ADVANCED SOIL MECHANICS

ADVANCED SOIL MECHANICS

Braja M. Das
The University of Texas at El Paso

Hemisphere Publishing Corporation
Washington New York London

McGraw-Hill Book Company
New York St. Louis San Francisco Auckland Bogotá
Hamburg Johannesburg London Madrid Mexico
Montreal New Delhi Panama Paris São Paulo
Singapore Sydney Tokyo Toronto

ADVANCED SOIL MECHANICS

1234567890 BCBC 898765432

This book was set in Press Roman by Communication Crafts Ltd. The editors were Julienne V. Brown, Brenda Munz Brienza, and Communication Crafts Ltd. BookCrafters, Inc., was printer and binder.

Library of Congress Cataloging in Publication Data

Das, Braja M., date

Advanced soil mechanics

Bibliography: p

Includes index.

1. Soil mechanics. I. Title.

TA710.D257 624.1'5136 81-6931

ISBN 0-07-015416-3 AACR2

To
Janice and Valerie

CONTENTS

4 Pore Water Pressure due to Undrained Loading

5 Consolidation

6 Evaluation of Soil Settlement

7 Shear Strength of Soils

PREFACE

This textbook is intended for use in an introductory graduate level course, and the general sequence followed is similar to that used in classrooms in various universities. The book has been developed primarily from class notes that I prepared for teaching initially at South Dakota State University and later at The University of Texas at El Paso.

The first chapter is on *Soil Aggregate* and is a general review of most of the materials to which students are introduced in the first course in soil mechanics offered at the undergraduate level. The remaining six chapters, dealing with permeability and seepage, stress distribution in a soil mass due to various types of loading conditions, development of pore water pressure due to undrained loading conditions, consolidation, methods of calculation of settlement of soils, and shear strength of soils, are presented in such a manner that readers who are unfamiliar with the subject will not face any serious problems in understanding. The basic concepts are presented in the earlier sections of each chapter and are then followed by more advanced topics.

The text has been extensively illustrated for better understanding. During the past ten to fifteen years, several new studies have been published in the geotechnical journals around the world. I have made an effort to include the important findings of most of these works as seem pertinent to the materials covered in this text.

A number of example problems are given in each chapter as well as a fairly large number of representative problems for solution by the students at the end of each chapter.

An extensive list of references is given at the end of each chapter which can be used by readers for in-depth review and/or research work.

Dual units—conventional English and SI—have been used throughout.

I am indebted to my wife, Janice, for her help in typing the manuscript. She also helped in preparing most of the figures and tables. Thanks are also due to Sands H. Figuers, graduate student at The University of Texas at El Paso, for his help in several stages during the preparation of the manuscript.

Braja M. Das

ADVANCED SOIL MECHANICS

CHAPTER

ONE

SOIL AGGREGATE

1.1 INTRODUCTION

Soils are aggregates of mineral particles, and together with air and/or water in the void spaces they form three-phase systems. A large portion of the earth's surface is covered by soils, and they are widely used as construction and foundation materials. Soil mechanics is the branch of engineering that deals with the engineering properties of soil and its behavior under stresses and strains.

This chapter is primarily designed to be a review of fundamentals to which the reader will already have been exposed in some detail. It is divided into eight major parts: weight–volume relations for the three-phase systems, grain-size distribution of soil particles, clay minerals, consistency, classification systems, compaction, volume change of soils, and the effective stress concept.

1.2 WEIGHT-VOLUME RELATIONSHIPS

1.2.1 Basic Definitions

Figure 1.1*a* shows a soil mass that has a total volume V and a total weight W. To develop the weight–volume relationships, the three phases of the soil mass, i.e., soil solids, air, and water, have been separated in Fig. 1.1*b*. Note that

$$W = W_s + W_w \tag{1.1}$$

and, also,

$$V = V_s + V_w + V_a \tag{1.2}$$

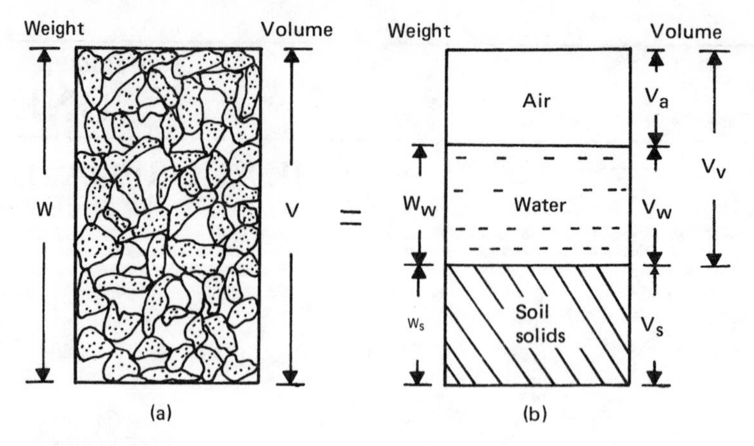

Fig. 1.1 Weight–volume relationships for soil aggregate.

$$V_v = V_w + V_a \tag{1.3}$$

where W_s = weight of soil solids
W_w = weight of water
V_s = volume of the soil solids
V_w = volume of water
V_a = volume of air

The weight of air is assumed to be zero. The volume relations commonly used in soil mechanics are void ratio, porosity, and degree of saturation.

Void ratio e is defined as the ratio of the volume of voids to the volume of solids:

$$e = \frac{V_v}{V_s} \tag{1.4}$$

Porosity n is defined as the ratio of the volume of voids to the total volume:

$$n = \frac{V_v}{V} \tag{1.5}$$

Also, $V = V_s + V_v$

and so

$$n = \frac{V_v}{V_s + V_v} = \frac{V_v/V_s}{V_s/V_s + V_v/V_s} = \frac{e}{1+e} \tag{1.6}$$

Degree of saturation S_r is the ratio of the volume of water to the volume of voids and is generally expressed as a percentage:

$$S_r\,(\%) = \frac{V_w}{V_v} \times 100 \tag{1.7}$$

The weight relations used are moisture content and unit weight. *Moisture content*

w is defined as the ratio of the weight of water to the weight of soil solids, generally expressed as a percentage:

$$w\,(\%) = \frac{W_w}{W_s} \times 100 \tag{1.8}$$

Unit weight γ is the ratio of the total weight to the total volume of the soil aggregate:

$$\gamma = \frac{W}{V} \tag{1.9}$$

This is sometimes referred to as moist unit weight since it includes the weight of water and the soil solids. If the entire void space is filled with water (i.e., $V_a = 0$), it is a saturated soil; Eq. (1.9) will then give use the saturated unit weight γ_{sat}.

The dry unit weight γ_d is defined as the ratio of the weight of soil solids to the total volume:

$$\gamma_d = \frac{W_s}{V} \tag{1.10}$$

Useful weight-volume relations can be developed by considering a soil mass in which the volume of soil solids is unity, as shown in Fig. 1.2. Since $V_s = 1$, from the definition of void ratio given in Eq. (1.4) the volume of voids is equal to the void ratio e. The weight of soil solids can be given by

$$W_s = G_s \gamma_w V_s = G_s \gamma_w \qquad \text{(since } V_s = 1\text{)}$$

where G_s is the specific gravity of soil solids, and γ_w is the unit weight of water ($62.4\ \text{lb/ft}^3$, or $9.81\ \text{kN/m}^3$).

From Eq. (1.8), the weight of water is $W_w = wW_s = wG_s\gamma_w$. So the moist unit weight is

$$\gamma = \frac{W}{V} = \frac{W_s + W_w}{V_s + V_v} = \frac{G_s\gamma_w + wG_s\gamma_w}{1 + e} = \frac{G_s\gamma_w(1 + w)}{1 + e} \tag{1.11}$$

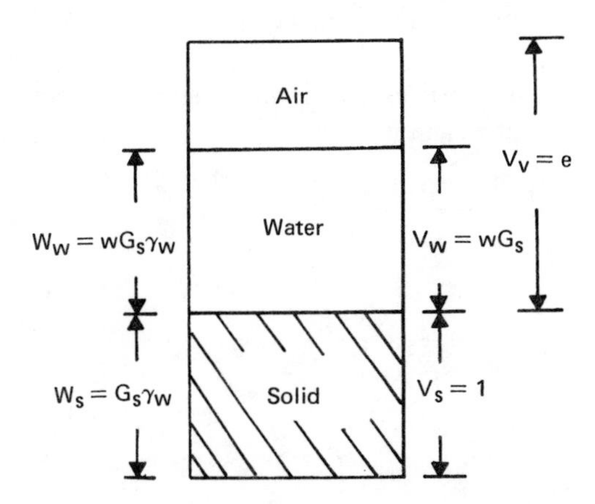

Fig. 1.2 Weight–volume relation for $V_s = 1$.

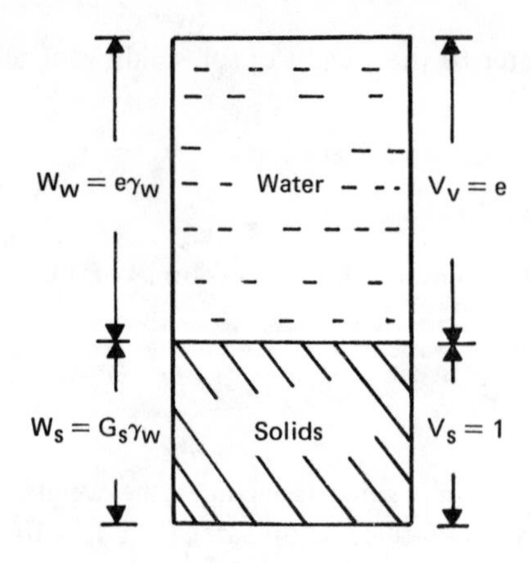

Fig. 1.3 Weight–volume relation for saturated soil with $V_s = 1$.

The dry unit weight can also be determined from Fig. 1.2 as

$$\gamma_d = \frac{W_s}{V} = \frac{G_s \gamma_w}{1+e} \tag{1.12}$$

The degree of saturation can be given by

$$S_r = \frac{V_w}{V_v} = \frac{W_w/\gamma_w}{V_v} = \frac{wG_s\gamma_w/\gamma_w}{e} = \frac{wG_s}{e} \tag{1.13}$$

For saturated soils, $S_r = 1$. So, from Eq. (1.13),

$$e = wG_s \tag{1.14}$$

By referring to Fig. 1.3, the relation for the unit weight of a saturated soil can be obtained as

$$\gamma_{\text{sat}} = \frac{W}{V} = \frac{W_s + W_w}{V} = \frac{G_s\gamma_w + e\gamma_w}{1+e} \tag{1.15}$$

Basic relations for unit weight such as Eqs. (1.11), (1.12), and (1.15) in terms of porosity n can also be derived by considering a soil mass that has a total volume of unity as shown in Fig. 1.4. In this case (for $V = 1$), from Eq. (1.5), $V_v = n$. So, $V_s = V - V_v = 1 - n$.

The weight of soil solids is equal to $(1-n)G_s\gamma_w$, and the weight of water $W_w = wW_s = w(1-n)G_s\gamma_w$. Thus the moist unit weight is

$$\gamma = \frac{W}{V} = \frac{W_s + W_w}{V} = \frac{(1-n)G_s\gamma_w + w(1-n)G_s\gamma_w}{1}$$

$$= G_s\gamma_w(1-n)(1+w) \tag{1.16}$$

The dry unit weight is

$$\gamma_d = \frac{W_s}{V} = (1-n)G_s\gamma_w \tag{1.17}$$

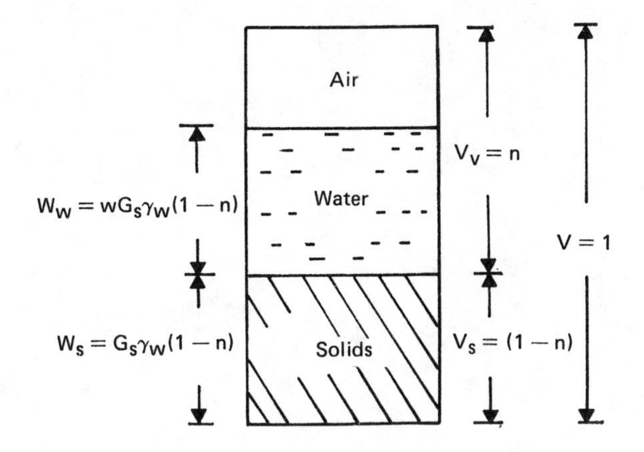

Fig. 1.4 Weight–volume relationship with $V = 1$.

If the soil is saturated (Fig. 1.5),

$$\gamma_{sat} = \frac{W_s + W_w}{V} = (1-n)G_s\gamma_w + n\gamma_w = [G_s - n(G_s - 1)]\gamma_w \tag{1.18}$$

Several other functional relationships are given in Table 1.1.

Example 1.1 For a soil in natural state, given $e = 0.8$, $w = 24\%$, and $G_s = 2.68$.

(*a*) Determine the moist unit weight, dry unit weight, and degree of saturation.

(*b*) If the soil is made completely saturated by adding water, what would its moisture content be at that time? Also find the saturated unit weight.

SOLUTION *Part (a):* From Eq. (1.11), the moist unit weight is

$$\gamma = \frac{G_s\gamma_w(1+w)}{1+e}$$

Since $\gamma_w = 9.81 \text{ kN/m}^3$,

$$\gamma = \frac{(2.68)(9.81)(1+0.24)}{1+0.8} = \mathit{18.11\ kN/m^3}$$

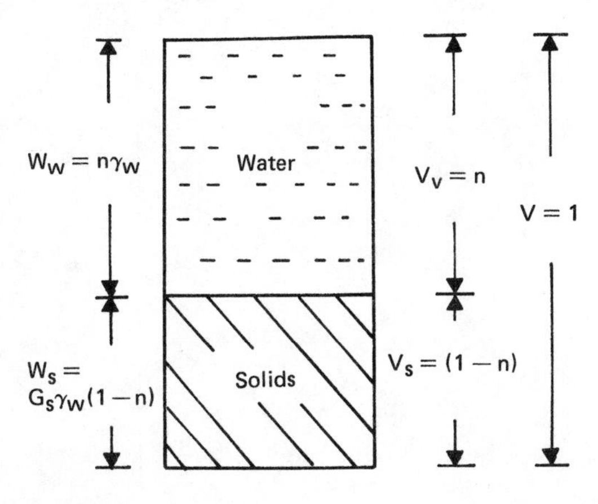

Fig. 1.5 Weight–volume relationship for saturated soil with $V = 1$.

Table 1.1 Functional relationships of various soil properties for saturated soils

Jumikis, A. R., *Soil Mechanics*, 1962, pp. 90–91, D. Van Nostrand Company, Inc., Princeton, New Jersey

	Sought quantities					
Quantities γ_w and:	Specific gravity G_s	Dry unit weight γ_d	Saturated unit weight γ_{sat}	Saturated moisture content, %	Porosity n	Void ratio e
$G_s; \gamma_d$			$\left(1 - \frac{1}{G_s}\right)\gamma_d + \gamma_w$	$\left(\frac{1}{\gamma_d} - \frac{1}{G_s\gamma_w}\right)\gamma_w$	$1 - \frac{\gamma_d}{G_s\gamma_w}$	$\frac{G_s\gamma_w}{\gamma_d} - 1$
$G_s; \gamma_{sat}$		$\frac{\gamma_{sat} - \gamma_w}{G_s - 1} G_s$		$\frac{G_s\gamma_w - \gamma_{sat}}{(\gamma_{sat} - \gamma_w) G_s}$	$\frac{G_s\gamma_w - \gamma_{sat}}{(G_s - 1)\gamma_w}$	$\frac{G_s\gamma_w - \gamma_{sat}}{\gamma_{sat} - \gamma_w}$
$G_s; w$		$\frac{G_s}{1 + wG_s}\gamma_w$	$\frac{1 + w}{1 + wG_s} G_s\gamma_w$		$\frac{wG_s}{1 + wG_s}$	wG_s
$G_s; n$		$G_s(1 - n)\gamma_w$	$[G_s - n(G_s - 1)]\gamma_w$	$\frac{n}{G_s(1 - n)}$		$\frac{n}{1 - n}$
$G_s; e$		$\frac{G_s}{1 + e}\gamma_w$	$\frac{G_s + e}{1 + e}\gamma_w$	$\frac{e}{G_s}$	$\frac{e}{1 + e}$	
$\gamma_d; \gamma_{sat}$	$\frac{\gamma_d}{\gamma_w + \gamma_d - \gamma_{sat}}$			$\frac{\gamma_{sat}}{\gamma_d} - 1$	$\frac{\gamma_{sat} - \gamma_d}{\gamma_w}$	$\frac{\gamma_{sat} - \gamma_d}{\gamma_w + \gamma_d - \gamma_{sat}}$
$\gamma_d; w$	$\frac{\gamma_d}{\gamma_w - w\gamma_d}$		$(1 + w)\gamma_d$		$w\frac{\gamma_d}{\gamma_w}$	$\frac{w\gamma_d}{\gamma_w - w\gamma_d}$

$\gamma_d; n$	$\dfrac{\gamma_d}{(1-n)\gamma_w}$		$\gamma_d + n\gamma_w$	$\dfrac{n\gamma_w}{\gamma_d}$		$\dfrac{n}{1-n}$
$\gamma_d; e$	$(1+e)\dfrac{\gamma_d}{\gamma_w}$		$\dfrac{e\gamma_w}{1+e} + \gamma_d$	$\dfrac{e}{1+e}\dfrac{\gamma_w}{\gamma_d}$	$\dfrac{e}{1+e}$	
$\gamma_{sat}; w$	$\dfrac{\gamma_{sat}}{\gamma_w - w(\gamma_{sat} - \gamma_w)}$	$\dfrac{\gamma_{sat}}{1+w}$			$\dfrac{w\gamma_{sat}}{(1+w)\gamma_w}$	$\dfrac{w\gamma_{sat}}{\gamma_w - w(\gamma_{sat} - \gamma_w)}$
$\gamma_{sat}; n$	$\dfrac{\gamma_{sat} - n\gamma_w}{(1-n)\gamma}$	$\gamma_{sat} - n\gamma_w$		$\dfrac{n\gamma_w}{\gamma_{sat} - n\gamma_w}$		$\dfrac{n}{1-n}$
$\gamma_{sat}; e$	$(1+e)\dfrac{\gamma_{sat}}{\gamma_w} - e$	$\gamma_{sat} - \dfrac{e}{1+e}\gamma_w$		$\dfrac{e\gamma_w}{\gamma_{sat} + e(\gamma_{sat} - \gamma_w)}$	$\dfrac{e}{1+e}$	
$w; n$	$\dfrac{n}{(1-n)w}$	$\dfrac{n}{w}\gamma_w$	$n\dfrac{1+w}{w}\gamma_w$			$\dfrac{n}{1-n}$
$w; e$	$\dfrac{e}{w}$	$\dfrac{e}{(1-e)w}\gamma_w$	$\dfrac{e}{w}\dfrac{1+w}{1+e}\gamma_w$		$\dfrac{e}{1+e}$	

From Eq. (1.12), the dry unit weight is

$$\gamma_d = \frac{G_s\gamma_w}{1+e} = \frac{(2.68)(9.81)}{1+0.8} = \mathit{14.61\ kN/m^3}$$

From Eq. (1.13), the degree of saturation is

$$S_r\,(\%) = \frac{wG_s}{e} \times 100 = \frac{(0.24)(2.68)}{0.8} \times 100 = \mathit{80.4\%}$$

Part (b): From Eq. (1.14), for saturated soils, $e = wG_s$, or

$$w\,(\%) = \frac{e}{G_s} \times 100 = \frac{0.8}{2.68} \times 100 = \mathit{29.85\%}$$

From Eq. (1.15), the saturated unit weight is

$$\gamma_{sat} = \frac{G_s\gamma_w + e\gamma_w}{1+e} = \frac{9.81\,(2.68+0.8)}{1+0.8} = \mathit{18.97\,kN/m^3}$$

1.2.2 General Range of Void Ratio and Dry Unit Weight Encountered in Granular Soils

For granular soils (sand and gravel), the range of void ratio generally encountered can be visualized by considering an ideal situation in which particles are spheres of equal size. The loosest and the densest possible arrangements that we can obtain from these equal spheres are, respectively, the simple cubic and the pyramidal type of packing as shown in Fig. 1.6. The void ratio corresponding to the simple cubic type of arrangement is 0.91; that for the pyramidal type of arrangement is 0.34. In the case of natural granular soils, particles are neither of equal size nor perfect spheres. The small-sized particles may occupy void spaces between the larger ones, which will tend to reduce the void ratio of natural soils as compared to that for equal spheres. On the other hand, the irregularity in the shape of the particles generally tends to increase the void ratio of soil as compared to ideal spheres. As a result of these two factors, the void ratios encountered in real soils are approximately in the same range as those obtained in the case of equal spheres.

Table 1.2 gives some typical values of void ratios and dry unit weights encountered in granular soils.

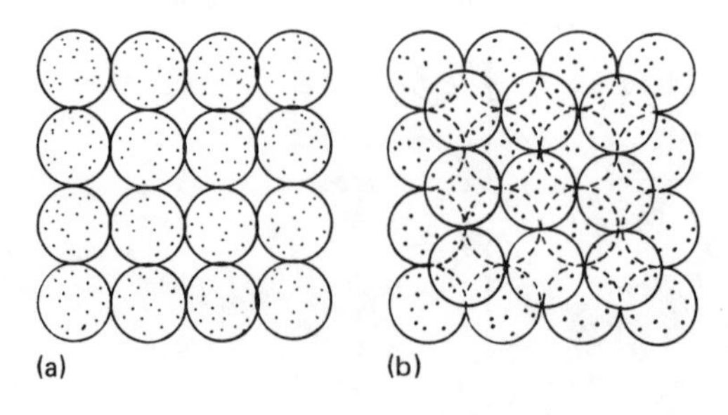

Fig. 1.6 Simple cubic (*a*) and pyramidal (*b*) types of arrangement of equal spheres.

Table 1.2 Typical values of void ratios and dry unit weights for granular soils

	Void ratio e		Dry unit weight γ_d			
			Minimum		Maximum	
Soil type	Maximum	Minimum	lb/ft³	kN/m³	lb/ft³	kN/m³
Gravel	0.6	0.3	103	16	127	20
Coarse sand	0.75	0.35	95	15	123	19
Fine sand	0.85	0.4	90	14	118	19
Standard Ottawa sand	0.8	0.5	92	14	110	17
Gravelly sand	0.7	0.2	97	15	138	22
Silty sand	1	0.4	83	13	118	19
Silty sand and gravel	0.85	0.15	90	14	144	23

1.2.3 Relative Density and Relative Compaction

Relative density is a term generally used to describe the degree of compaction of coarse-grained soils. Relative density D_r is defined as

$$D_r = \frac{e_{max} - e}{e_{max} - e_{min}} \tag{1.19}$$

where e_{max} = maximum possible void ratio
e_{min} = minimum possible void ratio
e = void ratio in natural state of soil

Equation (1.19) can also be expressed in terms of dry unit weight of the soil:

$$\gamma_{d(max)} = \frac{G_s \gamma_w}{1 + e_{min}}$$

or

$$e_{min} = \frac{G_s \gamma_w}{\gamma_{d(max)}} - 1 \tag{1.20}$$

Similarly,

$$e_{max} = \frac{G_s \gamma_w}{\gamma_{d(min)}} - 1 \tag{1.21}$$

and

$$e = \frac{G_s \gamma_w}{\gamma_d} - 1 \tag{1.22}$$

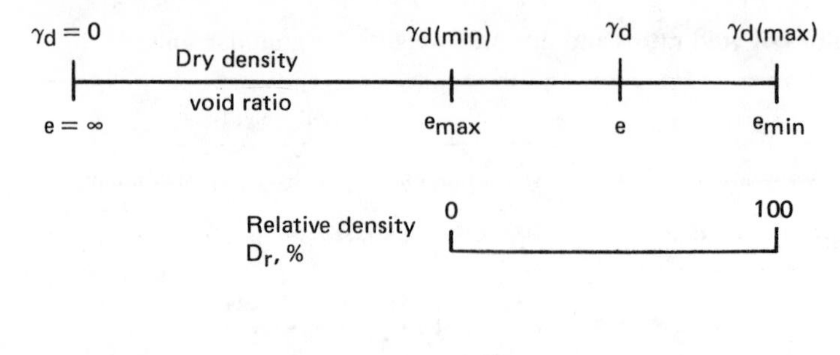

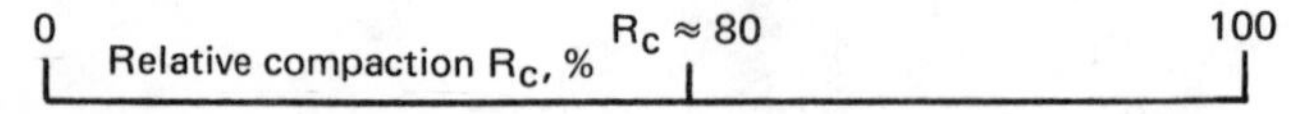

Fig. 1.7 Relative density and relative compaction concepts. *(After K. L. Lee and A. Singh, Relative Density and Relative Compaction,* J. Soil Mech. Found. Div., ASCE, *vol. 97, no. SM7, 1971.)*

where $\gamma_{d(max)}$, $\gamma_{d(min)}$, and γ_d are the maximum, minimum, and natural-state dry unit weights of the soil. Substitution of Eqs. (1.20), (1.21), and (1.22) into Eq. (1.19) yields

$$D_r = \frac{\gamma_{d(max)}}{\gamma_d} \frac{\gamma_d - \gamma_{d(min)}}{\gamma_{d(max)} - \gamma_{d(min)}} \tag{1.23}$$

Relative density is generally expressed as a percentage. It has been used by several investigators to correlate the angle of friction of soil, the soil liquefication potential, etc.

Another term occasionally used in regard to the degree of compaction of coarse-grained soils is *relative compaction*, R_c, which is defined as

$$R_c = \frac{\gamma_d}{\gamma_{d(max)}} \tag{1.24}$$

The difference between relative density and relative compaction is shown in Fig. 1.7. Comparing Eqs. (1.23) and (1.24),

$$R_c = \frac{R_o}{1 - D_r(1 - R_o)} \tag{1.25}$$

where $R_o = \gamma_{d(min)}/\gamma_{d(max)}$.

Lee and Singh (1971) reviewed 47 different soils and gave the approximate relation between relative compaction and relative density as

$$R_c = 80 + 0.2D_r \tag{1.26}$$

where D_r is in percent.

1.2.4 Specific Gravity of Soil Solids

The specific gravity of soil solids, G_s, has been used in the weight-volume relations derived in Sec. 1.2.1. The value of G_s for most natural soils falls in the general range

Table 1.3 Typical values of G_s

Soil type	G_s
Gravel	2.65–2.68
Sand	2.65–2.68
Silt	2.66–2.7
Clay	2.68–2.8

of 2.65 to 2.75. For organic soils, the value may fall below 2. Some typical values of G_s are given in Table 1.3.

1.3 GRAIN-SIZE DISTRIBUTION OF SOILS

1.3.1 Sieve Analysis and Hydrometer Analysis

For a basic understanding of the nature of soils that are generally encountered, the type of distribution of grain sizes in a given soil mass must be known.

The grain-size distribution of coarse-grained soils (gravelly and/or sandy) is usually determined by sieve analysis. Oven-dried soil with the lumps thoroughly broken down is passed through a number of sieves. The weight of the dry soil retained on each sieve is determined, and based on these weights the cumulative percent passing a given sieve is determined. This is generally referred to as *percent finer.* The numbers of the standard sieves used in the United States and their corresponding openings are given in Table 1.4.

Table 1.5 shows the results of a typical sieve analysis. It is customary to plot the grain-size distribution on semilogarithmic graph paper with the percent finer on the

Table 1.4 U.S. standard sieves

Sieve no.	Opening size, mm
3	6.35
4	4.76
6	3.36
8	2.38
10	2.00
16	1.19
20	0.84
30	0.59
40	0.42
50	0.297
60	0.25
70	0.21
100	0.149
140	0.105
200	0.074
270	0.053

Table 1.5 Results of a sieve analysis*

U.S. standard sieve no.	Sieve opening, mm	Mass of soil retained on each sieve, g	Cumulative mass of soil retained, g	Cumulative mass of soil passing each sieve, g	Percent finer†
4	4.76	10	10	640	98.5
10	2.00	30	40	610	93.8
16	1.19	52	92	558	85.8
30	0.59	80	172	478	73.4
40	0.42	141	313	337	51.8
60	0.25	96	409	241	37.1
100	0.149	105	514	136	20.9
200	0.074	85	599	51	7.8
Pan		51			

* Mass of total dry soil = 650 g

† Percent finer = $\dfrac{\text{cumulative mass of soil passing each sieve}}{\text{mass of the total dry soil}} \times 100$

arithmetic scale and the sieve openings on the logarithmic scale. The results of the sieve analysis given in Table 1.5 are plotted in Fig. 1.8.

The grain-size distribution can be used to determine some of the basic soil parameters such as the effective size, the uniformity coefficient, and the coefficient of gradation. The *effective size* of a soil is the diameter through which 10% of the total soil mass is passing and is referred to as D_{10}. The *uniformity coefficient* C_u is defined as

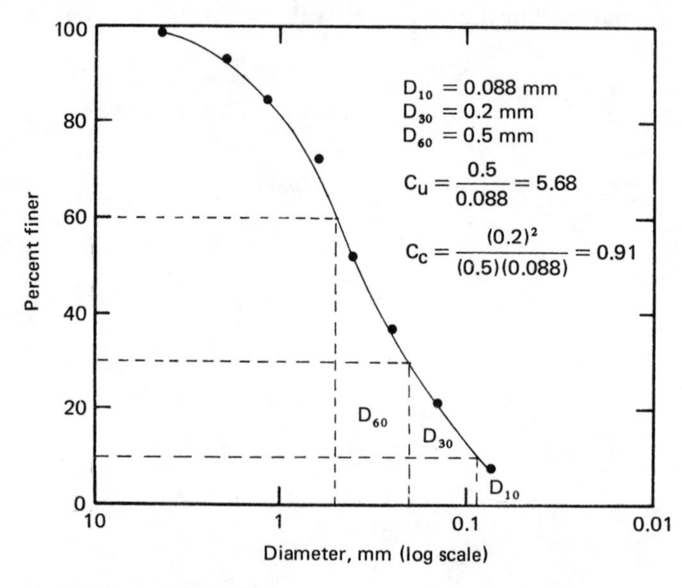

Fig. 1.8 Grain-size distribution.

$$C_u = \frac{D_{60}}{D_{10}} \tag{1.27}$$

where D_{60} is the diameter through which 60% of the total soil mass is passing.

The *coefficient of gradation* C_c is defined as

$$C_c = \frac{(D_{30})^2}{(D_{60})(D_{10})} \tag{1.28}$$

where D_{30} is the diameter through which 30% of the total soil mass is passing.

The uniformity coefficient and the coefficient of gradation for the sieve analysis shown in Table 1.5 are also shown in Fig. 1.8.

A soil is called a *well-graded* soil if the distribution of the grain sizes extends over a rather large range. In that case, the value of the uniformity coefficient is large. Generally, a soil is referred to as well graded if C_u is larger than about 4 to 6 and C_c is between 1 and 3. When most of the grains in a soil mass are of approximately the same size—i.e., C_u is close to 1—the soil is called *poorly graded.* A soil might have a combination of two or more well-graded soil fractions, and this type of soil is referred to as a *gap-graded* soil. Figure 1.9 shows the comparison of the general nature of the grain-size distributions for well-graded, poorly graded, and gap-graded soils.

The sieve analysis technique described above is applicable for soil grains larger than No. 200 (0.074 mm) sieve size. For fine-grained soils, the technique used for determination of the grain sizes is hydrometer analysis. This is based on the principle of sedimentation of soil grains. When soil particles are dispersed in water, they will

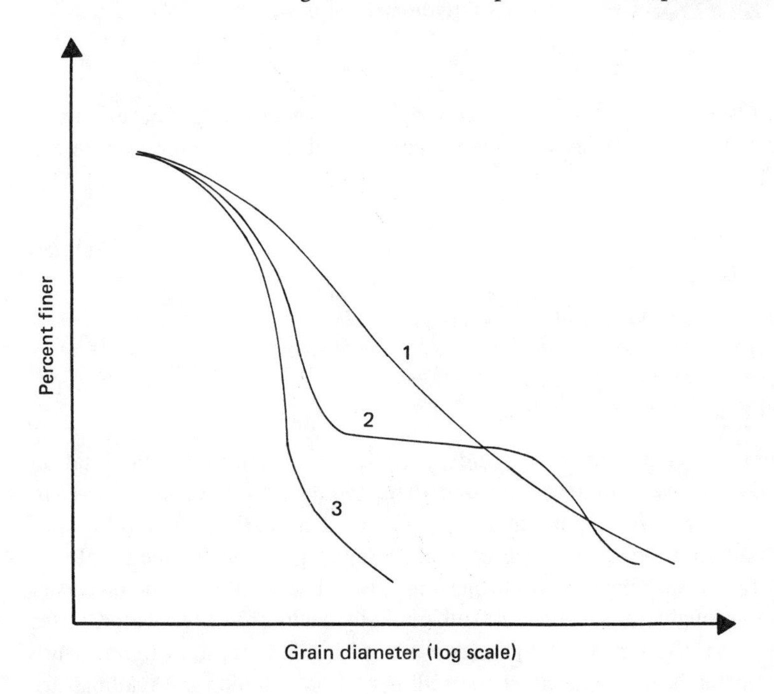

Fig. 1.9 Nature of grain-size distribution for (1) well-graded, (2) poorly graded, and (3) gap-graded soils.

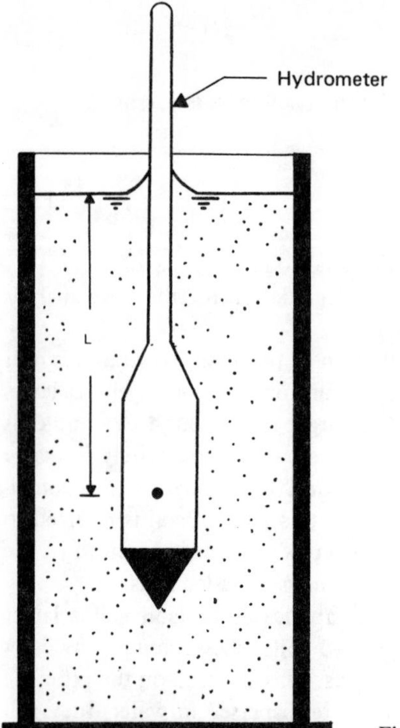

Fig. 1.10 Hydrometer analysis.

settle at different velocities depending on their weights, shapes, and sizes. For simplicity, it is assumed that all soil particles are spheres, and the velocity of a soil particle can be given by Stokes law as

$$V = \frac{\gamma_s - \gamma_w}{18\eta} D^2 \tag{1.29}$$

where V = velocity = distance/time = L/t
γ_w, γ_s = unit weight of water and soil particles, respectively
η = absolute viscosity of water
D = diameter of the soil particles

In the laboratory, hydrometer tests are generally conducted in a sedimentation cylinder, and 50 g of oven-dried soil is used. The sedimentation cylinder is 18 in (457.2 m) high and 2.5 in (63.5 mm) in diameter, and it is marked for a volume of 1000 ml. A 125-ml solution of 4% sodium hexametaphosphate in distilled water is generally added to the specimen as the dispersing agent. The volume of the dispersed soil suspension is brought up to the 1000 ml mark by adding distilled water. After thorough mixing, the sedimentation cylinder is placed inside a constant-temperature bath. The hydrometer is then placed in the sedimentation cylinder and readings are taken to the top of the meniscus (Fig. 1.10) at various elapsed times.

When the hydrometer is placed in the soil suspension at a time t after the start of sedimentation, it measures the liquid density in the vicinity of its bulb at a depth L (Fig. 1.10). The liquid density is a function of the amount of soil particles present per unit volume of the suspension at that depth. ASTM 152H hydrometers are calibrated to read the amount in grams of soil particles in suspension per 1000 ml (for $G_s = 2.65$ at a temperature of 20°C). Also, at a time t the soil particles in suspension at depth L will have diameters smaller than those calculated by Eq. (1.29), since the larger particles would have settled beyond the zone of measurement. Hence, the percent of soil finer than a given diameter D can be calculated. Since the actual conditions under which the test is conducted may be different from those for which the hydrometers are calibrated ($G_s = 2.65$, temperature of 20°C), it may be necessary to make corrections to the observed hydrometer readings. For further details regarding the corrections, the reader should refer to a soils laboratory manual (e.g., Bowles, 1978).

1.3.2 Soil-Separate Size Limits

Based on the grain size of soil particles, a number of agencies have developed the size limits of gravel, sand, silt, and clay. Some of these size limits are presented in Fig. 1.11. Note that some agencies classify clay as particles smaller than 0.005 mm in size, and others classify it as particles smaller than 0.002 mm in size. However, it needs to be realized that particles defined as clay on the basis of their size are not necessarily clay minerals. Clay particles possess the tendency to develop plasticity when mixed with water; these are clay minerals. Fine particles of quartz, feldspar, or mica may be present in a soil in the size range defined for clay, but these will not develop plasticity when mixed with water. It appears to be more appropriate for soil particles with sizes less than 2 μm or 5 μm as defined under various systems to be called *clay-size particles* rather than *clay*. True clay particles are mostly of colloidal size range (less than 1 μm), and 2 μm is probably the upper limit.

Some fundamentals of clay minerals are presented in the following sections.

1.4 CLAY MINERALS

1.4.1 Composition and Structure of Clay Minerals

Clay minerals are complex silicates of aluminum, magnesium, and iron. Two basic crystalline units form the clay minerals: (1) a silicon-oxygen tetrahedron, and (2) an aluminum or magnesium octahedron. A silicon-oxygen tetrahedron unit, shown in Fig. 1.12*a*, consists of four oxygen atoms surrounding a silicon atom. The tetrahedron units combine to form a *silica sheet* as shown in Fig. 1.13*a*. Note that the three oxygen atoms located at the base of each tetrahedron are shared by neighboring tetrahedra. Each silicon atom with a positive valance of 4 is linked to four oxygen atoms with a total negative valance of 8. However, each oxygen atom at the base of the tetrahedron is linked to two silicon atoms. This leaves one negative valance charge of the top oxygen atom of each tetrahedron to be counterbalanced. Figure 1.12*b* shows an octahedral

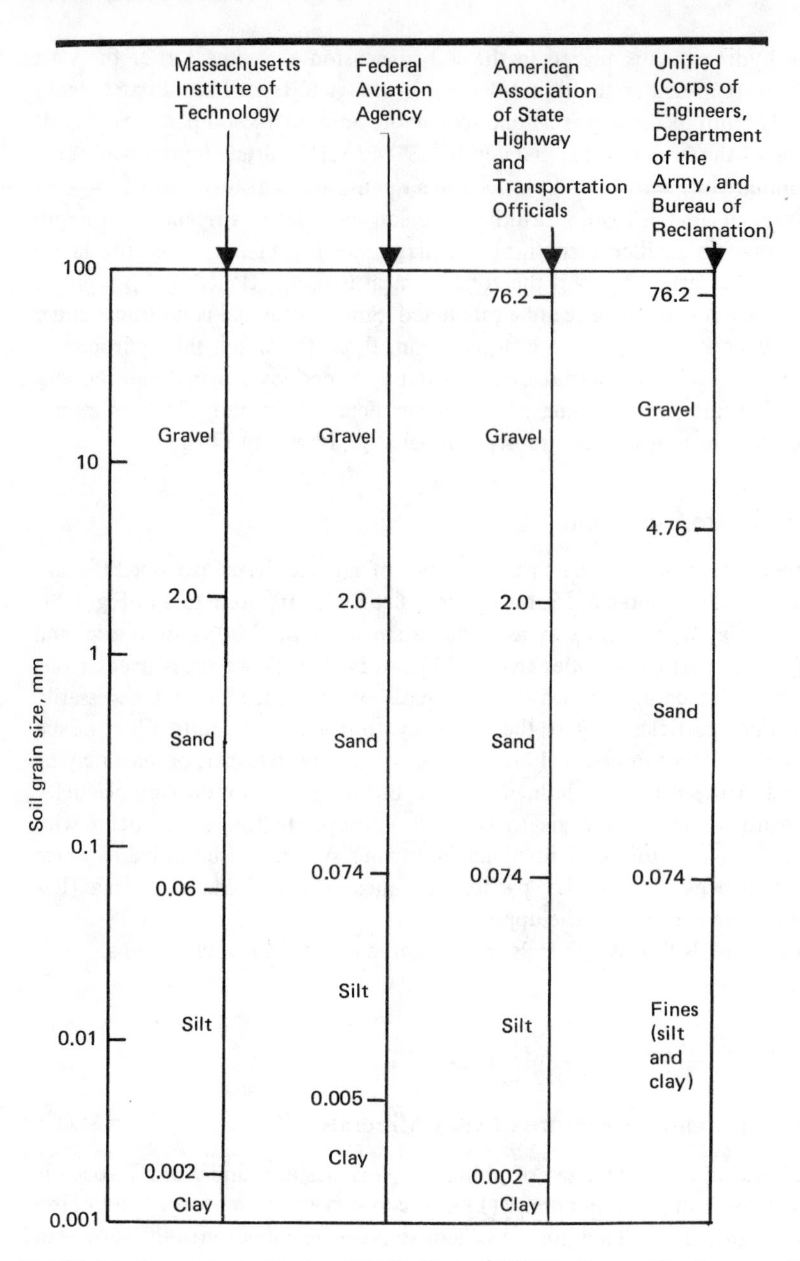

Fig. 1.11 Soil-separate size limits of M.I.T., FAA, AASHTO, Corps of Engineers, and USBR.

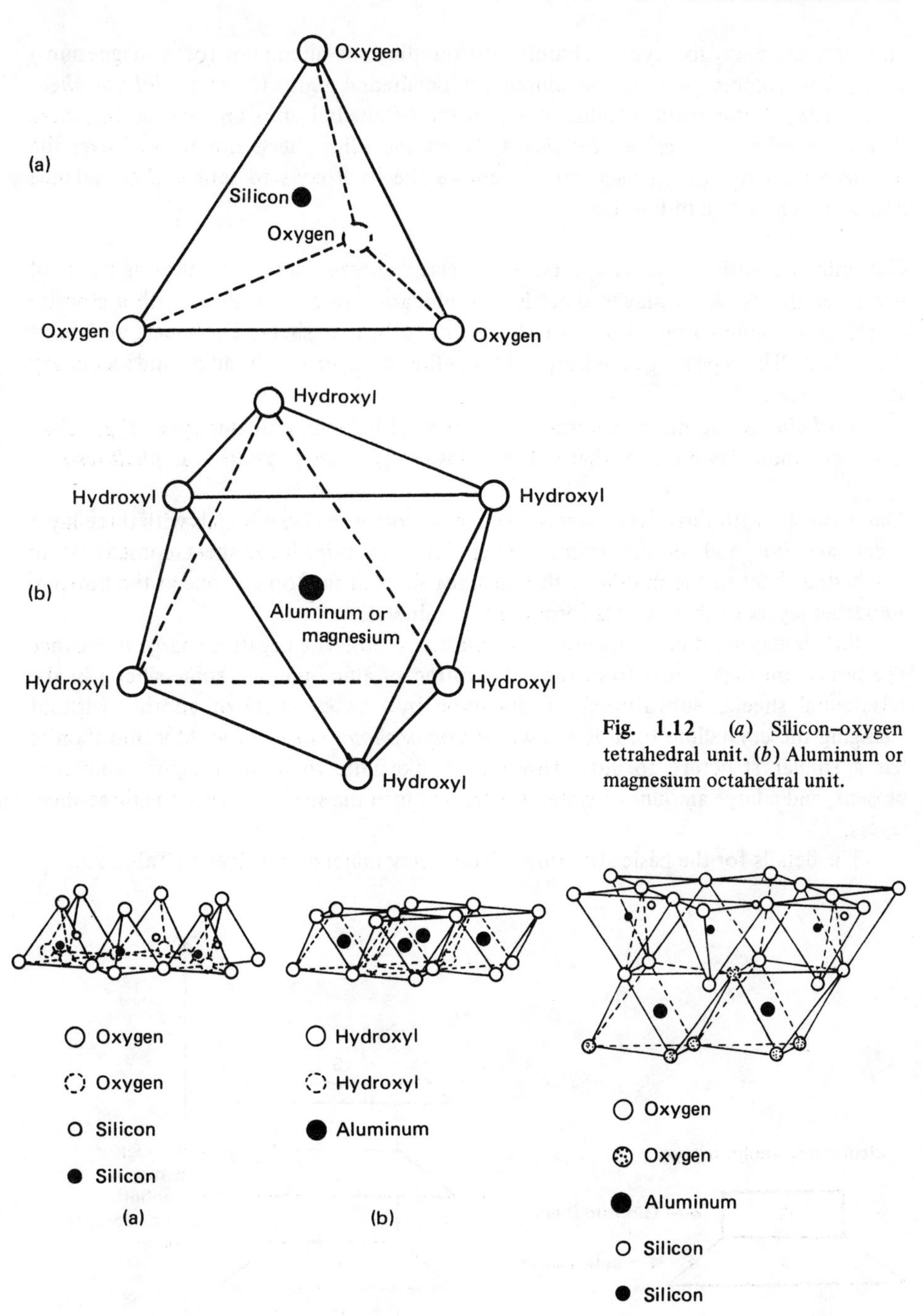

Fig. 1.12 (*a*) Silicon–oxygen tetrahedral unit. (*b*) Aluminum or magnesium octahedral unit.

Fig. 1.13 (*a*) Silica sheet. (*b*) Gibbsite sheet. (*c*) Silica–gibbsite sheet. *(After R. E. Grim, Physio-chemical Properties of Soils: Clay Minerals,* J. Soil Mech. Found. Div., ASCE, *vol. 85, no. SM2, 1959.)*

unit consisting of six hydroxyl units surrounding an aluminum (or a magnesium) atom. The combination of the aluminum octahedral units forms a *gibbsite sheet* (Fig. 1.13*b*). If the main metallic atoms in the octahedral units are magnesium, these sheets are referred to as *brucite sheets.* When the silica sheets are stacked over the octahedral sheets, the oxygen atoms replace the hydroxyls to satisfy their valance bonds. This is shown in Fig. 1.13*c*.

Clay minerals with two-layer sheets. Some clay minerals consist of repeating layers of two-layer sheets. A two-layer sheet is a combination of a silica sheet with a gibbsite sheet, or a combination of a silica sheet with a brucite sheet. The sheets are about 7.2 Å thick. The repeating layers are held together by hydrogen bonding and secondary valence forces.

Kaolinite is the most important clay mineral belonging to this type (Fig. 1.14). Other common clay minerals that fall into this category are *serpentine* and *halloysite.*

Clay minerals with three-layer sheets. The most common clay minerals with three-layer sheets are *illite* and *montmorillonite* (Fig. 1.15). A three-layer sheet consists of an octahedral sheet in the middle with one silica sheet at the top and one at the bottom. Repeated layers of these sheets form the clay minerals.

Illite layers are bonded together by potassium ions. The negative charge to balance the potassium ions comes from the substitution of aluminum for some silicon in the tetrahedral sheets. Substitution of this type by one element for another without changing the crystalline form is known as *isomorphous substitution.* Montmorillonite has a similar structure to illite. However, unlike illite there are no potassium ions present, and a large amount of water is attracted into the space between the three-sheet layers.

The details for the basic structure of some clay minerals are given in Table 1.6.

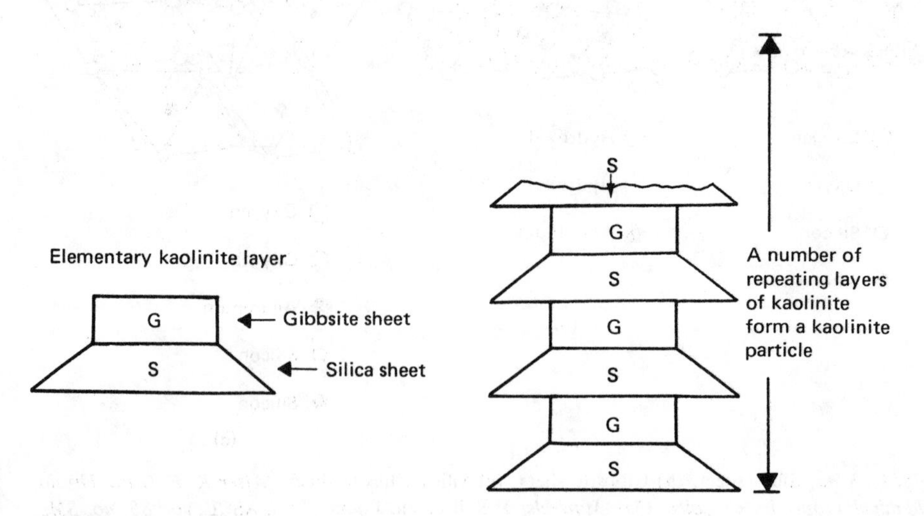

Fig. 1.14 Symbolic structure for kaolinite.

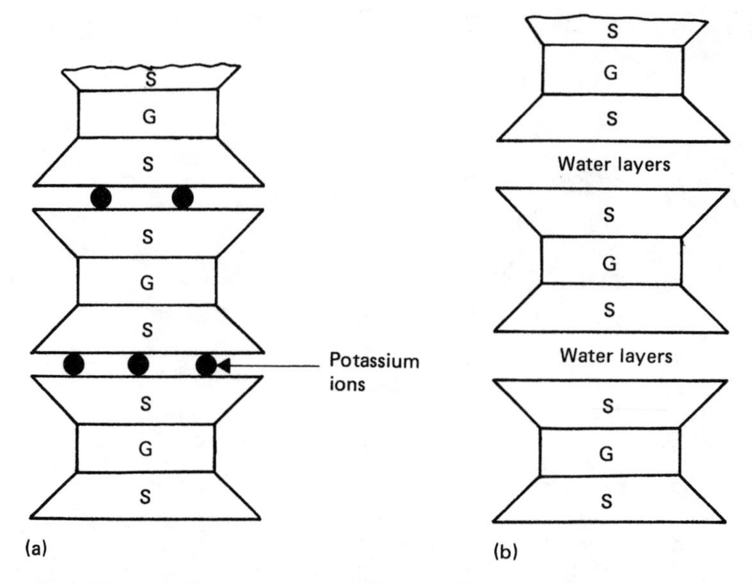

Fig. 1.15 Symbolic structures of (*a*) illite and (*b*) montmorillonite.

1.4.2 Specific Surface of Clay Minerals

The surface area of clay particles per unit mass is generally referred to as *specific surface.* The lateral dimensions of kaolinite platelets are about 1000 to 20,000 Å with thicknesses of 100 to 1000 Å. Illite particles have lateral dimensions of 1000 to 5000 Å and thicknesses of 50 to 500 Å. Similarly, montmorillonite particles have lateral dimensions of 1000 to 5000 Å with thicknesses of 10 to 50 Å. If we consider several clay samples all having the same mass, the highest surface area will be in the sample in which the particle sizes are the smallest. So it is easy to realize that the specific surface of kaolinite will be small compared to that of montmorillonite. The specific surfaces of kaolinite, illite, and montmorillonite are about 15, 90, and 800 m^2/g, respectively. Table 1.6 lists the specific surfaces of some clay minerals.

1.4.3 Cation Exchange Capacity

Clay particles carry a net negative charge. In an ideal crystal, the positive and negative charges would be balanced. However, isomorphous substitution and broken continuity of structures result in a net negative charge at the faces of the clay particles. (There are also some positive charges at the edges of these particles.) To balance the negative charge, the clay particles attract positively charged ions from salts in their pore water. These are referred to as exchangeable ions. Some are more strongly attracted than others, and the cations can be arranged in a series in terms of their affinity for attraction as follows:

$$Al^{3+} > Ca^{2+} > Mg^{2+} > NH_4^+ > K^+ > H^+ > Na^+ > Li^+$$

Table 1.6 Clay minerals

Mineral	Structure symbol*	Isomorphous substitution (nature and amount)	Linkage between sheets (type and strength)	Specific surface, m^2/g	Potential exchange capacity, mE/100 g	Actual exchange capacity, mE/100 g
Serpentine	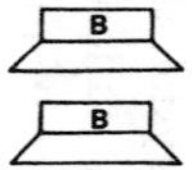 B / B	None	H-Bonding + secondary valence		1	1
Halloysite ($4H_2O$)	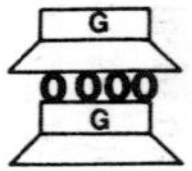 G / OOOO / G	Al for Si, 1 in 100	Secondary valence	40	12	12
Halloysite ($2H_2O$)	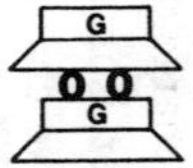 G / OO / G	Al for Si, 1 in 100	Secondary valence	40	12	12
Kaolinite	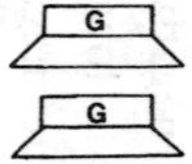 G / G	Al for Si, 1 in 400	H-Bonding + secondary valence	10–20	3	3
Talc	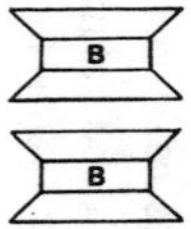 B / B	None	Secondary valence		1	1
Pyrophyllite	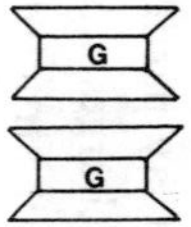 G / G	None	Secondary valence		1	1

Muscovite	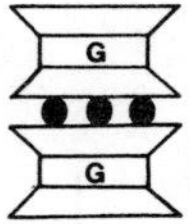 Al for Si, 1 in 4	Secondary valence + K linkage		250	5–20
Vermiculite	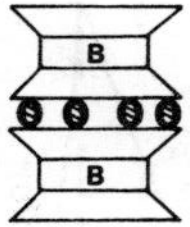 Al, Fe for Mg Al for Si	Secondary valence + Mg linkage	5–400	150	150
Illite	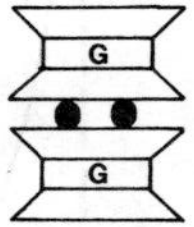 Al for Si, 1 in 7 Mg, Fe for Al Fe, Al for Mg	Secondary valence + K linkage	80–100	150	25
Montmorillonite	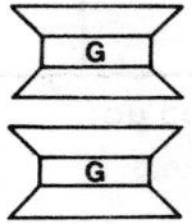 Mg for Al, 1 in 6	Secondary valence + exchangeable ion linkage	800	100	100
Nontronite	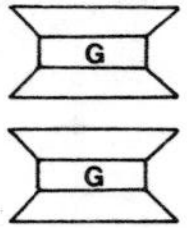 Al for Si, 1 in 6	Secondary valence + exchangeable ion linkage	800	100	100
Chlorite	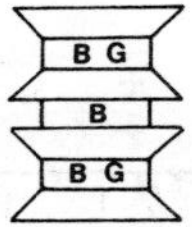 Al for Si, Fe Al for Mg	Secondary valence + brucite linkage	5–50	20	20

After T. W. Lambe and R. V. Whitman, "Soil Mechanics," Wiley, New York, 1969.
* Symbols: O H_2O ● K ● Mg [G] Gibbsite [B] Brucite.

This series indicates that, for example, Al^{3+} ions can replace Ca^{2+} ions, and Ca^{2+} ions can replace Na^{+} ions. The process is called *cation exchange.* For example,

$$Na_{clay} + CaCl_2 \rightarrow Ca_{clay} + NaCl$$

Cation exchange capacity (CEC) of a clay is defined as the amount of exchangeable ions, expressed in milliequivalents, per 100 g of dry clay. Table 1.6 gives the cation exchange capacity of some clays. The laboratory procedure for determination of CEC is given in ASTM *Special Technical Publication No. 479* (1970).

1.4.4 Nature of Water in Clay

The presence of exchangeable cations on the surface of clay particles was discussed in the preceding section. Some salt precipitates (cations in excess of the exchangeable ions and their associated anions) are also present on the surface of dry clay particles. When water is added to clay, these cations and anions float around the clay particles (Fig. 1.16).

At this point, it must be pointed out that water molecules are dipolar, since the hydrogen atoms are not symmetrically arranged around the oxygen atoms (Fig. 1.17*a*). This means that a molecule of water is like a rod with positive and negative charges at opposite ends (Fig. 1.17*b*). There are three general mechanisms by which these dipolar water molecules, or *dipoles,* can be electrically attracted toward the surface of the clay particles (Fig. 1.18):

a. Attraction between the negatively charged faces of clay particles and the positive ends of dipoles.
b. Attraction between cations in the double layer and the negatively charged ends of dipoles. The cations are in turn attracted by the negatively charged faces of clay particles.

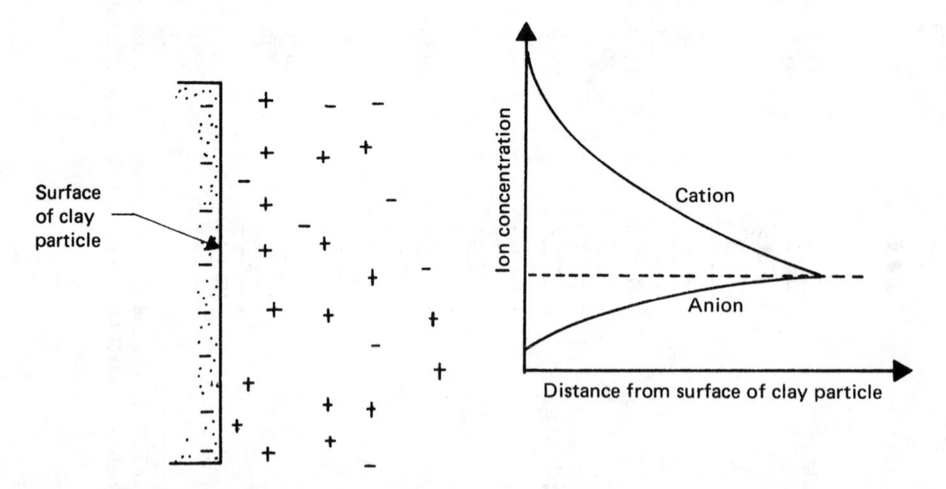

Fig. 1.16 Diffuse double layer.

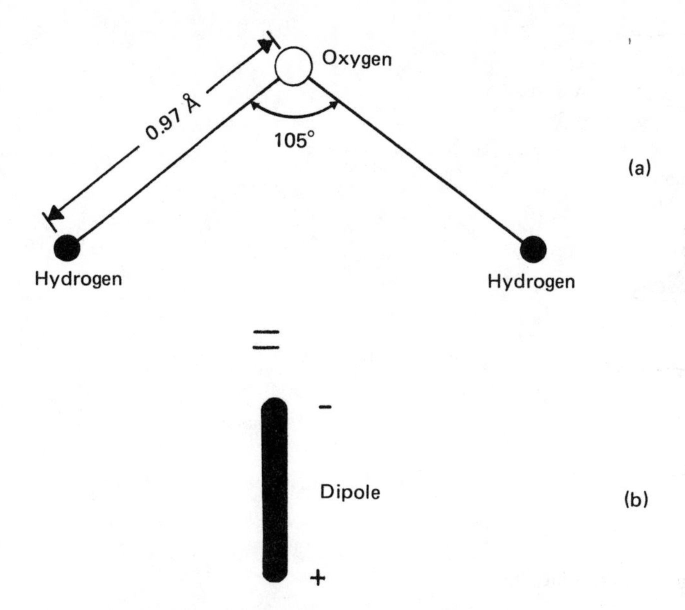

Fig. 1.17 Dipolar nature of water.

c. Sharing of the hydrogen atoms in the water molecules by hydrogen bonding between the oxygen atoms in the clay particles and the oxygen atoms in the water molecules.

The electrically attracted water that surrounds the clay particles is known as *double-layer water.* The plastic property of clayey soils is due to the existence of

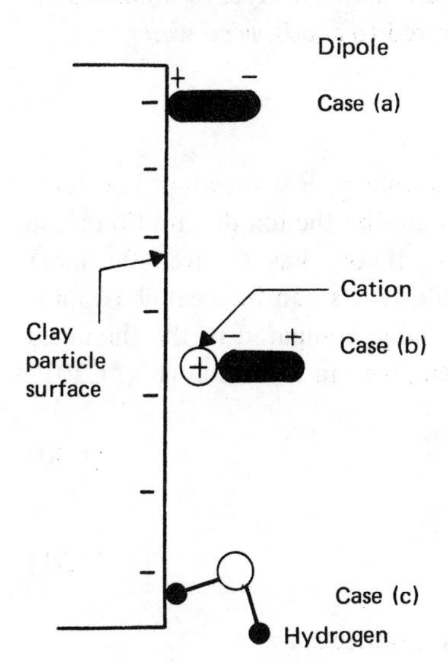

Fig. 1.18 Dipolar water molecules in diffuse double layer.

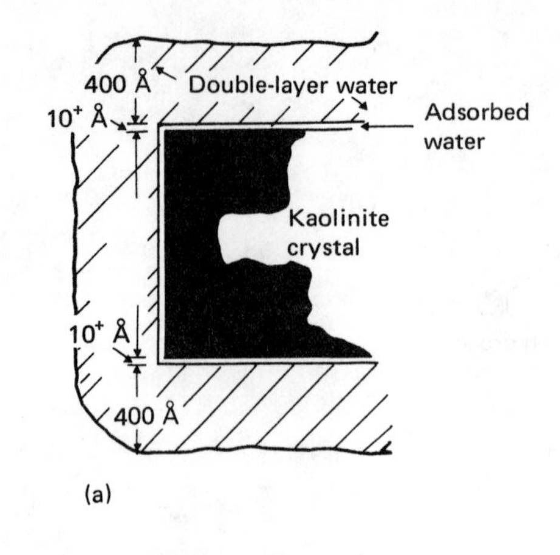

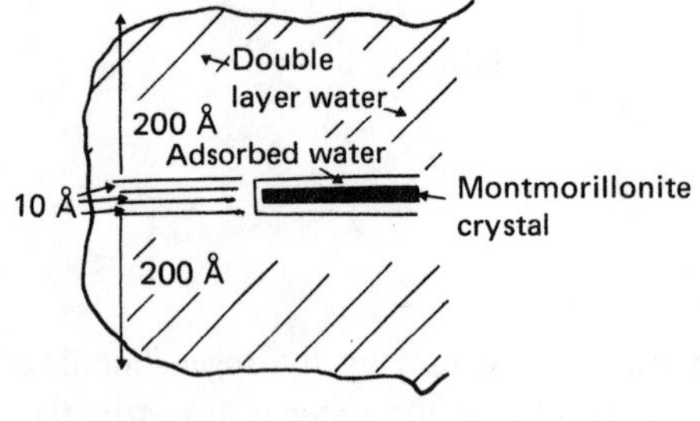

Fig. 1.19 Clay water (*a*) Typical kaolinite particle, 10,000 by 1000 Å. (*b*) Typical montmorillonite particle, 1000 by 10 Å. *(After T. W. Lambe, Compacted Clay: Structure,* Trans. ASCE, *vol. 125, 1960.)*

double-layer water. Thicknesses of double-layer water for typical kaolinite and montmorillonite crystals are shown in Fig. 1.19. Since the innermost layer of double-layer water is very strongly held by a clay particle, it is referred to as *adsorbed water.*

1.4.5 Repulsive Potential

The nature of the distribution of ions in the diffuse double layer is shown in Fig. 1.16. Several theories have been presented in the past to describe the ion distribution close to a charged surface. Of these, the Gouy-Chapman theory has received the most attention. Let us assume that the ions in the double layers can be treated as point charges, and that the surface of the clay particles is large compared to the thickness of the double layer. According to Boltzmann's theorem, we can write that (Fig. 1.20)

$$n_+ = n_{+(0)} \exp \frac{-v_+ e\Phi}{KT} \tag{1.30}$$

$$n_- = n_{-(0)} \exp \frac{v_- e\Phi}{KT} \tag{1.31}$$

where n_+ = local concentration of positive ions at a distance x
n_- = local concentration of negative ions at a distance x

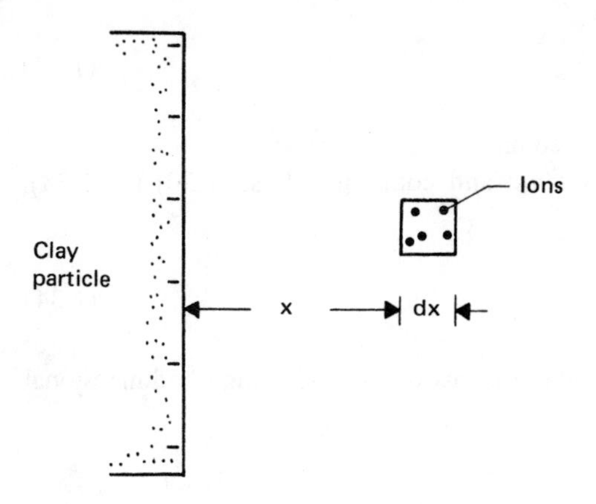

Fig. 1.20 Derivation of repulsive potential equation.

$n_{+(0)}, n_{-(0)}$ = concentration of positive and negative ions away from clay surface in equilibrium liquid

Φ = average electric potential at a distance x (Fig. 1.21)

v_+, v_- = ionic valences

e = unit electrostatic charge, 4.8×10^{-10} esu

K = Boltzmann's constant, 1.38×10^{-16} erg/K

T = absolute temperature

The charge density ρ at a distance x is given by

$$\rho = v_+ e n_+ - v_- e n_- \tag{1.32}$$

According to Poisson's equation,

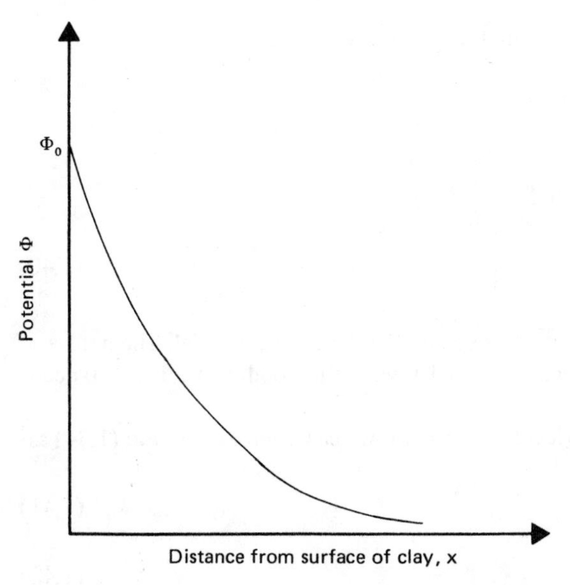

Fig. 1.21 Nature of variation of potential Φ with distance from the clay surface.

$$\frac{d^2\Phi}{dx^2} = \frac{-4\pi\rho}{\lambda} \tag{1.33}$$

where λ is the dielectric constant of the medium.

Assuming $\upsilon_+ = \upsilon_-$ and $n_{+(0)} = n_{-(0)} = n_0$, and combining Eqs. (1.30) to (1.33), we obtain

$$\frac{d^2\Phi}{dx^2} = \frac{8\pi n_0 \upsilon e}{\lambda} \sinh \frac{\upsilon e\Phi}{KT} \tag{1.34}$$

It is convenient to rewrite Eq. (1.34) in terms of the following nondimensional quantities:

$$y = \frac{\upsilon e\Phi}{KT} \tag{1.35}$$

$$z = \frac{\upsilon e\Phi_0}{KT} \tag{1.36}$$

and $$\xi = \kappa x \tag{1.37}$$

where Φ_0 is the potential at the surface of the clay particle and

$$\kappa^2 = \frac{8\pi n_0 e^2 \upsilon^2}{\lambda KT} \qquad (\text{cm}^{-2}) \tag{1.38}$$

Thus, from Eq. (1.34),

$$\frac{d^2 y}{d\xi^2} = \sinh y \tag{1.39}$$

The boundary conditions for solving Eq. (1.39) are:

1. At $\xi = \infty, y = 0$ and $dy/d\xi = 0$.
2. At $\xi = 0, y = z$, i.e., $\Phi = \Phi_0$.

The solution yields the relation

$$e^{y/2} = \frac{(e^{z/2} + 1) + (e^{z/2} - 1)e^{-\xi}}{(e^{z/2} + 1) - (e^{z/2} - 1)e^{-\xi}} \tag{1.40}$$

Equation (1.40) gives an approximately exponential decay of potential. The nature of the variation of the nondimensional potential y with the nondimensional distance is given in Fig. 1.22.

For a small surface potential (less than 25 mV), we can approximate Eq. (1.34) as

$$\frac{d^2\Phi}{dx^2} = \kappa^2\Phi \tag{1.41}$$

$$\Phi = \Phi_0 e^{-\kappa x} \tag{1.42}$$

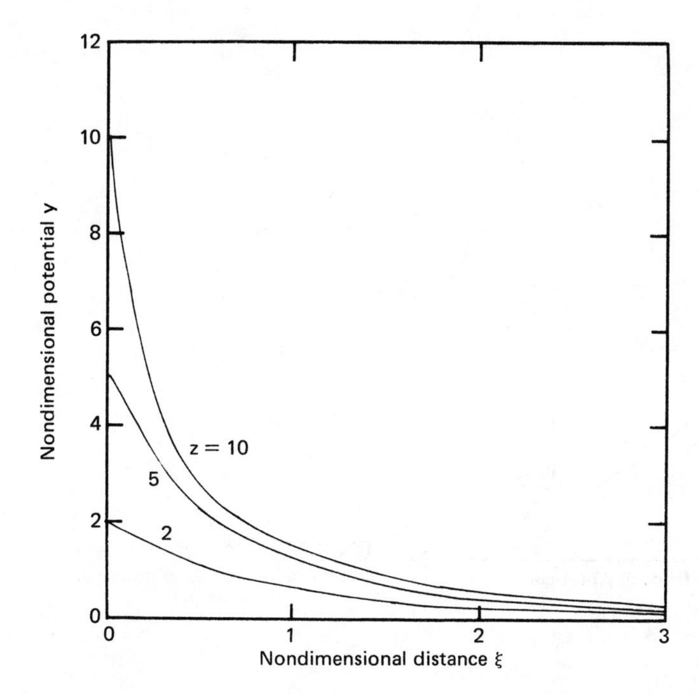

Fig. 1.22 Variation of nondimensional potential with nondimensional distance.

Equation (1.42) describes a purely exponential decay of potential. For this condition, the center of gravity of the diffuse charge is located at a distance of $x = 1/\kappa$. The term $1/\kappa$ is generally referred to as the double-layer *thickness.*

There are several factors that will affect the variation of the repulsive potential with distance from the surface of the clay layer. The effect of the cation concentration and ionic valence is shown in Figs. 1.23 and 1.24, respectively. For a given value of Φ_0 and x, the repulsive potential Φ decreases with the increase of ion concentration n_0 and ionic valence v.

When clay particles are close to and parallel to each other, the nature of variation of the potential will be as shown in Fig. 1.25. Note for this case that at $x = 0, \Phi = \Phi_0$, and at $x = d$ (midway between the plates), $\Phi = \Phi_d$ and $d\Phi/dx = 0$. Numerical solutions for the nondimensional potential $y = y_d$ (i.e., $\Phi = \Phi_d$) for various values of z and $\xi = \kappa d$ (i.e., $x = d$) are given by Verwey and Overbeek (1948) (see also Fig. 1.26).

1.4.6 Repulsive Pressure

The repulsive pressure midway between two parallel clay plates (Fig. 1.27) can be given by the Langmuir equation

$$p = 2n_0 KT \left(\cosh \frac{ve\phi_d}{KT} - 1\right) \tag{1.43}$$

where p is the repulsive pressure, i.e., the difference between the osmotic pressure

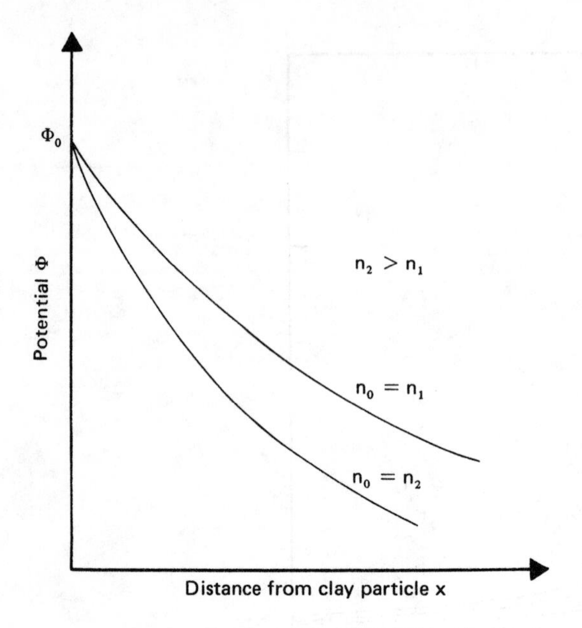

Fig. 1.23 Effect of cation concentration on the repulsive potential.

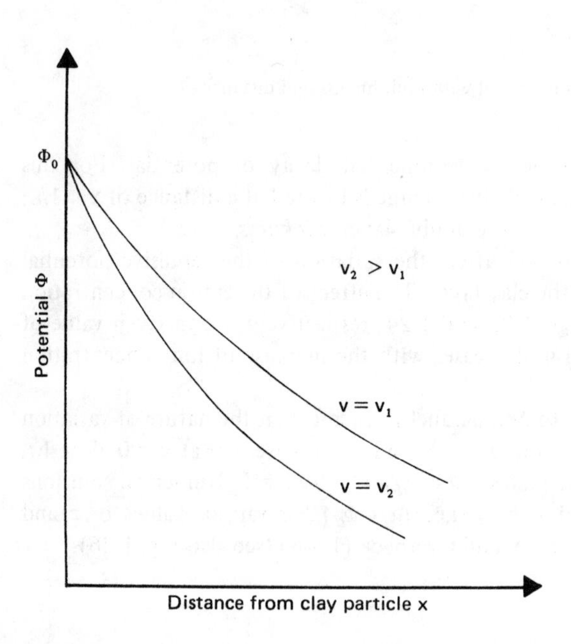

Fig. 1.24 Effect of ionic valence on the repulsive potential.

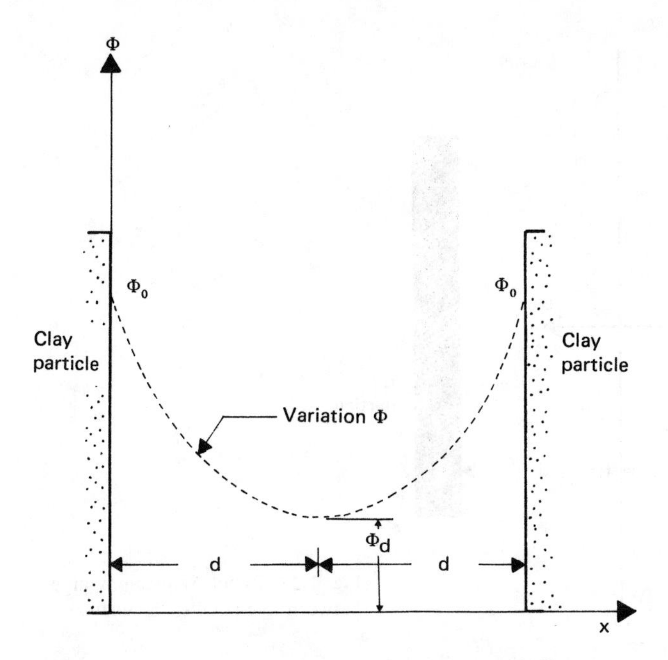

Fig. 1.25 Variation of Φ between two parallel clay particles.

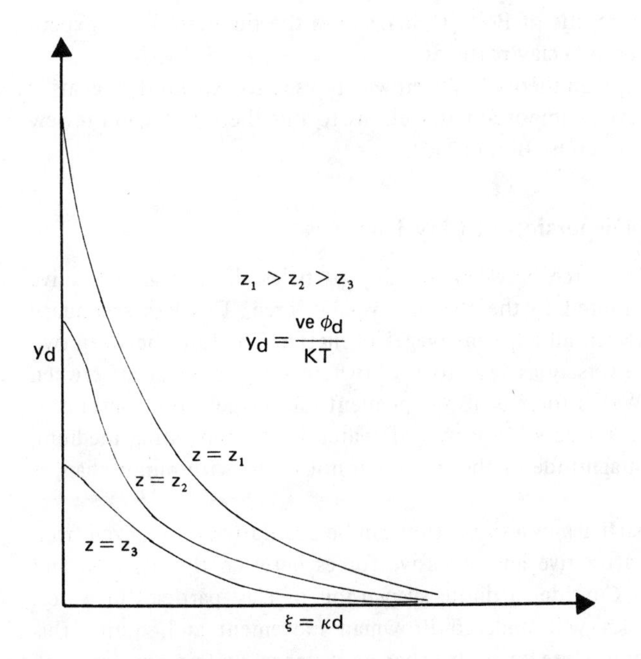

Fig. 1.26 Nature of variation of the nondimensional midplane potential for two parallel plates.

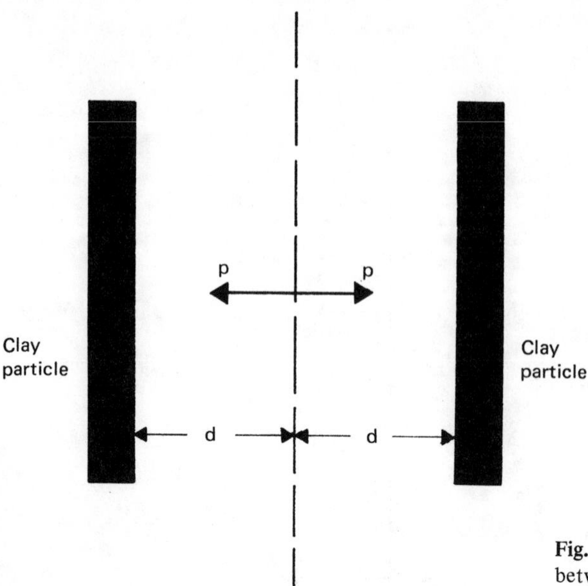

Fig. 1.27 Repulsive pressure midway between two parallel clay plates.

midway between the plates in relation to that in the equilibrium solution. Figure 1.28, which is based on the results of Bolt (1956), shows the theoretical and experimental variation of p between two clay particles.

Although the Guoy-Chapman theory has been widely used to explain the behavior of clay, there have been several important objections to this theory. A good review of these objections has been given by Bolt (1955).

1.4.7 Flocculation and Dispersion of Clay Particles

In addition to the repulsive force between the clay particles there is an attractive force, which is largely attributed to the Van der Waal's force. This is a secondary bonding force that acts between all adjacent pieces of matter. The force between two flat parallel surfaces varies inversely as $1/x^3$ to $1/x^4$, where x is the distance between the two surfaces. Van der Waal's force is also dependent on the dielectric constant of the medium separating the surfaces. However, if water is the separating medium, substantial changes in the magnitude of the force will not occur with minor changes in the constitution of water.

The behavior of clay particles in a suspension can be qualitatively visualized from our understanding of the attractive and repulsive forces between the particles and with the aid of Fig. 1.29. Consider a dilute suspension of clay particles in water. These colloidal clay particles will undergo Brownian movement and, during this random movement, will come close to each other at distances within the range of interparticle forces. The forces of attraction and repulsion between the clay particles vary at different rates with respect to the distance of separation. The force of repulsion decreases exponentially with distance, whereas the force of attraction decreases as the inverse third or fourth power of distance, as shown in Fig. 1.29. Depending on the distance of separation, if the magnitude of the repulsive force is greater than the magnitude of the attractive force, the net result will be repulsion. The clay particles will settle individually and form a dense layer at the bottom; however, they will remain separate from their neighbors (Fig. 1.30*a*). This is referred to as the *dispersed state* of the soil. On the other hand, if the net force between the particles is attraction, flocs will be formed and these flocs will settle to the bottom. This is called *flocculated* clay (Fig. 1.30*b*).

Salt flocculation and non-salt flocculation. We saw in Fig. 1.23 the effect of salt concentration, n_0, on the repulsive potential of clay particles. High salt concentration will depress the double layer of clay particles and hence the force of repulsion. We noted earlier in this section that the Van der Waal's force largely contributes to the

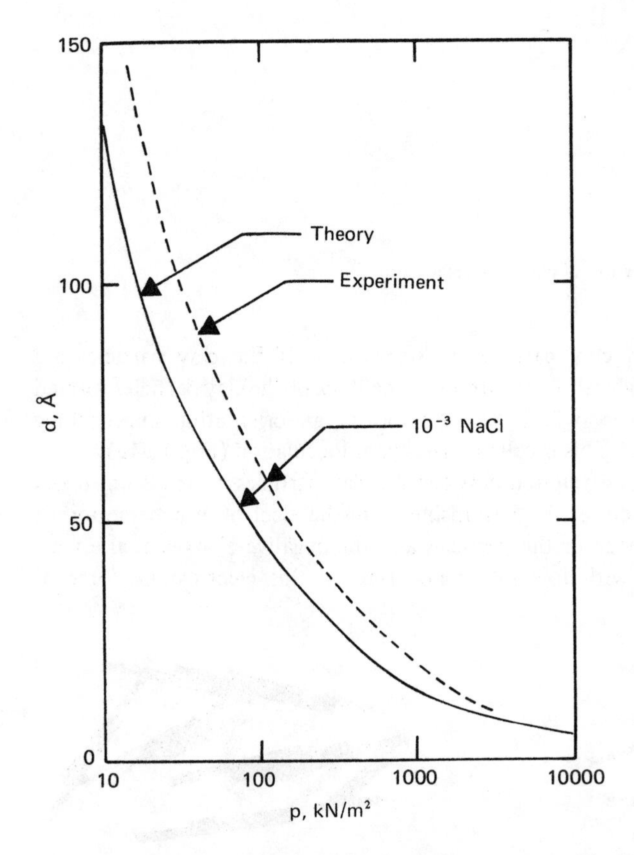

Fig. 1.28 Repulsive pressure between sodium montmorillonite clay particles. *(Redrawn after G. H. Bolt, Physical and Chemical Analysis of Compressibility of Pure Clay,* Geotechnique, *vol. 6, p. 86, 1956. By permission of the publisher, The Institution of Civil Engineers.)*

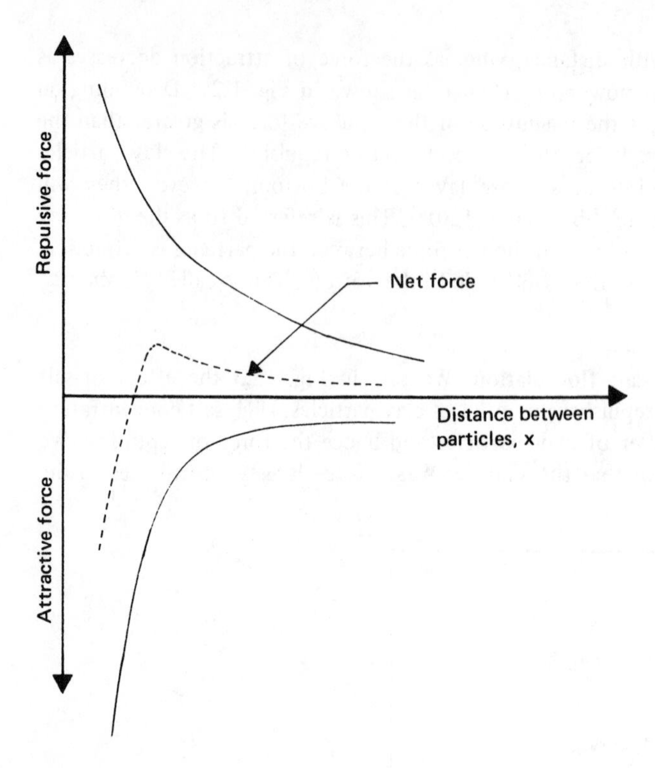

Fig. 1.29 Dispersion and flocculation of clay in a suspension.

force of attraction between clay particles in suspension. If the clay particles are suspended in water with a high salt concentration, the flocs of the clay particles formed by dominant attractive forces will give them mostly an orientation approaching parallelism (face-to-face type). This is called a salt-type flocculation (Fig. 1.31*a*).

Another type of force of attraction between the clay particles, which is not taken into account in colloidal theories, is that arising from the electrostatic attraction of the positive charges at the edge of the particles and the negative charges at the face. In a soil-water suspension with low salt concentration, this electrostatic force of

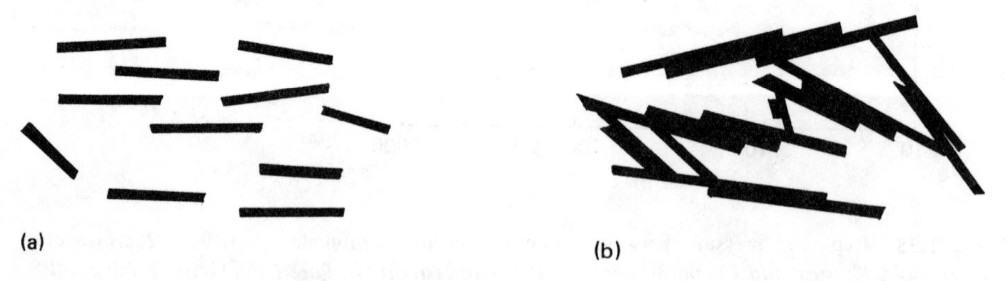

Fig. 1.30 (*a*) Dispersion and (*b*) flocculation of clay.

Fig. 1.31 (*a*) Salt and (*b*) non-salt flocculation of clay particles. *(After T. W. Lambe, Compacted Clay: Structure,* Trans. ASCE, *vol. 125, 1960.)*

attraction may produce a flocculation with an orientation approaching a perpendicular array. This is shown in Fig. 1.31*b* and is referred to as non-salt flocculation.

1.5 CONSISTENCY OF COHESIVE SOILS

1.5.1 Atterberg Limits

The presence of clay minerals in a fine-grained soil will allow it to be remolded in the presence of some moisture without crumbling. If a clay slurry is dried, the moisture content will gradually decrease and the slurry will pass from a liquid state to a plastic state. With further drying, it will change to a semisolid state and finally to a solid state as shown in Fig. 1.32. In about 1911, a Swedish scientist, A. Atterberg, developed a method for describing the limit consistency of fine-grained soils on the basis of moisture content. These limits are the *liquid limit,* the *plastic limit,* and the *shrinkage limit.*

The *liquid limit* is defined as the moisture content, in percent, at which the soil changes from a liquid state to a plastic state. The liquid limit is now generally determined by the standard Casagrande device (Casagrande, 1932, 1948). The moisture

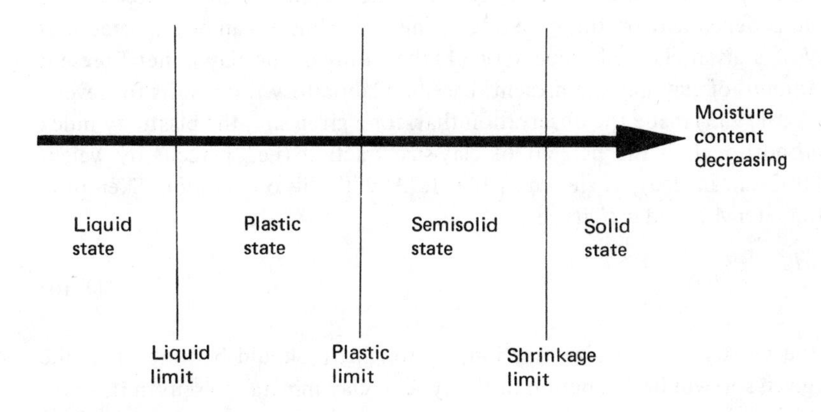

Fig. 1.32 Consistency of cohesive soils.

contents (in percent) at which the soil changes from a plastic to a semisolid state and from a semisolid to a solid state are defined, respectively, as the *plastic limit* and the *shrinkage limit.* These limits are generally referred to as the *Atterberg limits.*

The Atterberg limits of cohesive soil depend on several factors, such as amount and type of clay minerals and type of adsorbed cation.

The difference between the liquid limit and the plastic limit of a soil is defined as the *plasticity index PI:*

$$PI = LL - PL \tag{1.44}$$

where LL is the liquid limit and PL the plastic limit.

1.5.2 Liquidity Index

The relative consistency of a cohesive soil can be defined by a ratio called the *liquidity index LI.* It is defined as

$$LI = \frac{w_N - PL}{LL - PL} = \frac{w_N - PL}{PI} \tag{1.45}$$

where w_N is the natural moisture content. It can be seen from Eq. (1.45) that, if $w_N = LL$, then the liquidity index is equal to 1. Again, if $w_N = PL$, the liquidity index is equal to 0. Thus, for a natural soil deposit which is in a plastic state (i.e., $LL \geqslant w_N \geqslant PL$), the value of the liquidity index varies between 1 and 0. A natural soil deposit with $w_N > LL$ will have a liquidity index greater than 1. In an undisturbed state, these soils may be stable; however, a sudden shock may transform them into a liquid state. Such soils are called *sensitive clays.*

1.5.3 Activity

It was pointed out in Sec. 1.4.4 that oriented water (adsorbed and double layer) gives rise to the plastic property of a clay soil. The thickness of the oriented water around a clay particle is dependent on the type of clay mineral. Thus, it can be expected that the plasticity of a given clay will depend on (1) the nature of the clay mineral present and (2) the amount of clay mineral present. Based on laboratory test results for several soils, Skempton (1953) made the observation that, for a given soil, the plasticity index is directly proportional to the percent of clay-size fraction (i.e., percent by weight finer than 0.002 mm in size), as shown in Fig. 1.33. With this observation, Skempton defined a parameter A called *activity* as

$$A = \frac{PI}{C} \tag{1.46}$$

where C is the percent of clay-size fraction, by weight. It should be noted that the activity of a given soil will be a function of the type of clay mineral present in it.

The activities of several sand–clay mineral mixtures have been evaluated by Seed

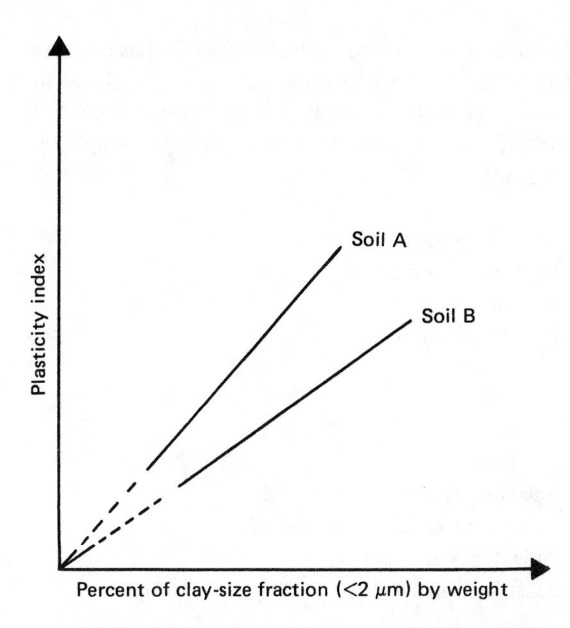

Fig. 1.33 Variation of plasticity index with the percent of clay-size fraction.

et al. (1964*b*). They concluded that although *PI* bears a linear relation to clay-size fractions, the line of correlation may not pass through the origin.

For practical purposes, it seems convenient to define activity as

$$A = \frac{\Delta(PI)}{\Delta C} \tag{1.47}$$

Activity has been used as an index property to determine the swelling potential of expansive clays (Sec. 1.8.2).

1.6 SOIL CLASSIFICATION

Soil classification is the arrangement of soils into various groups or subgroups to provide a common language to express briefly the general usage characteristics without detailed descriptions. At the present time, two major soil classification systems are available for general engineering use. They are the unified system, which is described below, and the AASHTO system. Both systems use simple index properties such as grain-size distribution, liquid limit, and plasticity index of soil.

1.6.1 Unified Soil Classification System

The unified system of soil classification was originally proposed by A. Casagrande in 1942 and was then revised in 1952 by the Corps of Engineers and the U.S. Bureau of Reclamation. In its present form, the system is widely used by various organizations, geotechnical engineers in private consulting business, and building codes.

Initially, there are two major divisions in the system. A soil is classified as a coarse-grained soil (gravelly and sandy) if more than 50% is retained on a No. 200 sieve and as a fine-grained soil (silty and clayey) if more than 50% is passing through a No. 200 sieve. The soil is then further classified by a number of subdivisions, as shown in Table 1.7. The following symbols are used:

G: gravel
S: sand
C: clay
M: silt
O: organic silt or clay
Pt: peat and highly organic soil

W: well-graded
P: poorly graded
H: high plasticity
L: low plasticity

Example 1.2 For a soil specimen, given:
passing No. 4 sieve = 92%
passing No. 40 sieve = 78%
passing No. 10 sieve = 81%
passing No. 200 sieve = 65%
liquid limit = 48
plasticity index = 32
Classify the soil by the unified classification system.

SOLUTION Since more than 50% is passing through a No. 200 sieve, it is a fine-grained soil, i.e., it could be *ML*, *CL*, *OL*, *MH*, *CH*, or *OH*. Now, if we plot $LL = 48$ and $PI = 32$ on the plasticity chart given in Table 1.7, it falls in the zone *CL*.

So the soil is classified as *CL*.

1.7 COMPACTION OF SOILS

1.7.1 Theory of Compaction and Proctor Compaction Test

Compaction of loose fills is a simple way of increasing the stability and load-bearing capacity of soils, and this is generally achieved by using smooth-wheel rollers, sheepsfoot rollers, rubber-tire rollers, and vibratory rollers.

In the compaction process, loose fills are placed in small lifts. Water is then added to the soil to serve as a lubricating agent on the soil particles. With the application of compacting effort, the soil particles slip over each other and move into a densely packed position. The effect of increasing the moisture content is demonstrated in Fig. 1.34. A silty clay when compacted dry with a compaction effort of 12,375 ft·lb/ft^3 (593 kJ/m^3) can be compacted to a unit weight of 85 lb/ft^3 (13.4 kN/m^3). However, as the moisture content is increased under the same compactive effort, the weight of soil solids in a unit volume gradually increases. A peak is reached with a moist unit weight of about 125 lb/ft^3 (19.65 kN/m^3) at a moisture content of about 20%. So the dry unit weight attained by adding water is

$$\gamma_d = \frac{\gamma}{1+w} = \frac{125}{1+0.2} = 104.17\,\text{lb/ft}^3\ (16.38\,\text{kN/m}^3)$$

If the moisture content is increased further, the soil becomes more workable. However, the moisture takes up the space that might have been occupied by soil solids; consequently, the moist unit weight decreases.

The degree of compaction of a soil is measured in terms of dry unit weight. For a given compactive effort, if the dry unit weight is plotted against the corresponding moisture content, the nature of the plot will be as shown in Fig. 1.35. The moisture content corresponding to the *maximum dry unit weight* is called the *optimum moisture content.*

Also plotted in Fig. 1.35 are the dry unit weights assuming the degree of saturation to be 100%. These are the *theoretical maximum* dry unit weights that can be attained for a given moisture content when there will be no air in the void spaces. With the degree of saturation as 100%,

$$e = wG_s \tag{1.14}$$

The maximum dry unit weight at a given moisture content with zero air voids can be given by [Eq. (1.12)]

$$\gamma_{\text{zav}} = \frac{G_s \gamma_w}{1+e} = \frac{G_s \gamma_w}{1 + wG_s} = \frac{\gamma_w}{1/G_s + w} \tag{1.48}$$

where γ_{zav} is the zero-air-void unit weight (dry).

The nature of the variation of γ_{zav} with moisture content is also shown in Fig. 1.35.

Proctor compaction test. A standard laboratory soil-compaction test was first developed by Proctor (1933), and this is usually referred to as the *standard Proctor test* (ASTM designation D-698, AASHTO designation T-99). The test is conducted by compaction of three layers of soil in a mold that is $\frac{1}{30}$ ft^3 (944 cm^3) in volume. Each layer of soil is subjected to 25 blows by a hammer weighing 5.5 lb (24.5 N) with a 12-in (304.8-mm) drop. From the known volume of the mold, weight of moist compacted soil in the mold, and moisture content of the compacted soil, the dry unit weight of compaction can be determined as

$$\gamma_{\text{moist}} = \frac{\text{weight of moist soil in the mold}}{\text{volume of mold}}$$

$$\gamma_d = \frac{\gamma_{\text{moist}}}{1 + w}$$

where w is the moisture content of soil.

The test can be repeated several times at various moisture contents of soil. By plotting γ_d against the corresponding moisture content, the optimum moisture content and the maximum dry unit weight can be obtained.

The plot of γ_d vs. moisture content of soils is influenced by several factors, the most important of which are the type and amount of compaction effort, the grain-size distribution, and the amount and type of clay minerals present. Figure 1.36 shows the γ_d vs. moisture content plots for eight different soils. Note that the compactive effort for all the soils is the same (standard Proctor compaction test) and is equal to

Table 1.7 Unified soil classification system

Major divisions	Group symbols	Typical names	Criteria of classification*
Course-grained soils (percent passing No. 200 sieve less than 50)			
Gravels (percent of coarse fraction passing No. 4 sieve less than 50)			
Gravels with little or no fines	*GW*	Well-graded gravels, gravel–sand mixtures (little or no fines)	$C_u = \frac{D_{60}}{D_{10}} > 4; C_c = \frac{(D_{30})^2}{D_{10} \times D_{60}}$ between 1 and 3
	GP	Poorly graded gravels, gravel–sand mixtures (little or no fines)	Not meeting the two criteria for *GW*
Gravels with fines	*GM*	Silty gravels, gravel–sand–silt mixtures	Atterburg limits below "A" line *or* plasticity index less than 4†
	GC	Clayey gravels, gravel–sand–clay mixtures	Atterburg limits above "A" line *with* plasticity index greater than 7†
Sands (percent of coarse fraction passing No. 4 sieve greater than 50)			
Clean sands (little or no fines)	*SW*	Well-graded sands, gravelly sands (little or no fines)	$C_u = \frac{D_{60}}{D_{10}} > 6; C_c = \frac{(D_{30})^2}{D_{10} \times D_{60}}$ between 1 and 3
	SP	Poorly graded sands, gravelly sands (little or no fines)	Not meeting the two criteria for *SW*
Sands with fines (appreciable amount of fines)	*SM*	Silty sands, sand–silt mixtures	Atterburg limits below "A" line *or* plasticity index less than 4†
	SC	Clayey sands, sand–clay mixtures	Atterburg limits above "A" line *with* plasticity index greater than 7†

Fine grained soils (percent passing No. 200 sieve greater than 50%)			
Silts and clay (liquid limit less than 50)	*ML*	Inorganic silts, very fine sands, rock flour, silty or clayey fine sands	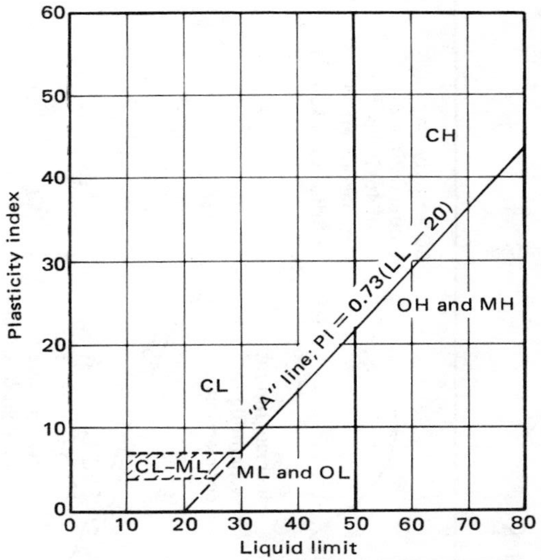
	CL	Inorganic clays (low to medium plasticity), gravelly clays, sandy clays, silty clays, lean clays	
	OL	Organic silts, organic silty clays (low plasticity)	
Silts and clay (liquid limit greater than 50)	*MH*	Inorganic silts, micaceous or diatomaceous fine sandy or silty soils, elastic silts	
	CH	Inorganic clays (high plasticity), fat clays	
	OH	Organic clays (medium to high plasticity), organic silts	
Highly organic soils	*Pt*	Peat, mulch, and other highly organic soils	

* Classification based on percentage of fines:

Percent passing No. 200	Classification
Less than 5	*GW*, *GP*, *SW*, *SP*
More than 12	*GM*, *GC*, *SM*, *SC*
5 to 12	Borderline – dual symbols required such as *GW–GM*, *GW–GC*, *GP–GM*, *GP–SC*, *SW–SM*, *SW–SC*, *SP–SM*, *SP–SC*

† Atterburg limits above "A" line and plasticity index between 4 and 7 are borderline cases. It needs dual symbols.

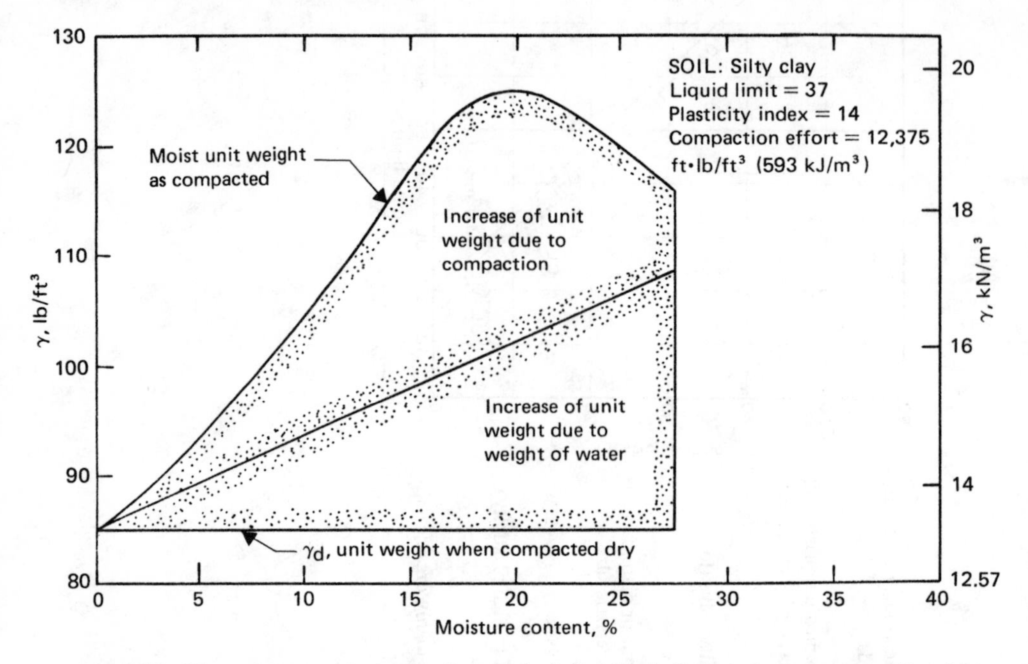

Fig. 1.34 The moisture content vs. unit weight relationship indicating the increased unit weight resulting from the addition of water and that due to the compaction effort applied. *(Redrawn after A. W. Johnson and J. R. Sallberg, Factors Influencing Compaction Test Results,* Highway Research Board, Bulletin 319, *1962.)*

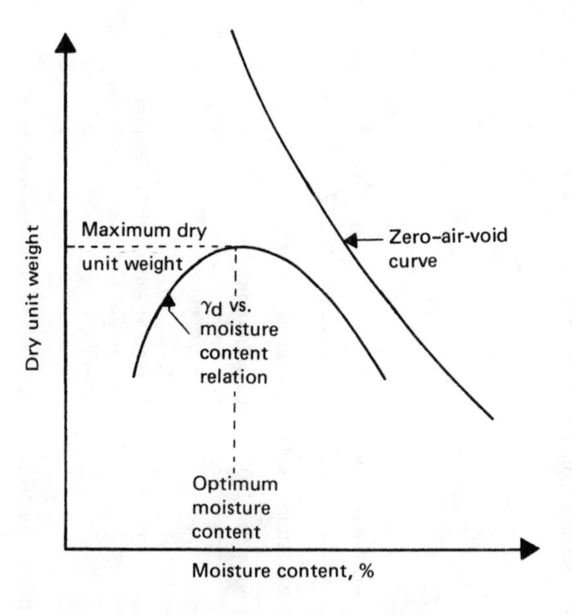

Fig. 1.35 Nature of the variation of dry unit weight of soil with moisture content in a compaction test.

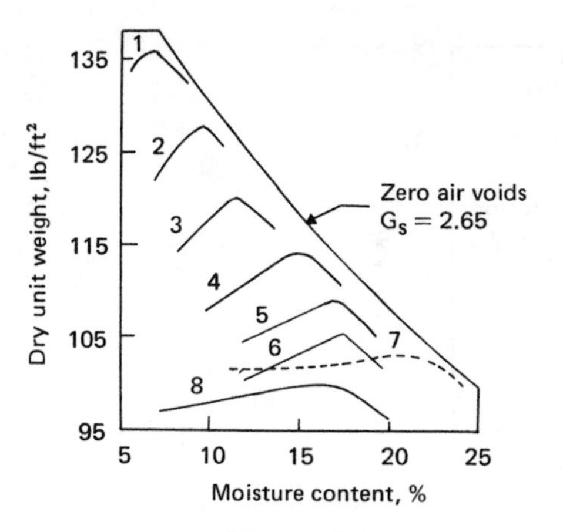

Soil Texture and Plasticity Data

No.	Description	Sand	Silt	Clay	LL	PI
1	Well-graded loamy sand	88	10	2	16	NP
2	Well-graded sandy loam	78	15	13	16	NP
3	Medium-graded sandy loam	73	9	18	22	4
4	Lean sandy silty clay	32	33	35	28	9
5	Lean silty clay	5	64	31	36	15
6	Loessial silt	5	85	10	26	2
7	Heavy clay	6	22	72	67	40
8	Poorly graded sand	94	−6−		NP	−

Fig. 1.36 Moisture content vs. dry unit weight relationships for eight soils according to AASHTO method T-99. (Note: $1\ \text{lb/ft}^3 = 157.21\ \text{N/m}^3$.) *(After A. W. Johnson and J. R. Sallberg, Factors Influencing Compaction Test Results,* Highway Research Board, Bulletin 319, *1962.)*

$$\frac{(5.5\ \text{lb/blow})\ (3\ \text{layers})\ (25\ \text{blows/layer})\ (1\text{-ft drop})}{\frac{1}{30}\ \text{ft}^3} = 12{,}375\ \text{ft}\cdot\text{lb/ft}^3$$

$$(\approxeq 593\ \text{kJ/m}^3)$$

The effect of compactive effort on the dry unit weight vs. moisture content relation is shown in Fig. 1.37. With increasing compactive effort the optimum moisture content decreases, and at the same time the maximum dry unit weight of compaction increases.

With the development of heavier compaction equipment, the standard Proctor test has been modified for better representation of field conditions. In the *modified Proctor test* (ASTM designation D-1577 and AASHTO designation T-180), the same mold as in the standard Proctor test is used. However, the soil is compacted in five layers with a 10-lb (44.5-N) hammer giving 25 blows to each layer. The height of drop of the hammer is 18 in (457.2 mm). Hence the compactive effort in the modified Proctor test is equal to

$$\frac{(25\ \text{blows/layer})\ (5\ \text{layers})\ 10\ \text{lb/blow})\ (1.5\text{-ft drop})}{\frac{1}{30}\ \text{ft}^3} = 56{,}250\ \text{ft}\cdot\text{lb/ft}^3$$

$$(\approxeq 2694\ \text{kJ/m}^3)$$

Conducting Proctor tests in sandy and gravelly soils is rather tedious because of lack of control over the moisture content. The nature of the dry unit weight vs. moisture content plot for a sand is shown in Fig. 1.38. With increasing moisture content, the dry unit weight gradually decreases and then increases up to the optimum

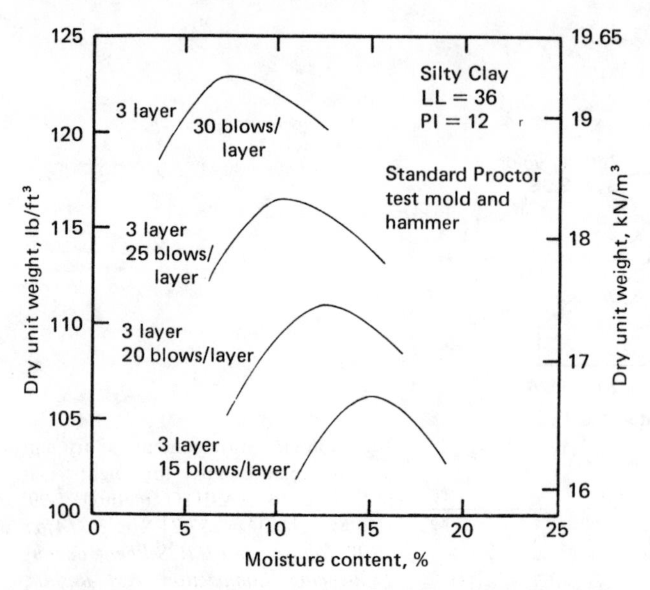

Fig. 1.37 Effect of compactive effort on dry unit weight vs. moisture content relation.

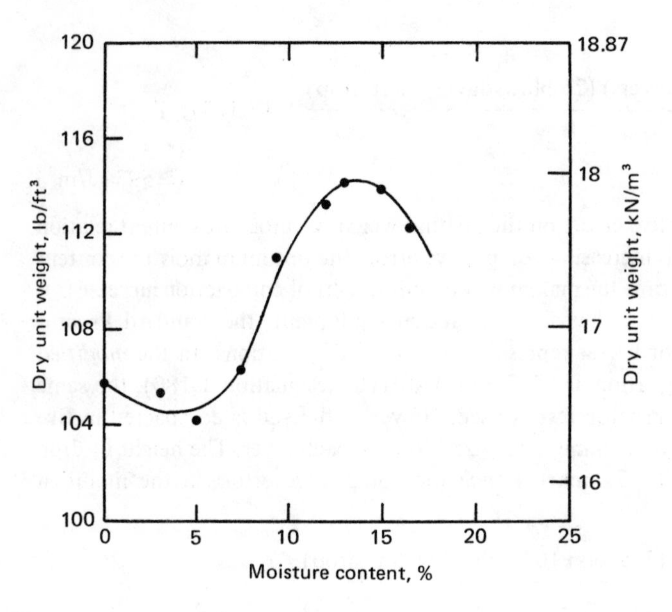

Fig. 1.38 Proctor compaction test results on a sand (AASHTO test designation T-99).

moisture content. The decrease of dry unit weights obtained at lower moisture contents is a result of the effect of capillary tension in the pore water. The capillary tension resists the movement of soil particles and thus prevents the soil from becoming densely packed.

1.7.2 Harvard Miniature Compaction Device

The Harvard miniature compaction device is used in the laboratory for compaction and preparation of soil specimens that are mostly used in research work. Unlike the Proctor test, the compaction is achieved by kneading. The volume of the mold of the Harvard miniature compaction device is $\frac{1}{454}$ ft^3 (62.4 cm^3). A tamper with a calibrated spring delivers the static pressure to the soil layers. The spring pressure may be 20 lb (89 N) or 40 lb (178 N). The number of layers of soil in the mold and the number of tamps can be varied, thus varying the energy of compaction per unit volume of soil.

1.7.3 Effect of Organic Content on Compaction of Soil

Soils with high percentages of organic content are often encountered during construction work. Increase of organic content in a soil tends to decrease the maximum dry unit weight of compaction and increase the compressibility of the soil, tendencies which are not desirable in the construction of foundations, embankments, etc. Franklin et al. (1973) studied the effect of organic contents on the strength and compaction characteristics of mechanical mixtures of inorganic soils and peat and of natural soil samples with the same organic content. The mineralogy of the inorganic fraction of these samples was reasonably the same. Samples for these tests were compacted in the Harvard miniature compaction device with three layers, 40-lb spring force, and 40 tamps per each layer. Figure 1.39 shows the variation of the maximum dry unit weight of compaction with the organic content, and the variation of the optimum moisture content with the organic content is shown in Fig. 1.40. The organic content O for these soils is defined as

$$O = \frac{\text{loss of dry weight due to heating the soil from 105 to 400°C}}{\text{dry weight (at 105°C)}} \tag{1.49}$$

Two major conclusions can be drawn from Figs. 1.39 and 1.40. (1) If the organic content in a given soil is more than about 10%, the maximum dry unit weight of compaction decreases considerably. (2) The optimum moisture content increases with the increase of organic contents of soil.

1.7.4 Field Compaction

The degree of compaction achieved in the field by a roller will depend on several factors that include:

1. Thickness of list.
2. Area over which the pressure is applied.

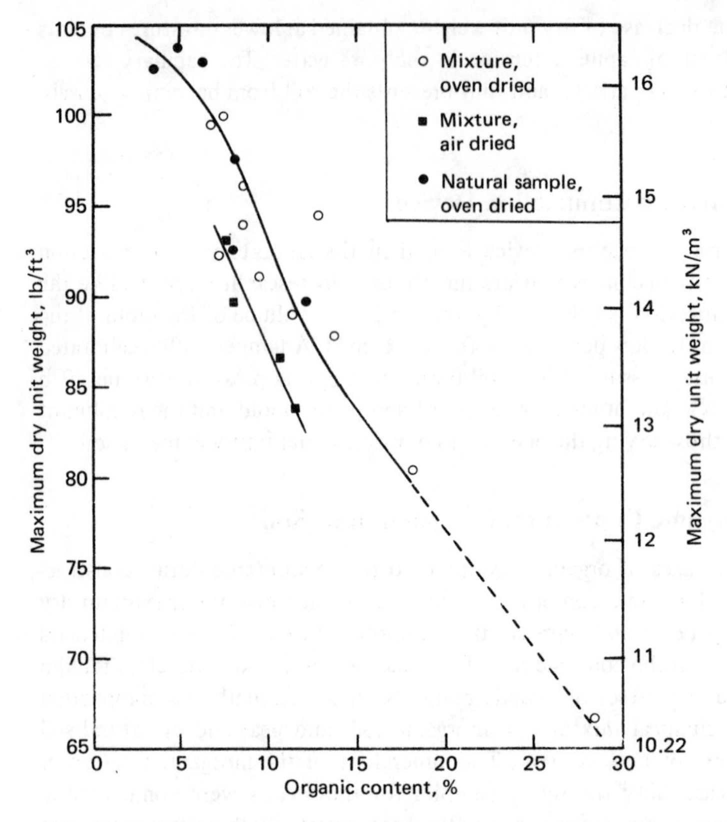

Fig. 1.39 Maximum dry unit weight vs. organic content for all compaction tests. (Note: 1 lb/ft^3 = 157.21 N/m^3.) *(Redrawn after A. F. Franklin, L. F. Orozco, and R. Semrau, Compaction of Slightly Organic Soils,* J. Soil Mech. Found. Div., ASCE, *vol. 99, no. SM7, 1973.)*

3. Intensity of pressure applied to the soil.
4. Type of roller.
5. Number of roller passes.

The growth curve during the field compaction of a lean clay is shown in Fig. 1.41. It can be seen that the dry unit weight of the soil gradually increases with the increase of the number of passes of a roller. After about eight to ten passes, the increase in γ_d of the soil is rather negligible. In most cases, the maximum economically attainable dry unit weight is achieved with about ten roller passes.

Vibratory rollers are particularly useful for compacting granular soils. Self-propelled and towed vibratory rollers of various sizes, weights, and vibration frequencies are available. The vibrations are generally produced by rotating an off-center weight. Vertical vibrations can be obtained by using two synchronized counter-rotating weights, as shown in Fig. 1.42. At any given instant, the horizontal forces developed by the counter-rotating weights are canceled out. Figure 1.43 shows the nature of compaction of a dune sand achieved by a vibratory roller after five passes. The low

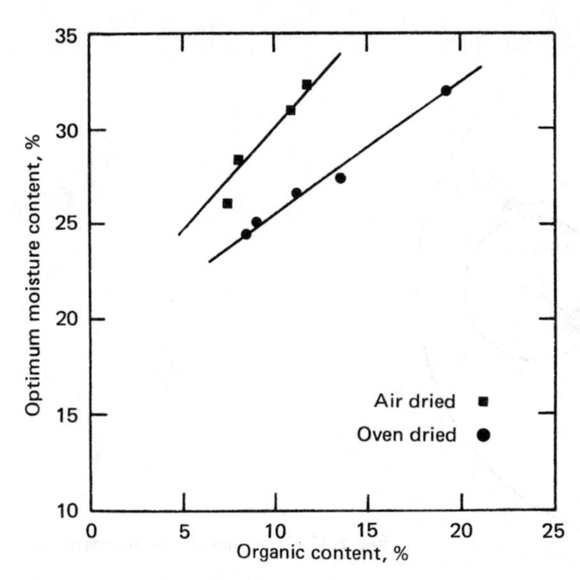

Fig. 1.40 Effect of drying history and organic content on optimum moisture content. *(After A. F. Franklin, L. F. Orozco, and R. Semrau, Compaction of Slightly Organic Soils,* J. Soil Mech. Found. Div., ASCE, *vol. 99, no. SM7, 1973.)*

unit weight that remains in the uppermost zone is due to vibration and lack of confinement in sand. Figure 1.44 shows the compacted unit-weight profiles for 8-ft lifts of the same dune sand (Fig. 1.43) for 2, 5, 15, and 45 roller passes. For field compaction work, the specification requires that the granular soil be compacted to a certain minimum relative density at all depths. Determination of the height of each lift depends on the type of roller and the economic number of passes. The method for determination of the lift height is shown in Fig. 1.45.

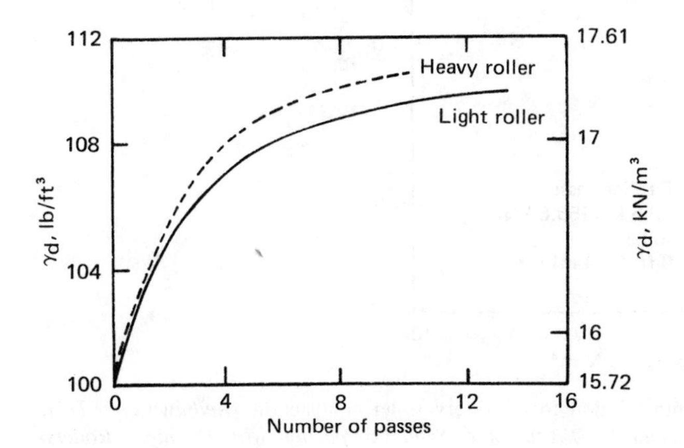

Fig. 1.41 Effect of number of passes on compaction of lean clay. *(After E. H. Yoder, "Principles of Pavement Design," Wiley, New York, 1959.)*

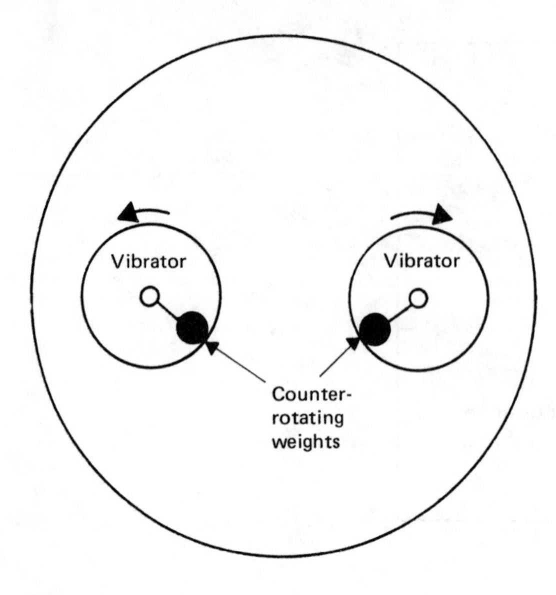

Fig. 1.42 Principles of vibratory rollers.

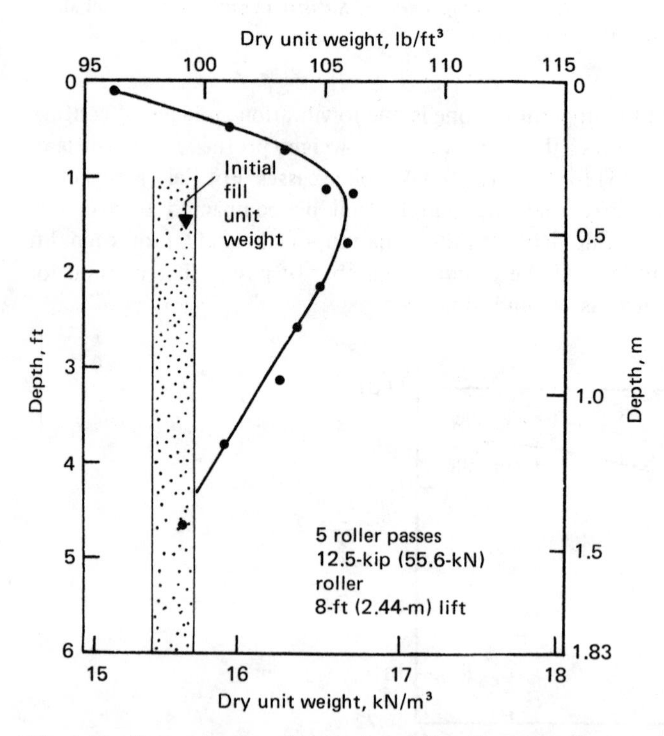

Fig. 1.43 Unit weight vs. depth relation for vibratory roller compaction. *(Redrawn after D. J. D'Appolonia, R. V. Whitman, and E. D'Appolonia, Sand Compaction with Vibratory Rollers,* J. Soil Mech. Found. Div., ASCE, *vol. 95, no. SM1, 1969.)*

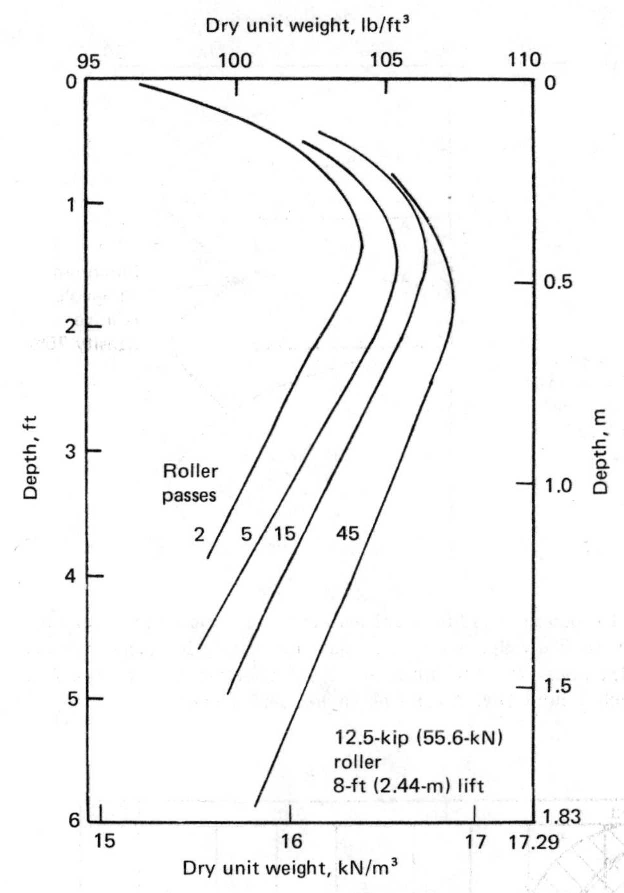

Fig. 1.44 Compacted unit weight profiles for 8-ft (2.44-m) lift heights for 2, 5, 15, and 45 vibratory roller passes. *(Redrawn after D. J. D'Appolonia, R. V. Whitman, and E. D'Appolonia, Sand Compaction with Vibratory Rollers,* J. Soil Mech. Found. Div., ASCE, *vol. 95, no. SM1, 1969.)*

1.7.5 In-Place Densification of Granular Soils

Several new techniques—such as the Terra-Proble method, Vibroflotation, blasting, and building sand compaction-piles—have been successfully used for compaction of in situ granular soils. In this section, the vibroflotation and blasting techniques will be treated in some detail.

Vibroflotation. The vibroflotation technique, which is patented by the Vibroflotation Foundation Company, is most suitable for the range of soil grain sizes shown in Fig. 1.46. The process involves the use of a device called a Vibroflot, which is a cylindrical piece of equipment about 6 ft ($\approx$2 m) long, 15 in ($\approx$400 mm) in diameter, and weighing about 4000 lb ($\approx$17.8 kN). An eccentric weight inside the cylinder develops a centrifugal force of about 10 tons (89 kN) at 1800 rpm. The device has water jets at the top

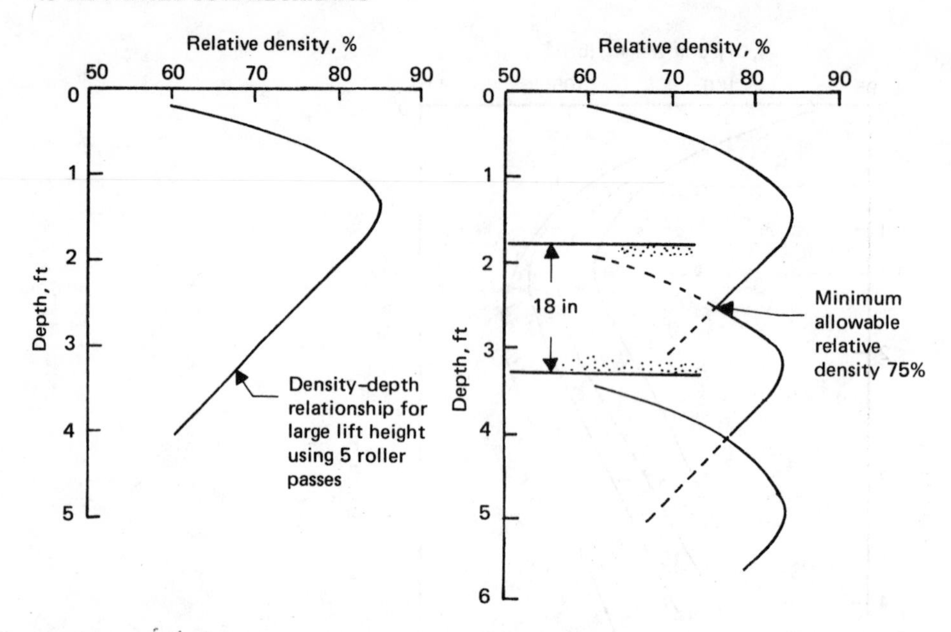

Fig. 1.45 Approximate method for determining lift height required to achieve a minimum compacted relative density of 75% with five roller passes using data for a large lift height. (Note: 1 ft = 0.3048 m.) *(After D. J. D'Appolonia, R. V. Whitman, and E. D'Appolonia, Sand Compaction with Vibratory Rollers,* J. Soil Mech. Found. Div., ASCE, *vol. 96, no. SM1, 1969.)*

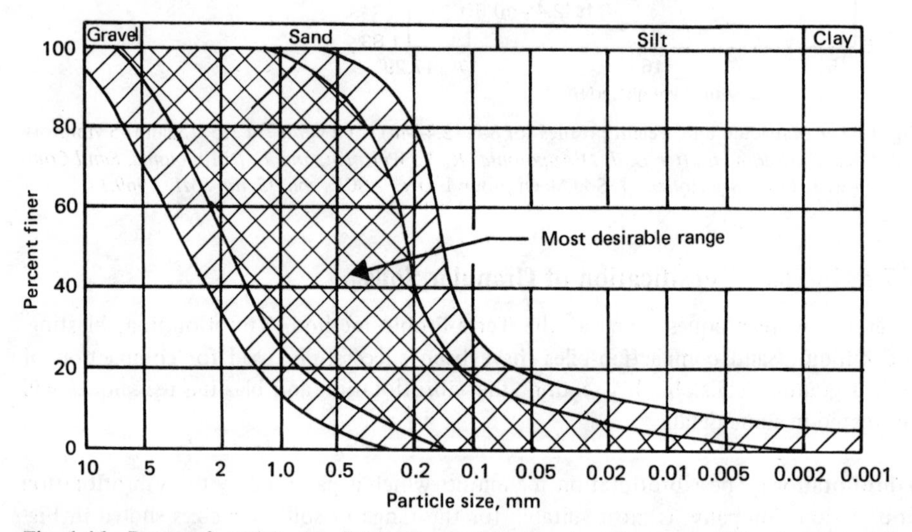

Fig. 1.46 Range of particle size distribution suitable for densification by Vibroflotation. *(After J. K. Mitchell, In-Place Treatment of Foundation Soils,* J. Soil Mech. Found. Div., ASCE, *vol. 96. no. SM1, 1970.)*

and bottom; these have a flow rate of 60 to 80 gal/min (0.2274 to 0.303 m^3/min) at a pressure of 60 to 80 lb/ft^2 (415 to 550 kN/m^2). The compaction process is illustrated in Fig. 1.47.

The Vibroflot sinks into the ground at the rate of 3 to 6 ft/min (≈1 to 2 m/min). When the desired depth is reached, the top jet is turned off. The device is then withdrawn at the rate of 1 ft/min (≈0.3 m/min) and sand is added from the top.

Examples of several Vibroflotation applications have been summarized by Mitchell (1970). In a regular working day, compaction of 10,000 to 20,000 yd^3 (≈2550 to 5100 m^3) is not uncommon.

Blasting. The range of soil grain sizes suitable for compaction by the blasting method is the same as for Vibroflotation (Fig. 1.46). In this method, compaction is achieved by successive detonations of small explosive charges in saturated soils. Relative densities of 70 to 80% up to a depth of 60 to 75 ft (≈20 to 25 m) can be achieved. In this method, explosive charges (60% dynamite, 30% special gelatin dynamite, and ammonite are most commonly used) are placed at about two-thirds the thickness of the stratum to be densified. The spacings of the charges vary from 10 to 25 ft (≈3 to 8 m). Three to five successive detonations of several spaced charges are usually required to achieve the desired compaction. Repeated shots are more effective than either a single large one or several small ones detonated simultaneously. The shock waves due to blasting cause liquefaction of the saturated sand, followed by densification. Practically no compaction is achieved in the top 3 ft (≈1 m), and so this zone usually needs recompaction by rollers. Mitchell (1970) has tabulated several projects where blasting has been used for densification.

The relation for the weight of charge and the sphere of influence for compaction can be given (Lyman, 1942; Mitchell, 1970) by the approximate relation

$$W = CR^3 \tag{1.50}$$

where W = weight of charge, lb
R = sphere of influence, ft
C = 0.0025 for 60% dynamite

If blasting is to be used in dry or partly saturated soils, preflooding is desirable.

1.8 VOLUME CHANGE OF SOILS

1.8.1 Shrinkage and Swelling of Clay

Clay soils undergo a volume change when the moisture content is changed; decrease of moisture content will cause shrinkage, increase of moisture content will result in swelling. The degree of change in volume depends on factors such as type and amount of clay minerals present in the soil, specific surface area of the clay, structure of the soil, pore-water salt concentration, valence of the exchangeable cation, etc. Large volume changes of soils have resulted in extensive structural damages.

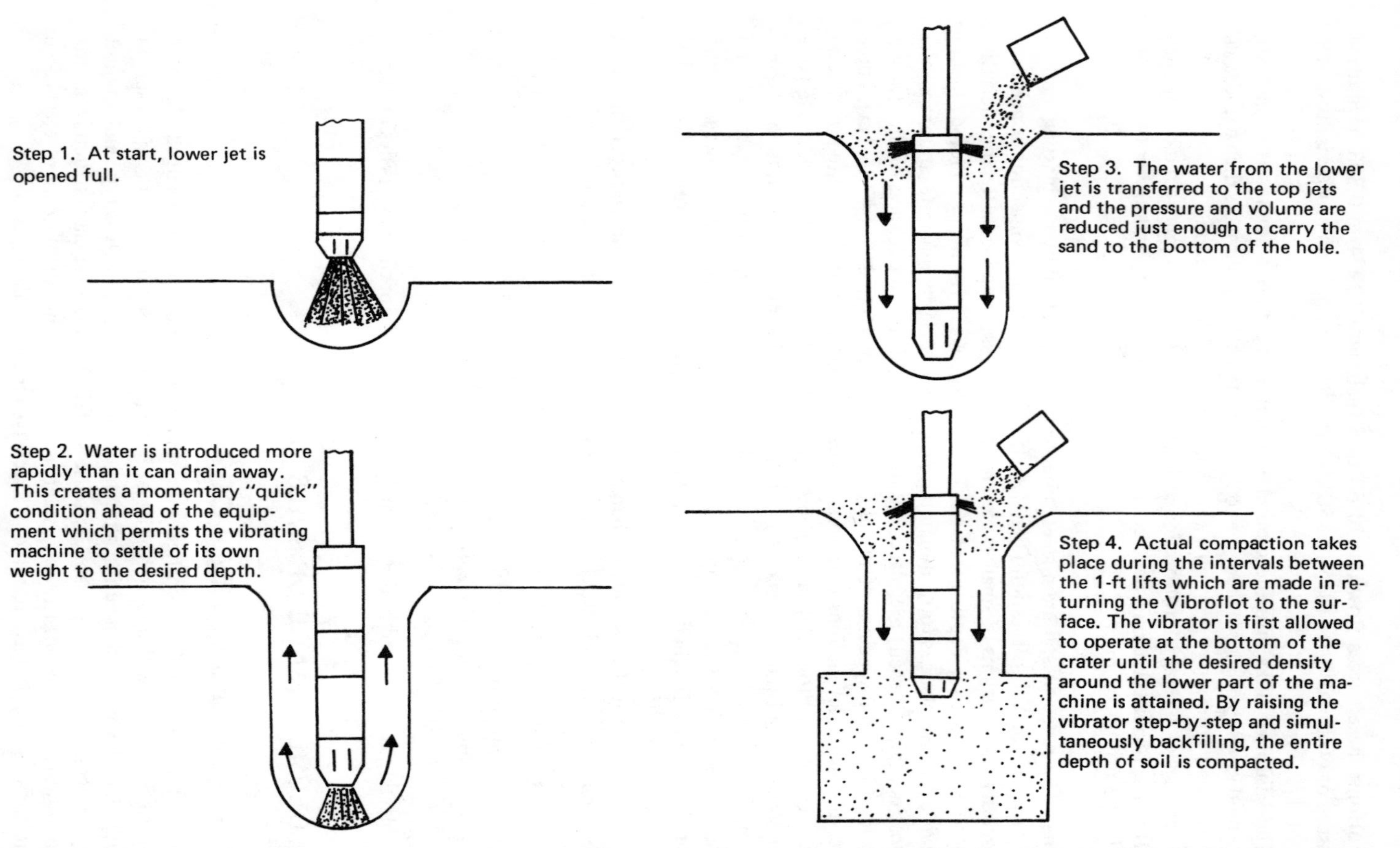

Fig. 1.47 Compaction process by Vibroflotation. *(Redrawn after J. K. Mitchell, In-Place Treatment of Foundation Soils,* J. Soil Mech. Found. Div., ASCE, *vol. 96, SM1, 1970.)*

The swelling of clays results from the increase in the thickness of the diffuse ion layer as water is supplied. Monovalent exchangeable sodium ions will cause greater swelling than divalent calcium ions.

Figure 1.48 shows the axial shrinkage of a silty clay soil. The shrinkage in this case was measured after compacting the samples at various moisture contents. Figure 1.49 shows the swelling pressures developed in a compacted sandy clay soil when samples were confined to approximately constant volume by means of compaction molds and pistons on their upper surfaces. For these tests, free access of water was given to the samples.

1.8.2 Swelling Potential of Clay Soils

Because of the potential hazards to foundations constructed on soils that undergo high volume change, there is a need to identify such soils. A number of attempts have been made by various investigators to develop a reliable method for their identification (e.g., Holtz and Gibbs, 1956; Bruijn, 1961; Jennings and Knight, 1958; Seed et al., 1962; Nayak and Christensen, 1971).

Seed et al. (1962) conducted several tests on laboratory-compacted sand–clay mineral mixtures to determine their swell potential. Swell potential is defined as the

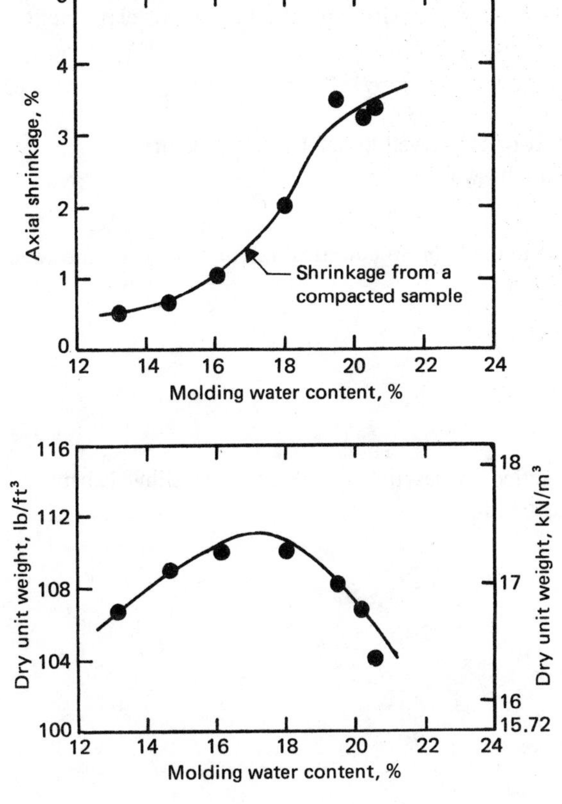

Fig. 1.48 Axial shrinkage of a silty clay. *(Redrawn after H. B. Seed and C. K. Chan, Structure and Strength Characteristics of Compacted Clays,* J. Soil Mech. Found. Div., ASCE, *vol. 85, no. SM5, 1959.)*

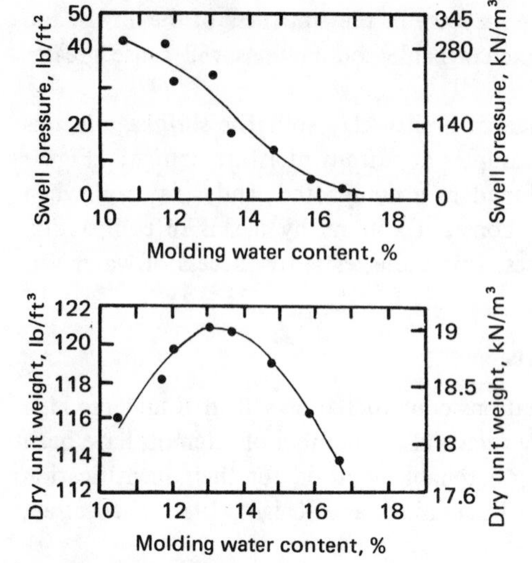

Fig. 1.49 Influence of molding water content on swell pressure of a sandy clay. *(After H. B. Seed and C. K. Chan, Structure and Strength Characteristics of Compacted Clays,* J. Soil Mech. Found. Div., ASCE, *vol. 95, no. SM5, 1959.)*

percentage of swell under a 1-psi (6.9-kN/m^2) surcharge of a laterally confined specimen compacted at optimum moisture content to maximum dry density in the standard AASHTO compaction test. Based on these tests, a well-defined relationship was established between the swell potential, the activity, and the percent of clay fraction (less than 2 μm size) in the soil:

$$S = (3.6 \times 10^{-5}) A^{2.44} C^{3.44} \tag{1.51}$$

where S = swell potential (percent of axial swell under 1-lb/ft^2 pressure)
C = percent of clay fraction, by weight
A = activity = $\Delta(PI)/\Delta C$

Equation (1.51) was further extended for an empirical relation between the swell potential and plasticity index of soil:

$$S = K(60)(PI)^{2.44}$$

where $K = 3.6 \times 10^{-5}$, or

$$S = (2.16 \times 10^{-3})(PI)^{2.44} \tag{1.52}$$

Conforming to the USBR practice to describe a soil on its swelling capability, Seed et al. made the following classification:

Degree of expansion	S
Low	0–1.5
Medium	1.5–5
High	5–25
Very high	>25

Figure 1.50 shows the classification chart for the swelling potential.

Ranganatham and Satyanarayana (1965) have suggested a correlation for the swell potential similar to that given by Seed et al. This is based on shrinkage index, swell activity, and the percentage of clay-size fraction in the soil. The *shrinkage index SI* is defined as the difference between the liquid limit and the plastic limit of soil. The shrinkage index bears a linear relation to the percent of the clay-size fraction by weight present in soil. This is shown in Fig. 1.51. *Swell activity SA* is defined as

$$SA = \frac{\Delta(SI)}{\Delta C} \tag{1.53}$$

The correlations for swell potential given by Ranganatham and Satyanarayana are

$$S = (4.57 \times 10^{-5})\,(SA)^{2.67}\,C^{3.44} \tag{1.54}$$

$$S = (41.13 \times 10^{-5})\,(SI)^{2.67} \tag{1.55}$$

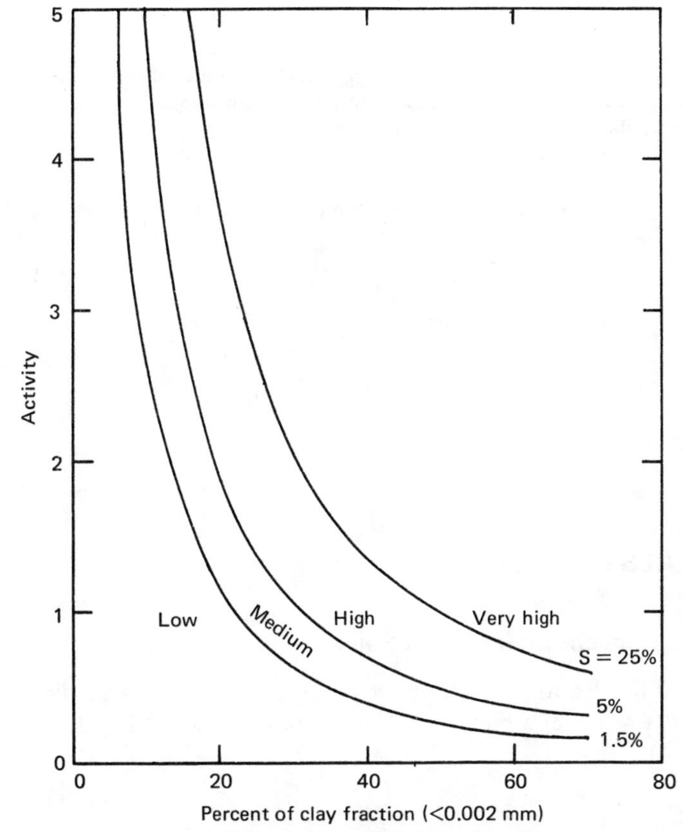

Fig. 1.50 Classification chart for swell potential. *(Redrawn after H. B. Seed, R. J. Woodward, and R. Lundgren, Prediction of Swelling Potential for Compacted Clays,* J. Soil Mech. Found. Div., ASCE, *vol. 88, no. SM3, 1962.)*

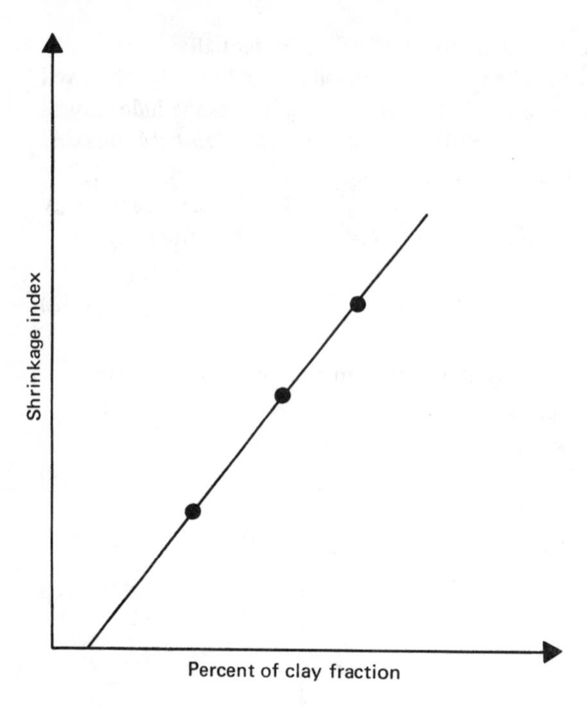

Fig. 1.51 Nature of variation of shrinkage index with percent of clay fraction.

The classifications for degree of soil expansion based on the shrinkage index are as follows:

Degree of expansion	*SI*
Low	0–20
Medium	20–30
High	30–60
Very high	>60

1.9 EFFECTIVE STRESS

1.9.1 Effective Stress Concept in Saturated Soils

Terzaghi (1925, 1936) was the first to suggest the principle of effective stress. According to this, the *total vertical stress* σ at a point O in a soil mass as shown in Fig. 1.52*a* can be given by

$$\sigma = h_1 \gamma + h_2 \gamma_{sat} \tag{1.56}$$

The total vertical stress σ consists of two parts. One part is carried by water and is continuous and acts with equal intensity in all directions. This is the *pore water pressure* or *neutral stress, u*. From Fig. 1.52*a*,

$$u = \gamma_w h_2 \tag{1.57}$$

The other part is the stress carried by the soil structure and is called the *effective stress* σ'. Thus

$$\sigma = \sigma' + u \tag{1.58}$$

Combining Eqs. (1.56) to (1.58),

$$\sigma' = \sigma - u = (h_1\gamma + h_2\gamma_{sat}) - h_2\gamma_w = h_1\gamma + h_2(\gamma_{sat} - \gamma_w)$$

$$= h_1\gamma + h_2\gamma' \tag{1.59}$$

where γ' is the submerged unit weight of soil, $\gamma_{sat} - \gamma_w$.

For dry soils, $u = 0$, so $\sigma = \sigma'$.

In general, if the normal total stresses at a point in a soil mass are σ_1, σ_2, and σ_3 as shown in Fig. 1.53, the effective stresses can be given as follows:

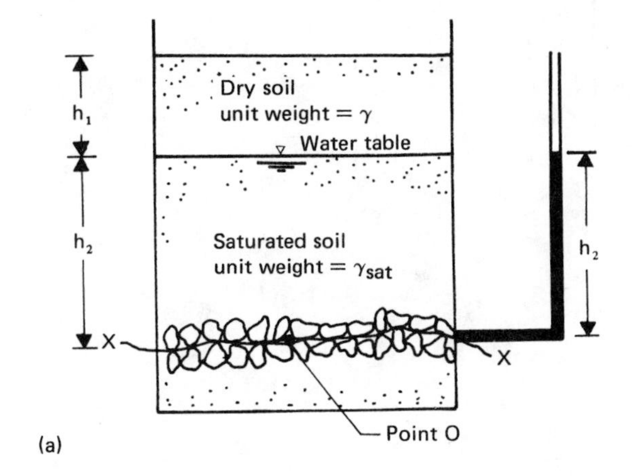

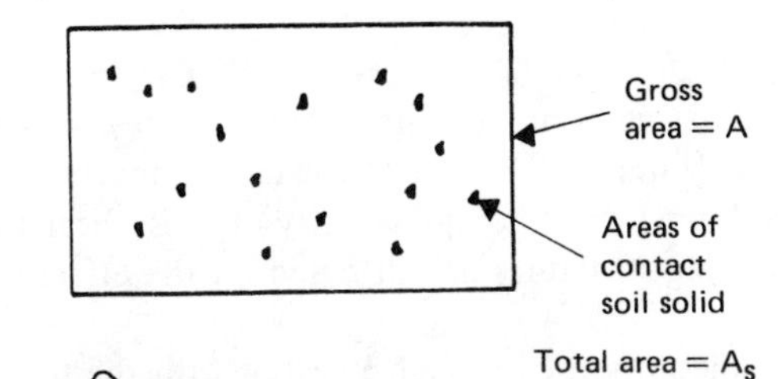

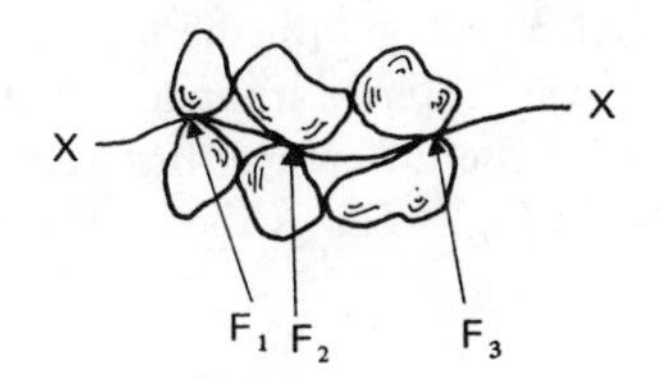

Fig. 1.52 Effective-stress concept. (*a*) Section. (*b*) Section at the level of *O*. (*c*) Forces carried by soil solids at their place of contact.

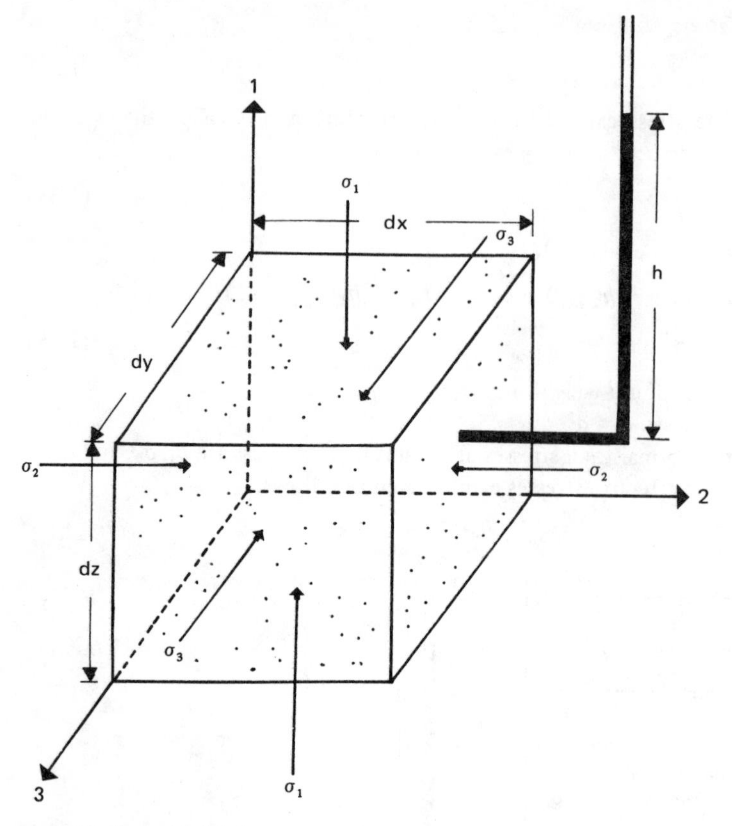

Fig. 1.53 Normal total stresses in a soil mass.

In direction 1, $\sigma_1' = \sigma_1 - u$

In direction 2, $\sigma_2' = \sigma_2 - u$

In direction 3, $\sigma_3' = \sigma_3 - u$

where σ_1', σ_2', and σ_3' are the effective stresses and u is the pore water pressure, $h\gamma_w$.

The principle of effective stress [Eq. (1.58)] is one of the most important findings in soil mechanics. The present developments on compressibility of soils, shear strength, and lateral earth pressure on retaining structures are all based on the effective stress concept.

The term effective stress is sometimes used interchangeably with the term intergranular stress by soils and foundation engineers. Although the terms are approximately the same, there is some difference. In order to visualize the difference, first refer to Fig. 1.52. The total vertical force F at the level of O in Fig. 1.52*a* is the sum of the following forces:

1. The force carried by soil solids at their point of contact, F_s. This can be seen by considering a wavy surface XX which passes through the point O and the points of contact of the solid particles. $F_1, F_2, F_3, \ldots$ are the resultant forces acting at the points of contact of the soil solids. So,

   $$F_s = F_{1(v)} + F_{2(v)} + F_{3(v)} + \cdots$$

   where $F_{1(v)}, F_{2(v)}, F_{3(v)}, \ldots$ are the vertical components of the forces $F_1, F_2, F_3, \ldots$.
2. The force carried by water, F_w,

   $$F_w = u(A - A_s)$$

   where u = pore water pressure $= \gamma_w h_2$
   A = gross area of cross section of soil (Fig. 1.52*b*)
   A_s = area occupied by soil solid-to-solid contact (Fig. 1.52*b*)
3. The electrical attractive force between the solid particles at the level of O, F_A
4. The electrical repulsive force between the solid particles at the level of O, F_R.

Thus, the total vertical force is

$$F = F_s + F_w - F_A + F_R$$

or

$$\sigma = \frac{F}{A} = \frac{F_s}{A} + \frac{F_w}{A} - \frac{F_A}{A} + \frac{F_R}{A}$$

where σ is the total stress at the level of O, and so

$$\sigma = \sigma_{ig} + u\left(1 - \frac{A_s}{A}\right) - A' + R' = \sigma_{ig} + u(1 - a) - A' + R'$$

where $\sigma_{ig} = F_s/A$ = intergranular stress
$a = A_s/A$
$A' = F_A/A$ = electrical attractive force per unit area of cross section of soil
$R' = F_R/A$ = electrical repulsive force per unit area of cross section of soil

Hence

$$\sigma_{ig} = \sigma - u(1 - a) + A' - R' \tag{1.60}$$

The value of a in the above equation is very small in the working stress range. We can thus approximate Eq. (1.60) as

$$\sigma_1 = \sigma - u - A' + R' \tag{1.61}$$

For granular soils, silts, and clays of low plasticity, the magnitudes of A' and R' are small; so, for all practical purposes, the intergranular stress becomes

$$\sigma_{ig} \approx \sigma - u \tag{1.62}$$

For this case, Eqs. (1.58) and (1.62) are similar and $\sigma' = \sigma_{ig}$. However, if $A' - R'$ is large, $\sigma_{ig} \neq \sigma'$. Such situations can be encountered in highly plastic, dispersed clays.

1.9.2 Critical Hydraulic Gradient and Boiling

Consider a condition where there is an upward flow of water through a soil layer, as shown in Fig. 1.54*a*. The total stress at a point O is

$$\sigma = h_1\gamma_w + h_2\gamma_{sat} \tag{1.63}$$

where γ_{sat} is the saturated unit weight of soil. The pore water pressure at O is

$$u = (h_1 + h_2 + x)\gamma_w \tag{1.64}$$

and the effective stress at O is

$$\sigma' = \sigma - u = (h_1\gamma_w + h_2\gamma_{sat}) - (h_1 + h_2 + x)\gamma_w$$

$$= h_2\gamma' - x\gamma_w \tag{1.65}$$

If the flow rate of water through the soil is continuously increased, the value of x will increase and will reach a condition where $\sigma' = 0$. This condition is generally referred to as *boiling.* Since the effective stress in the soil is zero, the soil will not be stable. Thus

$$\sigma' = 0 = h_2\gamma' - x\gamma_w$$

or

$$i_{cr} = \frac{x}{h_2} = \frac{\gamma'}{\gamma_w} \tag{1.66}$$

where i_{cr} is the critical hydraulic gradient.

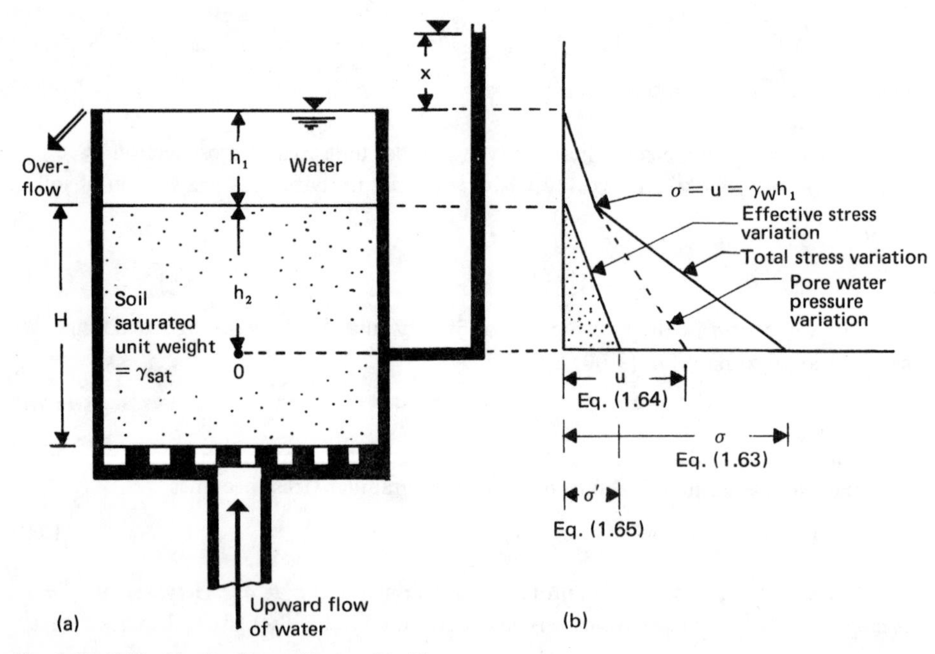

Fig. 1.54 Critical hydraulic gradient and boiling.

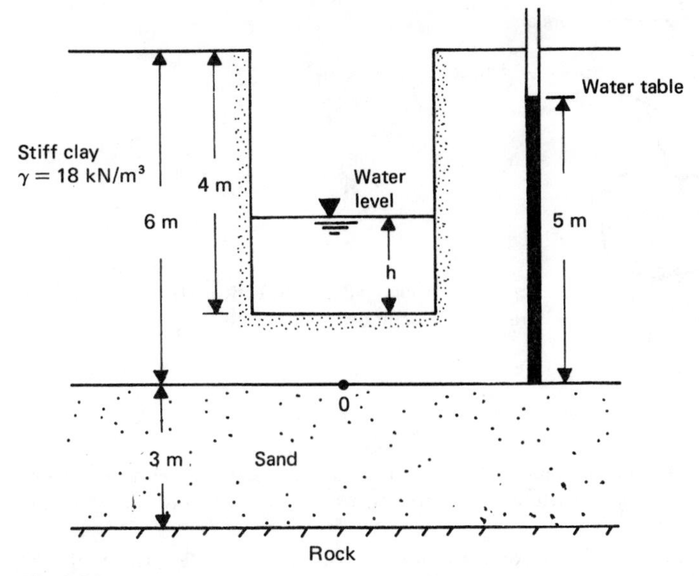

Fig. 1.55

Example 1.3 The sand layer of the soil profile shown in Fig. 1.55 is under artesian pressure. A trench is to be excavated in the clay up to a depth of 4 m. Determine the depth of water h to avoid boiling.

SOLUTION Referring to Fig. 1.55, at O the total stress is $\sigma = h\gamma_w + 2\gamma_{clay}$ and the pore water pressure is $u = 5\gamma_w$. For boiling, the effective stress $\sigma' = 0$. So, $h\gamma_w + 2\gamma_{clay} - 5\gamma_w = 0$, and

$$h = \frac{5\gamma_w - 2\gamma_{clay}}{\gamma_w} = \frac{5(9.81) - 2(18)}{9.81} = 1.33\,m$$

1.9.3 Effective Stress in Unsaturated Soils

The effective stress relation for saturated soils was given in Eq. (1.58) as $\sigma' = \sigma - u$. In unsaturated soils, the void spaces are occupied both by water and by air (Fig. 1.56). For such cases, the equation for the effective stress may be modified to read as (Bishop et al., 1960).

$$\sigma' = \sigma - u_a + \chi(u_a - u_w) \tag{1.67}$$

where u_a = pore air pressure
u_w = pore water pressure
χ = fraction of unit cross-sectional area of soil occupied by water

For saturated soils $\chi = 1$, and for dry soils $\chi = 0$. Bishop et al. (1960) have determined the nature of the variation of χ with the degree of saturation for several soils, based on their triaxial test results for unsaturated soil specimens.

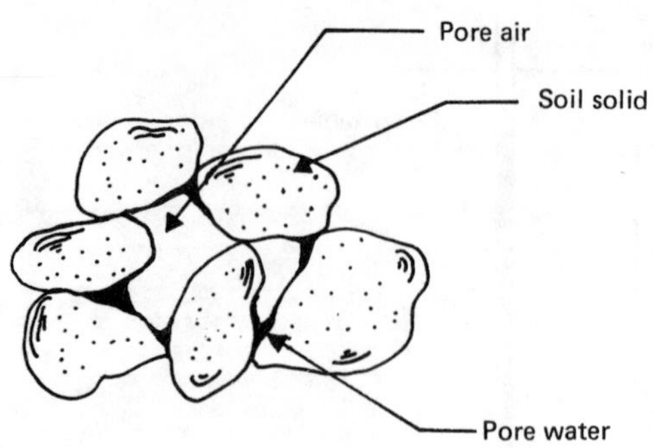

Fig. 1.56 Effective stress in partially saturated soil.

PROBLEMS

1.1 For a given soil, the in situ void ratio is 0.72 and $G_s = 2.61$. Calculate the porosity, dry unit weight (lb/ft³ and kN/m³), and the saturated unit weight. What would the moist unit weight be when the soil is 60% saturated?

1.2 A saturated clay soil has a moisture content of 40%. Given that $G_s = 2.78$, calculate the dry unit weight and saturated unit weight of the soil. Calculate the porosity of the soil.

1.3 For an undisturbed soil, the total volume is 5.1 ft³, the moist weight is 601 lb, the dry weight is 522 lb, and the void ratio is 0.6. Calculate the moisture content, dry unit weight, moist unit weight, degree of saturation, porosity, and G_s.

1.4 If a granular soil is compacted to a moist unit weight of 20.45 kN/m³ at a moisture content of 18%, what is the relative density of the compacted soil, given $e_{max} = 0.85$, $e_{min} = 0.42$, and $G_s = 2.65$?

1.5 For Prob. 1.4, what is the relative compaction?

1.6 From the results of a sieve analysis given below plot a graph for percent finer vs. grain size and then determine (*a*) the effective size, (*b*) the uniformity coefficient, and (*c*) the coefficient of gradation.

U.S. sieve No.	Mass of soil retained on each sieve, g
4	12.0
10	48.4
20	92.5
40	156.5
60	201.2
100	106.8
200	162.4
Pan	63.2

Would you consider this to be a well-graded soil?

1.7 The grain-size distribution curve for a soil is given in Fig. P1.1. Determine the percent of gravel, sand, silt, and clay present in this sample according to the M.I.T. soil-separate size limits (Fig. 1.11).

1.8 For a natural silty clay, the liquid limit is 55, the plastic limit is 28, and the percent finer than 0.002 mm is 29%. Estimate its activity.

1.9 Classify the following soils according to the unified soil classification system.

Soil	Percent passing U.S. sieve							*LL*	*PL*
	No. 4	No. 10	No. 20	No. 40	No. 60	No. 100	No. 200		
A	94	63	21	10	7	5	3		NP
B	98	80	65	55	40	35	30	28	18
C	98	86	50	28	18	14	20		NP
D	100	49	40	38	26	18	10		NP
E	80	60	48	31	25	18	8		NP
F	100	100	98	93	88	83	77	63	48

1.10 The results of a standard laboratory Proctor test are given below. Plot the curve for dry unit weight vs. moisture content. Determine the optimum moisture content and the maximum dry unit weight.

Moisture content, %	Weight of moist soil in the Proctor mold, lb
10.4	3.57
12.5	3.69
13.8	3.79
15.0	3.92
16.0	3.98
17.1	3.95
18.2	3.83
19.8	3.72

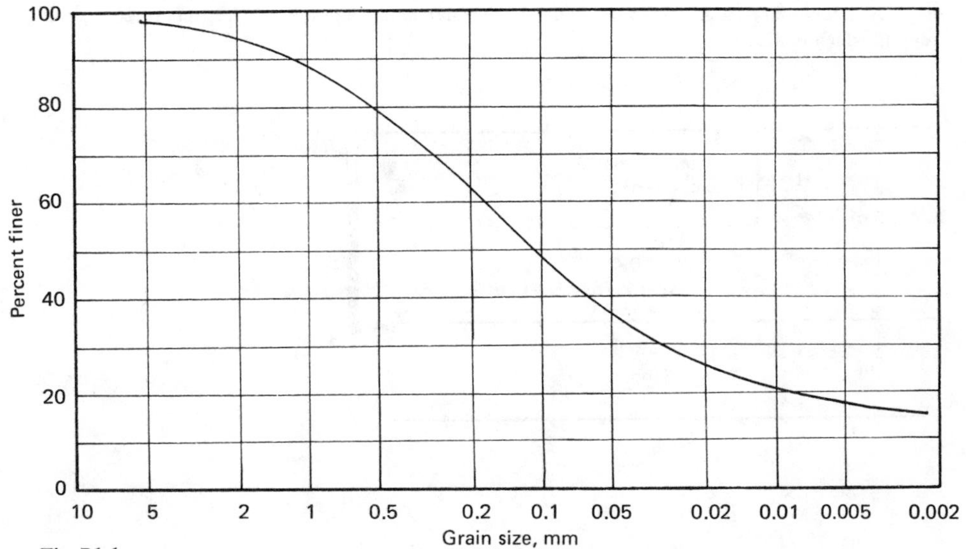

Fig. P1.1

1.11 Calculate the zero-air-void unit weights γ_{zav} for $G_s = 2.68$ and with the moisture content w varying from 0 to 20%. Plot your results on a graph (γ_{zav} vs. w).

1.12 For a clayey soil, $LL = 65$ and $PL = 31$. What would be its swell potential? How would you classify this, based on the degree of expansion—low, medium, high, or very high?

1.13 A soil profile is shown in Fig. P1.2. Draw the variation of total stress, pore water pressure, and effective stress with depth. Use SI units for your calculations.

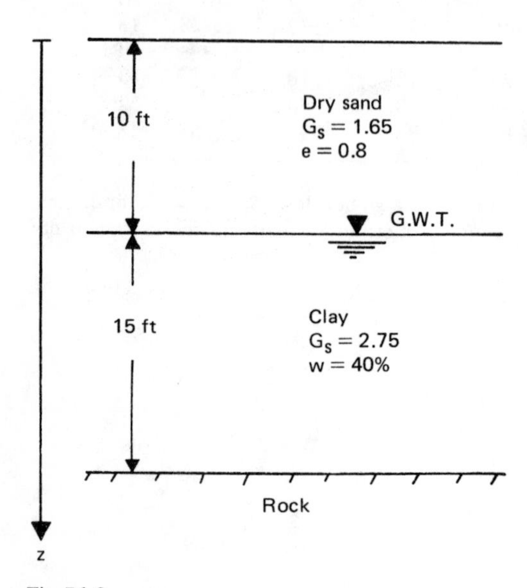

Fig. P1.2

1.14 A trench is to be excavated in a clay layer as shown in Fig. P1.3. The sand layer below the clay layer is under artesian pressure. What is the maximum depth H of the trench excavation beyond which boiling would occur?

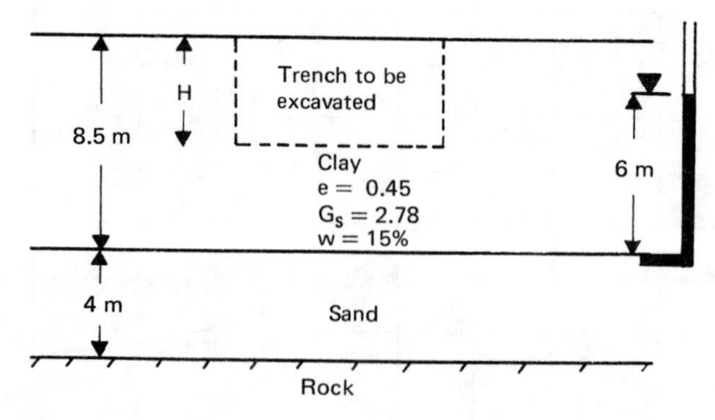

Fig. P1.3

REFERENCES

American Society for Testing and Materials, "Special Procedure for Testing Soil and Rock for Engineering Purposes," STP 479, 5th ed., 1970.

Atterberg, A., Uber die Physikalische Bodenuntersuschung und Uber die Plastizitat der Tone, *Int. Mitt. Bodenkunde,* vol. 1, p. 5, 1911.

Bishop, A. W., I. Alpan, G. E. Blight, and J. B. Donald, Factors Controlling the Strength of Partially Saturated Cohesive Soils, *Proc. Res. Conf. Shear Strength Cohesive Soils, ASCE,* pp. 503–532, 1960.

Bolt, G. H., Analysis of Validity of Gouy-Chapman Theory of the Electric Double Layer, *J. Colloid Sci.,* vol. 10, p. 206, 1955.

Bolt, G. H., Physical Chemical Analysis of Compressibility of Pure Clay, *Geotechnique,* vol. 6, p. 86, 1956.

Bowles, J. E., "Engineering Properties of Soils and Their Measurement," 2d ed., McGraw-Hill, New York, 1978.

Bruijn, C. J., Swelling Characteristics of a Transported Soil Profile of Leehof Vereenigning (Transvall), *Proc. 5th Int. Conf. Soil Mech. Found. Eng.,* vol. 1, p. 43, 1961.

Casagrande, A., Research on Atterberg Limits of Soils, *Public Roads,* vol. 13, pp. 121–130, 136, 1932.

Casagrande, A., "Classification and Identification of Soils," *Trans. ASCE,* vol. 113, 1948.

Casagrande, A., Notes on the Design of the Liquid Limit Device, *Geotechnique,* vol. 8, no. 2, pp. 84–91, 1958.

D'Appolonia, D. J., R. V. Whitman, and E. D'Appolonia, Sand Compaction with Vibratory Rollers, *J. Soil Mech. Found. Div., ASCE,* vol. 95, no. SM1, pp. 263–284, 1969.

Franklin, A. F., L. F. Orozco, and R. Semrau, Compaction of Slightly Organic Soils, *J. Soil Mech. Found. Div., ASCE,* vol. 99, no. SM7, pp. 541–557, 1973.

Grim, R. E., Physico-Chemical Properties of Soils, *J. Soil Mech. Found. Div., ASCE,* vol. 85, no. SM2, pp. 1–17, 1959.

Holtz, W. G., and H. T. Gibbs, Engineering Properties of Expansive Soils, *Trans. ASCE,* p. 641, 1956.

Jennings, J. E. B., and K. Knight, "The Prediction of Total Heave from Double Odometer Test," Symposium on Expansive Clays, South African Institute of Civil Engineers, p. 13, 1958.

Johnson, A. W., and J. R. Sallberg, Factors that Influence Field Compaction of Soils, *Highway Research Board Bull. 272,* 1960.

Johnson, A. W., and J. R. Sallberg, "Factors Influencing Compaction Test Results," *Highway Research Board Bull.* 319, 1962.

Jumikis, A. R., "Soil Mechanics," Van Nostrand, Princeton, N.J, 1962.

Lambe, T. W., Compacted Clay: Structure, *Trans. ASCE,* vol. 125, pp. 682–717, 1960.

Lambe, T. W., and R. V. Whitman, "Soil Mechanics," Wiley, New York, 1969.

Lee, K. L., and A. Singh, Relative Density and Relative Compaction, *J. Soil Mech. Found. Div., ASCE,* vol. 97, no. SM7, pp. 1049–1052, 1971.

Lyman, A. K. B., Compaction of Cohesionless Foundation Soils by Explosives, *Trans. ASCE,* vol. 107, pp. 1330–1348, 1942.

Mitchell, J. K., In-Place Treatment of Foundation Soils, *J. Soil Mech. Found. Div., ASCE,* vol. 96, no. SM1, pp. 73–110, 1970.

Nayak, N. V., and R. W. Christensen, Swelling Characteristics of Compacted Expansive Soils, *Clay and Clay Min.,* vol. 19, pp. 251–261, 1971.

Proctor, E. R., Design and Construction of Rolled Earth Dams, *Eng. News Record,* pp. 245–284, 286–289, 348–351, 372–376, 1933.

Ranganatham, B. V., and B. Satyanarayana, A Rational Method of Predicting Swelling Potential for Compacted Expansive Clays, *Proc. 6th Int. Conf. Soil Mech. Found. Eng.,* vol. 1, pp. 92–96, 1965.

Seed, H. B., and C. K. Chan, Structure and Strength Characteristics of Compacted Clay, *J. Soil Mech. Found. Eng. Div., ASCE,* vol. 85, no. SM5, p. 87, 1959.

Seed, H. B., R. J. Woodward, and R. Lundgren, Prediction of Swelling Potential for Compacted Clays, *J. Soil Mech. Found. Div., ASCE,* vol. 88, no. SM3, pp. 53–87, 1962.

Seed, H. B., R. J. Woodward, and R. Lundgren, Clay Minerological Aspects of the Atterberg Limits, *J. Soil Mech. Found. Eng. Div., ASCE,* vol. 90, no. SM4, pp. 107–131, 1964*a*.

Seed, H. B., R. J. Woodward, and R. Lundgren, Fundamental Aspects of Atterberg Limits, *J. Soil Mech. Found. Eng. Div., ASCE,* vol. 90, no. SM6, pp. 75–105, 1964*b*.

Skempton, A. W., The Colloidal Activity of Clay, *Proc. 3d Int. Conf. Soil Mech. Found. Eng.,* vol. 1, pp. 57–61, 1953.

Terzaghi, K., "Erdbaumechanik auf Bodenphysikalischer Grundlage," Deuticke, Vienna, 1925.

Terzaghi, K., Relation between Soil Mechanics and Foundation Engineering: Presidential Address, *Proc. 1st Int. Conf. Soil Mech. Found. Eng., Cambridge, Mass.,* vol. 3, pp. 13–18, 1936.

Verwey, E. J. W., and J. Th. G. Overbeek, "Theory of Stability of Lyophobic Colloids," Elsevier North Holland, New York, Amsterdam, 1948.

Yoder, E. J., "Principles of Pavement Design," Wiley, New York, 1959.

CHAPTER

TWO

PERMEABILITY AND SEEPAGE

Any given mass of soil consists of solid particles of various sizes with interconnected void spaces. The continuous void spaces in a soil permit water to flow from a point of high energy to a point of low energy. Permeability is defined as the property of a soil which allows the seepage of fluids through its interconnected void spaces. This chapter is devoted to the study of the basic parameters involved in the flow of water through soils.

2.1 PERMEABILITY

2.1.1 Darcy's Law

In order to obtain a fundamental relation for the quantity of seepage through a soil mass under a given condition, consider a case as shown in Fig. 2.1. The cross-sectional area of the soil is equal to A and the rate of seepage is q.

According to Bernoulli's theorem, the total head for flow at any section in the soil can be given by

$$\text{Total head} = \text{elevation head} + \text{pressure head} + \text{velocity head} \tag{2.1}$$

The velocity head for flow through soil is very small and can be neglected. So, the total heads at sections A and B can be given by

$$\text{Total head at } A = z_A + h_A$$

$$\text{Total head at } B = z_B + h_B$$

where z_A and z_B are the elevation heads, and h_A and h_B are the pressure heads.

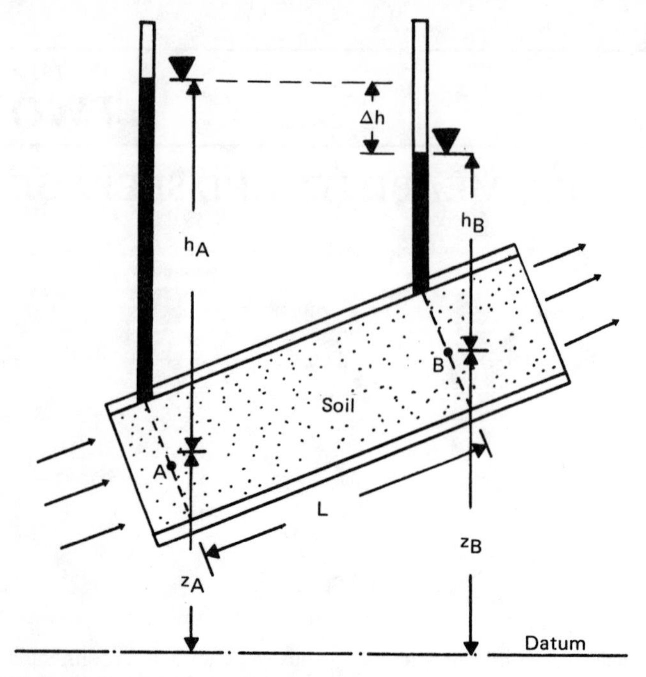

Fig. 2.1 Development of Darcy's law.

The loss of head Δh between sections A and B is

$$\Delta h = (z_A + h_A) - (z_B + z_B) \tag{2.2}$$

The hydraulic gradient i can be written as

$$i = \frac{\Delta h}{L} \tag{2.3}$$

where L is the distance between sections A and B.

Darcy (1856) published a simple relation between the discharge velocity and the hydraulic gradient:

$$v = ki \tag{2.4}$$

where v = discharge velocity
i = hydraulic gradient
k = coefficient of permeability

Hence, the rate of seepage q can be given by

$$q = kiA \tag{2.5}$$

Note that A is the cross section of the soil perpendicular to the direction of flow.

The coefficient of permeability k has the units of velocity, such as cm/s or mm/s, and is a measure of the resistance of the soil to flow of water. When the properties of water affecting the flow are included, we can express k by the relation

$$k\,(\text{cm/s}) = \frac{K\rho g}{\mu} \tag{2.6}$$

where K = intrinsic permeability, cm^2
ρ = mass density of the fluid, g/cm^3
g = acceleration due to gravity, cm/sec^2
μ = absolute viscosity of the fluid, poise [that is, g/(cm · s)]

It must be pointed out that the velocity v given by Eq. (2.4) is the discharge velocity calculated on the basis of the gross cross-sectional area. Since water can flow only through the interconnected pore spaces, the actual velocity of seepage through soil, v_s, can be given by

$$v_s = \frac{v}{n} \tag{2.7}$$

where n is the porosity of the soil.

Some typical values of the coefficient of permeability are given in Table 2.1.

The coefficient of permeability of soils is generally expressed at a temperature of 20°C. At any other temperature T, the coefficient of permeability can be obtained from Eq. (2.6) as

$$\frac{k_{20}}{k_T} = \frac{(\rho_{20})(\mu_T)}{(\rho_T)(\mu_{20})}$$

where k_T, k_{20} = coefficient of permeability at T°C and 20°C, respectively
ρ_T, ρ_{20} = mass density of the fluid at T°C and 20°C, respectively
μ_T, μ_{20} = coefficient of viscosity at T°C and 20°C, respectively

Since the value of ρ_{20}/ρ_T is approximately 1, we can write

$$k_{20} = k_T \frac{\mu_T}{\mu_{20}} \tag{2.8}$$

Table 2.2 gives the values of μ_T/μ_{20} for a temperature T varying from 10 to 30°C.

Table 2.1 Typical values of coefficient of permeability for various soils

Material	Coefficient of permeability, mm/s
Coarse	10 to 10^3
Fine gravel, coarse and medium sand	10^{-2} to 10
Fine sand, loose silt	10^{-4} to 10^{-2}
Dense silt, clayey silt	10^{-5} to 10^{-4}
Silty clay, clay	10^{-8} to 10^{-5}

Table 2.2 Values of μ_T/μ_{20}

Temperature T, °C	μ_T/μ_{20}	Temperature T, °C	μ_T/μ_{20}
10	1.298	21	0.975
11	1.263	22	0.952
12	1.228	23	0.930
13	1.195	24	0.908
14	1.165	25	0.887
15	1.135	26	0.867
16	1.106	27	0.847
17	1.078	28	0.829
18	1.051	29	0.811
19	1.025	30	0.793
20	1.000		

2.1.2 Validity of Darcy's Law

Darcy's law given by Eq. (2.4), $v = ki$, is true for laminar flow of water through the void spaces. Several studies have been made to investigate the range over which Darcy's law is valid, and an excellent summary of these works was given by Muskat (1937). A criterion for investigating the range can be furnished by Reynolds number. For flow through soils, Reynolds number R_n can be given by the relation

$$R_n = \frac{vD\rho}{\mu} \tag{2.9}$$

where v = discharge (superficial) velocity, cm/s
D = average diameter of the soil particle, cm
ρ = density of the fluid, g/cm^3
μ = coefficient of viscosity, g/(cm · s)]

For laminar flow conditions in soils, experimental results show that

$$R_n = \frac{vD\rho}{\mu} \leqslant 1 \tag{2.10}$$

With coarse sand, assuming $D = 0.45$ mm and making use of Eq. (2.70), we have $k \approx 100D^2 = 100(0.045)^2 = 0.203$ cm/s. Assuming $i = 1$, then $v = ki = 0.203$ cm/s. Also, $\rho_{\text{water}} \approx 1$ g/cm^3, and $\mu_{20°\text{C}} = (10^{-5})(981)$ g/(cm · s). Hence

$$R_n = \frac{(0.203)(0.045)(1)}{(10^{-5})(981)} = 0.931 < 1$$

From the above calculations, we can conclude that, for flow of water through all types of soil (sand, silt, and clay), the flow is laminar and Darcy's law is valid. With coarse sands, gravels, and boulders, turbulent flow of water can be expected, and the hydraulic gradient can be given by the relation

$$i = av + bv^2 \tag{2.11}$$

where a and b are experimental constants (see Forchheimer 1902, for example).

Leps (1973) summarized a number of works concerned with the determination of the velocity of flow through clean gravel and rocks. All investigators appear to agree that the average velocity of flow through the void spaces can be given by the relation

$$v_v = CR_H^{0.5} i^{0.54} \tag{2.12}$$

where v_v = average velocity of flow through voids
C = a constant which is a function of shape and roughness of rock particles
R_H = hydraulic mean radius
i = hydraulic gradient

2.1.3 Factors Affecting the Coefficient of Permeability

The coefficient of permeability depends on several factors, most of which are listed below:

1. Shape and size of the soil particles.
2. Void ratio. Permeability increases with increase of void ratio.
3. Degree of saturation. Permeability increases with increase of degree of saturation. The variation of the value of k with degree of saturation for Madison sand is shown in Fig. 2.2. Figure 2.3 shows the effect of the degree of saturation on the

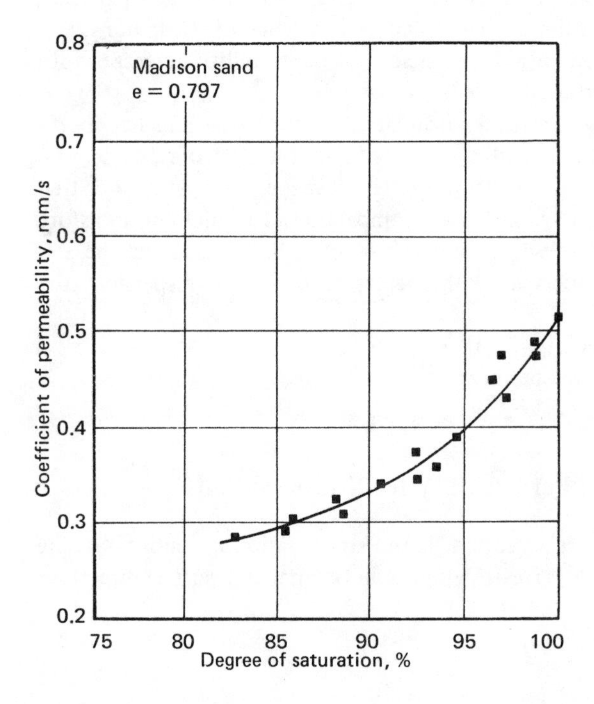

Fig. 2.2 Influence of degree of saturation on permeability of Madison sand.

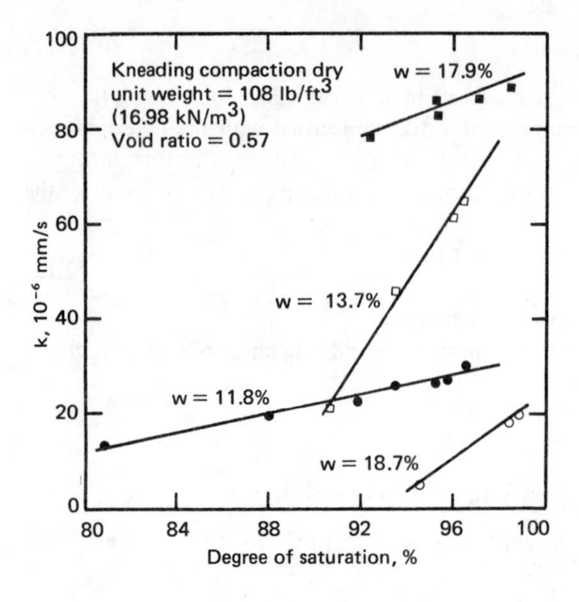

Fig. 2.3 Influence of degree of saturation on permeability of compacted silty clay. (Note: Samples aged 21 days at constant water content and unit weight after compaction prior to test.) *(Redrawn after J. K. Mitchell, D. R. Hooper, and R. G. Campanella, Permeability of Compacted Clay,* J. Soil Mech. Found. Div., ASCE, *vol. 91, no. SM4, 1965.)*

value of k for a silty clay. Note that the silty clay specimens were prepared by kneading compaction to a dry unit weight of 108 lb/ft^3 (16.98 kN/m^3). The molding moisture contents w were varied.

4. Composition of soil particles. For sands and silts this is not important; however, for soils with clay minerals this is one of the most important factors. Permeability in this case depends on the thickness of water held to the soil particles, which is a function of the cation exchange capacity, valence of the cations, etc. Other factors remaining the same, the coefficient of permeability decreases with increasing thickness of the diffuse double layer.
5. Soil structure. Fine-grained soils with a flocculated structure have a higher coefficient of permeability than those with a dispersed structure. This fact is demonstrated in Fig. 2.4 for the case of a silty clay. The test specimens were prepared to a constant dry unit weight by kneading compaction. The molding moisture content was varied. Note that with the increase of moisture content the soil becomes more and more dispersed. With increasing degree of dispersion, the permeability decreases.
6. Viscosity of the permeant [see Eq. (2.6)].
7. Density and concentration of the permeant.

2.1.4 Effective Coefficient of Permeability for Stratified Soils

In general, natural soil deposits are stratified. If the stratification is continuous, the effective coefficients of permeability for flow in the horizontal and vertical directions can be readily calculated.

Flow in the horizontal direction. Figure 2.5 shows several layers of soil with horizontal stratification. Due to fabric anisotropy, the coefficient of permeability of each soil layer may vary depending on the direction of flow. So, let us assume that $k_{h_1}, k_{h_2}, k_{h_3}, \ldots,$ are the coefficients of permeability of layers 1, 2, 3, . . ., respectively, for flow in the horizontal direction. Similarly, let $k_{v_1}, k_{v_2}, k_{v_3}, \ldots,$ be the coefficients of permeability for flow in the vertical direction.

Considering unit width of the soil layers as shown in Fig. 2.5, the rate of seepage in the horizontal direction can be given by

$$q = q_1 + q_2 + q_3 + \cdots + q_n \tag{2.13}$$

where q is the flow rate through the stratified soil layers combined, and $q_1, q_2, q_3, \ldots$ is the rate of flow through soil layers 1, 2, 3, . . ., respectively. Note that for flow in the horizontal direction (which is the direction of stratification of the soil layers),

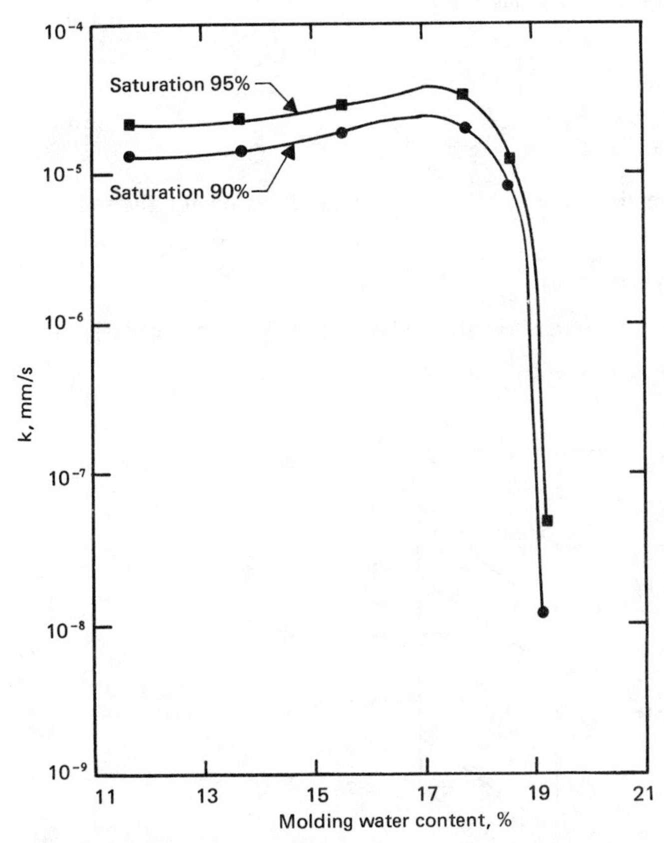

Fig. 2.4 Dependence of permeability on the structure of a silty clay. *(Redrawn after J. K. Mitchell, D. R. Hooper, and R. G. Campanella, Permeability of Compacted Clay,* J. Soil Mech. Found. Div., ASCE, *vol. 91, no. SM4, 1965.)*

the hydraulic gradient is the same for all layers. So,

$$\begin{aligned} q_1 &= k_{h_1} i H_1 \\ q_2 &= k_{h_2} i H_2 \\ q_3 &= k_{h_3} i H_3 \\ &\vdots \end{aligned} \tag{2.14}$$

and

$$q = k_{e(h)} i H \tag{2.15}$$

where i = hydraulic gradient
$k_{e(h)}$ = effective coefficient of permeability for flow in horizontal direction
H_1, H_2, H_3 = thicknesses of layers 1, 2, 3, respectively
$H = H_1 + H_2 + H_3 + \cdots$

Substitution of Eqs. (2.14) and (2.15) into Eq. (2.13) yields

$$k_{e(h)} H = k_{h_1} H_1 + k_{h_2} H_2 + k_{h_3} H_3 + \cdots$$

Hence,

$$k_{e(h)} = \frac{1}{H}(k_{h_1} H_1 + k_{h_2} H_2 + k_{h_3} H_3 + \cdots) \tag{2.16}$$

Flow in the vertical direction. For flow in the vertical direction for the soil layers shown in Fig. 2.6,

$$v = v_1 = v_2 = v_3 = \cdots = v_n \tag{2.17}$$

where $v_1, v_2, v_3, \ldots$ are the discharge velocities in layers $1, 2, 3, \ldots$, respectively; or

$$v = k_{e(v)} i = k_{v_1} i_1 = k_{v_2} i_2 = k_{v_3} i_3 = \cdots \tag{2.18}$$

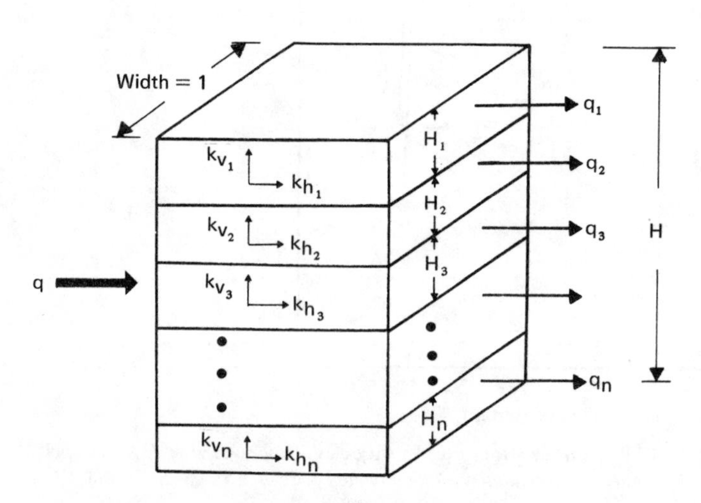

Fig. 2.5 Flow in horizontal direction in stratified soil deposit.

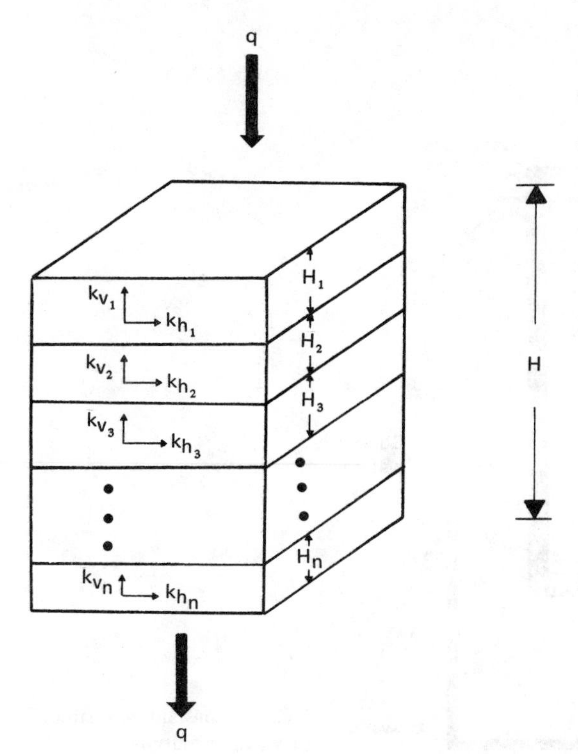

Fig. 2.6 Flow in vertical direction in stratified soil deposit.

where $k_{e(v)}$ = effective coefficient of permeability for flow in vertical direction

$k_{v_1}, k_{v_2}, k_{v_3}, \ldots$ = coefficients of permeability of layers 1, 2, 3, . . ., respectively, for flow in vertical direction

$i_1, i_2, i_3, \ldots$ = hydraulic gradient in soil layers 1, 2, 3, . . ., respectively

For flow at right angles to the direction of stratification,

Total head loss = (head loss in layer 1) + (head loss in layer 2) + $\cdots$

or
$$iH = i_1H_1 + i_2H_2 + i_3H_3 + \cdots \tag{2.19}$$

Combining Eqs. (2.18) and (2.19)

$$\frac{v}{k_{e(v)}} H = \frac{v}{k_{v_1}} H_1 + \frac{v}{k_{v_2}} H_2 + \frac{v}{k_{v_3}} H_3 + \cdots$$

or
$$k_{e(v)} = \frac{H}{H_1/k_{v_1} + H_2/k_{v_2} + H_3/k_{v_3} + \cdots} \tag{2.20}$$

2.1.5 Determination of Coefficient of Permeability in the Laboratory

The four most common laboratory methods for determining the coefficient of permeability of soils are the following:

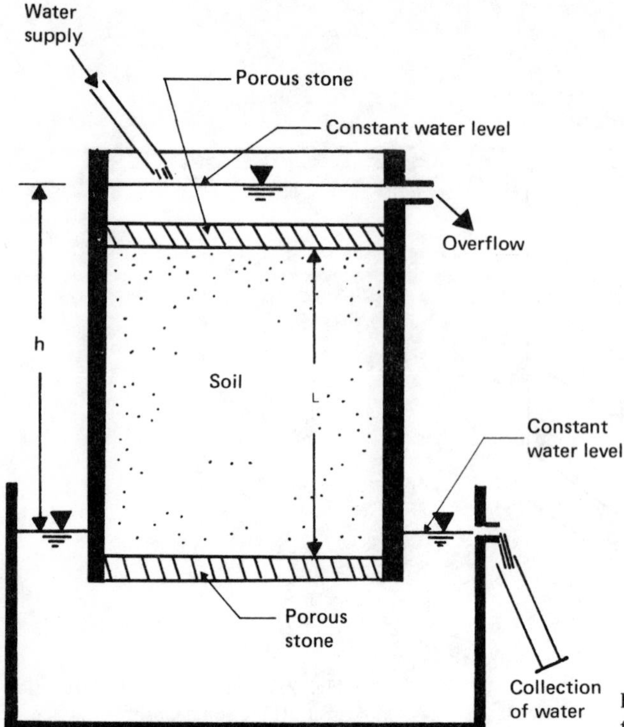

Fig. 2.7 Constant-head laboratory permeability test.

1. Constant-head test.
2. Falling-head test.
3. Indirect determination from consolidation test.
4. Indirect determination by horizontal capillary test.

The general principles of these methods are given below.

Constant-head test. The constant-head test is suitable for more permeable granular materials. The basic laboratory test arrangement is shown in Fig. 2.7. The soil specimen is placed inside a cylindrical mold, and the constant head loss, h, of water flowing through the soil is maintained by adjusting the supply. The outflow water is collected in a measuring cylinder, and the duration of the collection period is noted. From Darcy's law, the total quantity of flow Q in time t can be given by

$$Q = qt = kiAt$$

where A is the area of cross section of the specimen. But $i = h/L$, where L is the length of the specimen, and so $Q = k(h/L)At$. Rearranging this gives

$$k = \frac{QL}{hAt} \tag{2.21}$$

Once all the quantities in the right-hand side of Eq. (2.21) have been determined from the test, the coefficient of permeability of the soil can be calculated.

Falling-head test. The falling-head permeability test is more suitable for fine-grained soils. Figure 2.8 shows the general laboratory arrangement for the test. The soil specimen is placed inside a tube, and a standpipe is attached to the top of the specimen. Water from the standpipe flows through the specimen. The initial head difference h_1 at time $t = 0$ is recorded, and water is allowed to flow through the soil such that the final head difference at time $t = t$ is h_2.

The rate of flow through the soil is

$$q = kiA = k\frac{h}{L}A = -a\frac{dh}{dt} \tag{2.22}$$

where h = head difference at any time t
A = area of specimen
a = area of standpipe
L = length of specimen

From Eq. (2.22),

$$\int_0^t dt = \int_{h_1}^{h_2} \frac{aL}{Ak}\left(-\frac{dh}{h}\right)$$

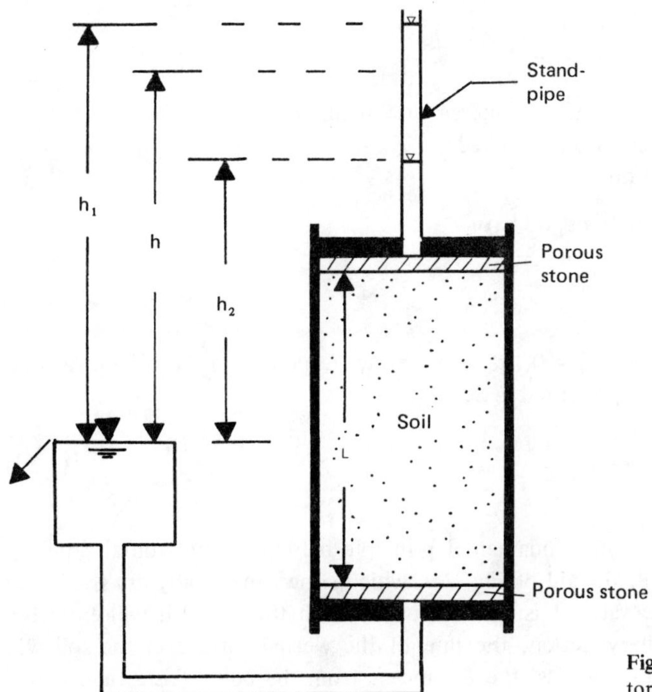

Fig. 2.8 Falling-head laboratory permeability test.

or $$k = 2.303 \frac{aL}{At} \log \frac{h_1}{h_2} \tag{2.23}$$

The values of a, L, A, t, h_1, and h_2 can be determined from the test, and then the coefficient of the permeability k for a soil can be calculated from Eq. (2.23).

Permeability from consolidation test. The coefficient of permeability of clay soils is often determined by the consolidation test, the procedures of which are explained in Sec. 5.1.6. From Eq. (5.25),

$$T_v = \frac{C_v t}{H^2}$$

where T_v = time factor
C_v = coefficient of consolidation
H = length of average drainage path
t = time

The coefficient of consolidation is [see Eq. (5.15)]

$$C_v = \frac{k}{\gamma_w m_v}$$

where γ_w = unit weight of water
m_v = volume coefficient of compressibility

Also, $$m_v = \frac{\Delta e}{\Delta\sigma\,(1+e)}$$

where Δe = change of void ratio for incremental loading
$\Delta\sigma$ = incremental pressure applied
e = initial void ratio

Combining these three equations, we have

$$k = \frac{T_v \gamma_w\,\Delta e\,H^2}{t\,\Delta\sigma\,(1+e)} \tag{2.24}$$

For 50% consolidation, $T_v = 0.198$; and the corresponding t_{50} can be estimated according to the procedure presented in Sec. 5.1.6.

Hence, $$k = \frac{0.198\gamma_w\,\Delta e\,H^2}{t_{50}\,\Delta\sigma\,(1+e)} \tag{2.25}$$

Horizontal capillary test. The fundamental principle behind the horizontal capillary test can be explained with the aid of Fig. 2.9, which shows an initially dry soil inside a horizontal tube. If the valve A is opened, water from the reservoir will enter the tube and, through capillary action, the line of the wetted surface in the soil will gradually advance—in other words, the distance x from the point 1 is a function of time t.

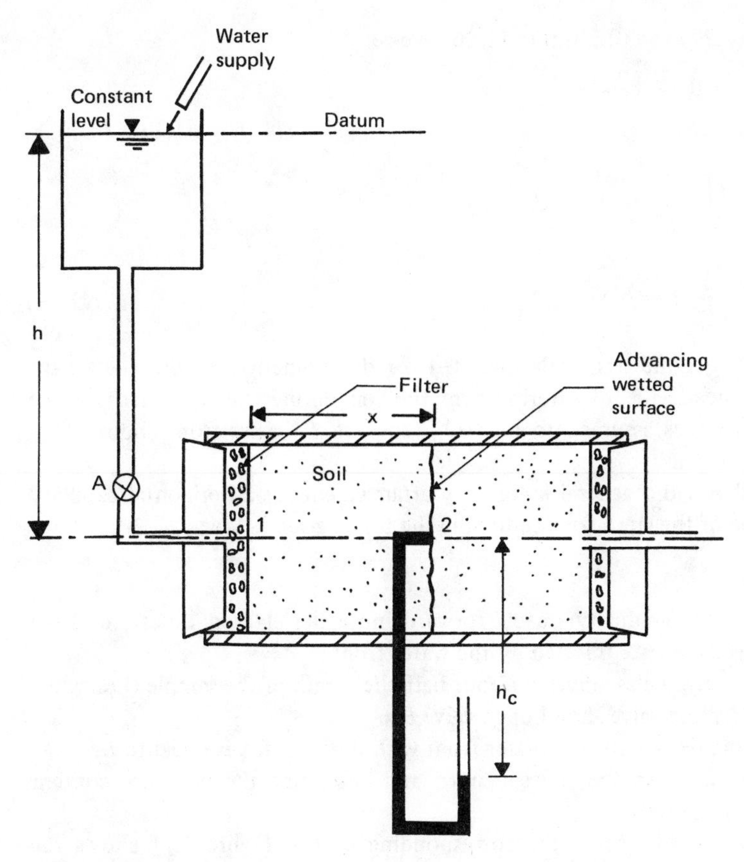

Fig. 2.9 Development of Eq. (2.29) for horizontal capillary test.

At point 1, the total head is zero (based on the datum shown in Fig. 2.9). At point 2 (immediately to the left of the wetted surface), the total head is $-(h + h_c)$. Using Darcy's law,

$$v = nS_r v_s = ki \tag{2.26}$$

where n = porosity
S_r = degree of saturation
v_s = seepage velocity

But
$$v_s = \frac{dx}{dt} \tag{2.27}$$

and
$$i = \frac{(\text{total head at 1}) - (\text{total head at 2})}{x}$$

or
$$i = \frac{0 - [-(h + h_c)]}{x} = \frac{h + h_c}{x} \tag{2.28}$$

Substituting Eqs. (2.27) and (2.28) into (2.26), we get

$$v_s = \frac{dx}{dt} = k \frac{1}{nS_r} \frac{h + h_c}{x}$$

$$\int_{x_1}^{x_2} x\, dx = \int_0^t \frac{k}{nS_r} (h + h_c)\, dt$$

$$\frac{x_2^2 - x_1^2}{t} = \frac{2k}{nS_r} (h + h_c) \tag{2.29}$$

Equation (2.29) is the basic relation used for determination of the coefficient of permeability. The degree of saturation of the soil during the movement of the water front is sometimes assumed to be 100%. In fact, S_r varies from about 75 to 95% for tests in most soils.

Figure 2.10 shows the general laboratory arrangement for a horizontal capillary test. A brief outline of the steps for conducting the test is given below.

1. Open the valve A.
2. As the water front gradually travels forward, note the elapsed times t and the corresponding distances x traveled by the water front.
3. When the water front has traveled about half the length of the sample (i.e., when x is about $L/2$), close valve A and open valve B.
4. Continue to note the advance of water front with time, until x is equal to L.
5. Close valve B. Remove the soil specimen and determine the moisture content and the degree of saturation.
6. Plot the values of x^2 against the corresponding time t. Figure 2.11 shows the nature of the plot, which consists of two straight lines. The portion Oa is for

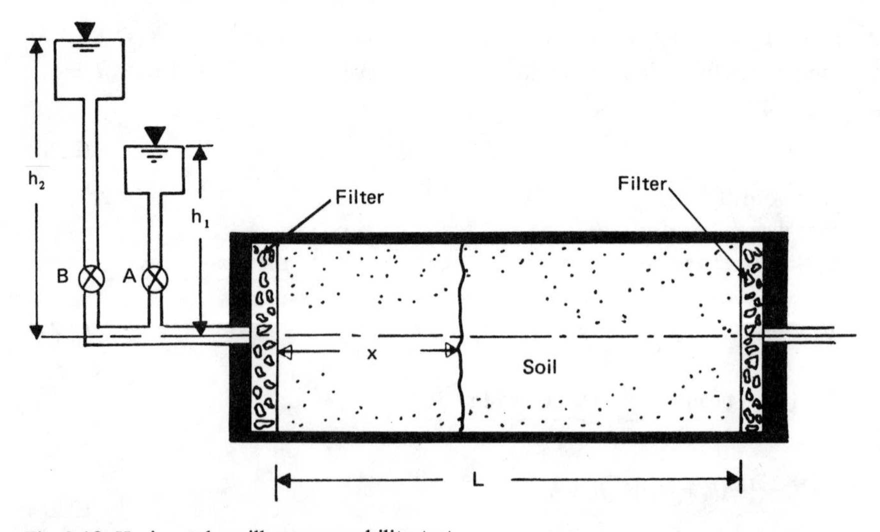

Fig. 2.10 Horizontal capillary permeability test.

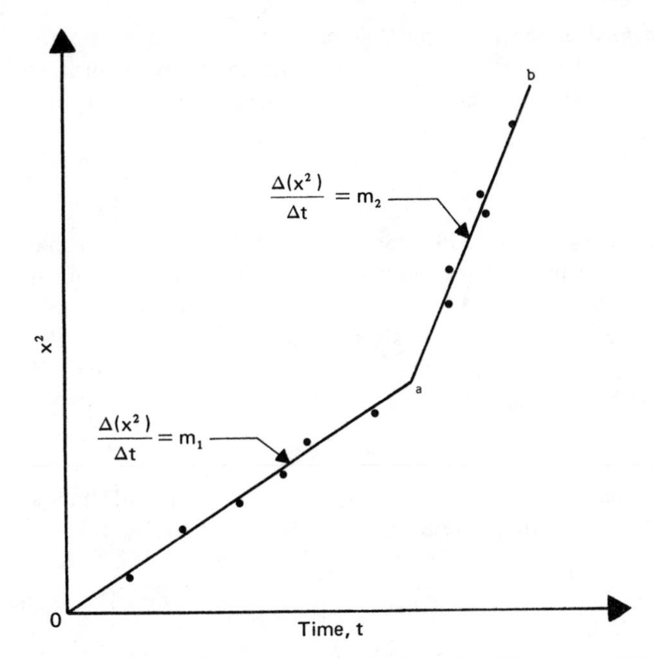

Fig. 2.11 Plot of x^2 against time t in horizontal capillary permeability test.

the readings taken in step 2, and the portion ab is for the readings taken in step 4.

7. From Eq. (2.29) we can write

$$\frac{\Delta(x^2)}{\Delta t} = \frac{2k}{nS_r}(h + h_c) \tag{2.30}$$

The left-hand side of Eq. (2.30) represents the slope of the straight-line plot of x^2 vs. t.

8. Determine the slopes of lines Oa and ab. Let these be m_1 and m_2, respectively. So,

$$m_1 = \frac{2k}{nS_r}(h_1 + h_c) \tag{2.31}$$

and

$$m_2 = \frac{2k}{nS_r}(h_2 + h_c) \tag{2.32}$$

Since n, S_r, h_1, h_2, m_1, and m_2 are determined from the test, the above two equations contain only two unknowns (k and h_c) and thus can be solved.

2.1.6 Determination of Coefficient of Permeability in the Field

It is sometimes difficult to obtain undisturbed soil specimens from the field. For large construction projects, it is advisable to conduct permeability tests in situ and compare the results with those obtained in the laboratory. Several techniques are presently available for determination of the coefficient of permeability in the field, such as pumping from wells, borehole tests, etc., and some of these methods will be treated briefly in this section.

Pumping from wells

Gravity wells. Figure 2.12 shows a permeable layer underlain by an impermeable stratum. The coefficient of permeability of the top permeable layer can be determined by pumping from a well at a constant rate and observing the steady-state water table in nearby observation wells. The steady state is established when the water level in the test well and the observation wells become constant. At the steady state, the rate of discharge due to pumping can be expressed as

$$q = kiA$$

From Fig. 2.12, i is approximately equal to dh/dr (this is referred to as Dupuit's assumption), and $A = 2\pi rh$. Substituting these in the above equation for rate of discharge, we get

$$q = k\frac{dh}{dr}\,2\pi rh$$

$$\int_{r_1}^{r_2}\frac{dr}{r} = \frac{2\pi k}{q}\int_{h_1}^{h_2} h\,dh$$

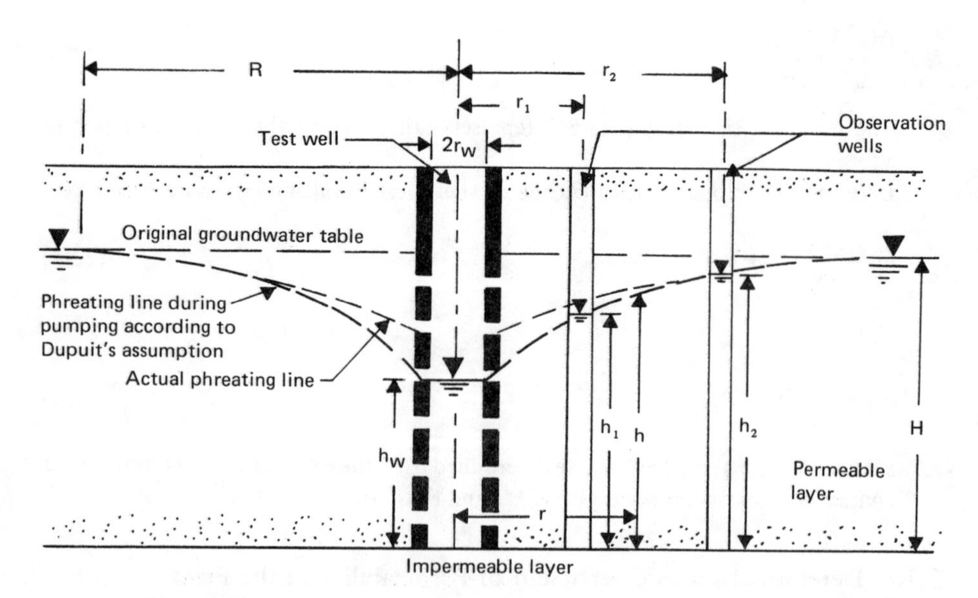

Fig. 2.12 Determination of coefficient of permeability by pumping from wells – gravity well.

So, $$k = \frac{2.303q\,[\log (r_2/r_1)]}{\pi(h_2^2 - h_1^2)} \tag{2.33}$$

If the values of r_1, r_2, h_1, h_2, and q are known from field measurements, the coefficient of permeability can be calculated from the simple relation given in Eq. (2.33).

According to Kozeny (1933), the maximum radius of influence, R (Fig. 2.12), for drawdown due to pumping can be given by

$$R = \sqrt{\frac{12t}{n}\sqrt{\frac{qk}{\pi}}} \tag{2.34}$$

where n = porosity
R = radius of influence
t = time during which discharge of water from well has been established

Also note that if we substitute $h_1 = h_w$ at $r_1 = r_w$ and $h_2 = H$ at $r_2 = R$, then

$$k = \frac{2.303q\,[\log (R/r_w)]}{\pi(H^2 - h_w^2)} \tag{2.35}$$

where H is the depth of the original groundwater table from the impermeable layer.

The depth h at any distance r from the well ($r_w \leqslant r \leqslant R$) can be determined from Eq. (2.33) by substituting $h_1 = h_w$ at $r_1 = r_w$ and $h_2 = h$ at $r_2 = r$. Thus

$$k = \frac{2.303q\,[\log (r/r_w)]}{\pi(h^2 - h_w^2)}$$

or $$h = \sqrt{\frac{2.303q}{\pi k}\log\frac{r}{r_w} + h_w^2} \tag{2.36}$$

It must be pointed out that Dupuit's assumption (i.e., that $i = dh/dr$) does introduce large errors regarding the actual phreatic line near the wells during steady-state pumping. This is shown in Fig. 2.12. For r greater than H to $1.5H$, the phreatic line predicted by Eq. (2.36) will coincide with the actual phreatic line.

The relation for the coefficient of permeability given by Eq. (2.33) has been developed on the assumption that the well fully penetrates the permeable layer. If the well partially penetrates the permeable layer as shown in Fig. 2.13, the coefficient of permeability can be better represented by the following relation (Mansur and Kaufman 1962):

$$q = \frac{\pi k\,[(H-s)^2 - t^2]}{2.303 \log (R/r_w)}\left[1 + \left(0.30 + \frac{10r_w}{H}\right)\sin\frac{1.8s}{H}\right] \tag{2.37}$$

The notations used in the right-hand side of Eq. (2.37) are shown in Fig. 2.13.

Artesian wells. The coefficient of permeability for a confined aquifier can also be determined from well pumping tests. Figure 2.14 shows an artesian well penetrating the full depth of an aquifier from which water is pumped out at a constant rate. Pumping is continued until a steady state is reached. The rate of water pumped out at steady state is given by

$$q = kiA = k\frac{dh}{dr}2\pi rT \tag{2.38}$$

where T is the thickness of the confined aquifier, or

$$\int_{r_1}^{r_2}\frac{dr}{r} = \int_{h_1}^{h_2}\frac{2\pi kT}{q}\,dh \tag{2.39}$$

Solution of Eq. (2.39) gives

$$k = \frac{q\log(r_2/r_1)}{2.727\,T(h_2 - h_1)} \tag{2.40}$$

Hence, the coefficient of permeability k can be determined by observing the drawdown in two observation wells, as shown in Fig. 2.14.

If we substitute $h_1 = h_w$ at $r_1 = r_w$ and $h_2 = H$ at $r_2 = R$ in Eq. (2.40) we get

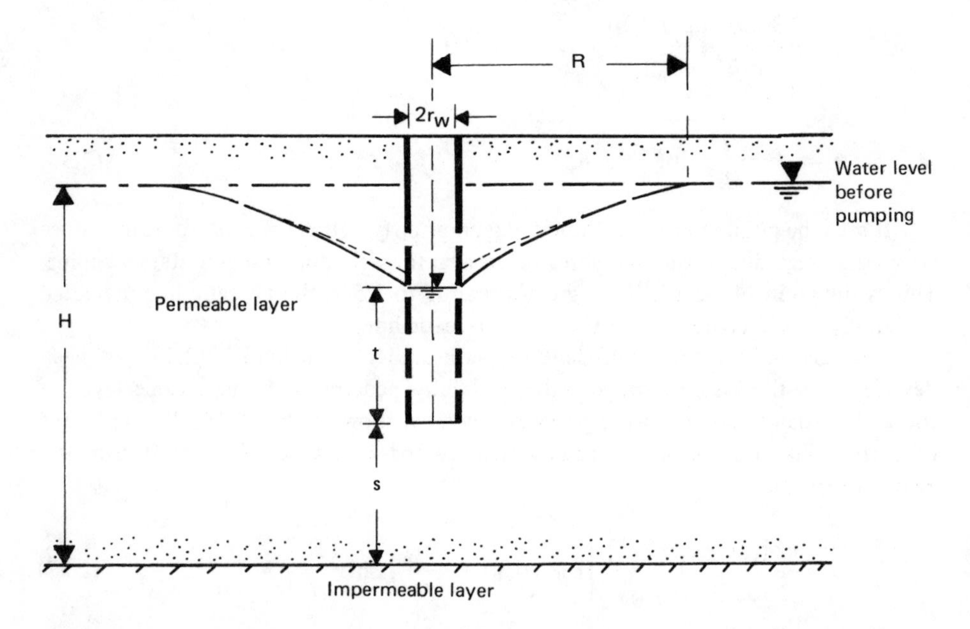

Fig. 2.13 Pumping from partially penetrating gravity wells.

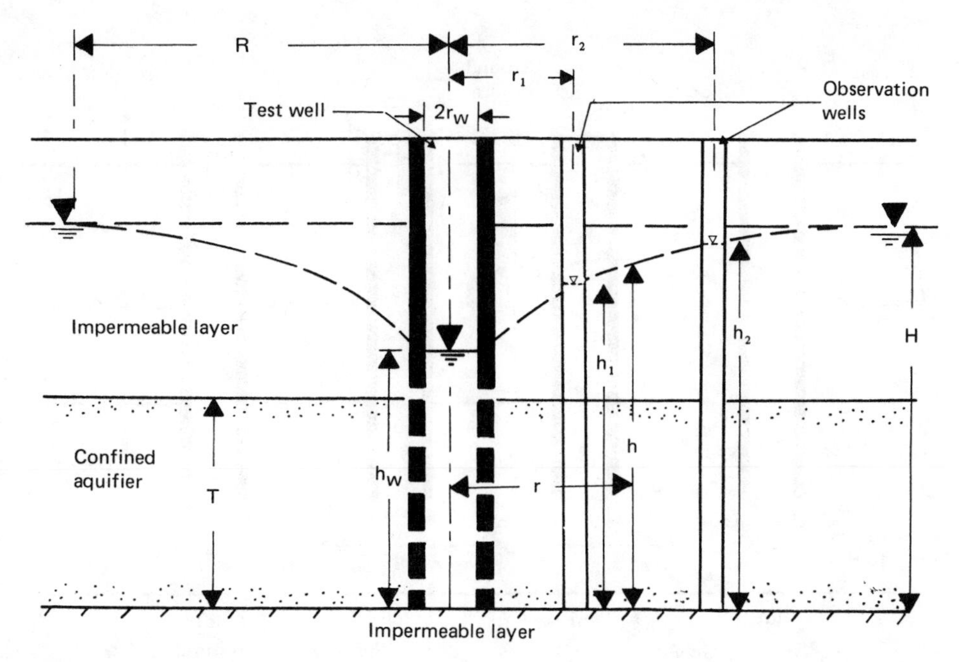

Fig. 2.14 Determination of coefficient of permeability by pumping from wells—confined aquifier case.

$$k = \frac{q \log (R/r_w)}{2.727\, T(H - h_w)} \tag{2.41}$$

Bore hole test recommended by USBR. Well pumping tests are costly, and so for economic reasons bore holes are used in many cases to estimate the coefficient of permeability of soils and rocks (U.S. Bureau of Reclamation, 1961).

Open-end test. Figure 2.15 shows the schematic diagram for determination of the permeability of soils by means of the open-end test. Casings are inserted in the bore holes, and they extend to the soil layers whose permeability needs to be determined. The groundwater table can be above or below the bottom of the casings.

Before conducting the permeability test, the hole is cleaned out. The test begins by adding clear water to maintain gravity flow at a constant head. The coefficient of permeability can be determined by the equation

$$k = \frac{q}{5.5rh} \tag{2.42}$$

where r = inside radius of the casing
h = differential head of water
q = rate of supply of water to maintain constant head

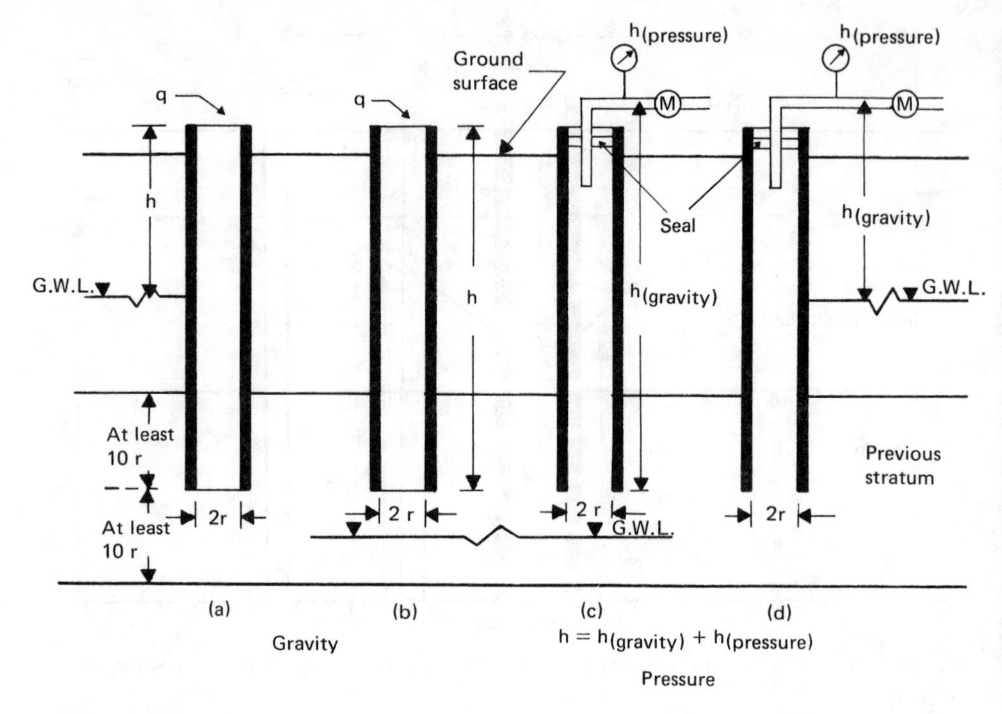

Fig. 2.15 Open-end test for soil permeability in the field. *(Redrawn after U.S. Bureau of Reclamation, 1961, by permission of the United States Department of the Interior, Water and Power Resources Service.)*

To ensure that Eq. (2.42) is used correctly, the following points should be noted:

1. In Fig. 2.15*a* and *b*, the constant water level maintained inside the casing coincides with the top of the casing.
2. In Fig. 2.15*b*, h is measured up to the bottom of the casing.
3. In soils that have low permeability, it may be desirable to apply some pressure to the water. This is the case in Fig. 2.15*c* and *d*. For these cases, the value of h used in Eq. (2.42) is given by

$$h = h_{\text{gravity}} + h_{\text{pressure}} \tag{2.43}$$

4. Any consistent set of units may be used in Eq. (2.42). If q is expressed in gal/min, then

$$k(\text{ft/yr}) = C_1 \frac{q \text{ (gal/min)}}{h \text{ (ft)}} \tag{2.44}$$

The values of C_1 for various sizes of standard casings used for field exploration are given in Table 2.3. The inside and outside diameters of these casings are given in Table 2.4.

Table 2.3 Values of C_1 for various sizes of casings

Size of casings	C_1
EX	204,000
AX	160,000
BX	129,000
NX	102,000

After U.S. Bureau of Reclamation (1961)

Packer test. The arrangement for carrying out permeability tests in a bore hole below the casing is shown in Fig. 2.16. These tests can be made above or below the ground-water table. However, the soils should be such that the bore hole will stay open without any casing. The test procedure is illustrated in Fig. 2.16. The equations for calculation of the coefficients of permeability are as follows:

$$k = \frac{q}{2\pi L h} \log \frac{L}{r} \qquad \text{for } L \geqslant 10r \tag{2.45}$$

and

$$k = \frac{q}{2\pi L h} \sinh^{-1} \frac{L}{2r} \qquad \text{for } 10r > L \geqslant r \tag{2.46}$$

where q = constant rate of flow into bore hole
L = length of test hole
r = radius of bore hole
h = differential head of water

For convenience,

$$k\,(\text{ft/yr}) = C_p \frac{q\,(\text{gal/min})}{H\,(\text{ft})} \tag{2.47}$$

The values of C_p are given in Table 2.5.

Table 2.4 Diameters of standard casings

Casing	Inside diameter, in	Outside diameter, in
EX	$1\frac{1}{2}$	$1\frac{13}{16}$
AX	$1\frac{29}{32}$	$2\frac{1}{4}$
BX	$2\frac{3}{8}$	$2\frac{7}{8}$
NX	3	$3\frac{1}{2}$

Note: 1 in = 25.4 mm.

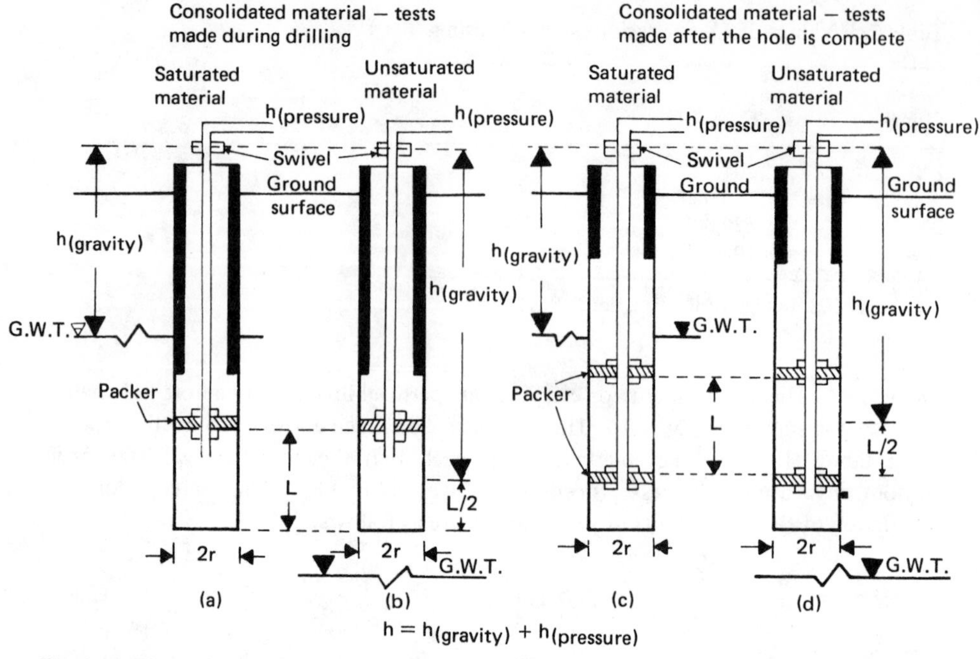

Fig. 2.16 The packer test for soil permeability. *(Redrawn after U.S. Bureau of Reclamation, 1961, by permission of the United States Department of the Interior, Water and Power Resources Service.)*

Variable-head tests by means of piezometer observation wells. The U.S. Department of the Navy (1971) has adopted some standard variable-head tests for determination of the coefficient of permeability by means of piezometer observation wells. These methods are described in Table 2.6. Careful attention should be paid to the notations. Figure 2.17 gives the shape factor coefficient S' used for condition (A). Figure 2.18 gives the shape factor coefficient C_S used for condition (F-1). Figure 2.19 shows the analysis of permeability by variable-head tests with reference to Table 2.6.

Example 2.1 Refer to Fig. 2.12 for the pumping test from a gravity well. If $r_1 = 50$ ft, $r_2 = 90$ ft, $h_1 = 92$ ft, and $h_2 = 96$ ft and the pumping rate from the well is 90 gal/min, determine the coefficient of permeability k.

SOLUTION $q = 90$ gal/min $= 90\,(0.1337) = 12.03$ ft^3/min. From Eq. (2.33),

$$k = \frac{2.303q\,[\log(r_2 - r_1)]}{\pi(h_2^2 - h_1^2)} = \frac{2.303\,(12.03)\,[\log(90/50)]}{\pi(96^2 - 92^2)} = \mathit{0.003\,ft/min}$$

Example 2.2 Refer to Fig. 2.13. For the steady state-condition, $r_w = 0.4$ m, $H = 28$ m, $s = 8$ m, and $t = 10$ m. The coefficient of permeability of the layer is 0.03 mm/s. For the steady-state pumping condition, estimate the rate of discharge q in m^3/min.

SOLUTION From Eq. (2.37),

$$q = \frac{\pi k\,[(H-s)^2 - t^2]}{2.303\ [\log (R/r_w)]}\left[1 + \left(0.30 + \frac{10 r_w}{H}\right)\sin\frac{1.8s}{H}\right]$$

$k = 0.03$ mm/s $= 0.0018$ m/min

So,

$$q = \frac{\pi(0.0018)[(28-8)^2 - 10^2]}{2.303\ [\log (R/0.4)]}\left\{1 + \left[0.30 + \frac{(10)\,(0.4)}{28}\right]\sin\frac{1.8(8)}{28}\right\}$$

$$= \frac{0.8976}{\log (R/0.4)}$$

From this equation for q, we can construct the following table:

R, m	q, m³
25	0.5
30	0.48
40	0.45
50	0.43
100	0.37

From the above table, the rate of discharge is approximately *0.45 m³/min.*

Table 2.5 Values of C_p [Eq. (2.47)]

L, ft	Diameter of test hole			
	EX	AX	BX	NX
1	31,000	28,500	25,800	23,300
2	19,400	18,100	16,800	15,500
3	14,400	13,600	12,700	11,800
4	11,600	11,000	10,300	9,700
5	9,800	9,300	8,800	8,200
6	8,500	8,100	7,600	7,200
7	7,500	7,200	6,800	6,400
8	6,800	6,500	6,100	5,800
9	6,200	5,900	5,600	5,300
10	5,700	5,400	5,200	4,900
15	4,100	3,900	3,700	3,600
20	3,200	3,100	3,000	2,800

After U. S. Bureau of Reclamation (1961)
Note: 1 ft = 0.3048 m.

Table 2.6 Computation of permeability from variable head tests (for observation well of constant cross section)

Condition	Diagram	Shape factor F	Permeability k by variable head test	Applicability
Observation well or piezometer in saturated isotropic stratum of infinite depth				
(A) Uncased hole	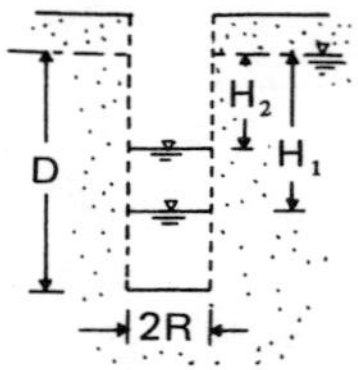 D, H_2, H_1, 2R	$F = 16\pi DS'R$	$k = \dfrac{R}{16DS'} \times \dfrac{H_2 - H_1}{t_2 - t_1}$ for $\dfrac{D}{R} < 50$	Simplest methods for permeability determination. Not applicable in stratified soils. For values of S', see Fig. 2.17.
(B) Cased hole, soil flush with bottom	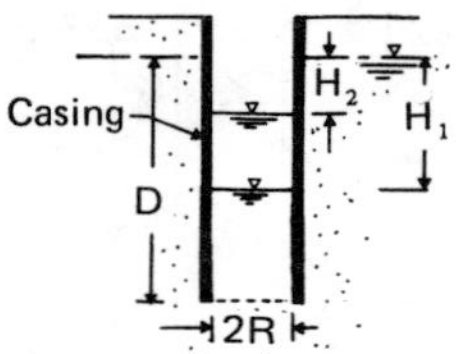 Casing, D, H_2, H_1, 2R	$F = \dfrac{11R}{2}$	$k = \dfrac{2\pi R}{11(t_2 - t_1)} \ln \dfrac{H_1}{H_2}$ for 6 in (0.1524 m) $\leqslant D \leqslant$ 60 in (1.524 m)	Used for permeability determination at shallow depths below the water table. May yield unreliable results in falling-head test with silting of bottom of hole.
(C) Cased hole, uncased or perforated extension of length L	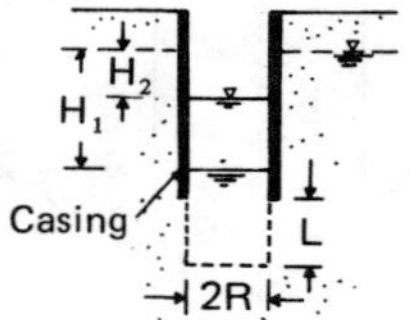 H_2, H_1, Casing, L, 2R	$F = \dfrac{2\pi L}{\ln (L/R)}$	$k = \dfrac{R^2}{2L(t_2 - t_1)} \ln \dfrac{L}{R} \ln \dfrac{H_1}{H_2}$ for $\dfrac{L}{R} > 8$	Used for permeability determination at greater depths below water table.
(D) Cased hole, column of soil inside casing to height L	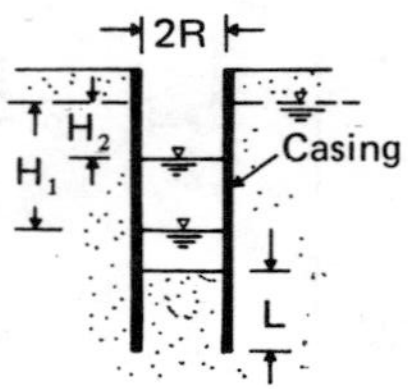 2R, H_2, H_1, Casing, L	$F = \dfrac{11\pi R^2}{2\pi R + 11L}$	$k = \dfrac{2\pi R + 11L}{11(t_2 - t_1)} \ln \dfrac{H_1}{H_2}$	Principal use is for permeability in vertical direction in anisotropic soils.

Observation well or piezometer in acquifer with impervious upper layer

Conditions	Diagram	Shape factor	Permeability	Applicability
(E) Cased hole, opening flush with upper boundary of aquifer of infinite depth	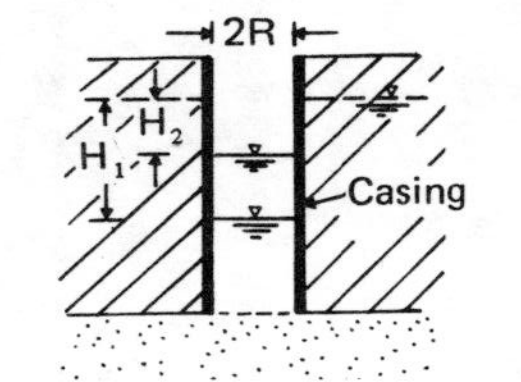	$F = 4R$	$k = \dfrac{\pi R}{4(t_2 - t_1)} \ln \dfrac{H_1}{H_2}$	Used for permeability determination when surface impervious layer is relatively thin. May yield unreliable results in falling-head test with silting of bottom of hole.
(F) Cased hole, uncased or perforated extension into aquifer of finite thickness: (1) $\dfrac{L_1}{T} \leqslant 0.20$	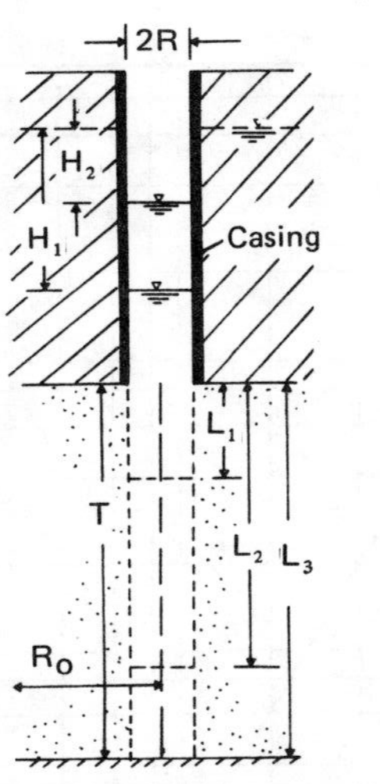	(1) $F = C_S R$	$k = \dfrac{\pi R}{C_S(t_2 - t_1)} \ln \dfrac{H_1}{H_2}$	Used for permeability determinations at depths greater than about 5 ft (1.524 m). For values of C_S, see Fig. 2.18.
(2) $0.2 < \dfrac{L_2}{T} < 0.85$		(2) $F = \dfrac{2\pi L_2}{\ln (L_2/R)}$	$k = \dfrac{R^2 \ln (L_2/R)}{2L_2(t_2 - t_1)} \ln \dfrac{H_1}{H_2}$ for $\dfrac{L}{R} > 8$	Used for permeability determinations at greater depths and for fine-grained soils using porous intake point of piezometer.
(3) $\dfrac{L_3}{T} = 1.00$ Note: R_O is the effective radius to source at constant head		(3) $F = \dfrac{2\pi L_3}{\ln (R_O/R)}$	$k = \dfrac{R^2 \ln (R_O/R)}{2L_3(t_2 - t_1)} \ln \dfrac{H_1}{H_2}$	Assume value of $R_O/R = 200$ for estimates unless observation wells are made to determine actual value of R_O.

After U.S. Navy, 1971.

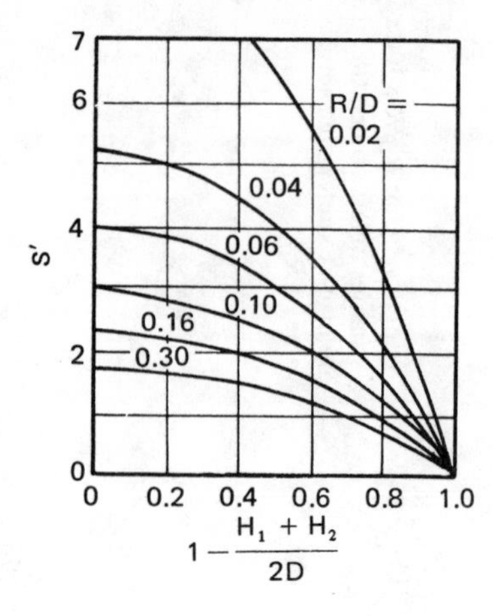

Fig. 2.17 Shape factor coefficient S' used for condition (A) of Table 2.6. *(After U.S. Navy 1971, based on Figure 11-11 (p. 555) from* Soil Engineering, *3d ed., by Merlin G. Spangler and Richard L. Hardy. Copyright 1951 © 1963, 1973 by Harper & Row, Publishers, Inc. Reprinted by permission of the publisher.)*

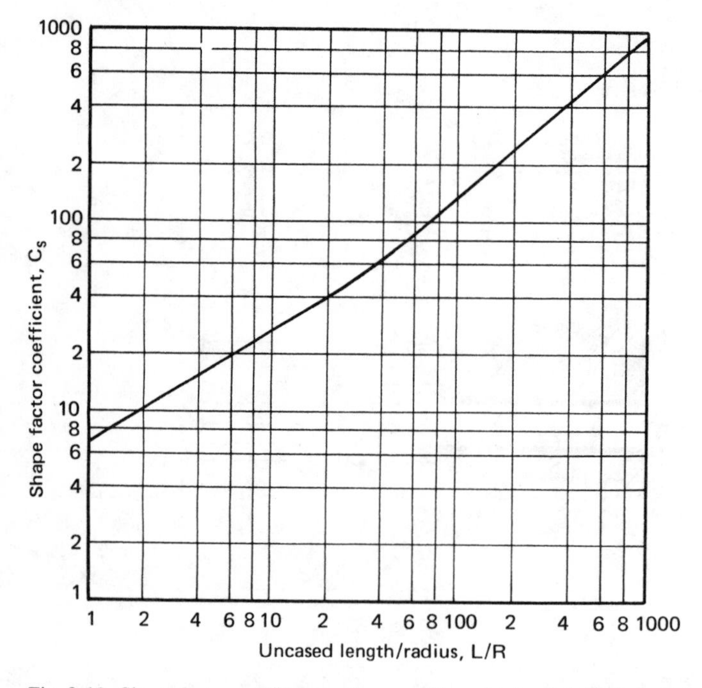

Fig. 2.18 Shape factor coefficient, C_S–condition (F-1) of Table 2.6. *(After U.S. Navy, 1971.)*

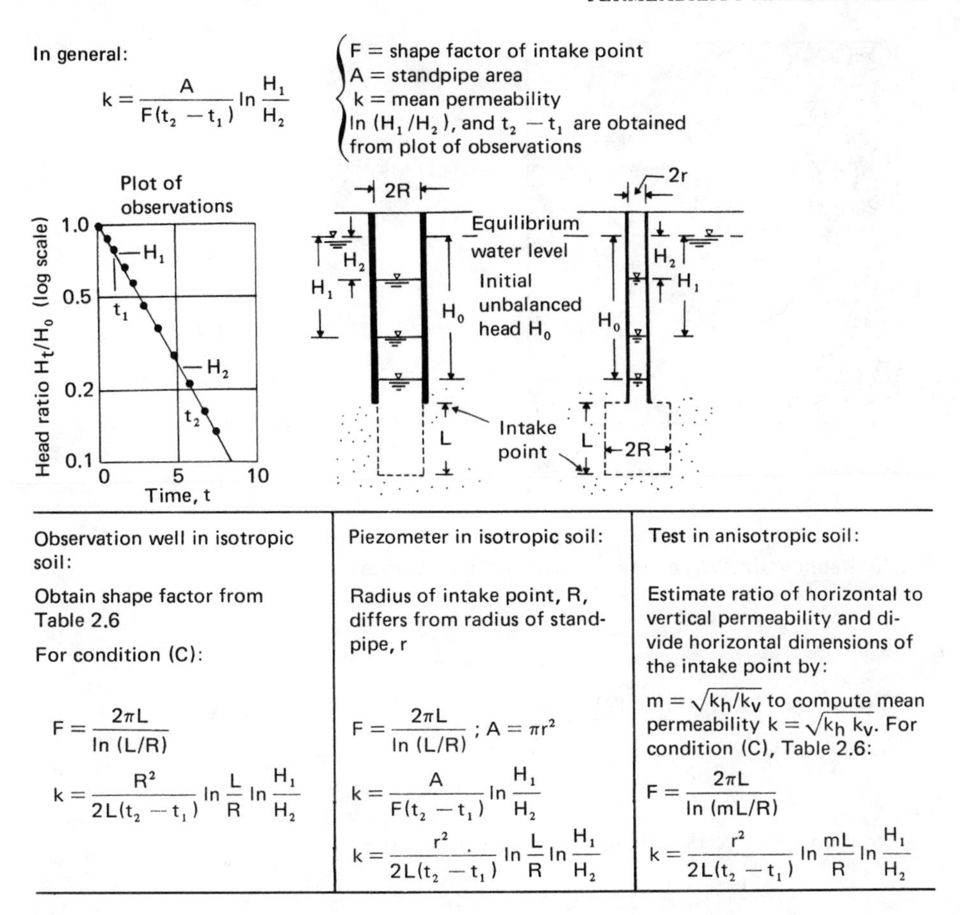

Observation well in isotropic soil:	Piezometer in isotropic soil:	Test in anisotropic soil:
Obtain shape factor from Table 2.6 For condition (C):	Radius of intake point, R, differs from radius of standpipe, r	Estimate ratio of horizontal to vertical permeability and divide horizontal dimensions of the intake point by: $m = \sqrt{k_h/k_v}$ to compute mean permeability $k = \sqrt{k_h k_v}$. For condition (C), Table 2.6:
$F = \dfrac{2\pi L}{\ln(L/R)}$ $k = \dfrac{R^2}{2L(t_2 - t_1)} \ln \dfrac{L}{R} \ln \dfrac{H_1}{H_2}$	$F = \dfrac{2\pi L}{\ln(L/R)}$; $A = \pi r^2$ $k = \dfrac{A}{F(t_2 - t_1)} \ln \dfrac{H_1}{H_2}$ $k = \dfrac{r^2}{2L(t_2 - t_1)} \ln \dfrac{L}{R} \ln \dfrac{H_1}{H_2}$	$F = \dfrac{2\pi L}{\ln(mL/R)}$ $k = \dfrac{r^2}{2L(t_2 - t_1)} \ln \dfrac{mL}{R} \ln \dfrac{H_1}{H_2}$

Fig. 2.19 Analysis of permeability by variable-head tests. *(After U.S. Navy, 1971.)*

2.1.7 Theoretical Solution for Coefficient of Permeability

It was pointed out earlier in this chapter that the flow through soils finer than coarse gravel is laminar. The interconnected voids in a given soil mass can be visualized as a number of capillary tubes through which water can flow (Fig. 2.20).

According to the Hagen-Poiseuille's equation, the quantity of flow of water in unit time, q, through a capillary tube of radius R can be given by

$$q = \frac{\gamma_w S}{8\mu} R^2 a \tag{2.48}$$

where γ_w = unit weight of water
μ = absolute coefficient of viscosity
a = area cross section of tube
S = hydraulic gradient

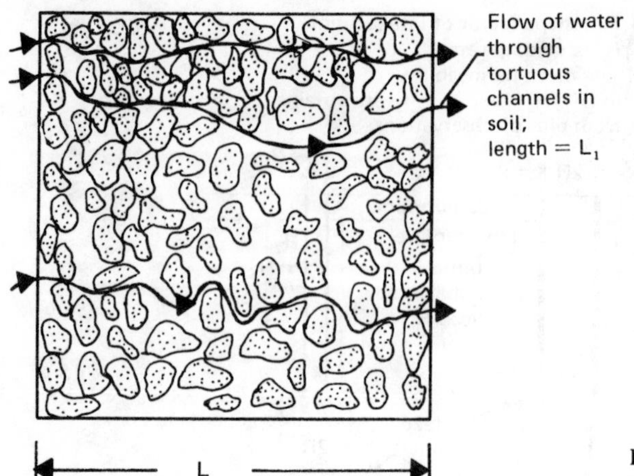

Fig. 2.20 Flow of water through tortuous channels in soil.

The hydraulic radius R_H of the capillary tube can be given by

$$R_H = \frac{\text{area}}{\text{wetted perimeter}} = \frac{\pi R^2}{2\pi R} = \frac{R}{2} \tag{2.49}$$

From Eqs. (2.48) and (2.49),

$$q = \frac{1}{2} \frac{\gamma_w S}{\mu} R_H^2 a \tag{2.50}$$

For flow through two parallel plates, we can also derive

$$q = \frac{1}{3} \frac{\gamma_w S}{\mu} R_H^2 a \tag{2.51}$$

So, for laminar flow conditions, the flow through any cross section can be given by a general equation:

$$q = \frac{\gamma_w S}{C_S \mu} R_H^2 a \tag{2.52}$$

where C_S is the shape factor. Also, the average velocity of flow v_a is given by

$$v_a = \frac{q}{a} = \frac{\gamma_w S}{C_S \mu} R_H^2 \tag{2.53}$$

For an actual soil, the interconnected void spaces can be assumed to be a number of tortuous channels (Fig. 2.20), and for these the term S in Eq. (2.53) is equal to $\Delta h/\Delta L_1$. Now,

$$R_H = \frac{\text{area}}{\text{perimeter}} = \frac{\text{(area) (length)}}{\text{(perimeter) (length)}} = \frac{\text{volume}}{\text{surface area}}$$

$$= \frac{1}{\text{(surface area)/(volume of pores)}} \tag{2.54}$$

If the total volume of soil is V, the volume of voids is $V_v = nV$, where n is porosity. Let S_V be equal to the surface area per unit volume of soil (bulk). From Eq. (2.54),

$$R_H = \frac{\text{volume}}{\text{surface area}} = \frac{nV}{S_V V} = \frac{n}{S_V} \tag{2.55}$$

Substituting Eq. (2.55) into Eq. (2.53) and taking $v_a = v_s$ (where v_s is the actual seepage velocity through soil), we get

$$v_s = \frac{\gamma_w}{C_S \mu} S \frac{n^2}{S_V^2} \tag{2.56}$$

It must be pointed out that the hydraulic gradient i used for soils is the macroscopic gradient. The factor S in Eq. (2.56) is the microscopic gradient for flow through soils. Referring to Fig. 2.20, $i = \Delta h/\Delta L$ and $S = \Delta h/\Delta L_1$. So,

$$i = \frac{\Delta h}{\Delta L_1} \frac{\Delta L_1}{\Delta L} = ST \tag{2.57}$$

or $$S = \frac{i}{T} \tag{2.58}$$

where T is tortuosity, $\Delta L_1/\Delta L$.

Again, the seepage velocity in soils is

$$v_s = \frac{v}{n} \frac{\Delta L_1}{\Delta L} = \frac{v}{n} T \tag{2.59}$$

where v is the discharge velocity. Substitution of Eqs. (2.59) and (2.58) into Eq. (2.56) yields

$$v_s = \frac{v}{n} T = \frac{\gamma_w}{C_S \mu} \frac{i}{T} \frac{n^2}{S_V^2}$$

or $$v = \frac{\gamma_w}{C_S \mu S_V^2} \frac{n^3}{T^2} i \tag{2.60}$$

In Eq. (2.60), S_V is the surface area per unit volume of soil. If we define S_s as the surface area per unit volume of soil solids, then

$$S_s V_s = S_V V \tag{2.61}$$

where V_s is the volume of soil solids in a bulk volume V, that is,

$$V_s = (1 - n)V$$

So, $$S_s = \frac{S_V V}{V_s} = \frac{S_V V}{(1 - n)V} = \frac{S_V}{1 - n} \tag{2.62}$$

Combining Eqs. (2.60) and (2.62), we obtain

$$v = \frac{\gamma_w}{C_S \mu S_s^2 T^2} \frac{n^3}{(1-n)^2} i$$

$$= \frac{1}{C_S S_s^2 T^2} \frac{\gamma_w}{\mu} \frac{e^3}{1+e} i \tag{2.63}$$

where e is the void ratio. This relation is the Kozeny-Carman equation (Kozeny, 1927; Carman, 1956). Comparing Eqs. (2.4) and (2.63), we find that the coefficient of permeability is

$$k = \frac{1}{C_S S_s^2 T^2} \frac{\gamma_w}{\mu} \frac{e^3}{1+e} \tag{2.64}$$

The absolute permeability was defined by Eq. (2.6) as

$$K = k \frac{\mu}{\gamma_w}$$

Comparing Eqs. (2.6) and (2.64),

$$K = \frac{1}{C_S S_s^2 T^2} \frac{e^3}{1+e} \tag{2.65}$$

The Kozeny-Carman equation works well for describing coarse-grained soils such as sand and some silts. For these cases, the coefficient of permeability bears a linear relation to $e^3/(1+e)$. However, serious discrepancies are observed when the Kozeny-Carman equation is applied to clayey soils.

For granular soils, the shape factor C_S is approximately 2.5 and the tortuosity factor T is about $\sqrt{2}$.

2.1.8 Variation of Permeability with Void Ratio in Sand

Based on Eq. (2.64), the coefficient of permeability can be written as

$$k \propto \frac{e^3}{1+e} \tag{2.66}$$

or

$$\frac{k_1}{k_2} = \frac{e_1^3/(1+e_1)}{e_2^3/(1+e_2)} \tag{2.67}$$

where k_1 and k_2 are the coefficients of permeability of a given soil at void ratios of e_1 and e_2, respectively.

Several other relations for the coefficient of permeability and void ratio have been suggested. They are of the form

$$k \propto \frac{e^2}{1+e} \tag{2.68}$$

$$k \propto e^2 \tag{2.69}$$

Table 2.7 Coefficient of permeability for a uniform Madison sand
Laboratory constant-head test; D_{10} = 0.2 mm

Test no.	Void ratio e	Coefficient of permeability at 20°C, k_{20}, mm/s	$\frac{e^3}{1+e}$	$\frac{e^2}{1+e}$	e^2
1	0.797	0.504	0.282	0.353	0.635
2	0.704	0.394	0.205	0.291	0.496
3	0.606	0.303	0.139	0.229	0.367
4	0.804	0.539	0.288	0.358	0.646
5	0.688	0.356	0.193	0.280	0.473
6	0.617	0.286	0.144	0.235	0.381
7	0.755	0.490	0.245	0.325	0.57
8	0.687	0.436	0.192	0.280	0.472
9	0.582	0.275	0.125	0.214	0.339

For comparison of the validity of the relations given in Eqs. (2.67) to (2.69), the experimental results (laboratory constant-head test) for a uniform Madison sand are given in Table 2.7. Based on the calculations given, the permeability functions have been plotted against the coefficient of permeability in Fig. 2.21. From the plot, it appears that all three relations are equally good.

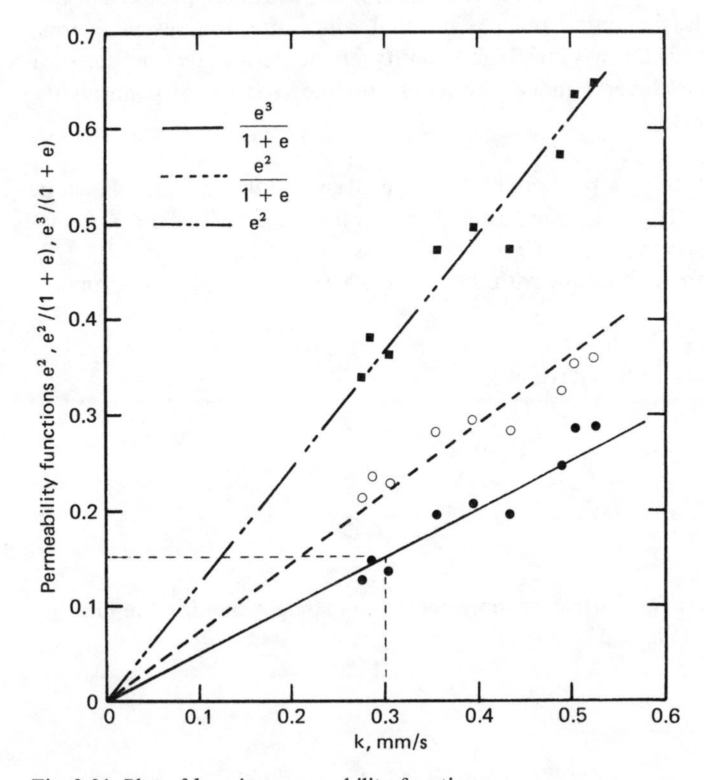

Fig. 2.21 Plot of k against permeability function.

A. Hazen (1911) gave an empirical relation for permeability of filter sands as

$$k = 100(D_{10})^2 \tag{2.70}$$

where k is in cm/s and D_{10} is the effective size of the soil in cm.

Equation (2.70) was obtained from the test results of Hazen where the effective size of soils varied from 0.1 to 3 mm and the uniformity coefficient for all soils was less than 5. The coefficient 100 is an average value. The individual test results showed a variation of the coefficient from 41 to 146. Although Hazen's relation is approximate, it shows a similarity to Eq. (2.69).

A. Casagrande has also given an empirical relation for k for fine or medium clean sands with bulky grain as

$$k = 1.4\, k_{0.85} e^2 \tag{2.71}$$

where $k_{0.85}$ is the coefficient of permeability at a void ratio of 0.85.

2.1.9 Variation of Permeability with Void Ratio in Clay

The Kozeny–Carman equation does not successfully explain the variation of the coefficient of permeability with void ratio for clayey soils. The discrepancies between the theoretical and experimental values are shown in Figs. 2.22 and 2.23. These results are based on consolidation-permeability tests (Olsen, 1961, 1962). The marked degree of variation between the theoretical and experimental values arise from several factors, including deviations from Darcy's law, high viscosity of the pore water, and unequal pore sizes. Olsen has developed a model to account for the variation of permeability due to unequal pore sizes.

Example 2.3 Results of a permeability test are given in Table 2.7 and drawn in Fig. 2.21. (*a*) Calculate the "composite shape factor," $C_S S_s^2 T^2$, of the Kozeny–Carmen equation, given $\mu_{20°C} = 10.09 \times 10^{-3}$ poise. (*b*) If $C_S = 2.5$ and $T = \sqrt{2}$, determine S_s. Compare this value with the theoretical value for a sphere of diameter D_{10}.

SOLUTION *Part (a):* From Eq. (2.64),

$$k = \frac{1}{C_S S_s^2 T^2} \frac{\gamma_w}{\mu} \frac{e^3}{1+e}$$

$$C_S S_s^2 T^2 = \frac{\gamma_w}{\mu} \frac{e^3/(1+e)}{k}$$

The value of $[e^3/(1+e)]/k$ is the slope of the straight line for the plot of $e^3/(1+e)$ against k (Fig. 2.21). So

$$\frac{e^3/(1+e)}{k} = \frac{0.15}{0.03\ \text{cm/s}} = 5$$

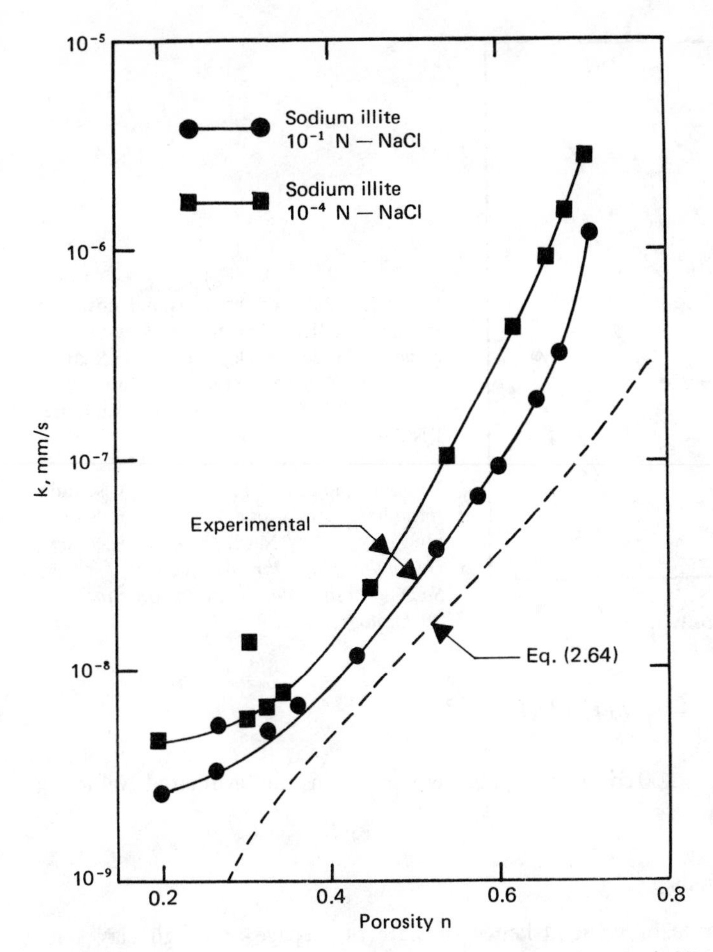

Fig. 2.22 Coefficient of permeability for sodium illite. *(After H. W. Olsen, Hydraulic Flow through Saturated Clays, Sc.D. Thesis, Massachusetts Institute of Technology, 1961.)*

$$C_S S_s^2 T^2 = \frac{(1\ \text{g/cm}^3)\,(981\ \text{cm/s}^2)}{10.09 \times 10^{-3}\ \text{poise}}\,(5) = 4.86 \times 10^5\ cm^{-2}$$

Part (b): (Note the units carefully.)

$$S_s = \sqrt{\frac{4.86 \times 10^5}{C_S T^2}} = \sqrt{\frac{4.86 \times 10^5}{2.5 \times (\sqrt{2})^2}} = 311.8\, cm^2/cm^3$$

For $D_{10} = 0.2$ mm,

$$S_s = \frac{\text{surface area of a sphere of radius 0.01 cm}}{\text{volume of sphere of radius 0.01 cm}}$$

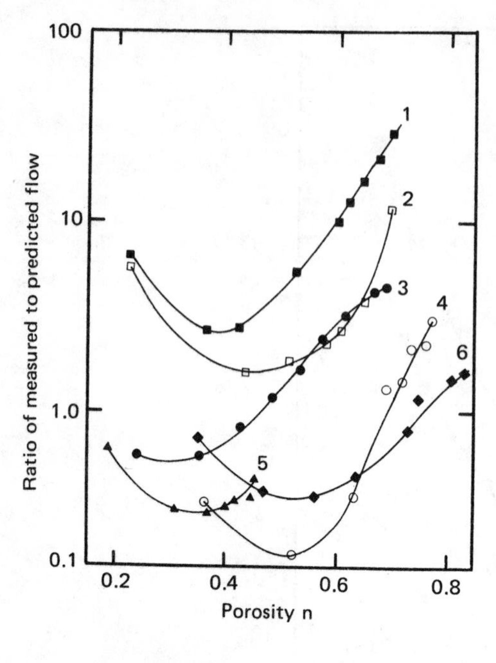

Fig. 2.23 Ratio of the measured flow rate to that predicted by the Kozeny-Carman equation for several clays. Curve 1: Sodium illite, $10^{-1}N$ NaCl. Curve 2: Sodium illite, $10^{-4}N$ NaCl. Curve 3: Natural kaolinite, Distilled water H_2O. Curve 4: Sodium Boston blue clay, $10^{-1}N$ NaCl. Curve 5: Sodium kaolinite, 1% (by Wt.) sodium tetraphosphate. Curve 6: Calcium Boston blue clay, $10^{-4}N$ NaCl. *(After H. W. Olsen, Hydraulic Flow through Saturated Clays, Sc.D. Thesis, Massachusetts Institute of Technology, 1961.)*

$$= \frac{4\pi(0.01)^2}{\frac{4}{3}\pi(0.01)^3} = \frac{3}{0.01} = 300\,cm^2/cm^3$$

This value of $S_S = 300\ \text{cm}^2/\text{cm}^3$ agrees closely with the estimated value of $S_s = 311.8\ \text{cm}^2/\text{cm}^3$.

2.1.10 Electroosmosis

The coefficient of permeability—and hence the rate of seepage—through clay soils is very small compared to that in granular soils, but the drainage can be increased by application of an external electric current. This phenomenon is a result of the exchangeable nature of the adsorbed cations in clay particles and the dipolar nature of the water molecules. The principle can be explained with the help of Fig. 2.24. When dc electricity is applied to the soil, the cations start to migrate to the cathode, which consists of a perforated metallic pipe. Since water is adsorbed on the cations, it is also dragged along. When the cations reach the cathode, they release the water, and the subsequent build up of pressure causes the water to drain out. This process is called *electroosmosis* and was first used by L. Casagrande in 1937 for soil stabilization in Germany.

Rate of drainage by electroosmosis. Figure 2.25 shows a capillary tube formed by clay particles. The surface of the clay particles have negative charges, and the cations are concentrated in a layer of liquid. According to the Helmholtz-Smoluchowski theory (Helmholtz, 1879; Smoluchowski, 1914; see also Mitchell, 1970, 1976), the flow velocity due to an applied dc voltage E can be given by

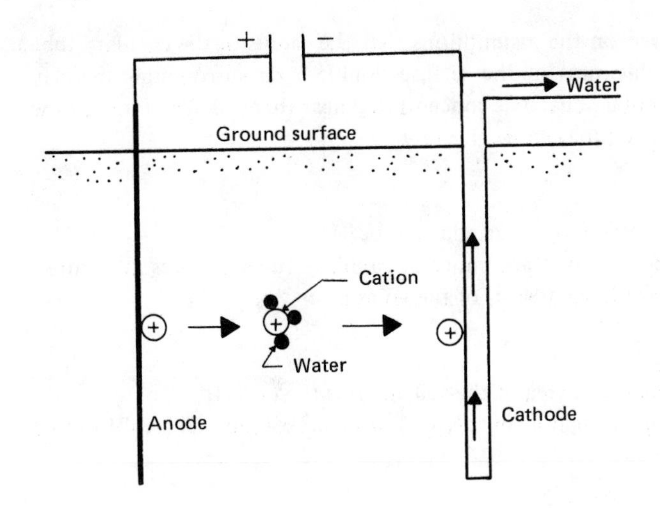

Fig. 2.24 Principles of electro-osmosis.

$$v_e = \frac{D\zeta}{4\pi\eta} \frac{E}{L} \tag{2.72}$$

where v_e = flow velocity due to applied voltage
D = dielectric constant
ζ = zeta potential
η = viscosity
L = electrode spacing

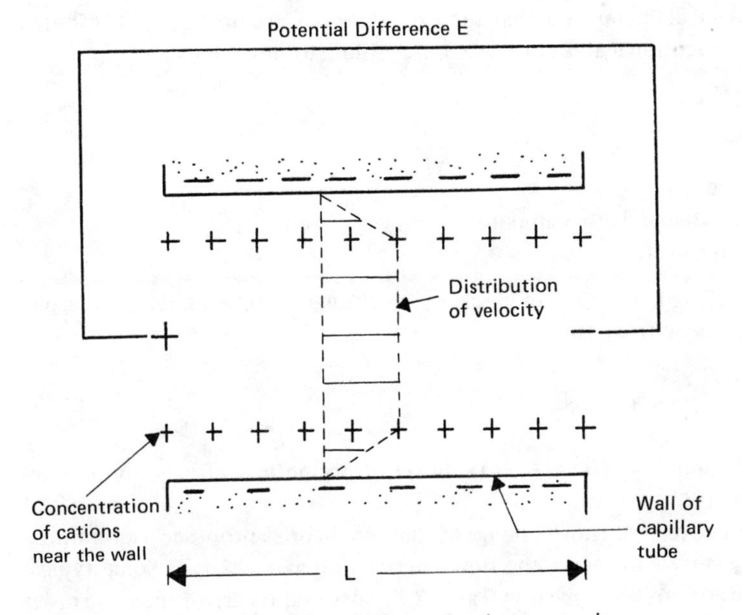

Fig. 2.25 Helmholtz–Smoluchowski theory for electroosmosis.

Equation (2.72) is based on the assumptions that the radius of the capillary tube is large compared to the thickness of the diffuse double layer surrounding the clay particles and that all the mobile charge is concentrated near the wall. The rate of flow of water through the capillary tube can be given by

$$q_c = a v_e \tag{2.73}$$

where a is area of the cross section of the capillary tube.

If a soil mass is assumed to have a number of capillary tubes as a result of interconnected voids, the cross-sectional area A_v of the voids is

$$A_v = nA$$

where A is the gross cross-sectional area of the soil and n is the porosity.

The rate of discharge q through a soil mass of gross cross-sectional area A can be expressed by the relation

$$q = A_v v_e = nAv_e = n \frac{D\zeta}{4\pi\eta} \frac{E}{L} A \tag{2.74}$$

$$= k_e i_e A \tag{2.75}$$

where $k_e = n(D\zeta/4\pi\eta)$ is the electroosmotic coefficient of permeability and i_e is the electrical potential gradient. The units of k_e can be $\text{cm}^2/(\text{s} \cdot \text{V})$ and the units of i_e can be V/cm. Note that Eq. (2.75) is similar in form to Eq. (2.5).

In contrast to the Helmholtz–Smoluchowski theory [Eq. (2.72)], which is based on flow through large capillary tubes, Schmid (1950, 1951) proposed a theory in which it was assumed that the capillary tubes formed by the pores between clay particles are small in diameter and that the excess cations are uniformly distributed across the pore cross-sectional area (Fig. 2.26). According to this theory,

$$v_e = \frac{r^2 A_o F}{8\eta} \frac{E}{L} \tag{2.76}$$

where r = pore radius
A_o = volume charge density in pore
F = Faraday constant

Based on Eq. (2.76), the rate of discharge q through a soil mass of gross cross-sectional area A can be written as

$$q = n \frac{r^2 A_o F}{8\eta} \frac{E}{L} A = k_e i_e A \tag{2.77}$$

where n is porosity and $k_e = n(r^2 A_o F/8\eta)$ is the electroosmotic coefficient of permeability.

Without arguing over the shortcomings of the two theories proposed, our purpose will be adequately served by using the flow-rate relation as $q = k_e i_e A$. Some typical values of k_e for several soils are given in Table 2.8. These values are of the same order

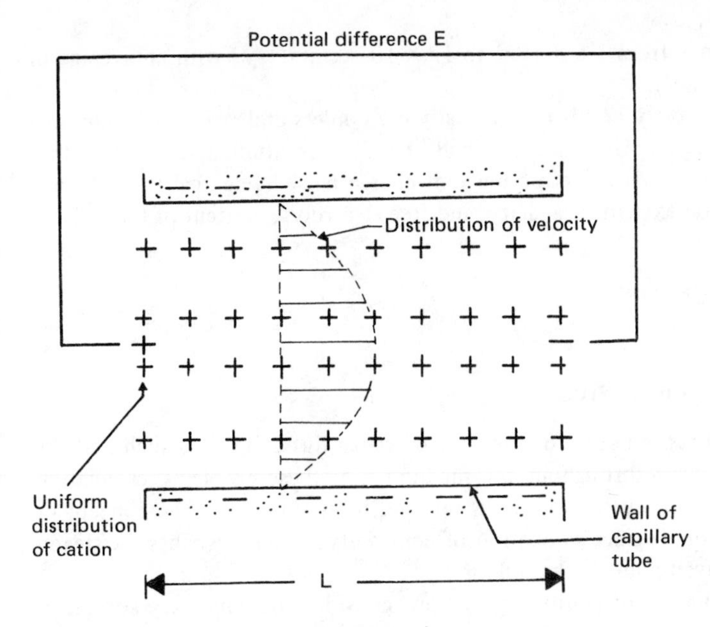

Fig. 2.26 Schmid theory for electroosmosis.

Table 2.8 Electroosmotic coefficient of permeability

Material	Water content, %	k_e, $cm^2/(sec \cdot V)$
London clay	52.3	5.8×10^{-5}
Boston blue clay	50.8	5.1×10^{-5}
Kaolin	67.7	5.7×10^{-5}
Clayey silt	31.7	5.0×10^{-5}
Rock flour	27.2	4.5×10^{-5}
Na-Montmorillonite	170	2.0×10^{-5}
Na-Montmorillonite	2000	12.0×10^{-5}
Mica powder	49.7	6.9×10^{-5}
Fine sand	26.0	4.1×10^{-5}
Quartz powder	23.5	4.3×10^{-5}
As quick clay	31.0	$2.0–2.5 \times 10^{-5}$
Bootlegger Cove clay	30.0	$2.4–5.0 \times 10^{-5}$
Silty clay, West Branch dam	32.0	$3.0–6.0 \times 10^{-5}$
Clay silt, Little Pic river, Ontario	26.0	1.5×10^{-5}

After Mitchell (1976)

of magnitude and range from 1.5×10^{-5} to 12×10^{-5} cm^2/(s·V) with an average of about 6×10^{-5} cm^2/(s·V).

Electroosmosis is costly and is not generally used unless drainage by conventional means cannot be achieved. Gray and Mitchell (1967) have studied the factors that affect the amount of water transferred per unit charge passed [e.g., gal/(h · A)], such as water content, cation exchange capacity, and free electrolyte content of the soil.

2.2 SEEPAGE

2.2.1 Equation of Continuity

In many practical cases, the nature of the flow of water through soil is such that the velocity and gradient vary throughout the medium. For these problems, calculation of flow is generally made by use of graphs referred to as *flow nets.* The concept of the flow net is based on Laplace's equation of continuity, which describes the steady flow condition for a given point in the soil mass.

To derive the equation of continuity of flow, consider an elementary soil prism at point A (Fig. 2.27b) for the hydraulic structure shown in Fig. 2.27(a). The flows entering the soil prism in the x, y, and z directions can be given from Darcy's law as

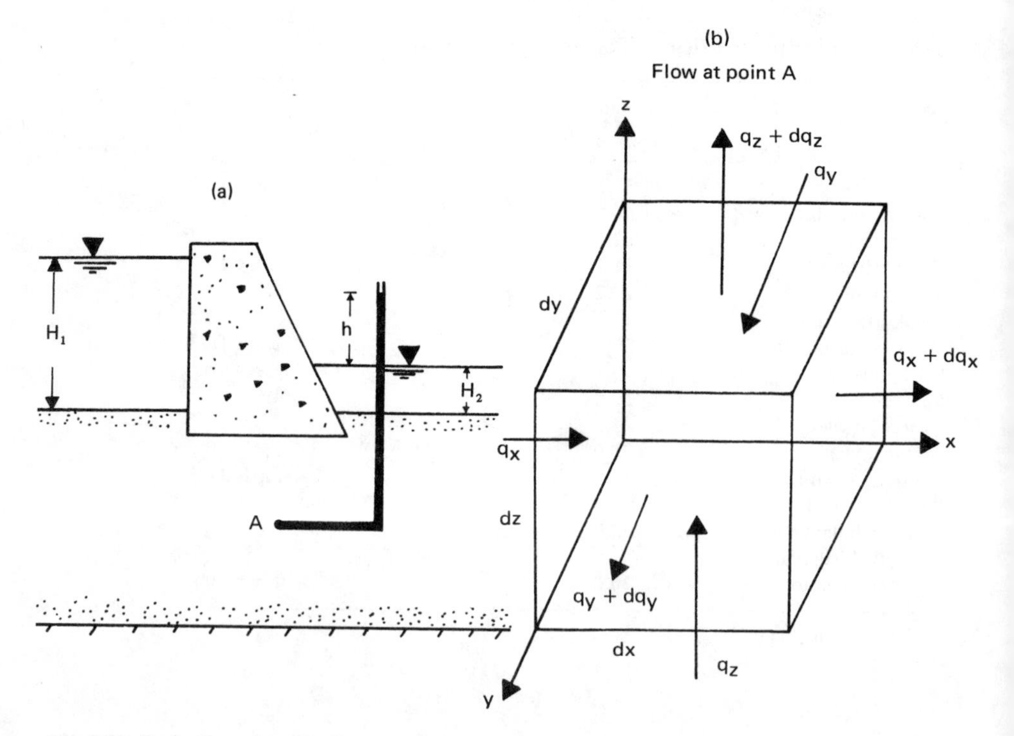

Fig. 2.27 Derivation of continuity equation.

$$q_x = k_x i_x A_x = k_x \frac{\partial h}{\partial x} dy\, dz \tag{2.78}$$

$$q_y = k_y i_y A_y = k_y \frac{\partial h}{\partial y} dx\, dz \tag{2.79}$$

$$q_z = k_z i_z A_z = k_z \frac{\partial h}{\partial z} dx\, dy \tag{2.80}$$

where q_x, q_y, q_z = flow entering in directions x, y, z, respectively
k_x, k_y, k_z = coefficients of permeability in directions x, y, z, respectively
h = hydraulic head at point A

The flows leaving the prism in the x, y, and z directions are, respectively,

$$q_x + dq_x = k_x(i_x + di_x)A_x$$

$$= k_x \left(\frac{\partial h}{\partial x} + \frac{\partial^2 h}{\partial x^2} dx\right) dy\, dz \tag{2.81}$$

$$q_y + dq_y = k_y \left(\frac{\partial h}{\partial y} + \frac{\partial^2 h}{\partial y^2} dy\right) dx\, dz \tag{2.82}$$

$$q_z + dq_z = k_z \left(\frac{\partial h}{\partial z} + \frac{\partial^2 h}{\partial z^2} dz\right) dx\, dy \tag{2.83}$$

For steady flow through an incompressible medium, the flow entering the elementary prism is equal to the flow leaving the elementary prism. So,

$$q_x + q_y + q_z = (q_x + dq_x) + (q_y + dq_y) + (q_z + dq_z) \tag{2.84}$$

Combining Eqs. (2.78) to (2.84), we obtain

$$k_x \frac{\partial^2 h}{\partial x^2} + k_y \frac{\partial^2 h}{\partial y^2} + k_z \frac{\partial^2 h}{\partial z^2} = 0 \tag{2.85}$$

For two-dimensional flow in the xz plane, Eq. (2.85) becomes

$$k_x \frac{\partial^2 h}{\partial x^2} + k_z \frac{\partial^2 h}{\partial z^2} = 0 \tag{2.86}$$

If the soil is isotropic with respect to permeability, $k_x = k_z = k$, and the continuity equation simplifies to

$$\frac{\partial^2 h}{\partial x^2} + \frac{\partial^2 h}{\partial z^2} = 0 \tag{2.87}$$

This is generally referred to as Laplace's equation.

Potential and stream functions. Consider a function $\phi(x, z)$ such that

$$\frac{\partial \phi}{\partial x} = v_x = -k \frac{\partial h}{\partial x} \tag{2.88}$$

and $$\frac{\partial \phi}{\partial z} = v_z = -k \frac{\partial h}{\partial z} \tag{2.89}$$

If we differentiate Eq. (2.88) with respect to x and Eq. (2.89) with respect to z and substitute in Eq. (2.87), we get

$$\frac{\partial^2 \phi}{\partial x^2} + \frac{\partial^2 \phi}{\partial z^2} = 0 \tag{2.90}$$

Therefore, $\phi(x, z)$ satisfies the Laplace equation. From Eqs. (2.88) and (2.89),

$$\phi(x, z) = -kh(x, z) + f(z) \tag{2.91}$$

and $$\phi(x, z) = -kh(x, z) + g(x) \tag{2.92}$$

Since x and z can be varied independently, $f(z) = g(x) = C$, a constant. So

$$\phi(x, z) = -kh(x, z) + C$$

or $$h(x, z) = \frac{1}{k}[C - \phi(x, z)] \tag{2.93}$$

If $h(x, z)$ is a constant equal to h_1, Eq. (2.93) represents a curve in the xz plane. For this curve, ϕ will have a constant value, ϕ_1. This is an *equipotential line*. So, by assigning to ϕ a number of values such as $\phi_1, \phi_2, \phi_3, \ldots$, we can get a number of equipotential lines along which $h = h_1, h_2, h_3, \ldots$, respectively. The slope along an equipotential line ϕ can now be derived:

$$d\phi = \frac{\partial \phi}{\partial x} dx + \frac{\partial \phi}{\partial z} dz \tag{2.94}$$

If ϕ is a constant along a curve, $d\phi = 0$. Hence,

$$\left(\frac{dz}{dx}\right)_\phi = -\frac{\partial \phi / \partial x}{\partial \phi / \partial z} = -\frac{v_x}{v_z} \tag{2.95}$$

Again, let $\psi(x, z)$ be a function such that

$$\frac{\partial \psi}{\partial z} = v_x = -k \frac{\partial h}{\partial x} \tag{2.96}$$

and $$-\frac{\partial \psi}{\partial x} = v_z = -k \frac{\partial h}{\partial z} \tag{2.97}$$

Combining Eqs. (2.88) and (2.96), we obtain

$$\frac{\partial \phi}{\partial x} = \frac{\partial \psi}{\partial z}$$

$$\frac{\partial^2 \psi}{\partial z^2} = \frac{\partial^2 \phi}{\partial x \, \partial z} \tag{2.98}$$

Again, combining Eqs. (2.89) and (2.97),

$$-\frac{\partial \phi}{\partial z} = \frac{\partial \psi}{\partial x}$$

$$-\frac{\partial^2 \phi}{\partial x\, \partial z} = \frac{\partial^2 \psi}{\partial x^2} \tag{2.99}$$

From Eqs. (2.98) and (2.99),

$$\frac{\partial^2 \psi}{\partial x^2} + \frac{\partial^2 \psi}{\partial z^2} = -\frac{\partial^2 \phi}{\partial x\, \partial z} + \frac{\partial^2 \phi}{\partial x\, \partial z} = 0$$

So $\psi(x, z)$ also satisfies Laplace's equation. If we assign to $\psi(x, z)$ various values $\psi_1, \psi_2, \psi_3, \ldots$, we get a family of curves in the xz plane. Now

$$d\psi = \frac{\partial \psi}{\partial x}\, dx + \frac{\partial \psi}{\partial z}\, dz \tag{2.100}$$

For a given curve, if ψ is constant, then $d\psi = 0$. Thus, from Eq. (2.100),

$$\left(\frac{dz}{dx}\right)_\psi = -\frac{\partial \psi/\partial x}{\partial \psi/\partial z} = \frac{v_z}{v_x} \tag{2.101}$$

Note that the slope, $(dz/dx)_\psi$, is in the same direction as the resultant velocity. Hence, the curves $\psi = \psi_1, \psi_2, \psi_3, \ldots$ are the *flow lines.*

From Eqs. (2.95) and (2.101), we can see that at a given point (x, z) the equipotential line and the flow line are orthogonal.

The functions $\phi(x, z)$ and $\psi(x, z)$ are called the *potential function* and the *stream function,* respectively.

2.2.2 Use of Continuity Equation for Solution of Simple Flow Problem

To understand the role of the continuity equation [Eq. (2.87)], consider a simple case of flow of water through two layers of soil as shown in Fig. 2.28. The flow is in one direction only, i.e., in the direction of the x axis. The lengths of the two soil layers (L_A and L_B) and their coefficients of permeability in the direction of the x axis (k_A and k_B) are known. The total heads at sections 1 and 3 are known. We are required to plot the total head at any other section for $0 < x < L_A + L_B$.

For one-dimensional flow, Eq. (2.87) becomes

$$\frac{\partial^2 h}{\partial x^2} = 0 \tag{2.102}$$

Integration of Eq. (2.102) twice gives

$$h = C_2 x + C_1 \tag{2.103}$$

where C_1 and C_2 are constants.

For flow through soil A, the boundary conditions are:

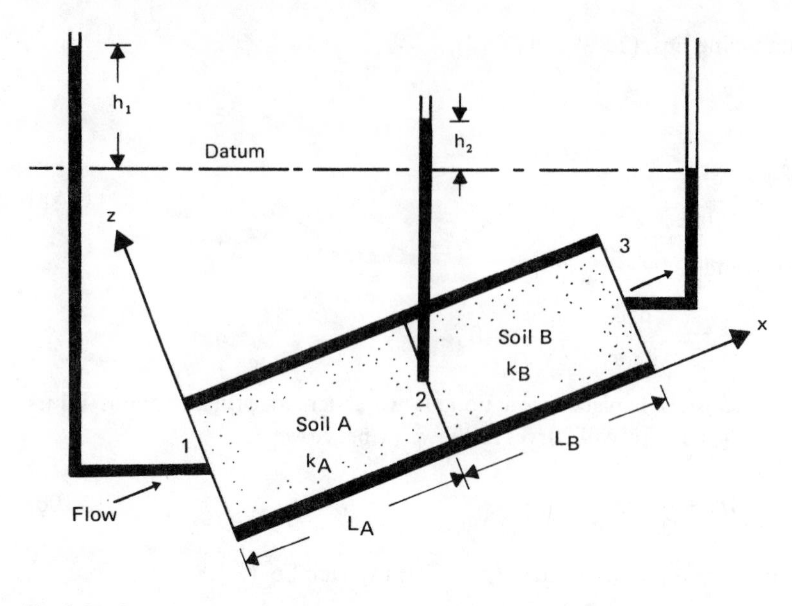

Fig. 2.28 One-directional flow through two layers of soil.

1. at $x = 0, h = h_1$
2. at $x = L_A, h = h_2$

However, h_2 is unknown ($h_1 > h_2$). From the first boundary condition and Eq. (2.103), $C_1 = h_1$. So,

$$h = C_2 x + h_1 \tag{2.104}$$

From the second boundary condition and Eq. (2.103),

$$h_2 = C_2 L_A + h_1 \qquad \text{or} \qquad C_2 = (h_2 - h_1)/L_A \qquad \text{So,}$$

So, $$h = -\frac{h_1 - h_2}{L_A} x + h_1 \qquad (\text{for } 0 \leqslant x \leqslant L_A) \tag{2.105}$$

For flow through soil B, the boundary conditions for solution of C_1 and C_2 in Eq. (2.102) are

1. at $x = L_A, h = h_2$
2. at $x = L_A + L_B, h = 0$

From the first boundary condition and Eq. (2.103), $h_2 = C_2 L_A + C_1$, or

$$C_1 = h_2 - C_2 L_A \tag{2.106}$$

Again, from the secondary boundary condition and Eq. (2.103), $0 = C_2(L_A + L_B) + C_1$, or

$$C_1 = -C_2(L_A + L_B) \tag{2.107}$$

Equating the right-hand sides of Eqs. (2.106) and (2.107),

$$h_2 - C_2 L_A = -C_2(L_A + L_B)$$

$$C_2 = -\frac{h_2}{L_B} \tag{2.108}$$

and then substituting Eq. (2.108) into Eq. (2.106), gives

$$C_1 = h_2 + \frac{h_2}{L_B} L_A = h_2 \left(1 + \frac{L_A}{L_B}\right) \tag{2.109}$$

Thus, for flow through soil B,

$$h = -\frac{h_2}{L_B} x + h_2 \left(1 + \frac{L_A}{L_B}\right) \qquad (\text{for } L_A \leqslant x \leqslant L_A + L_B) \tag{2.110}$$

With Eqs. (2.105) and (2.110), we can solve for h for any value of x from 0 to $L_A + L_B$, provided that h_2 is known. However,

$$q = \text{rate of flow through soil } A = \text{rate of flow through soil } B$$

So,

$$q = k_A \left(\frac{h_1 - h_2}{L_A}\right) A = k_B \left(\frac{h_2}{L_B}\right) A \tag{2.111}$$

where k_A and k_B are the coefficients of permeability of soils A and B, respectively, and A is the area of cross section of soil perpendicular to the direction of flow.

From Eq. (2.111),

$$h_2 = \frac{k_A h_1}{L_A(k_A/L_A + k_B/L_B)} \tag{2.112}$$

Substitution of Eq. (2.112) into Eqs. (2.105) and (2.110) yields, after simplification,

$$h = h_1 \left(1 - \frac{k_B x}{k_A L_B + k_B L_A}\right) \qquad (\text{for } x = 0 \text{ to } L_A) \tag{2.113}$$

$$h = h_1 \left[\frac{k_A}{k_A L_B + k_B L_A} (L_A + L_B - x)\right] \qquad (\text{for } x = L_A \text{ to } L_A + L_B) \tag{2.114}$$

2.2.3 Flow Nets

A set of flow lines and equipotential lines is called a *flow net.* As discussed in Sec. 2.2.1, a flow line is a line along which a water particle will travel. An equipotential line is a line joining the points that show the same piezometric elevation (i.e., hydraulic head $= h(x, z) =$ constant). Figure 2.29 shows an example of a flow net for a single row of sheet piles. The permeable layer is isotropic with respect to the coefficient of permeability, i.e., $k_x = k_z = k$. Note that the solid lines in Fig. 2.29 are the flow lines, and the broken lines are the equipotential lines. In drawing a flow net, the boundary conditions must be kept in mind. For example, in Fig. 2.29,

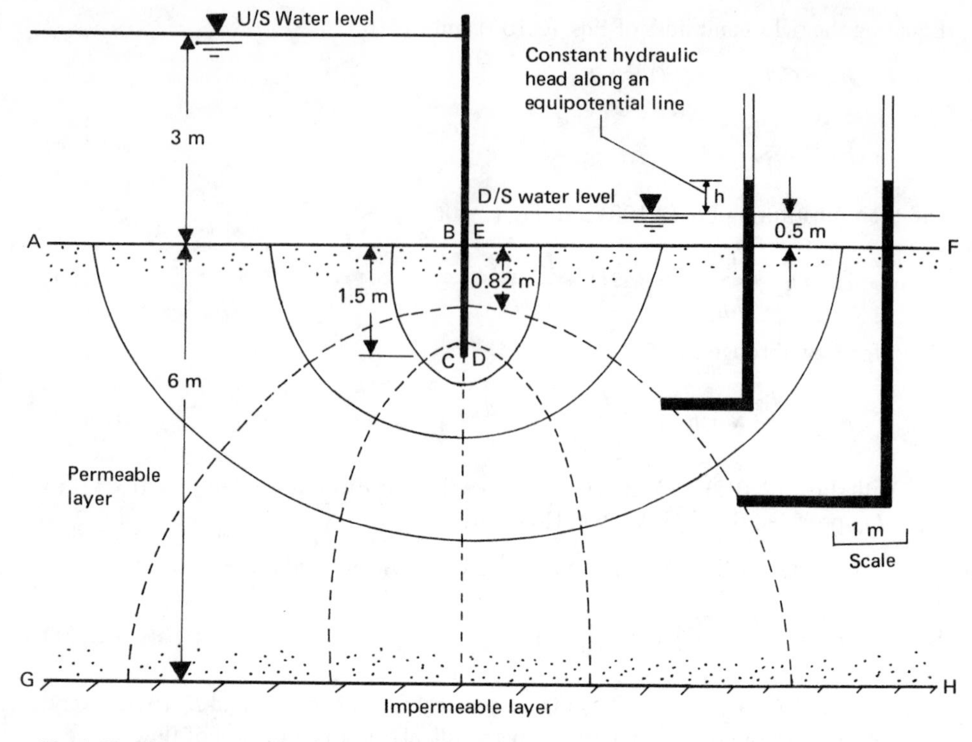

Fig. 2.29 Flow net around a single row of sheet piles.

1. *AB* is an equipotential line
2. *EF* is an equipotential line
3. *BCDE* (i.e., the sides of the sheet pile) is a flow line
4. *GH* is a flow line

The flow lines and the equipotential lines are drawn by trial and error. It must be remembered that the flow lines intersect the equipotential lines at right angles. The flow and equipotential lines are usually drawn in such a way that the flow elements are approximately squares. Drawing a flow net is time consuming and tedious because of the trial-and-error process involved. Once a satisfactory flow net has been drawn, it can be traced out.

Some other examples of flow nets are shown in Figs. 2.30 and 2.31 for flow under dams.

Calculation of seepage from a flow net under a hydraulic structure. A *flow channel* is the strip located between two adjacent flow lines. To calculate the seepage under a hydraulic structure, consider a flow channel as shown in Fig. 2.32. The equipotential lines crossing the flow channel are also shown, along with their corresponding hydraulic heads. Let Δq be the flow through the flow channel per unit length of the hydraulic

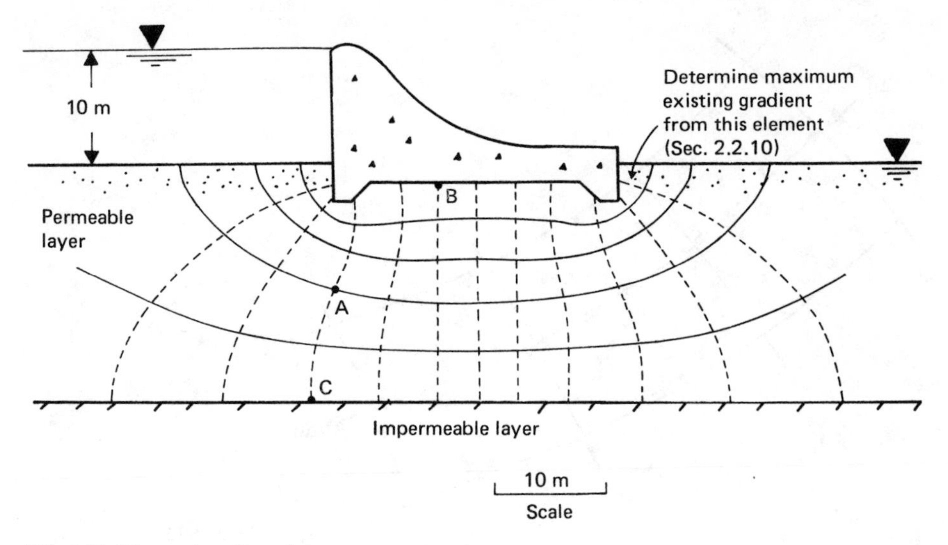

Fig. 2.30 Flow net under a dam.

structure (i.e., perpendicular to the section shown). According to Darcy's law,

$$\Delta q = kiA = k\left(\frac{h_1 - h_2}{l_1}\right)(b_1 \times 1) = k\left(\frac{h_2 - h_3}{l_2}\right)(b_2 \times 1)$$

$$= k\left(\frac{h_3 - h_4}{l_3}\right)(b_3 \times 1) = \cdots \tag{2.115}$$

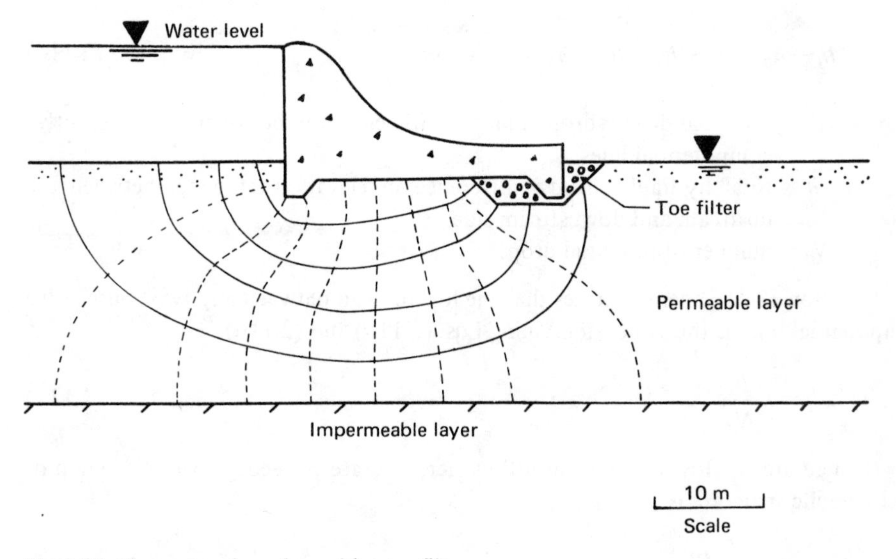

Fig. 2.31 Flow net under a dam with a toe filter.

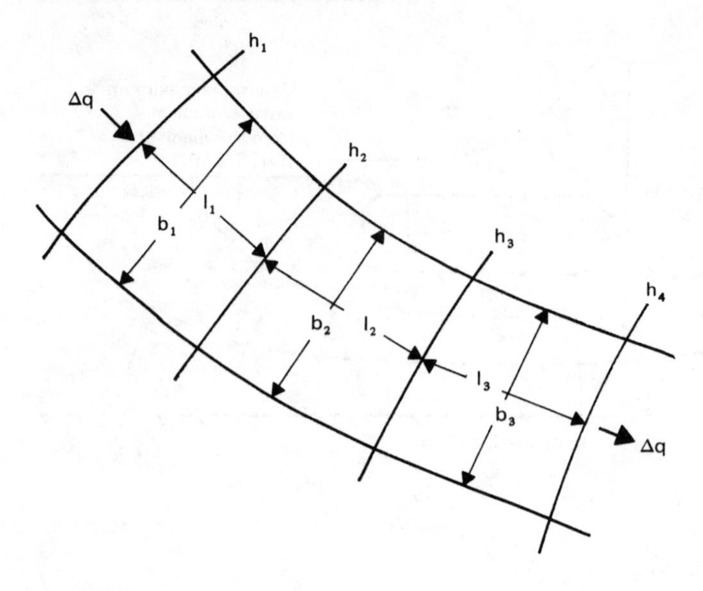

Fig. 2.32.

If the flow elements are drawn as squares, then

$$l_1 = b_1$$
$$l_2 = b_2$$
$$l_3 = b_3$$
$$\vdots$$

So, from Eq. (2.115), we get

$$h_1 - h_2 = h_2 - h_3 = h_3 - h_4 = \cdots = \Delta h = \frac{h}{N_d} \tag{2.116}$$

where Δh = potential drop = drop in piezometric elevation between two consecutive equipotential lines

h = total hydraulic head = difference in elevation of water between the upstream and downstream side

N_d = number of potential drops

Equation (2.116) demonstrates that the loss of head between any two consecutive equipotential lines is the same. Combining Eqs. (2.115) and (2.116),

$$\Delta q = k \frac{h}{N_d} \tag{2.117}$$

If there are N_f flow channels in a flow net, the rate of seepage per unit length of the hydraulic structure is

$$q = N_f \, \Delta q = kh \frac{N_f}{N_d} \tag{2.118}$$

Although flow nets are usually constructed in such a way that all flow elements are approximately squares, that need not always be the case. We could construct flow nets with all the flow elements drawn as rectangles. In that case, the length-to-width ratio of the flow nets has to be a constant, i.e.,

$$\frac{b_1}{l_1} = \frac{b_2}{l_2} = \frac{b_3}{l_3} = \cdots = n \tag{2.119}$$

For such flow nets, the rate of seepage per unit length of hydraulic structure can be given by

$$q = kh \frac{N_f}{N_d} n \tag{2.120}$$

Example 2.4 For the flow net shown in Fig. 2.30:
(*a*) How high would water rise if a piezometer is placed at (i) A, (ii) B, (iii) C?
(*b*) If $k = 0.01$ mm/s, determine the seepage loss of the dam in $\text{m}^3/(\text{day} \cdot \text{m})$.

SOLUTION The maximum hydraulic head h is 10 m. In Fig. 2.30, $N_d = 12$, $\Delta h = h/N_d = 10/12 = 0.833$.

Part (a), (i): To reach A, water has to go through three potential drops. So head lost is equal to $3 \times 0.833 = 2.5$ m. Hence the elevation of the water level in the piezometer at A will be $10 - 2.5 = 7.5\ m$ above the ground surface.

Part (a), (ii): The water level in the piezometer above the ground level is $10 - 5(0.833) = 5.84\ m$.

Part (a), (iii): Points A and C are located on the same equipotential line. So water in a piezometer at C will rise to the same elevation as at A, i.e., *7.5 m* above the ground surface.

Part (b): The seepage loss is given by $q = kh\,(N_f/N_d)$. From Fig. 2.30, $N_f = 5$ and $N_d = 12$. Since

$$k = 0.01 \text{ mm/s} = \left(\frac{0.01}{1000}\right)(60 \times 60 \times 24) = 0.864 \text{ m/day}$$

$$q = 0.864(10)(5/12) = 3.6\ m^3/(day \cdot m)$$

2.2.4 Hydraulic Uplift Force under a Structure

Flow nets can be used to determine the hydraulic uplifting force under a structure. The procedure can best be explained through a numerical example. Consider the dam section shown in Fig. 2.30, the cross section of which has been replotted in Fig. 2.33. To find the pressure head at point D (Fig. 2.33), we refer to the flow net shown in Fig. 2.30; the pressure head is equal to $(10 + 3.34\text{ m})$ minus the hydraulic head loss. Point D coincides with the third equipotential line beginning with the upstream side, which means that the hydraulic head loss at that point is $2(h/N_d) = 2(10/12) = 1.67$ m. So,

$$\text{Pressure head at } D = 13.34 - 1.67 = 11.67 \text{ m}$$

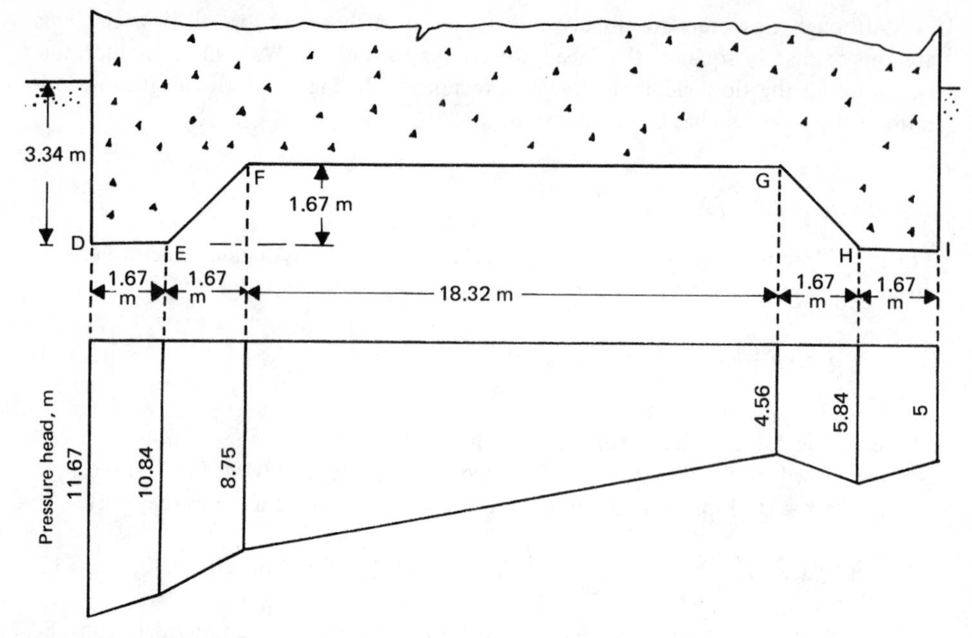

Fig. 2.33 Pressure head under the dam section shown in Fig. 2.30.

Similarly,

$$\text{Pressure head at } E = (10 + 3.34) - 3(10/12) = 10.84 \text{ m}$$

$$\text{Pressure head at } F = (10 + 1.67) - 3.5(10/12) = 8.75 \text{ m}$$

(Note that point F is approximately midway between the fourth and fifth equipotential lines starting from the upstream side.)

$$\text{Pressure head at } G = (10 + 1.67) - 8.5(10/12) = 4.56 \text{ m}$$

$$\text{Pressure head at } H = (10 + 3.34) - 9(10/12) = 5.84 \text{ m}$$

$$\text{Pressure head at } I = (10 + 3.34) - 10(10/12) = 5 \text{ m}$$

The pressure heads calculated above are plotted in Fig. 2.33. Between points F and G, the variation of pressure heads will be approximately linear. The hydraulic uplift force per unit length of the dam, U, can now be calculated as

$$U = \gamma_w \text{ (area of the pressure head diagram) } (1)$$

$$= 9.81 \left[\left(\frac{11.67 + 10.84}{2} \right)(1.67) + \left(\frac{10.84 + 8.75}{2} \right)(1.67) \right.$$

$$+ \left(\frac{8.75 + 4.56}{2} \right)(18.32) + \left(\frac{4.56 + 5.84}{2} \right)(1.67)$$

$$\left. + \left(\frac{5.84 + 5}{2} \right)(1.67) \right]$$

$$= 9.81(18.8 + 16.36 + 121.92 + 8.68 + 9.05)$$

$$= 1714.9 \text{ kN/m}$$

2.2.5 Flow Nets in Anisotropic Material

In developing the procedure described in Sec. 2.2.3 for plotting flow nets, we assumed that the permeable layer is isotropic, i.e., $k_{\text{horizontal}} = k_{\text{vertical}} = k$. Let us now consider the case of constructing flow nets for seepage through soils that show anisotropy with respect to permeability. For two-dimensional flow problems, we refer to Eq. (2.86):

$$k_x \frac{\partial^2 h}{\partial x^2} + k_z \frac{\partial^2 h}{\partial z^2} = 0$$

where $k_x = k_{\text{horizontal}}$ and $k_z = k_{\text{vertical}}$. This equation can be rewritten as

$$\frac{\partial^2 h}{(k_z/k_x)\partial x^2} + \frac{\partial^2 h}{\partial z^2} = 0 \tag{2.121}$$

Let $x' = \sqrt{k_z/k_x}\, x$; then

$$\frac{\partial^2 h}{(k_z/k_x)\partial x^2} = \frac{\partial^2 h}{\partial x'^2} \tag{2.122}$$

Substituting Eq. (2.122) into Eq. (2.121), we obtain

$$\frac{\partial^2 h}{\partial x'^2} + \frac{\partial^2 h}{\partial z^2} = 0 \tag{2.123}$$

Equation (2.123) is of the same form as Eq. (2.87), which governs the flow in isotropic soils and should represent two sets of orthogonal lines in the $x'z$ plane. The steps for construction of a flow net in an anisotropic medium are as follows:

1. To plot the section of the hydraulic structure, adopt a *vertical scale.*
2. Determine $\sqrt{\dfrac{k_z}{k_x}} = \sqrt{\dfrac{k_{\text{vertical}}}{k_{\text{horizontal}}}}$
3. Adopt a horizontal scale such that

   $$\text{Scale}_{\text{horizontal}} = \sqrt{\frac{k_z}{k_x}}\,(\text{scale}_{\text{vertical}})$$

4. With the scales adopted in steps 1 and 3, plot the cross section of the structure.
5. Draw the flow net for the transformed section plotted in step 4 in the same manner as is done for seepage through isotropic soils.
6. Calculate the rate of seepage as

$$q = \sqrt{k_x k_z}\, h \frac{N_f}{N_d} \tag{2.124}$$

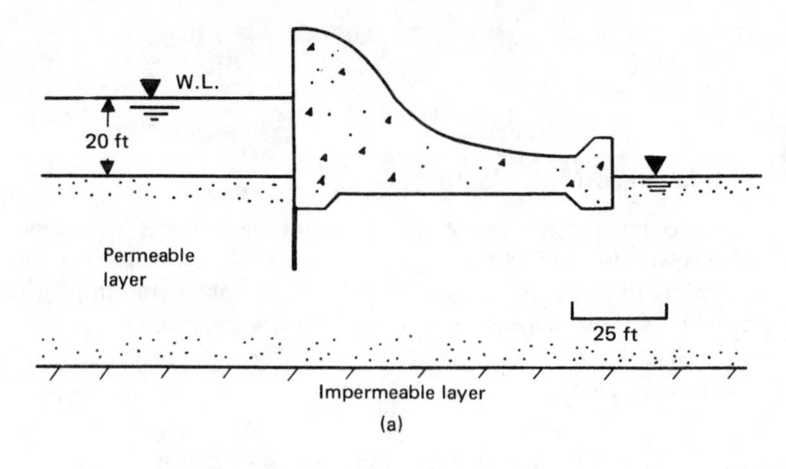

(a)

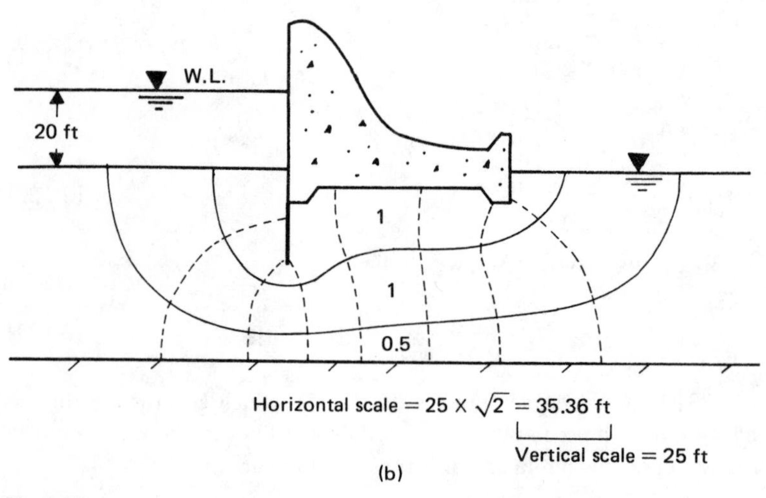

(b)

Fig. 2.34

Compare Eqs. (2.117) and (2.124). Both equations are similar, except for the fact that k in Eq. (2.117) is replaced by $\sqrt{k_x k_z}$ in Eq. (2.124).

Example 2.5 A dam section is shown in Fig. 2.34*a*. The coefficients of permeability of the permeable layer in the vertical and horizontal directions are 2×10^{-2} mm/s and 4×10^{-2} mm/s, respectively. Draw a flow net and calculate the seepage loss of the dam in $\text{ft}^3/(\text{day} \cdot \text{ft})$.

SOLUTION From the given data,

$k_z = 2 \times 10^{-2}$ mm/s $= 5.67$ ft/day

$k_x = 4 \times 10^{-2}$ mm/s $= 11.34$ ft/day

and

$h = 20$ ft

For drawing the flow net,

$$\text{Horizontal scale} = \sqrt{\frac{2 \times 10^{-2}}{4 \times 10^{-2}}}\,(\text{vertical scale})$$

$$= \frac{1}{\sqrt{2}}(\text{vertical scale})$$

On the basis of this, the dam section is replotted and the flow net drawn as in Fig. 2.34*b*. The rate of seepage is given by $q = \sqrt{k_x k_z}\, h(N_f/N_d)$. From Fig. 2.34*b*, $N_d = 8$ and $N_f = 2.5$ (the lowermost flow channel has a width-to-length ratio of 0.5). So,

$$q = \sqrt{(5.67)(11.34)}\,(20)\,(2.5/8) = 50.12\ ft^3/(day \cdot ft)$$

Example 2.6 A single row of sheet pile structure is shown in Fig. 2.35*a*. Draw a flow net for the transformed section. Replot this flow net in the natural scale also. The relationship between the permeabilities is given as $k_x = 6k_z$.

SOLUTION For the transformed section,

$$\text{Horizontal scale} = \sqrt{\frac{k_z}{k_x}}\,(\text{vertical scale})$$

$$= \frac{1}{\sqrt{6}}\,(\text{vertical scale})$$

The transformed section and the corresponding flow net are shown in Fig. 2.35*b*.

Figure 2.35*c* shows the flow net constructed to the natural scale. One important fact to be noticed from this is that when the soil is anisotropic with respect to permeability, *the flow and equipotential lines are not necessarily orthogonal.*

2.2.6 Construction of Flow Nets for Hydraulic Structures on Nonhomogeneous Subsoils

The flow-net construction technique described in Sec. 2.2.3 is for the condition where the subsoil is homogeneous. Rarely in nature do such ideal conditions occur; in most cases, we encounter stratified soil deposits such as those shown in Fig. 2.38. When a flow net is constructed across the boundary of two soils with different permeabilities, the flow net deflects at the boundary. This is called a *transfer condition.* Figure 2.36 shows a general condition where a flow channel crosses the boundary of two soils. Soil layers 1 and 2 have permeabilities of k_1 and k_2, respectively. The broken lines drawn across the flow channel are the equipotential lines.

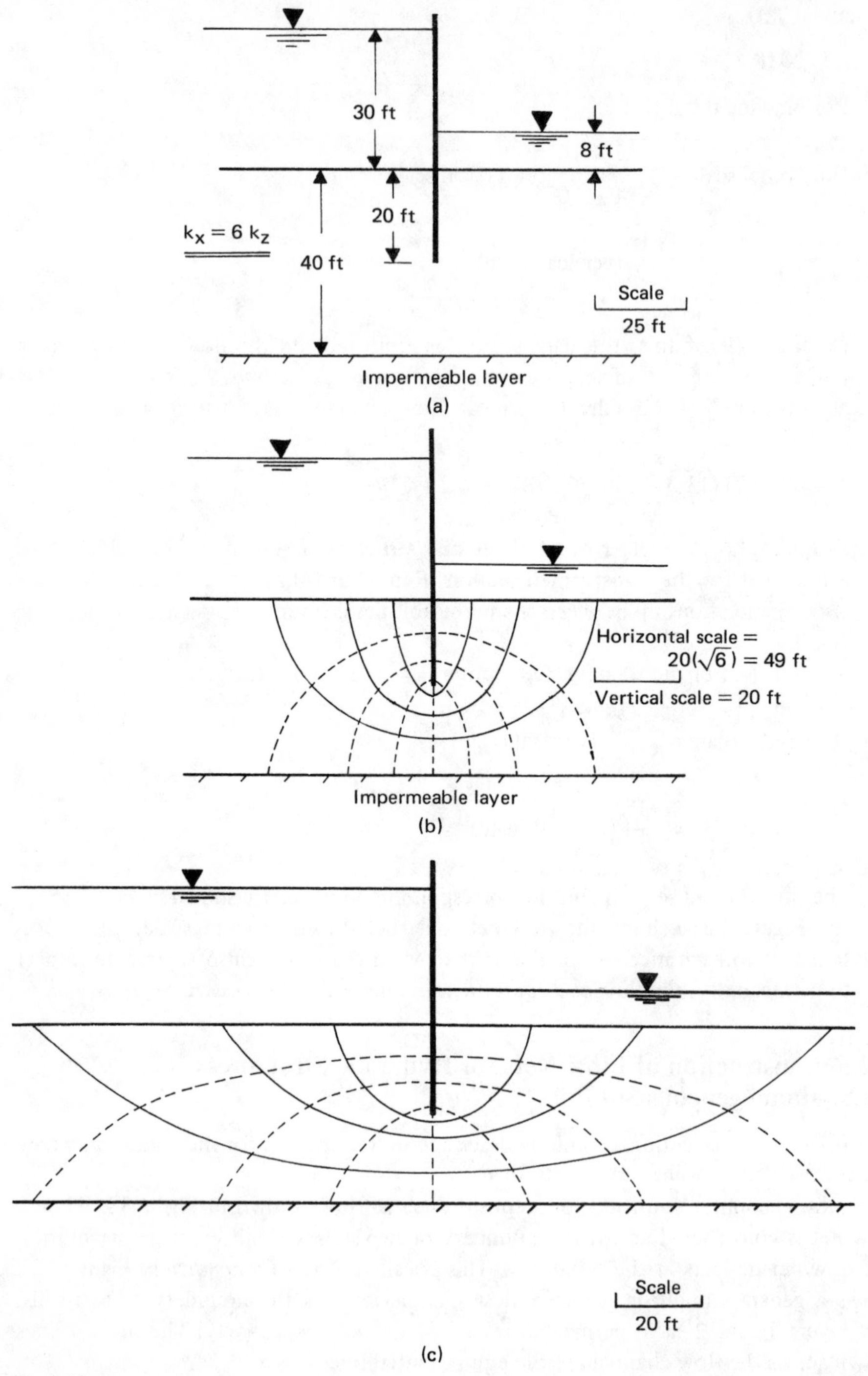
30 ft
8 ft
20 ft
$k_x = 6\ k_z$
40 ft
Scale
25 ft
Impermeable layer
(a)
Horizontal scale =
$20(\sqrt{6}) = 49$ ft
Vertical scale = 20 ft
Impermeable layer
(b)
Scale
20 ft
(c)

Fig. 2.35

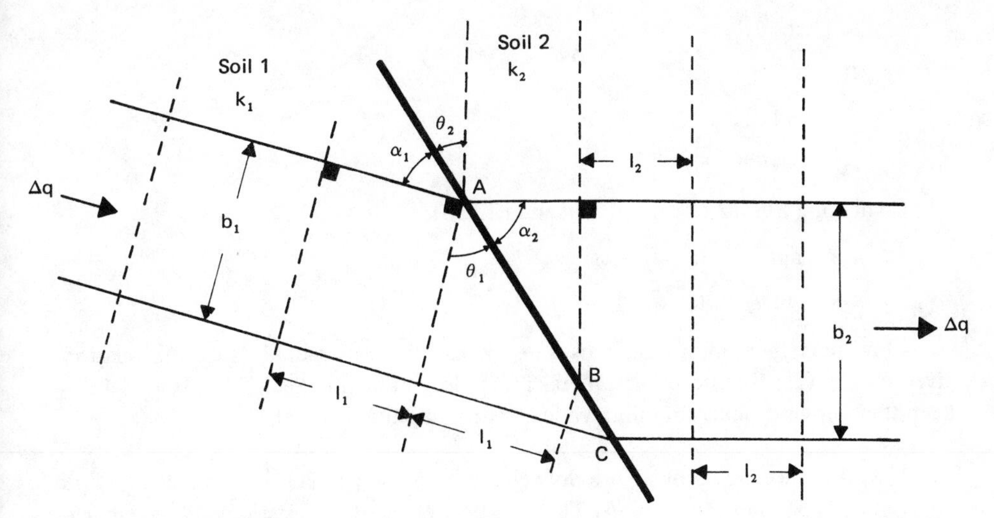

Fig. 2.36 Transfer condition.

Let Δh be the loss of hydraulic head between two consecutive equipotential lines. Considering a unit length perpendicular to the section shown, the rate of seepage through the flow channel is

$$\Delta q = k_1 \frac{\Delta h}{l_1} (b_1 \times 1) = k_2 \frac{\Delta h}{l_2} (b_2 \times 1)$$

or
$$\frac{k_1}{k_2} = \frac{b_2/l_2}{b_1/l_1} \tag{2.125}$$

where l_1 and b_1 are the length and width of the flow elements in soil layer 1, and l_2 and b_2 are the length and width of the flow elements in soil layer 2.

Referring again to Fig. 2.36,

$$l_1 = AB \sin \theta_1 = AB \cos \alpha_1 \tag{2.126a}$$

$$l_2 = AB \sin \theta_2 = AB \cos \alpha_2 \tag{2.126b}$$

$$b_1 = AC \cos \theta_1 = AC \sin \alpha_1 \tag{2.126c}$$

$$b_2 = AC \cos \theta_2 = AC \sin \alpha_2 \tag{2.126d}$$

From Eqs. (2.126*a*) and (2.126*c*),

$$\frac{b_1}{l_1} = \frac{\cos \theta_1}{\sin \theta_1} = \frac{\sin \alpha_1}{\cos \alpha_1}$$

or
$$\frac{b_1}{l_1} = \frac{1}{\tan \theta_1} = \tan \alpha_1 \tag{2.127}$$

Also, from Eqs. (2.126*b*) and (2.126*d*),

$$\frac{b_2}{l_2} = \frac{\cos \theta_2}{\sin \theta_2} = \frac{\sin \alpha_2}{\cos \alpha_2}$$

or
$$\frac{b_2}{l_2} = \frac{1}{\tan \theta_2} = \tan \alpha_2 \tag{2.128}$$

Combining Eqs. (2.125), (2.127), and (2.128),

$$\frac{k_1}{k_2} = \frac{\tan \theta_1}{\tan \theta_2} = \frac{\tan \alpha_2}{\tan \alpha_1} \tag{2.129}$$

Flow nets in nonhomogeneous subsoils can be constructed using the relations given by Eq. (2.129) and other general principles outlined in Sec. 2.2.3. It is useful to keep the following points in mind while constructing the flow nets:

1. If $k_1 > k_2$, we may plot square flow elements in layer 1. This means that $l_1 = b_1$ in Eq. (2.125). So $k_1/k_2 = b_2/l_2$. Thus the flow elements in layer 2 will be rectangles and their width-to-length ratios will be equal to k_1/k_2. This is shown in Fig. 2.37*a*.
2. If $k_1 < k_2$, we may plot square flow elements in layer 1 (i.e., $l_1 = b_1$). From Eq. (2.125), $k_1/k_2 = b_2/l_2$. So the flow elements in layer 2 will be rectangles. This is shown in Fig. 2.37*b*.

An example of the construction of a flow net for a dam section resting on a

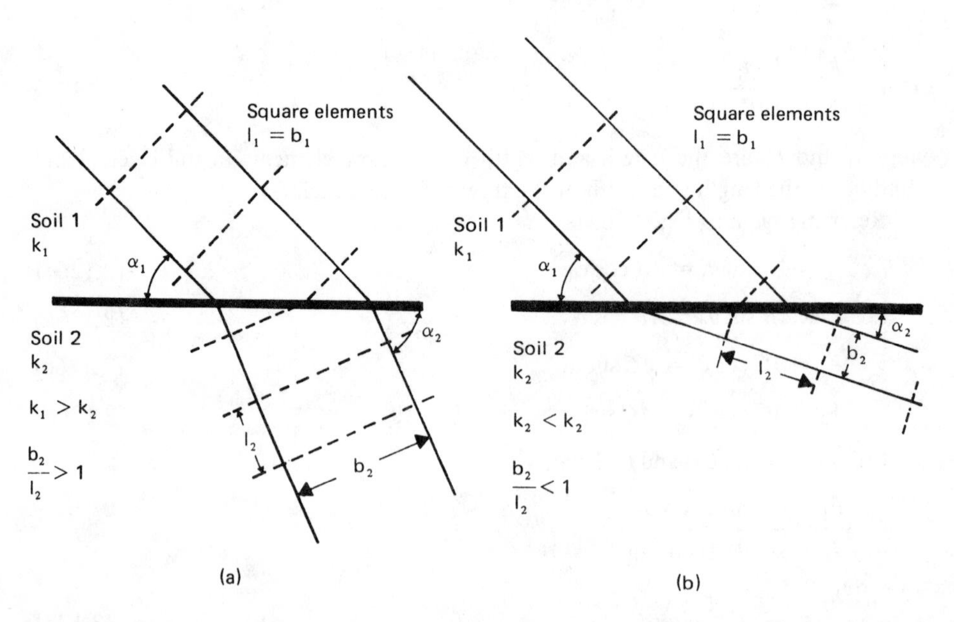

Fig. 2.37 Flow channel at the boundary between two soils with different coefficients of permeability.

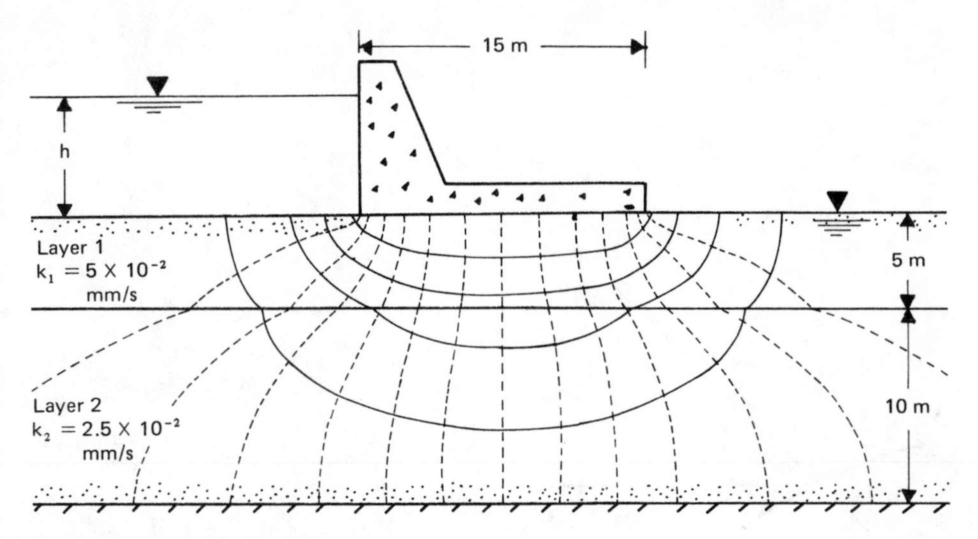

Fig. 2.38 Flow net under a dam.

two-layered soil deposit is given in Fig. 2.38. Note that $k_1 = 5 \times 10^{-2}$ mm/s and $k_2 = 2.5 \times 10^{-2}$ mm/s. So,

$$\frac{k_1}{k_2} = \frac{5.0 \times 10^{-2}}{2.5 \times 10^{-2}} = 2 = \frac{\tan \alpha_2}{\tan \alpha_1} = \frac{\tan \theta_1}{\tan \theta_2}$$

In soil layer 1, the flow elements are plotted as squares; and, since $k_1/k_2 = 2$, the length-to-width ratio of the flow elements in soil layer 2 is 1/2.

2.2.7 Directional Variation of Permeability in Anisotropic Medium

In anisotropic soils, the directions of the maximum and minimum permeabilities are generally at right angles to each other. However, the equipotential lines and the flow lines are not necessarily orthogonal, as was shown in Fig. 2.35*c*.

Figure 2.39 shows a flow line and an equipotential line. m is the direction of the tangent drawn to the flow line at O, and thus that is the direction of the resultant discharge velocity. Direction n is perpendicular to the equipotential line at O, and so it is the direction of the resultant hydraulic gradient. Using Darcy's law,

$$v_x = -k_{\max} \frac{\partial h}{\partial x} \tag{2.130}$$

$$v_z = -k_{\min} \frac{\partial h}{\partial z} \tag{2.131}$$

$$v_m = -k_\alpha \frac{\partial h}{\partial m} \tag{2.132}$$

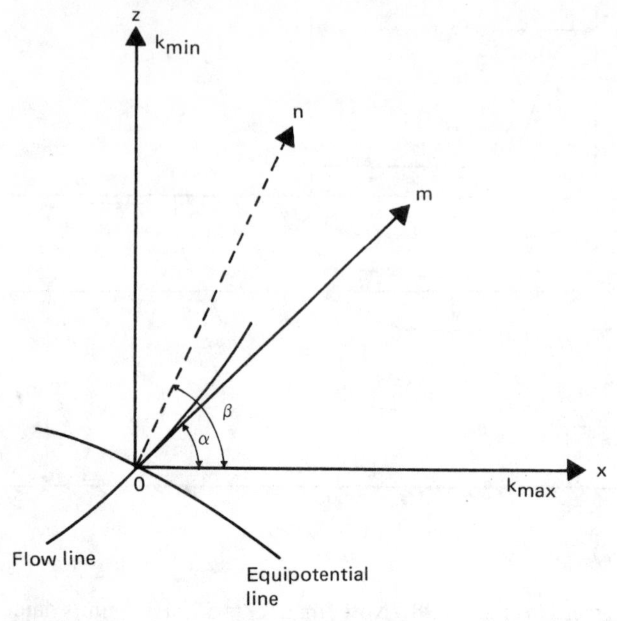

Fig. 2.39 Directional variation of the coefficient of permeability.

$$v_n = -k_\beta \frac{\partial h}{\partial n} \tag{2.133}$$

where $k_{\max}$ = maximum coefficient of permeability (in the horizontal x direction)
$k_{\min}$ = minimum coefficient of permeability (in the vertical z direction)
k_α, k_β = coefficients of permeability in m, n directions, respectively

Now, we can write

$$\frac{\partial h}{\partial m} = \frac{\partial h}{\partial x} \cos \alpha + \frac{\partial h}{\partial z} \sin \alpha \tag{2.134}$$

From Eqs. (2.130), (2.131), and (2.132), we have

$$\frac{\partial h}{\partial x} = -\frac{v_x}{k_{\max}} \qquad \frac{\partial h}{\partial x} = -\frac{v_z}{k_{\min}} \qquad \frac{\partial h}{\partial m} = -\frac{v_m}{k_\alpha}$$

Also, $v_x = v_m \cos \alpha$ and $v_z = v_m \sin \alpha$.

Substitution of these into Eq. (2.134) gives

$$-\frac{v_m}{k_\alpha} = -\frac{v_x}{k_{\max}} \cos \alpha - \frac{v_z}{k_{\min}} \sin \alpha$$

or

$$\frac{v_m}{k_\alpha} = \frac{v_m}{k_{\max}} \cos^2 \alpha + \frac{v_m}{k_{\min}} \sin^2 \alpha$$

so
$$\frac{1}{k_\alpha} = \frac{\cos^2 \alpha}{k_{max}} + \frac{\sin^2 \alpha}{k_{min}} \tag{2.135}$$

The nature of the variation of k_α with α as determined by Eq. (2.135) is shown in Fig. 2.40.

Again, we can say that

$$v_n = v_x \cos \beta + v_z \sin \beta \tag{2.136}$$

Combining Eqs. (2.130), (2.131), and (2.133),

$$k_\beta \frac{\partial h}{\partial n} = k_{max} \frac{\partial h}{\partial x} \cos \beta + k_{min} \frac{\partial h}{\partial z} \sin \beta \tag{2.137}$$

But
$$\frac{\partial h}{\partial x} = \frac{\partial h}{\partial n} \cos \beta \tag{2.138}$$

and
$$\frac{\partial h}{\partial z} = \frac{\partial h}{\partial n} \sin \beta \tag{2.139}$$

Substitution of Eqs. (2.138) and (2.139) into Eq. (2.137) yields

$$k_\beta = k_{max} \cos^2 \beta + k_{min} \sin^2 \beta \tag{2.140}$$

The variation of k_β with β is also shown in Fig. 2.40. It can be seen that, for given values of k_{max} and k_{min}, Eqs. (2.135) and (2.140) yield slightly different values of the directional permeability. However, the maximum difference will not be more than 25%.

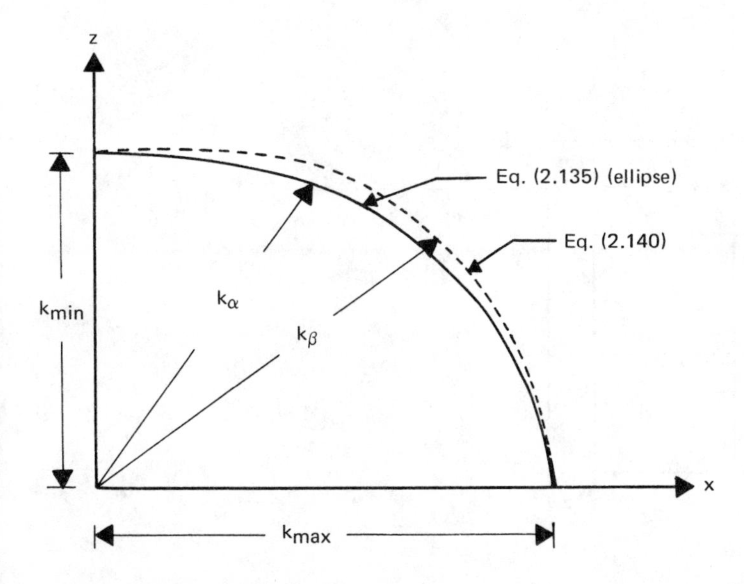

Fig. 2.40 Directional variation of permeability.

2.2.8 Numerical Analysis of Seepage

In this section, we develop some approximate finite-difference equations for solving seepage problems. We start from Laplace's equation, which was derived in Sec. 2.2.1; for two-dimensional seepage

$$k_x \frac{\partial^2 h}{\partial x^2} + k_z \frac{\partial^2 h}{\partial z^2} = 0 \tag{2.86}$$

Figure 2.41 shows a part of a region in which flow is taking place. For flow in the horizontal direction, using Taylor's series we can write

$$h_1 = h_0 + \Delta x \left(\frac{\partial h}{\partial x}\right)_0 + \frac{(\Delta x)^2}{2!}\left(\frac{\partial^2 h}{\partial x^2}\right)_0 + \frac{(\Delta x)^3}{3!}\left(\frac{\partial^3 h}{\partial x^3}\right)_0 + \cdots \tag{2.141}$$

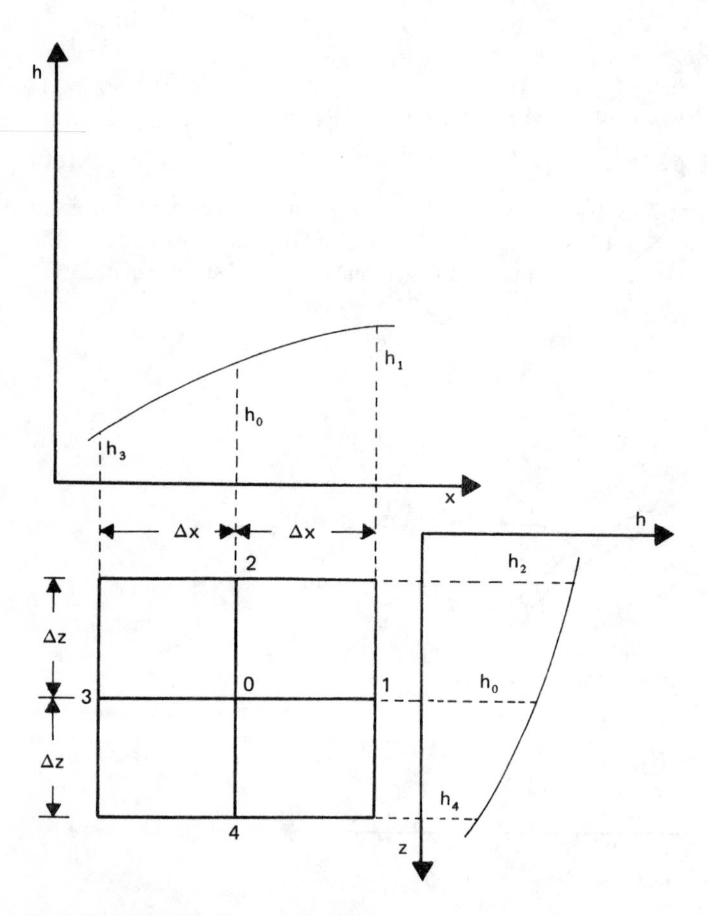

Fig. 2.41 Hydraulic heads for flow in a region.

and $$h_3 = h_0 - \Delta x \left(\frac{\partial h}{\partial x}\right)_0 + \frac{(\Delta x)^2}{2!}\left(\frac{\partial^2 h}{\partial x^2}\right)_0 - \frac{(\Delta x)^3}{3!}\left(\frac{\partial^3 h}{\partial x^3}\right)_0 + \cdots \tag{2.142}$$

Adding Eqs. (2.141) and (2.142), we obtain

$$h_1 + h_3 = 2h_0 + \frac{2(\Delta x)^2}{2!}\left(\frac{\partial^2 h}{\partial x^2}\right)_0 + \frac{2(\Delta x)^4}{4!}\left(\frac{\partial^4 h}{\partial x^4}\right)_0 + \cdots \tag{2.143}$$

Assuming Δx to be small, we can neglect the third and subsequent terms on the right-hand side of Eq. (2.143). Thus

$$\left(\frac{\partial^2 h}{\partial x^2}\right)_0 = \frac{h_1 + h_3 - 2h_0}{(\Delta x)^2} \tag{2.144}$$

Similarly, for flow in the z direction we can obtain

$$\left(\frac{\partial^2 h}{\partial z^2}\right)_0 = \frac{h_2 + h_4 - 2h_0}{(\Delta z)^2} \tag{2.145}$$

Substitution of Eqs. (2.144) and (2.145) into Eq. (2.86) gives

$$k_x \frac{h_1 + h_3 - 2h_0}{(\Delta x)^2} + k_z \frac{h_2 + h_4 - 2h_0}{(\Delta z)^2} = 0 \tag{2.146}$$

If $k_x = k_y = k$ and $\Delta x = \Delta z$, Eq. (2.146) simplifies to

$$h_1 + h_2 + h_3 + h_4 - 4h_0 = 0$$

or $$h_0 = \tfrac{1}{4}(h_1 + h_2 + h_3 + h_4) \tag{2.147}$$

Equation (2.147) can also be derived by considering Darcy's law, $q = kiA$. For the rate of flow from point 1 to point 0 through the channel shown hatched in Fig. 2.42*a*, we have

$$q_{1\text{-}0} = k \frac{h_1 - h_0}{\Delta x} \Delta z \tag{2.148}$$

Similarly,

$$q_{0\text{-}3} = k \frac{h_0 - h_3}{\Delta x} \Delta z \tag{2.149}$$

$$q_{2\text{-}0} = k \frac{h_2 - h_0}{\Delta z} \Delta x \tag{2.150}$$

$$q_{0\text{-}4} = k \frac{h_0 - h_4}{\Delta z} \Delta x \tag{2.151}$$

Since the total rate of flow into point 0 is equal to the total rate of flow out of point 0, $q_{in} - q_{out} = 0$. Hence

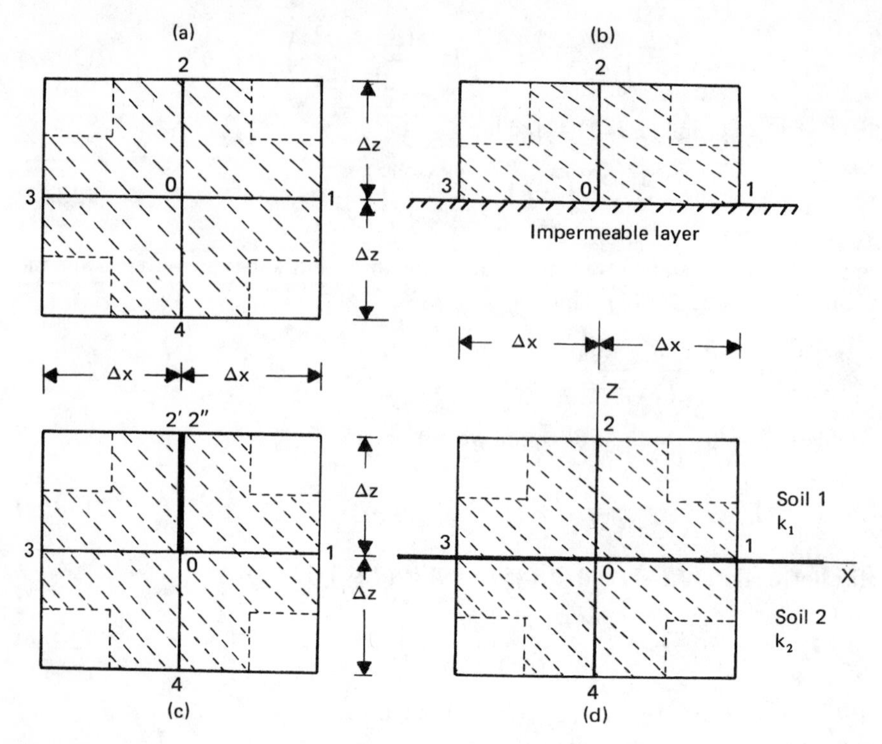

Fig. 2.42.

$$(q_{1\text{-}0} + q_{2\text{-}0}) - (q_{0\text{-}3} + q_{0\text{-}4}) = 0 \tag{2.152}$$

Taking $\Delta x = \Delta z$ and substituting Eqs. (2.148) to (2.151) into Eq. (2.152), we get

$$h_0 = \tfrac{1}{4}(h_1 + h_2 + h_3 + h_4)$$

If the point 0 is located on the boundary of a pervious and an impervious layer as shown in Fig. 2.42*b*, Eq. (2.147) must be modified as follows:

$$q_{1\text{-}0} = k\,\frac{h_1 - h_0}{\Delta x}\,\frac{\Delta z}{2} \tag{2.153}$$

$$q_{0\text{-}3} = k\,\frac{h_0 - h_3}{\Delta x}\,\frac{\Delta z}{2} \tag{2.154}$$

$$q_{0\text{-}2} = k\,\frac{h_0 - h_2}{\Delta z}\,\Delta x \tag{2.155}$$

For continuity of flow,

$$q_{1\text{-}0} - q_{0\text{-}3} - q_{0\text{-}2} = 0 \tag{2.156}$$

With $\Delta x = \Delta z$, combining Eqs. (2.153) to (2.156) gives

$$\frac{h_1 - h_0}{2} - \frac{h_0 - h_3}{2} - (h_0 - h_2) = 0$$

$$\frac{h_1}{2} + \frac{h_3}{2} + h_2 - 2h_0 = 0$$

or
$$h_0 = \tfrac{1}{4}(h_1 + 2h_2 + h_3) \tag{2.157}$$

When point 0 is located at the bottom of a piling (Fig. 2.42*c*), the equation for the hydraulic head for flow continuity can be given by

$$q_{1\text{-}0} + q_{4\text{-}0} - q_{0\text{-}3} - q_{0\text{-}2'} - q_{0\text{-}2''} = 0 \tag{2.158}$$

Note that 2′ and 2″ are two points at the same elevation on the opposite sides of the sheet pile with hydraulic heads of $h_{2'}$ and $h_{2''}$, respectively. For this condition we can obtain (for $\Delta x = \Delta z$), through a similar procedure to that above,

$$h_0 = \tfrac{1}{4}[h_1 + \tfrac{1}{2}(h_{2'} + h_{2''}) + h_3 + h_4] \tag{2.159}$$

Seepage in layered soils. Equation (2.147), which we derived above, is valid for seepage in homogeneous soils. However, for the case of flow across the boundary of one homogeneous soil layer to another, Eq. (2.147) must be modified. Referring to Fig. 2.42*d*, since the flow region is located half in soil 1 with a coefficient of permeability k_1 and half in soil 2 with a coefficient of permeability k_2, we can say that

$$k_x = \tfrac{1}{2}(k_1 + k_2) \tag{2.160}$$

Now, if we replace soil 2 by soil 1, it will have a hydraulic head of $h_{4'}$ in place of h_4. For the velocity to remain the same,

$$k_1 \frac{h_{4'} - h_0}{\Delta z} = k_2 \frac{h_4 - h_0}{\Delta z} \tag{2.161}$$

or
$$h_{4'} = \frac{k_2}{k_1}(h_4 - h_0) + h_0 \tag{2.162}$$

Thus, based on Eq. (2.86), we can write

$$\frac{k_1 + k_2}{2} \frac{h_1 + h_3 - 2h_0}{(\Delta x)^2} + k_1 \frac{h_2 + h_{4'} - 2h_0}{(\Delta z)^2} = 0 \tag{2.163}$$

Taking $\Delta x = \Delta z$ and substituting Eq. (2.162) into Eq. (2.163),

$$\frac{1}{2}(k_1 + k_2)\left[\frac{h_1 + h_3 - 2h_0}{(\Delta x)^2}\right] + \frac{k_1}{(\Delta x)^2}\left\{h_2 + \left[\frac{k_2}{k_1}(h_4 - h_0) + h_0\right] - 2h_0\right\} = 0 \tag{2.163a}$$

or
$$h_0 = \frac{1}{4}\left(h_1 + \frac{2k_1}{k_1 + k_2} h_2 + h_3 + \frac{2k_2}{k_1 + k_2} h_4\right) \tag{2.164}$$

The application of the equations developed in this section can best be demonstrated by the use of a numerical example. Consider the problem of determining the hydraulic heads at various points below the dam shown in Fig. 2.38. Let $\Delta x = \Delta z = 1.25$ m. Since the flow net below the dam will be symmetrical, we will consider only the left-hand half. The steps for determining the values of h at various points in the permeable soil layers are as follows:

1. Roughly sketch out a flow net.
2. Based on the rough flow net (step 1), assign some values for the hydraulic heads at various grid points. These are shown in Fig. 2.43*a*. Note that the values of h assigned here are in *percent*.
3. Consider the heads for row 1 (i.e., $i = 1$). The $h_{(i,j)}$ for $i = 1$ and $j = 1, 2, \ldots, 22$ are 100 in Fig. 2.43*a*; these are correct values based on the boundary conditions. The $h_{(i,j)}$ for $i = 1$ and $j = 23, 24, \ldots, 28$ are estimated values. The flow condition for these grid points is similar to that shown in Fig. 2.42*b*; and, according to Eq. (2.157), $(h_1 + 2h_2 + h_3) - 4h_0 = 0$, or

$$(h_{(i,j+1)} + 2h_{(i+1,j)} + h_{(i,j-1)}) - 4h_{(i,j)} = 0 \tag{2.165}$$

Since the hydraulic heads in Fig. 2.43 are assumed values, Eq. (2.165) will not be satisified. For example, for the grid point $i = 1$ and $j = 23$, $h_{(i,j-1)} = 100$, $h_{(i,j)} = 84$, $h_{(i,j+1)} = 68$, and $h_{(i+1,j)} = 78$. If these values are substituted in Eq. (2.165), we get $[68 + 2(78) + 100] - 4(84) = -12$, instead of zero. If we set -12 equal to R (where R stands for *residual*) and add $R/4$ to $h_{(i,j)}$, Eq. (2.165) will be satisfied. So the new, corrected value of $h_{(i,j)}$ is equal to $84 + (-3) = 81$, as shown in Fig. 2.43*b*. This is called the *relaxation process*. Similarly, the corrected head for the grid point $i = 1$ and $j = 24$ can be found as follows:

$[84 + 2(67) + 61] - 4(68) = 7 = R$;

So, $h_{(1,24)} = 68 + 7/4 = 69.75 \approx 69.8$.

The corrected values of $h_{(1,25)}$, $h_{(1,26)}$, and $h_{(1,27)}$ can be determined in a similar manner. Note that $h_{(1,28)} = 50$ is correct, based on the boundary condition. These are shown in Fig. 2.43*b*.

4. Consider the rows $i = 2$, 3, and 4. $h_{(i,j)}$ for $i = 2, \ldots, 4$ and $j = 2, 3, \ldots, 27$ should follow Eq. (2.147), $(h_1 + h_2 + h_3 + h_4) - 4h_0 = 0$; or

$$(h_{(i,j+1)} + h_{(i-1,j)} + h_{(i,j-1)} + h_{(i+1,j)}) - 4h_{(i,j)} = 0 \tag{2.166}$$

To find the corrected heads $h_{(i,j)}$, we proceed as in step 3. The residual R is calculated by substituting values into Eq. (2.166), and the corrected head is then given by $h_{(i,j)} + R/4$. Due to symmetry, the corrected values of $h_{(1,28)}$ for $i = 2$, 3, and 4, are all 50 as originally assumed. The corrected heads are shown in Fig. 2.43*b*.

5. Consider row $i = 5$ (for $j = 2, 3, \ldots, 27$). According to Eq. (2.164),

$$h_1 + \frac{2k_1}{k_1 + k_2} h_2 + h_3 + \frac{2k_2}{k_1 + k_2} h_4 - 4h_0 = 0 \tag{2.167}$$

7.5 m

j = 1 → j

i = 1 ↓ i

k_1 ↑ k_2 ↓

100	100	100	100	100	100	100	100	100	100	100	100	100	100	100	100	100	100	100	100	100	100	84	68	61	57	54	50
99	99	99	99	99	99	99	99	99	99	98	98	97	97	97	96	95	94	94	94	90	90	78	67	61	57	54	50
99	99	99	99	99	99	99	99	98	98	97	97	96	96	95	94	93	92	89	86	83	80	73	67	62	58	54	50
99	99	98	98	98	98	97	97	97	97	96	96	96	94	94	92	90	89	86	83	80	77	72	66	62	58	54	50
99	99	98	98	97	97	96	96	95	95	94	94	93	93	92	90	88	86	83	81	77	74	70	60	62	58	54	50
99	98	98	97	96	96	95	95	94	94	93	93	90	90	90	87	85	83	80	80	75	72	68	65	61	57	53	50
99	98	97	97	96	96	95	95	94	94	93	92	89	89	86	84	82	80	78	77	73	70	66	64	60	57	53	50
97	97	96	96	95	95	94	94	92	92	92	90	88	87	84	83	81	79	77	74	72	69	66	64	60	57	53	50
96	95	94	94	93	93	92	92	91	91	89	88	86	84	83	81	80	77	75	73	70	68	66	64	60	57	53	50
95	94	93	93	92	92	91	91	90	90	88	87	85	83	82	80	79	76	74	73	70	68	66	64	60	57	53	50
95	94	93	93	92	92	91	91	90	90	88	86	85	83	82	80	78	76	74	73	70	68	66	64	60	57	53	50
94	93	92	92	91	91	90	90	89	89	88	86	84	82	81	80	78	76	74	72	70	68	66	64	60	57	53	50
94	93	92	92	91	91	90	90	89	89	86	86	84	82	80	79	78	76	74	72	70	68	66	64	60	57	53	50

(a)

Fig. 2.43 Hydraulic head calculation by numerical method. (*a*) Initial assumption; (*b*) at the end of first iteration; (*c*) at the end of tenth iteration.

j = 1 → j

7.5 m

k_1 / k_2

i = 1																											
100	100	100	100	100	100	100	100	100	100	100	100	100	100	100	100	100	100	100	100	100	100	81	69.8	61.8	57.3	53.8	50
99	99.3	99.3	99.3	99.3	99.3	99.3	99.3	99.0	98.8	98.5	98.0	97.8	97.5	97.0	96.5	95.8	95.3	94.3	92.5	91.8	87.0	78.5	68.5	61.8	57.5	53.8	50
99	99.0	98.8	98.8	98.8	98.8	98.5	98.3	98.3	97.8	97.3	96.8	96.5	95.5	95.3	94.0	92.8	91.3	89.5	87.3	84.0	80.8	74.3	67.0	62.0	57.8	54.0	50
99	98.8	98.5	98.3	98.0	97.8	97.5	97.3	96.8	96.5	96.0	95.8	94.8	94.8	93.3	92.0	90.5	88.5	86.0	83.3	80.0	76.5	71.5	66.8	62.0	58.0	54.0	50
99	98.6	98.3	97.6	97.4	96.9	96.4	95.9	95.7	95.2	94.7	94.2	93.7	92.6	92.1	90.2	88.2	86.2	83.7	81.0	77.9	74.4	70.3	65.8	61.8	57.8	53.8	50
99	98.5	97.5	97.3	96.5	96.0	95.5	95.0	94.5	94.0	93.5	92.3	91.3	90.5	88.8	87.3	85.0	82.8	81.0	78.3	75.5	71.8	68.3	64.8	61.0	57.3	53.5	50
99	97.8	97.3	96.5	96.0	95.5	95.0	94.5	93.8	93.3	92.8	91.3	89.8	88.0	86.8	84.5	82.5	80.5	78.5	76.3	73.5	70.0	67.0	63.8	60.5	56.8	53.3	50
97	96.5	96.0	95.5	95.0	94.5	94.0	93.3	92.8	92.3	91.0	90.0	88.0	86.3	84.8	82.5	81.0	78.8	76.5	74.8	71.5	69.0	66.3	63.5	60.3	56.8	53.3	50
96	95.3	94.5	94.6	93.5	93.0	92.5	92.0	91.3	90.5	89.8	88.0	86.3	84.8	82.8	81.5	79.5	77.5	75.3	73.0	70.8	68.3	66.0	63.5	60.3	56.8	53.3	50
95	94.3	93.5	93.0	92.5	92.0	91.5	91.0	90.5	89.8	88.5	86.8	85.3	83.5	82.0	80.5	78.5	76.5	74.5	72.5	70.3	68.0	66.0	63.5	60.3	56.8	53.3	50
95	93.8	93.0	92.5	92.0	91.5	91.0	90.5	90.0	89.3	88.0	86.5	84.5	83.0	81.5	80.8	78.3	76.0	74.3	72.3	70.3	68.0	66.0	63.5	60.3	56.8	53.3	50
94	93.3	92.5	92.0	91.5	91.0	90.5	90.0	89.5	89.0	87.8	86.0	84.3	82.5	81.0	79.5	78.0	76.0	74.0	72.3	70.0	68.0	66.0	63.5	60.3	56.8	53.3	50
94	93.0	92.3	91.8	91.3	90.8	90.3	89.8	89.3	88.8	87.8	86.0	84.0	82.0	80.8	79.5	77.8	76.0	74.0	72.0	70.0	68.0	66.0	63.5	60.3	56.8	53.3	50

i ↓

(b)

7.5 m

$j = 1$ → j

$i = 1$ ↓ i

100	100	100	100	100	100	100	100	100	100	100	100	100	100	100	100	100	100	100	100	100	100	82.7	72.4	65.2	59.4	54.5	50
99.0	99.4	99.4	99.4	99.3	99.3	99.2	99.1	99.0	98.8	98.7	98.4	98.3	97.9	97.6	97.2	96.7	96.0	95.1	94.0	92.0	88.3	79.4	71.4	64.7	59.2	54.4	50
99.0	99.0	99.0	98.9	98.7	98.6	98.4	98.2	97.9	97.7	97.4	97.1	96.5	96.1	95.3	94.6	93.6	92.4	90.9	88.9	86.1	81.9	75.9	69.6	63.9	58.9	54.3	50
99.0	98.8	98.6	98.4	98.2	97.9	97.7	97.3	97.1	96.6	96.3	95.7	95.2	94.3	93.5	92.3	91.9	89.5	87.5	85.2	82.0	78.1	73.2	68.1	63.1	58.4	54.1	50
99.0	98.6	98.3	98.0	97.6	97.3	97.0	96.7	96.2	95.8	95.2	94.7	93.8	93.0	91.8	90.6	89.0	87.3	85.2	82.6	79.6	75.8	72.5	67.0	62.4	58.1	54.0	50
99.0	98.2	97.6	97.1	96.7	96.2	95.8	95.4	94.9	94.2	93.6	92.7	91.8	90.5	89.3	87.7	86.0	84.1	81.8	79.3	76.4	73.0	69.4	65.4	61.5	57.6	53.8	50
99.0	97.7	96.8	96.3	95.7	95.3	94.7	94.2	93.6	92.9	92.0	91.0	89.8	88.6	87.0	85.4	83.6	81.5	79.3	76.8	74.1	71.1	67.8	64.4	60.8	57.2	53.6	50
97.0	96.5	95.9	95.3	94.8	94.3	93.7	93.1	92.5	91.6	90.7	89.5	88.3	86.8	85.2	83.5	81.6	79.6	77.4	75.0	72.5	69.7	66.8	63.6	60.3	56.9	53.5	50
96.0	95.6	95.0	94.5	93.9	93.4	92.8	92.2	91.5	90.6	89.6	88.4	86.9	85.4	83.8	82.0	80.2	78.2	76.1	73.9	71.4	68.9	66.1	63.1	60.0	56.7	53.4	50
95.0	94.7	94.2	93.7	93.1	92.6	92.0	91.4	90.7	89.7	88.7	87.4	85.9	84.4	82.7	81.0	79.1	77.2	75.2	73.0	70.8	68.3	65.7	62.8	59.8	56.6	53.3	50
95.0	94.3	93.7	93.1	92.5	92.0	91.4	90.8	90.0	89.1	88.0	86.7	85.2	83.6	82.0	80.2	78.5	76.6	74.6	72.5	70.3	68.0	65.4	62.7	59.7	56.6	53.3	50
94.0	93.8	93.3	92.7	92.1	91.6	91.0	90.4	89.6	88.7	87.6	86.3	84.8	83.2	81.5	79.9	78.1	76.2	74.3	72.2	70.1	67.8	65.3	62.6	59.7	56.6	53.3	50
94.0	93.6	93.1	92.6	92.0	91.5	90.9	90.3	89.5	88.6	87.5	86.1	84.7	83.0	81.4	79.7	78.0	76.1	74.2	72.2	70.0	67.8	65.3	62.6	59.7	56.6	53.3	50

k_1

k_2

(c)

Fig. 2.43 *(Continued)*

Since $k_1 = 5 \times 10^{-2}$ mm/s and $k_2 = 2.5 \times 10^{-2}$ mm/s,

$$\frac{2k_2}{k_1 + k_2} = \frac{2(5) \times 10^{-2}}{(5 + 2.5) \times 10^{-2}} = 1.33$$

$$\frac{2k_2}{k_1 + k_2} = \frac{2(2.5) \times 10^{-2}}{(5 + 2.5) \times 10^{-2}} = 0.667$$

Using the above values, Eq. (2.167) can be rewritten as

$$h_{(i,j+1)} + 1.333\, h_{(i-1,j)} + h_{(i,j-1)} + 0.667\, h_{(i+1,j)} - 4h_{(i,j)} = 0$$

As in step 4, calculate the residual R by using the heads in Fig. 2.43*a*. The corrected values of the heads are given by $h_{(i,j)} + R/4$. These are shown in Fig. 2.43*b*. Note that, due to symmetry, the head at the grid point $i = 5$ and $j = 28$ is 50, as assumed initially.

6. Consider the rows $i = 6, 7, \ldots, 12$. $h_{(i,j)}$ for $i = 6, 7, \ldots, 12$ and $j = 2, 3, \ldots, 27$ can be found by using Eq. (2.147). Find the corrected head in a similar manner as done in step 4. The heads at $j = 28$ are all 50, as assumed. These values are shown in Fig. 2.43*b*.
7. Consider row $i = 13$. $h_{(i,j)}$ for $i = 13$ and $j = 2, 3, \ldots, 27$ can be found from Eq. (2.157), $(h_1 + 2h_2 + h_3) - 4h_0 = 0$, or

   $$h_{(i,j+1)} + 2h_{(i-1,j)} + h_{(i,j-1)} - 4h_{(i,j)} = 0$$

   With proper values of the head given in Fig. 2.43*a*, find the residual and the corrected heads as in step 3. Note that $h_{(13,28)} = 50$ due to symmetry. These values are given in Fig. 2.43*b*.
8. With the new heads, repeat steps 3 through 7. This iteration must be carried out several times until the residuals are negligible.

   Figure 2.43*c* shows the corrected hydraulic heads after ten iterations. With these values of h, the equipotential lines can now easily be drawn.

2.2.9 Seepage Force per Unit Volume of Soil Mass

Flow of water through a soil mass results in some force being exerted on the soil itself. To evaluate the *seepage force* per unit volume of soil, consider a soil mass bounded by two flow lines *ab* and *cd* and two equipotential lines *ef* and *gh*, as shown in Fig. 2.44. The soil mass has unit thickness at right angles to the section shown. The self-weight of the soil mass is (length) (width) (thickness) $(\gamma_{sat}) = (L)(L)(1)(\gamma_{sat}) = L^2\gamma_{sat}$. The hydrostatic force on the side *ef* of the soil mass is (pressure head) $(L)(1) = h_1\gamma_w L$. The hydrostatic force on the side *gh* of the soil mass is $h_2 L\gamma_w$. For equilibrium,

$$\Delta F = h_1\gamma_w L + L^2\gamma_{sat}\sin\alpha - h_2\gamma_w L \tag{2.168}$$

But $h_1 + L\sin\alpha = h_2 + \Delta h$, so

$$h_2 = h_1 + L\sin\alpha - \Delta h \tag{2.169}$$

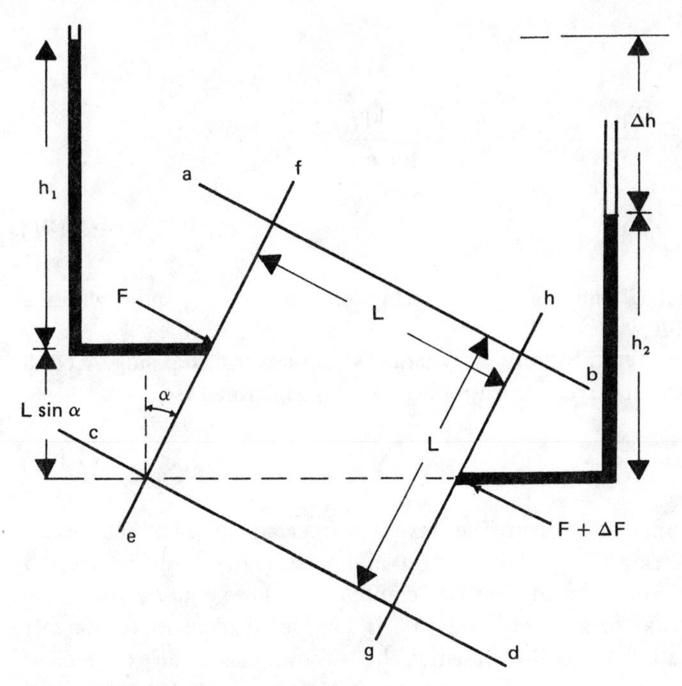

Fig. 2.44 Seepage force determination.

Combining Eqs. (2.168) and (2.169),

$$\Delta F = h_1 \gamma_w L + L^2 \gamma_{sat} \sin \alpha - (h_1 + L \sin \alpha - \Delta h)\gamma_w L$$

or

$$\Delta F = L^2 (\gamma_{sat} - \gamma_w) \sin \alpha + \Delta h \gamma_w L = \underbrace{L^2 \gamma' \sin \alpha}_{\text{submerged unit weight of soil}} + \underbrace{\Delta h \gamma_w L}_{\text{seepage force}} \tag{2.170}$$

where $\gamma' = \gamma_{sat} - \gamma_w$. From Eq. (2.170) we can see that the seepage force on the soil mass considered is equal to $\Delta h \gamma_w L$. Therefore,

$$\text{Seepage force per unit volume of soil mass} = \frac{\Delta h \gamma_w L}{L^2}$$

$$= \gamma_w \frac{\Delta h}{L} = \gamma_w i \tag{2.171}$$

where i is the hydraulic gradient.

2.2.10 Safety of Hydraulic Structures against Piping

We saw in Sec. 1.9.2 that when upward seepage occurs and the hydraulic gradient is equal to i_{cr}, *piping* or *heaving* originates in the soil mass:

$$i_{cr} = \frac{\gamma'}{\gamma_w}$$

$$\gamma' = \gamma_{sat} - \gamma_w = \frac{G_s\gamma_w + e\gamma_w}{1+e} - \gamma_w = \frac{(G_s - 1)\gamma_w}{1+e}$$

So,
$$i_{cr} = \frac{\gamma'}{\gamma_w} = \frac{G_s - 1}{1+e} \tag{2.171}$$

For the combinations of G_s and e generally encountered in soils, i_{cr} varies within a range of about 0.85 to 1.1.

Harza (1935) investigated the safety of hydraulic structures against piping. According to his work, the factor of safety against piping, F_S, can be defined as

$$F_S = \frac{i_{cr}}{i_{exit}} \tag{2.172}$$

where i_{exit} is the maximum exit gradient. The maximum exit gradient can be determined from the flow net. Referring to Fig. 2.30, the maximum exit gradient can be given by $\Delta h/l$ (Δh is the head lost between the last two equipotential lines, and l is the length of the flow element). A factor of safety of 3 to 4 is considered adequate for the safe performance of the structure. Harza also presented charts for the maximum exit gradient of dams constructed over deep homogeneous deposits (see Fig. 2.45). Using the notations shown in Fig. 2.45, the maximum exit gradient can be given by

$$i_{exit} = C\frac{h}{B} \tag{2.173}$$

A theoretical solution for the determination of the maximum exit gradient for

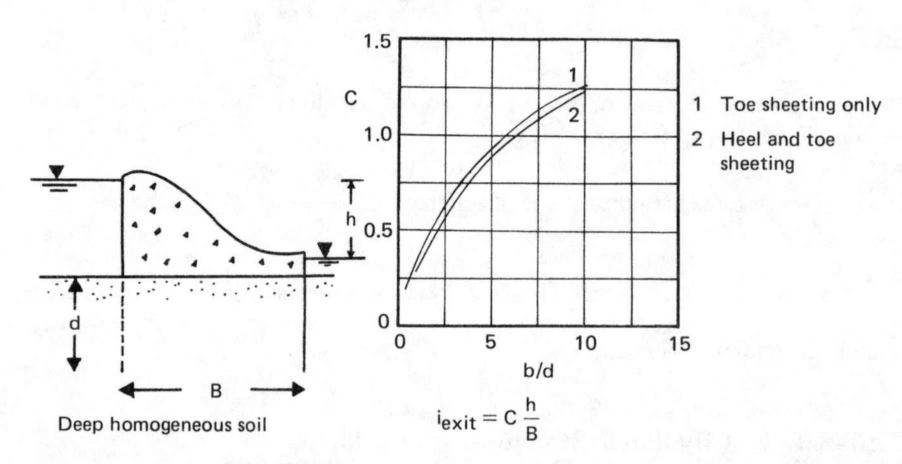

Fig. 2.45 Critical exit gradient. *(After L. F. Harza, Uplift and Seepage under Dam in Sand.,* Trans. ASCE, *vol. 100, 1935.)*

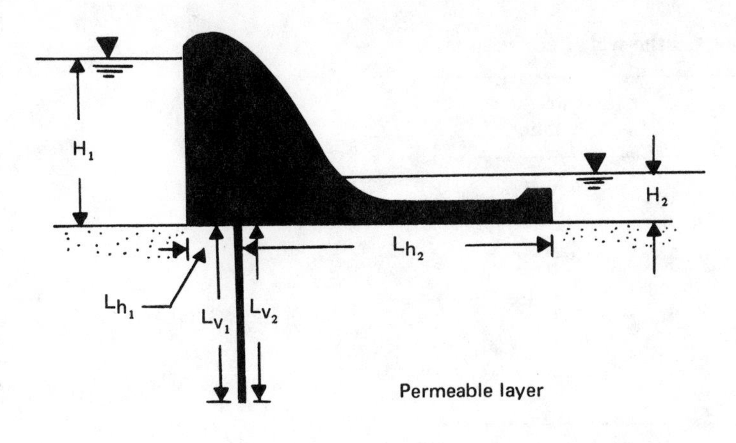

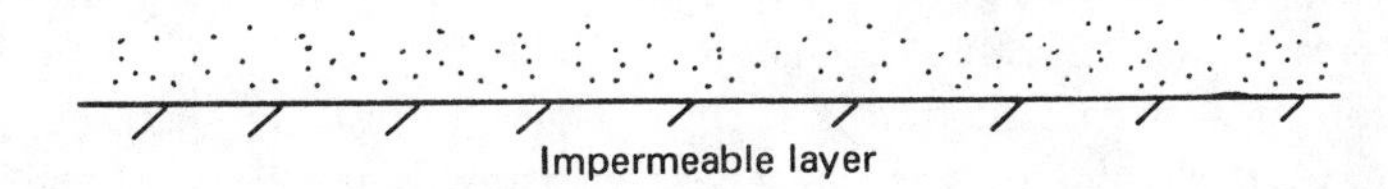

Fig. 2.46 Calculation of weighted creep distance.

a single row of sheet pile structures as shown in Fig. 2.29 is available (see Harr, 1962, p. 111) and is of the form

$$i_{\text{exit}} = \frac{1}{\pi} \frac{\text{maximum hydraulic head}}{\text{depth of penetration of sheet pile}} \tag{2.174}$$

Lane (1935) also investigated the safety of dams against piping and suggested an empirical approach to the problem. He introduced a term called *weighted creep distance* which is determined from the shortest flow path:

$$L_w = \frac{\Sigma L_h}{3} + \sum L_v \tag{2.175}$$

where L_w = weighted creep distance

$\Sigma L_h = L_{h_1} + L_{h_2} + \cdots$ = sum of horizontal distance along shortest flow path (see Fig. 2.46)

$\Sigma L_v = L_{v_1} + L_{v_2} + \cdots$ = sum of vertical distances along shortest flow path (see Fig. 2.46)

Once the weighted creep length has been calculated, the weighted creep ratio can be determined as (Fig. 2.46)

$$\text{Weighted creep ratio} = \frac{L_w}{H_1 - H_2} \tag{2.176}$$

Table 2.9 Safe values for the weighted creep ratio

Material	Safe weighted creep ratio
Very fine sand or silt	8.5
Fine sand	7.0
Medium sand	6.0
Coarse sand	5.0
Fine gravel	4.0
Coarse gravel	3.0
Soft to medium clay	2.0–3.0
Hard clay	1.8
Hard pan	1.6

For a structure to be safe against piping, Lane suggested that the weighted creep ratio should be equal to or greater than the safe values shown in Table 2.9.

If the cross section of a given structure is such that the shortest flow path has a slope steeper than 45°, it should be taken as a vertical path. If the slope of the shortest flow path is less than 45°, it should be considered as a horizontal path.

Terzaghi (1922) conducted some model tests with a single row of sheet piles as shown in Fig. 2.47 and found that the failure due to piping takes place within a distance of $D/2$ from the sheet piles (D is the depth of penetration of the sheet pile).

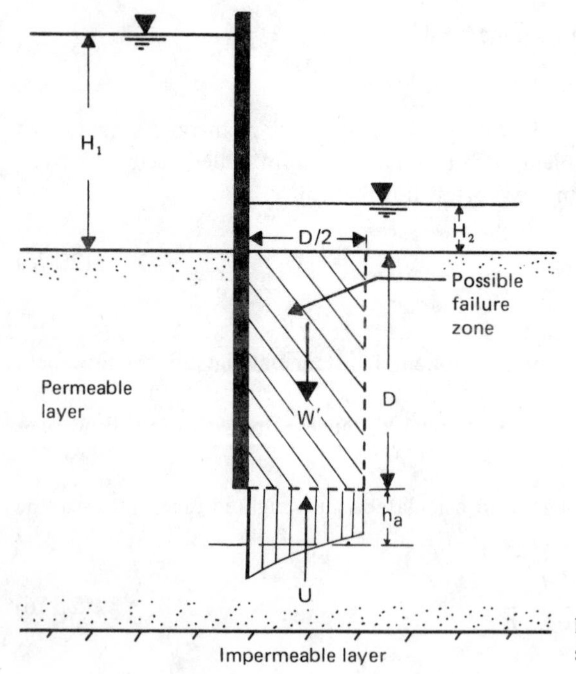

Fig. 2.47 Failure due to piping for a single-row sheet pile structure.

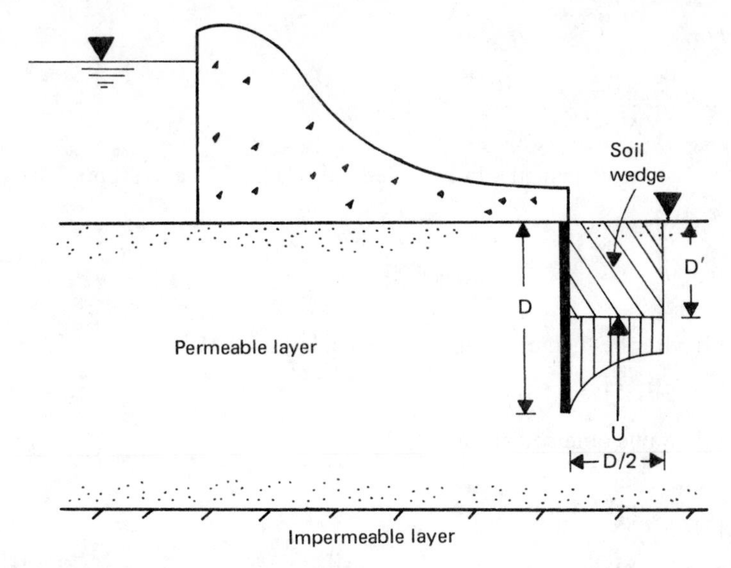

Fig. 2.48 Safety against piping under a dam.

Therefore, the stability of this type of structure can be determined by considering a soil prism on the downstream side of unit thickness and of section $D \times D/2$. Using the flow net, the hydraulic uplifting pressure can be determined as

$$U = \tfrac{1}{2}\gamma_w D h_a \tag{2.177}$$

where h_a is the average hydraulic head at the base of the soil prism. The submerged weight of the soil prism acting vertically downwards can be given by

$$W' = \tfrac{1}{2}\gamma' D^2 \tag{2.178}$$

Hence, the factor of safety against heave is

$$F_S = \frac{W'}{U} = \frac{\frac{1}{2}\gamma' D^2}{\frac{1}{2}\gamma_w D h_a} = \frac{D\gamma'}{h_a \gamma_w} \tag{2.179}$$

A factor of safety of about 4 is generally considered adequate.

For structures other than a single row of sheet piles, such as that shown in Fig. 2.48, Terzaghi (1943) recommended that the stability of several soil prisms of size $D/2 \times D' \times 1$ be investigated to find the minimum factor of safety. Note that $0 < D' \leqslant D$. However, Harr (1962, p. 125) suggested that a factor of safety of 4 to 5 with $D' = D$ should be sufficient for safe performance of the structure.

Example 2.7 A flow net for a single row of sheet piles is given in Fig. 2.29.

(*a*) Determine the factor of safety against piping by Harza's method.

(*b*) Determine the factor of safety against piping by Terzaghi's method [Eq. (2.179)]. Assume $\gamma' = 10.2\ \text{kN/m}^3$.

SOLUTION *Part (a):*

$$i_{exit} = \frac{\Delta h}{L} \qquad \Delta h = \frac{3-0.5}{N_d} = \frac{3-0.5}{6} = 0.417\,\text{m}$$

The length of the last flow element can be scaled out of Fig. 2.29 and is approximately 0.82 m. So

$$i_{exit} = \frac{0.417}{0.82} = 0.509$$

(We can check this with the theoretical equation given in Eq. (2.174):

$$i_{exit} = (1/\pi)\,[(3-0.5)/1.5] = 0.53$$

which is close to the value obtained above.)

$$i_{cr} = \frac{\gamma'}{\gamma_w} = \frac{10.2\,\text{kN/m}^3}{9.81\,\text{kN/m}^3} = 1.04$$

So, the factor of safety against piping is

$$\frac{i_{cr}}{i_{exit}} = \frac{1.04}{0.509} = 2.04$$

Part (b): A soil prism of cross section $D \times D/2$, where $D = 1.5$ m, on the downstream side adjacent to the sheet pile is plotted in Fig. 2.49*a*. The approximate hydraulic heads at the bottom of the prism can be evaluated by using the flow net. Referring to Fig. 2.29 (note that $N_d = 6$),

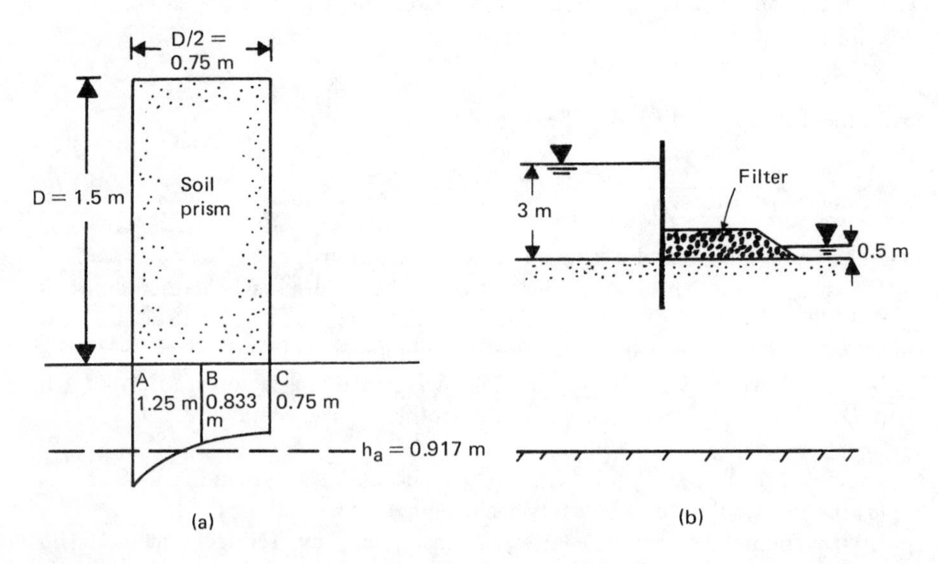

Fig. 2.49

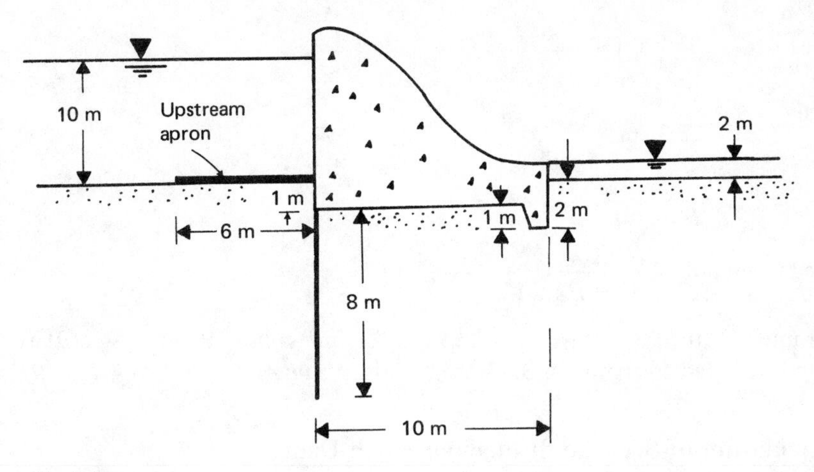

Fig. 2.50

$$h_A = \frac{3}{6}(3 - 0.5) = 1.25\text{ m}$$

$$h_B = \frac{2}{6}(3 - 0.5) = 0.833\text{ m}$$

$$h_C = \frac{1.8}{6}(3 - 0.5) = 0.75\text{ m}$$

$$h_a = \frac{0.375}{0.75}\left(\frac{1.25 + 0.75}{2} + 0.833\right) = 0.917\text{ m}$$

$$F_S = \frac{D\gamma'}{h_a \gamma_w} = \frac{1.5 \times 10.2}{0.917 \times 9.81} = 1.7$$

The factor of safety calculated here is rather low. However, it can be increased by placing some filter material (Sec. 2.2.15) on the downstream side above the ground surface as shown in Fig. 2.49*b*. This will increase the weight of the soil prism [W', see Eq. (2.178)].

Example 2.8 A dam section is shown in Fig. 2.50. The subsoil is fine sand. Using Lane's method, determine whether the structure is safe against piping.

SOLUTION From Eq. (2.175),

$$L_w = \frac{\Sigma L_h}{3} + \sum L_v$$

$$\sum L_h = 6 + 10 = 16\text{ m}$$

$$\sum L_v = 1 + (8 + 8) + 1 + 2 = 20\,\text{m}$$

$$L_w = \frac{16}{3} + 20 = 25.33\,\text{m}$$

From Eq. (2.176),

$$\text{Weighted creep ratio} = \frac{L_w}{H_1 - H_2} = \frac{25.33}{10 - 2} = 3.17$$

From Table 2.9, the safe weighted creep ratio for fine sand is about 7. Since the calculated weighted creep ratio is 3.17, the structure is *unsafe.*

2.2.11 Calculation of Seepage through an Earth Dam Resting on an Impervious Base

Several solutions have been proposed for determination of the quantity of seepage through a homogeneous earth dam. In this section, some of these solutions will be considered.

Dupuit's solution. Figure 2.51 shows the section of an earth dam in which *ab* is the *phreatic surface,* i.e., the uppermost line of seepage. The quantity of seepage through a unit length at right angles to the cross section can be given by Darcy's law as $q = kiA$.

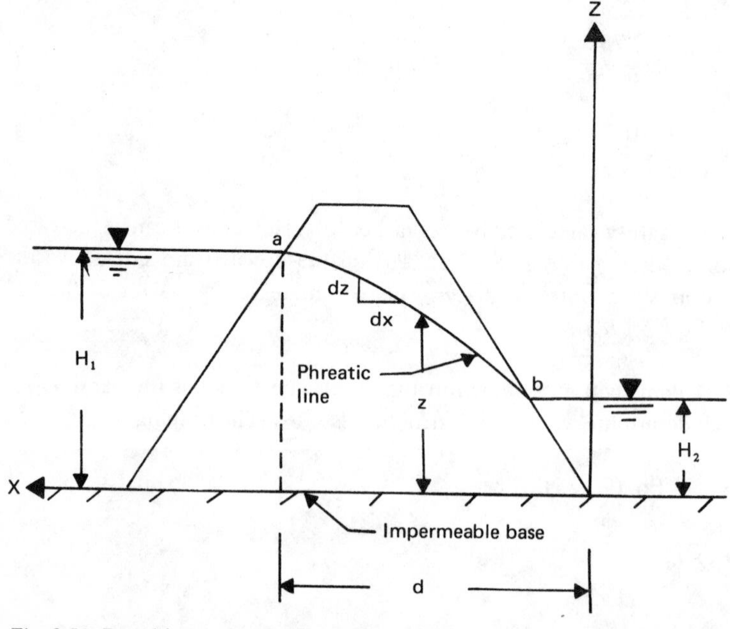

Fig. 2.51 Dupuit's solution for flow through an earth dam.

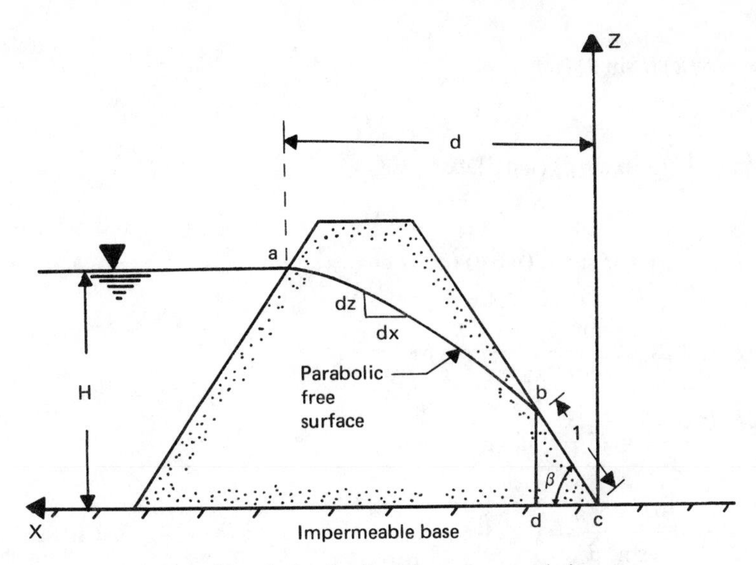

Fig. 2.52 Schaffernak's solution for flow through an earth dam.

Dupuit (1863) assumed that the hydraulic gradient i is equal to the slope of the free surface and is constant with depth, i.e., $i = dz/dx$. So,

$$q = k\frac{dz}{dx}[(z)(1)] = k\frac{dz}{dx}z$$

$$\int_0^d q\,dx = \int_{H_2}^{H_1} kz\,dz$$

$$qd = \frac{k}{2}(H_1^2 - H_2^2)$$

or
$$q = \frac{k}{2d}(H_1^2 - H_2^2) \tag{2.180}$$

Equation (2.180) represents a parabolic free surface. However, in the derivation of the equation, no attention has been paid to the entrance or exit conditions. Also note that if $H_2 = 0$ the phreatic line would intersect the impervious surface.

Schaffernak's solution. For calculation of seepage through a homogeneous earth dam, Schaffernak (1917) proposed that the phreatic surface will be like line *ab* in Fig. 2.52, i.e., it will intersect the downstream slope at a distance l from the impervious base. The seepage per unit length of the dam can now be determined by considering the triangle *bcd* in Fig. 2.52:

$$q = kiA \qquad A = (\overline{bd})(l) = l\sin\beta$$

From Dupuit's assumption, the hydraulic gradient is given by $i = dz/dx = \tan\beta$. So,

$$q = kz\frac{dz}{dx} = (k)(l \sin\beta)(\tan\beta) \tag{2.181}$$

or $$\int_{l \sin\beta}^{H} z\,dz = \int_{l \cos\beta}^{d} (l \sin\beta)(\tan\beta)\,dx$$

$$\frac{1}{2}(H^2 - l^2 \sin^2\beta) = (l \sin\beta)(\tan\beta)(d - l\cos\beta)$$

$$\frac{1}{2}(H^2 - l^2 \sin^2\beta) = l\frac{\sin^2\beta}{\cos\beta}(d - l\cos\beta)$$

$$\frac{H^2\cos\beta}{2\sin^2\beta} - \frac{l^2\cos\beta}{2} = ld - l^2\cos\beta$$

$$l^2\cos\beta - 2ld + \frac{H^2\cos\beta}{\sin^2\beta} = 0 \tag{2.182}$$

$$l = \frac{2d \pm \sqrt{4d^2 - 4[(H^2\cos^2\beta)/\sin^2\beta]}}{2\cos\beta}$$

so $$l = \frac{d}{\cos\beta} - \sqrt{\frac{d^2}{\cos^2\beta} - \frac{H^2}{\sin^2\beta}} \tag{2.183}$$

Once the value of l is known, the rate of seepage can be calculated from the equation $q = kl \sin\beta \tan\beta$.

Schaffernak suggested a graphical procedure to determine the value of l. This procedure can be explained with the aid of Fig. 2.53.

1. Extend the downstream slope line *bc* upwards.
2. Draw a vertical line *ae* through the point *a*. This will intersect the projection of line *bc* (step 1) at point *f*.
3. With *fc* as diameter, draw a semicircle *fhc*.

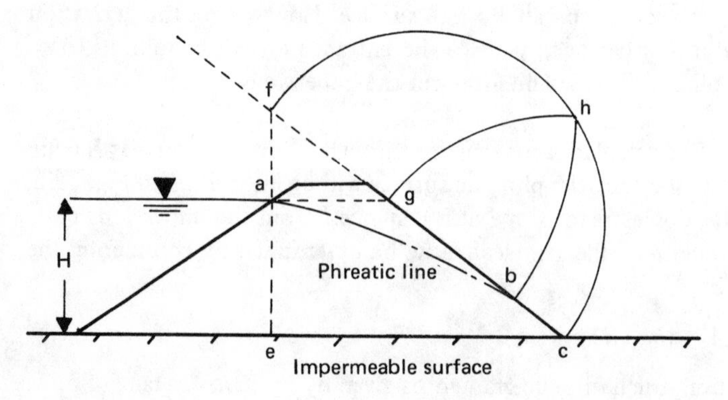

Fig. 2.53 Graphical construction for Schaffernak's solution.

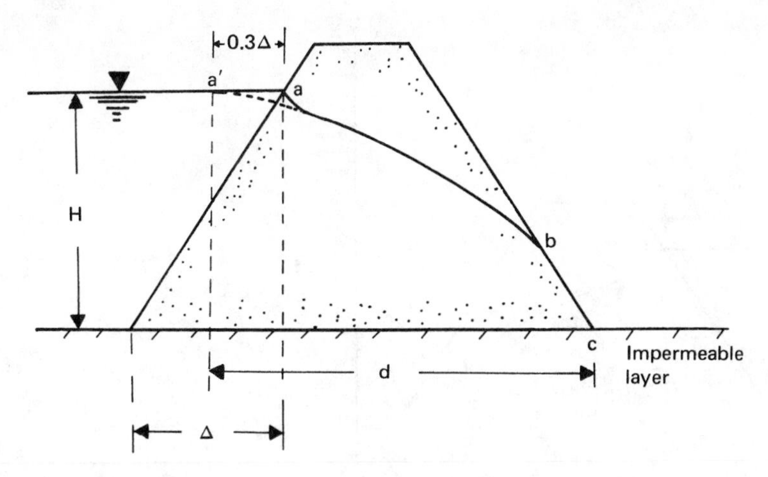

Fig. 2.54 Modified distance d for use in Eq. (2.183).

4. Draw a horizontal line ag.
5. With c as the center and cg as the radius, draw an arc of a circle, gh.
6. With f at the center and fh as the radius, draw an arc of a circle, hb.
7. Measure $bc = l$.

A Casagrande (1937) showed experimentally that the parabola ab shown in Fig. 2.52 should actually start from the point a' as shown in Fig. 2.54. Note that $aa' = 0.3\Delta$. So, with this modification, the value of d for use in Eq. (2.183) will be the horizontal distance between points a' and c.

L. Casagrande's solution. Equation (2.183) was obtained on the basis of Dupuit's assumption that the hydraulic gradient i is equal to dz/dx. L. Casagrande (1932) suggested that this relation is an approximation to the actual condition. In reality (see Fig. 2.55),

$$i = \frac{dz}{ds} \tag{2.184}$$

For a downstream slope of β greater than 30°, the deviations from Dupuit's assumption become more noticeable. Based on this assumption [Eq. (2.184)], the rate of seepage is $q = kiA$. Considering the triangle bcd in Fig. 2.55,

$$i = \frac{dz}{ds} = \sin\beta \qquad A = (bd)(l) = l \sin\beta$$

So
$$q = k\frac{dz}{dx}z = kl\sin^2\beta \tag{2.185}$$

or
$$\int_{l\sin\beta}^{H} z\,dz = \int_{l}^{s} (l\sin^2\beta)\,ds$$

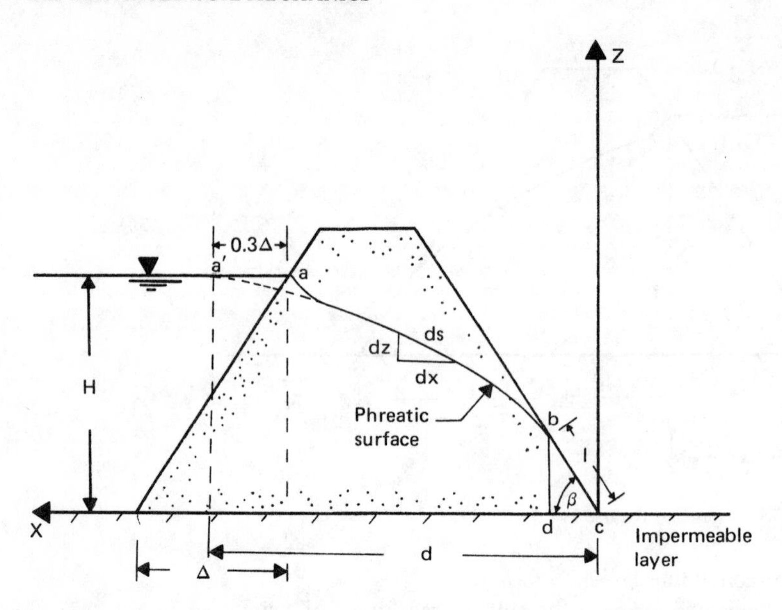

Fig. 2.55 L. Casagrande's solution for flow through an earth dam. (Note: length of the curve $a'bc = S$.)

where s is the length of the curve $a'bc$. Hence,

$$\frac{1}{2}(H^2 - l^2 \sin^2 \beta) = l \sin^2 \beta (s - l)$$

$$H^2 - l^2 \sin^2 \beta = 2ls \sin^2 \beta - 2l^2 \sin^2 \beta$$

$$l^2 - 2ls + \frac{H^2}{\sin^2 \beta} = 0 \tag{2.186}$$

The solution to Eq. (2.186) is

$$l = s - \sqrt{s^2 - \frac{H^2}{\sin^2 \beta}} \tag{2.187}$$

With about a 4 to 5% error, we can approximate s as the length of the straight line $a'c$. So,

$$s = \sqrt{d^2 + H^2} \tag{2.188}$$

Combining Eqs. (2.187) and (2.188),

$$l = \sqrt{d^2 + H^2} - \sqrt{d^2 - H^2 \cot^2 \beta} \tag{2.189}$$

Once l is known, the rate of seepage can be calculated from the equation

$$q = kl \sin^2 \beta.$$

A solution that avoids the approximation introduced in Eq. (2.189) was given by Gilboy (1934) and put into graphical form by Taylor (1948), as shown in Fig. 2.56. To use the graph,

1. Determine d/H.
2. For given values of d/H and β, determine m.
3. Calculate $l = mH/\sin\beta$.
4. Calculate $q = kl\sin^2\beta$.

Pavlovsky's solution. Pavlovsky (1931; also see Harr, 1962) also gave a solution for calculation of seepage through an earth dam. This can be explained with reference to Fig. 2.57. The dam section can be divided into three zones, and the rate of seepage through each zone can be calculated as follows.

Zone I (area agof) In this zone, the seepage lines are actually curved, but Pavlovsky assumed that they can be replaced by horizontal lines. The rate of seepage through an elementary strip dz can then be given by

$$dq = ki\,dA \qquad dA = (dz)(1) = dz$$

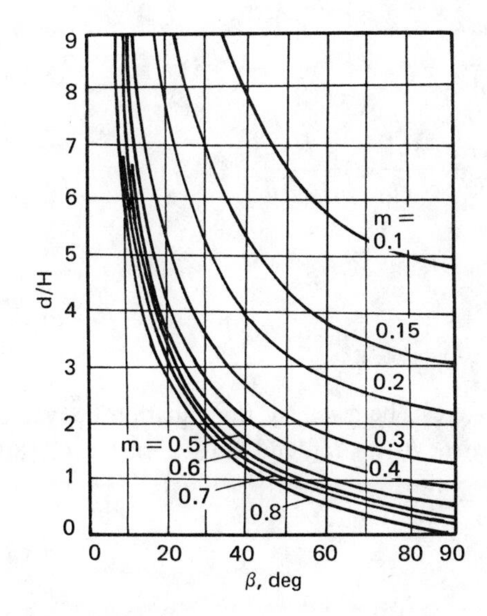

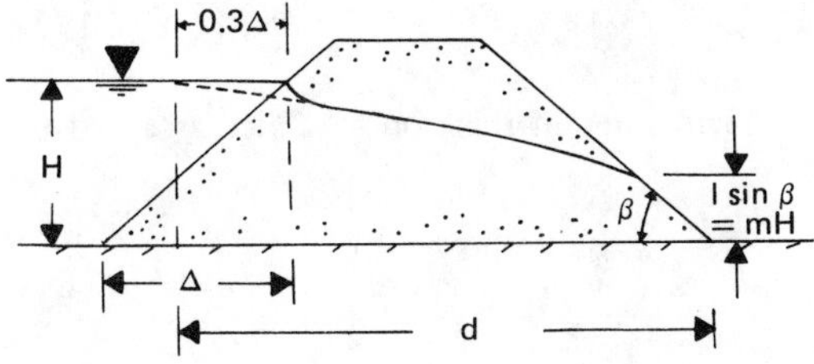

Fig. 2.56 Chart for solution by L. Casagrande's method based on Gilboy's solution. *(After D. W. Taylor, "Fundamentals of Soil Mechanics," Wiley, New York, 1948.)*

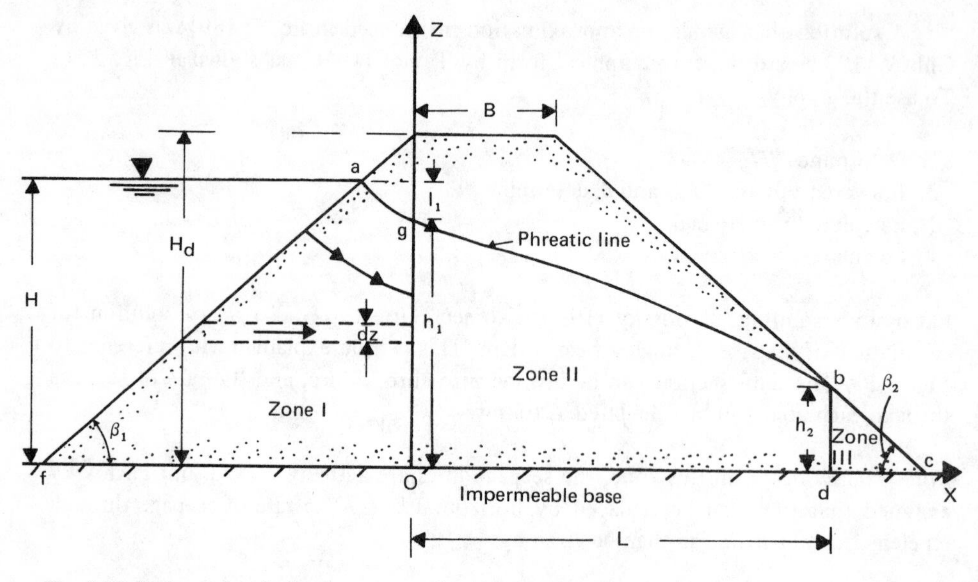

Fig. 2.57 Pavlovsky's solution for seepage through an earth dam.

$$i = \frac{\text{loss of head, } l_1}{\text{length of flow}} = \frac{l_1}{(H_d - z)\cot\beta_1}$$

So, $$q = \int dq = \int_0^{h_1} \frac{kl_1}{(H_d - z)\cot\beta_1} dz = \frac{kl_1}{\cot\beta_1} \ln \frac{H_d}{H_d - h_1}$$

But $l_1 = H - h_1$. So

$$q = \frac{k(H - h_1)}{\cot\beta_1} \ln \frac{H_d}{H_d - h_1} \tag{2.190}$$

Zone II (area ogbd) The flow in this zone can be given by the equation derived by Dupuit [Eq. (2.180)]. Substituting h_1 for H_1, h_2 for H_2, and L for d in Eq. (2.180), we get

$$q = \frac{k}{2L}(h_1^2 - h_2^2) \tag{2.191}$$

where $$L = B + (H_d - h_2)\cot\beta_2 \tag{2.192}$$

Zone III (area bcd) As in zone I, the stream lines in this zone are also assumed to be horizontal:

$$q = k\int_0^{h_2} \frac{dz}{\cot\beta_2} = \frac{kh_2}{\cot\beta_2} \tag{2.193}$$

Combining Eqs. (2.191), (2.192), and (2.193),

$$h_2 = \frac{B}{\cot \beta_2} + H_d - \sqrt{\left(\frac{B}{\cot \beta_2} + H_d\right)^2 - h_1^2} \tag{2.194}$$

From Eqs. (2.190) and (2.193),

$$\frac{H - h_1}{\cot \beta_1} \ln \frac{H_d}{H_d - h_1} = \frac{h_2}{\cot \beta_2} \tag{2.195}$$

Equations (2.194) and (2.195) contain two unknowns, h_1 and h_2, which can be solved graphically (see Ex. 2.9). Once these are known, the rate of seepage per unit length of the dam can be obtained from any one of the equations (2.190), (2.191), and (2.193).

Seepage through earth dams with $k_x \neq k_z$. If the soil in a dam section shows anisotropic behavior with respect to permeability, the dam section should first be plotted according to the transformed scale (as explained in Sec. 2.2.5):

$$x' = \sqrt{\frac{k_z}{k_x}}\, x$$

All calculations should be based on this transformed section. Also, for calculating the rate of seepage, the term k in the corresponding equations should be equal to $\sqrt{k_x k_z}$.

Example 2.9 The cross section of an earth dam is shown in Fig. 2.58. Calculate the rate of seepage through the dam [q in $m^3/(min \cdot m)$] by (*a*) Dupuit's method; (*b*) Schaffernak's method; (*c*) L. Casagrande's method; and (*d*) Pavlovsky's method.

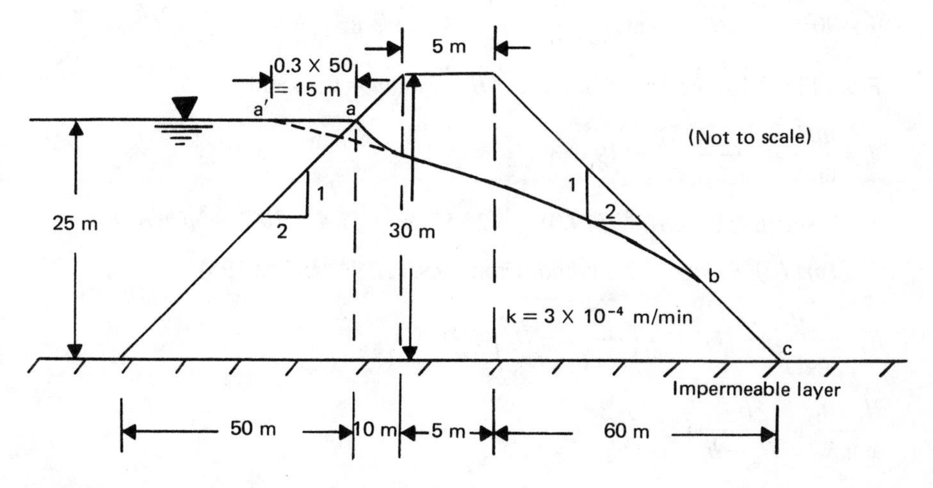

Fig. 2.58

SOLUTION *Part (a), Dupuit's method:* From Eq. (2.180),

$$q = \frac{k}{2d}(H_1^2 - H_2^2)$$

From Fig. 2.58, $H_1 = 25$ m and $H_2 = 0$; also, d (the horizontal distance between points a and c) is equal to 60 + 5 + 10 = 75 m. Hence

$$q = \frac{3 \times 10^{-4}}{2 \times 75}(25)^2 = 12.5 \times 10^{-4}\ m^3/(min \cdot m)$$

Part (b), Schaffernak's method: From Eqs. (2.181) and (2.183),

$$q = (k)(l \sin\beta)(\tan\beta) \qquad l = \frac{d}{\cos\beta} - \sqrt{\frac{d^2}{\cos^2\beta} - \frac{H^2}{\sin^2\beta}}$$

Using Casagrande's correction (Fig. 2.54), d (the horizontal distance between a' and c) is equal to 60 + 5 + 10 + 15 = 90 m. Also,

$$\beta = \tan^{-1}\frac{1}{2} = 26.57^\circ \qquad H = 25\ \text{m}$$

So,

$$l = \frac{90}{\cos 26.57^\circ} - \sqrt{\left(\frac{90}{\cos 26.57^\circ}\right)^2 - \left(\frac{25}{\sin 26.57^\circ}\right)^2}$$

$$= 100.63 - \sqrt{(100.63)^2 - (55.89)^2} = 16.95\ \text{m}$$

$$q = (3 \times 10^{-4})(16.95)(\sin 26.57^\circ)(\tan 26.57^\circ) = 11.37 \times 10^{-4}\ m^3/(min \cdot m)$$

Part (c), L. Casagrande's method: We will use the graph given in Fig. 2.56.

$$d = 90\ \text{m} \qquad H = 25\ \text{m} \qquad \frac{d}{H} = \frac{90}{25} = 3.6 \qquad \beta = 26.57^\circ$$

From Fig. 2.56, for $\beta = 26.57^\circ$ and $d/H = 3.6$, $m = 0.34$ and

$$l = \frac{mH}{\sin\beta} = \frac{0.34(25)}{\sin 26.57^\circ} = 19.0\ \text{m}$$

$$q = kl\sin^2\beta = (3 \times 10^{-4})(19.0)(\sin 26.57^\circ)^2 = 11.4 \times 10^{-4}\ m^3/(min \cdot m)$$

Part (d), Pavlovsky's method: From Eqs. (2.194) and (2.195),

$$h_2 = \frac{B}{\cot\beta_2} + H_d - \sqrt{\left(\frac{B}{\cot\beta_2} + H_d\right)^2 - h_1^2}$$

$$\frac{H - h_1}{\cot\beta_1}\ln\frac{H_d}{H_d - h_1} = \frac{h_2}{\cot\beta_2}$$

From Fig. 2.58, $B = 5$ m, $\cot\beta_2 = \cot 26.57^\circ = 2$, $H_d = 30$ m, $H = 25$ m. Substituting these values in Eq. (2.194), we get

$$h_2 = \frac{5}{2} + 30 - \sqrt{\left(\frac{5}{2} + 30\right)^2 - h_1^2}$$

or

$$h_2 = 32.5 - \sqrt{1056.25 - h_1^2} \tag{a}$$

Similarly, from Eq. (2.195),

$$\frac{25 - h_1}{2} \ln \frac{30}{30 - h_1} = \frac{h_2}{2}$$

or

$$h_2 = (25 - h_1) \ln \frac{30}{30 - h_1} \tag{b}$$

Eqs. (*a*) and (*b*) must be solved by trial and error:

h_1, m	h_2 from Eq. (*a*), m	h_2 from Eq. (*b*), m
2	0.062	1.587
4	0.247	3.005
6	0.559	4.240
8	1.0	5.273
10	1.577	6.082
12	2.297	6.641
14	3.170	6.915
16	4.211	6.859
18	5.400	6.414
20	6.882	5.493

Using the values of h_1 and h_2 calculated in the preceding table, we can plot the graph as shown in Fig. 2.59; and from that, $h_1 = 18.9$ m and $h_2 = 6.06$ m. From Eq. (2.193),

$$q = \frac{kh_2}{\cot \beta_2} = \frac{(3 \times 10^{-4})(6.06)}{2} = 9.09 \times 10^{-4} \, m^3/(min \cdot m)$$

2.2.12 Plotting of Phreatic Line for Seepage through Earth Dams

For construction of flow nets for seepage through earth dams, the phreatic line needs to be established first. This is usually done by the method proposed by Casagrande (1937) and is shown in Fig. 2.60*a*. Note that *aefb* in Fig. 2.60*a* is the actual phreatic line. The curve *a'efb'c'* is a parabola with its focus at *c*; the phreatic line coincides with this parabola, but with some deviations at the upstream and the downstream

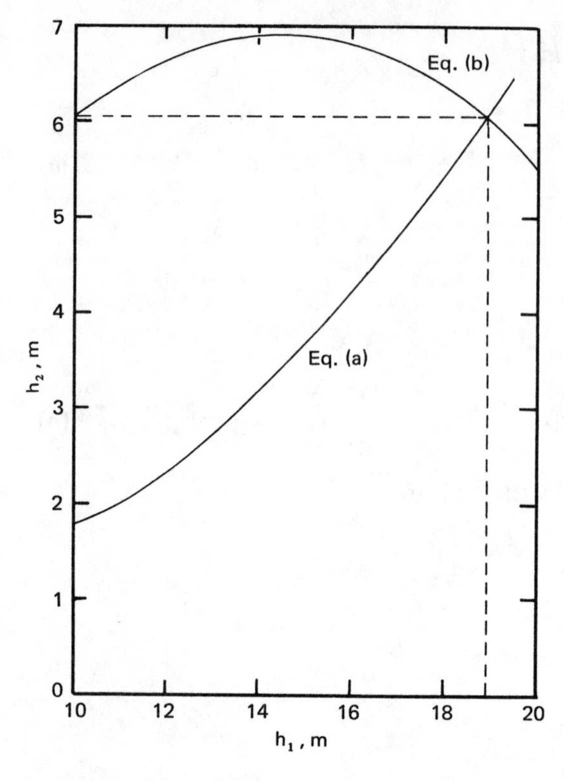

Fig. 2.59

faces. At a point a, the phreatic line starts at an angle of 90° to the upstream face of the dam and $aa' = 0.3\Delta$.

The parabola $a'efb'c'$ can be constructed as follows:

1. Let the distance cc' be equal to p. Now, referring to Fig. 2.60b, $Ac = AD$ (based on the properties of a parabola), $Ac = \sqrt{x^2 + z^2}$, and $AD = 2p + x$. Thus,

$$\sqrt{x^2 + z^2} = 2p + x \tag{2.196}$$

At $x = d$, $z = H$. Substituting these conditions into Eq. (2.196) and rearranging, we obtain

$$p = \tfrac{1}{2}\left(\sqrt{d^2 + H^2} - d\right) \tag{2.197}$$

Since d and H are known, the value of p can be calculated.

2. From Eq. (2.196),

$$x^2 + z^2 = 4p^2 + x^2 + 4px$$

$$x = \frac{z^2 - 4p^2}{4p} \tag{2.198}$$

With p known, the values of x for various values of z can be calculated from Eq. (2.198) and the parabola can be constructed.

To complete the phreatic line, the portion *ae* has to be approximated and drawn by hand. When $\beta < 30°$, the value of l can be calculated from Eq. (2.183) as

$$l = \frac{d}{\cos\beta} - \sqrt{\frac{d^2}{\cos^2\beta} - \frac{H^2}{\sin^2\beta}}$$

Note that $l = bc$ in Fig. 2.60*a*. Once point *b* has been located, the curve *fb* can be approximately drawn by hand.

If $\beta \geqslant 30°$, Casagrande proposed that the value of l can be determined by using the graph given in Fig. 2.61. In Fig. 2.60*a*, $b'b = \Delta l$, and $bc = l$. After locating the point *b* on the downstream face, the curve *fb* can be approximately drawn by hand.

Example 2.10 An earth dam section is shown in Fig. 2.62. Plot the phreatic line for seepage. For the earth dam section, $k_x = k_z$.

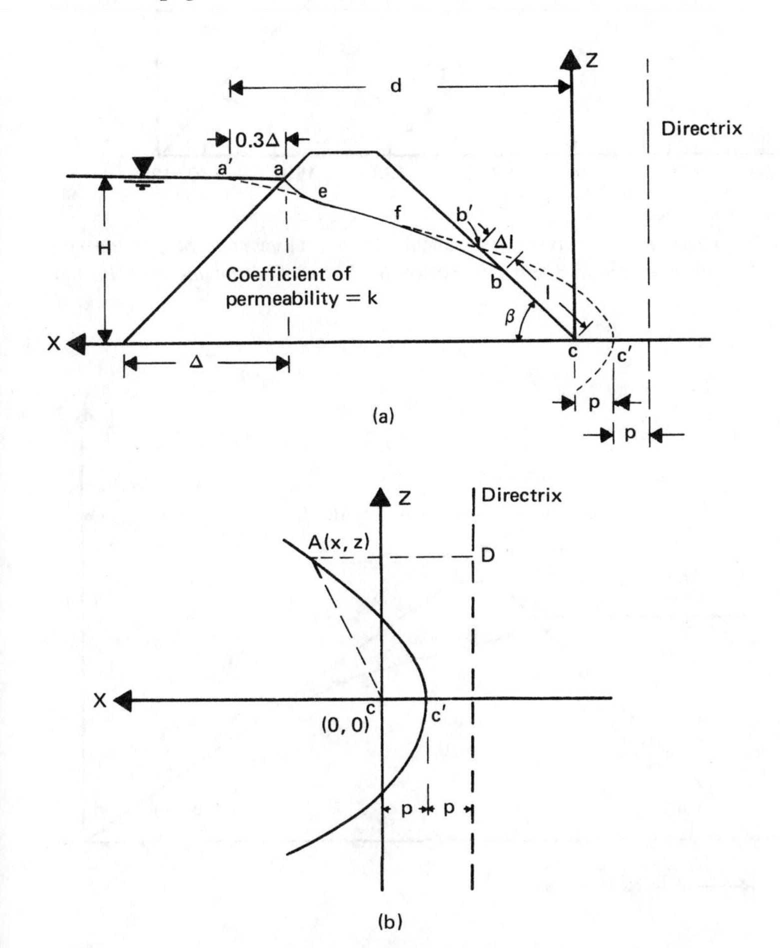

Fig. 2.60 Determination of phreatic line for seepage through an earth dam.

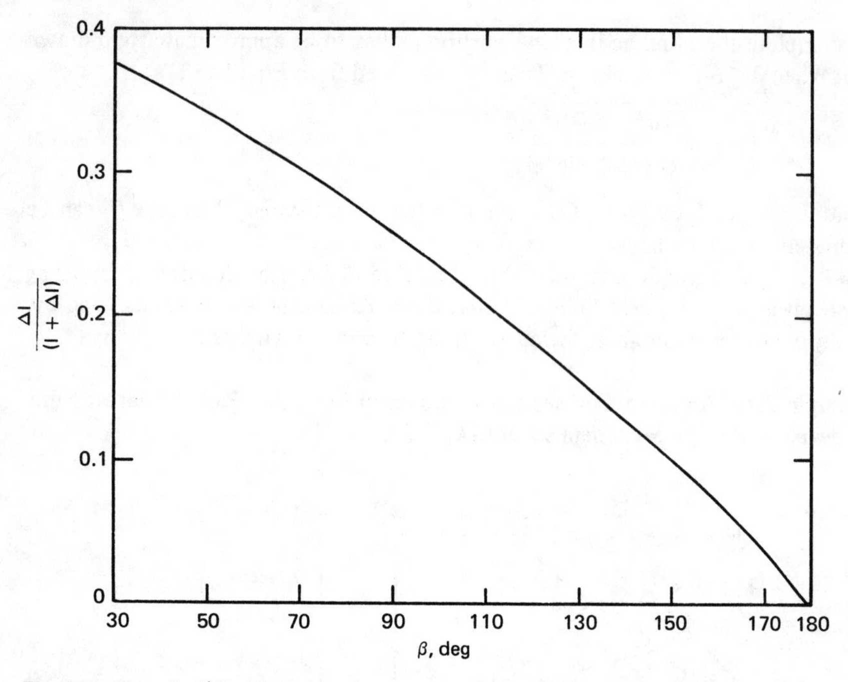

Fig. 2.61 Plot of $\Delta l/(l + \Delta l)$ against downstream slope angle. *(After A. Casagrande, Seepage through Dams,* Contribution to Soil Mechanics, 1925–1940, *Boston Society of Civil Engineering, Boston, 1937.)*

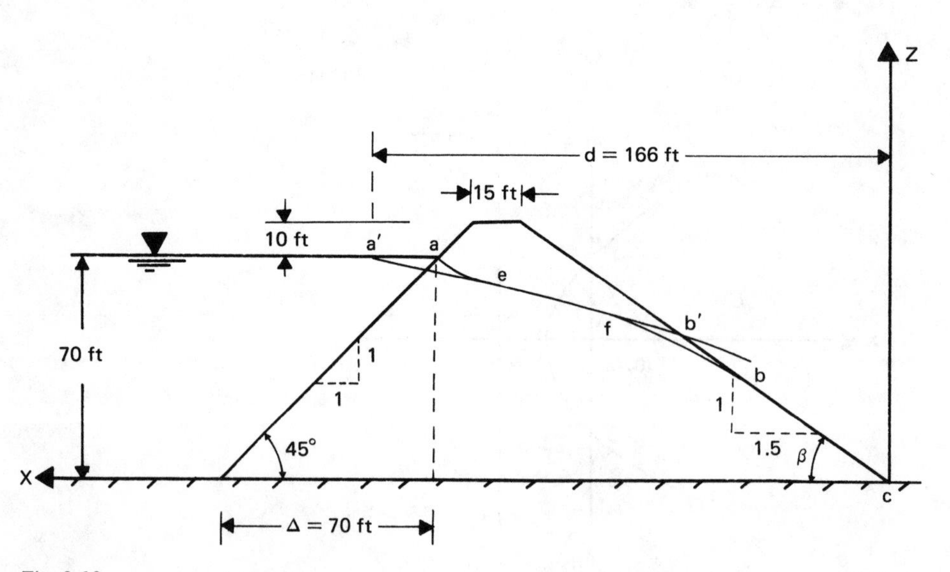

Fig. 2.62

SOLUTION

$$\beta = \tan^{-1}(1/1.5) = 33.69°$$

$$\Delta = 70 \cot 45° = 70 \text{ ft}$$

$$aa' = 0.3\Delta = 0.3(70) = 21 \text{ ft}$$

and

$$d = 80 \cot 33.69° + 15 + 10 \cot 45° + 21 = 120 + 15 + 10 + 21 = 166 \text{ ft}$$

From Eq. (2.197),

$$p = \tfrac{1}{2}(\sqrt{d^2 + H^2} - d) = \tfrac{1}{2}(\sqrt{166^2 + 70^2} - 166)$$
$$= \tfrac{1}{2}(180.16 - 166) = 7.08 \text{ ft}$$

Using Eq. (2.198), we can now determine the coordinates of several points of the parabola $a'efb'c'$:

z, ft	x from Eq. (2.198), ft
70	166
65	142.1
60	120.04
55	99.73
50	81.2
45	64.42

Using the values of x and corresponding z calculated in the above table, the basic parabola has been plotted in Fig. 2.62.

We calculate l as follows. The equation of the line cb' can be given by $z = x \tan\beta$, and the equation of the parabola [Eq. (2.198)] is $x = (z^2 - 4p^2)/4p$. The coordinates of point b' can be determined by solving the above two equations:

$$x = \frac{z^2 - 4p^2}{4p} = \frac{(x \tan \beta)^2 - 4p^2}{4p}$$

or

$$x^2 \tan^2 \beta - 4px - 4p^2 = 0$$

Hence

$$x^2 \tan^2 33.69° - 4(7.08)x - 4(7.08)^2 = 0$$

$$0.444x^2 - 28.32x - 200.5 = 0$$

The solution of the above equation gives $x = 70.22$ ft. So

$$cb' = \sqrt{70.22^2 + (70.22 \tan 33.69°)^2} = 84.39 \text{ ft} = l + \Delta l$$

From Fig. 2.61, for $\beta = 33.69°$,

$$\frac{\Delta l}{l + \Delta l} = 0.366 \qquad \Delta l = (0.366)(84.39) = 30.9 \text{ ft}$$

$$l = (l + \Delta l) - (\Delta l)$$

$$= 84.39 - 30.9 = 53.49 \text{ ft} \approx 54 \text{ ft}$$

So $l = cb = 54$ ft.

The curve portions *ae* and *fb* can now be approximately drawn by hand, which completes the phreatic line *aefb* (Fig. 2.62).

2.2.13 Entrance, Discharge, and Transfer Conditions of Line of Seepage through Earth Dams

A. Casagrande (1937) analyzed the entrance, discharge, and transfer conditions for the line of seepage through earth dams. When we consider the flow from a free-draining material (coefficient of permeability very large; $k_1 \approx \infty$) into a material of permeability k_2, it is called an *entrance*. Similarly, when the flow is from a material of permeability k_1 into a free-draining material ($k_2 \approx \infty$) it is referred to as *discharge*. Figure 2.63 shows various entrance, discharge, and transfer conditions. The transfer conditions show the nature of deflection of the line of seepage when passing from a material of permeability k_1 to a material of permeability k_2.

Using the conditions given in Fig. 2.63, we can determine the nature of the phreatic lines for various types of earth dam sections.

2.2.14 Flow-net Construction for Earth Dams

With a knowledge of the nature of the phreatic line and the entrance, discharge, and transfer conditions, we can now proceed to draw flow nets for earth dam sections. Figure 2.64 shows an earth dam section that is homogeneous with respect to permeability. To draw the flow net, the following steps must be followed:

1. Draw the phreatic line, since this is known.
2. Note that *ag* is an equipotential line and that *gc* is a flow line.
3. It is important to realize that the pressure head at any point on the phreatic line is zero; hence, the difference of total head between any two equipotential lines should be equal to the difference in elevation between the points where these equipotential lines intersect the phreatic line.

   Since loss of hydraulic head between any two consecutive equipotential lines is the same, determine the number of equipotential drops, N_d, the flow net needs to have and calculate $\Delta h = h/N_d$.
4. Draw the *head lines* for the cross section of the dam. The points of intersection

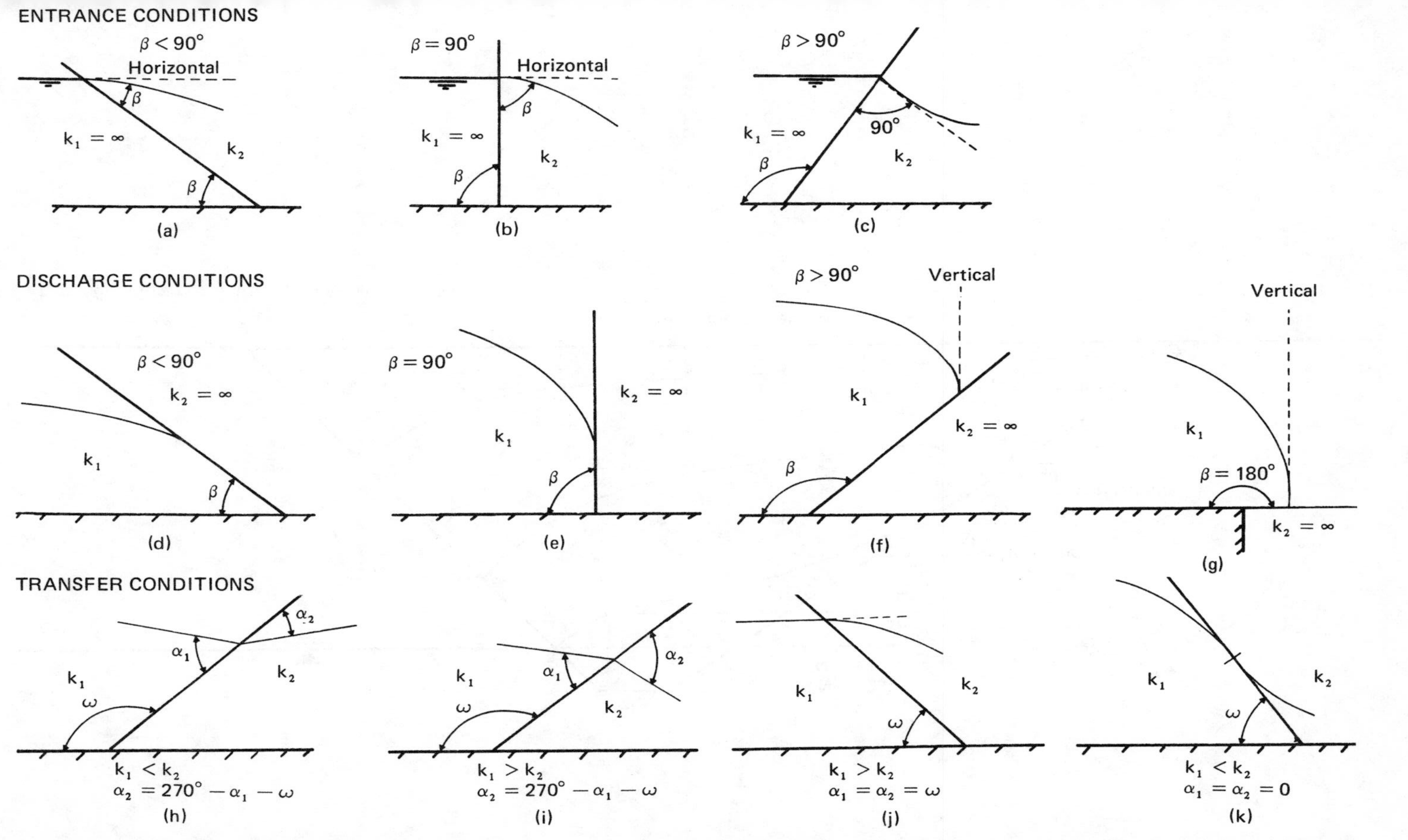

Fig. 2.63 Entrance, discharge, and transfer conditions. *(After A. Casagrande, Seepage through Dams,* Contribution to Soil Mechanics, 1925–1940, *Boston Society of Civil Engineers, Boston, 1937.)*

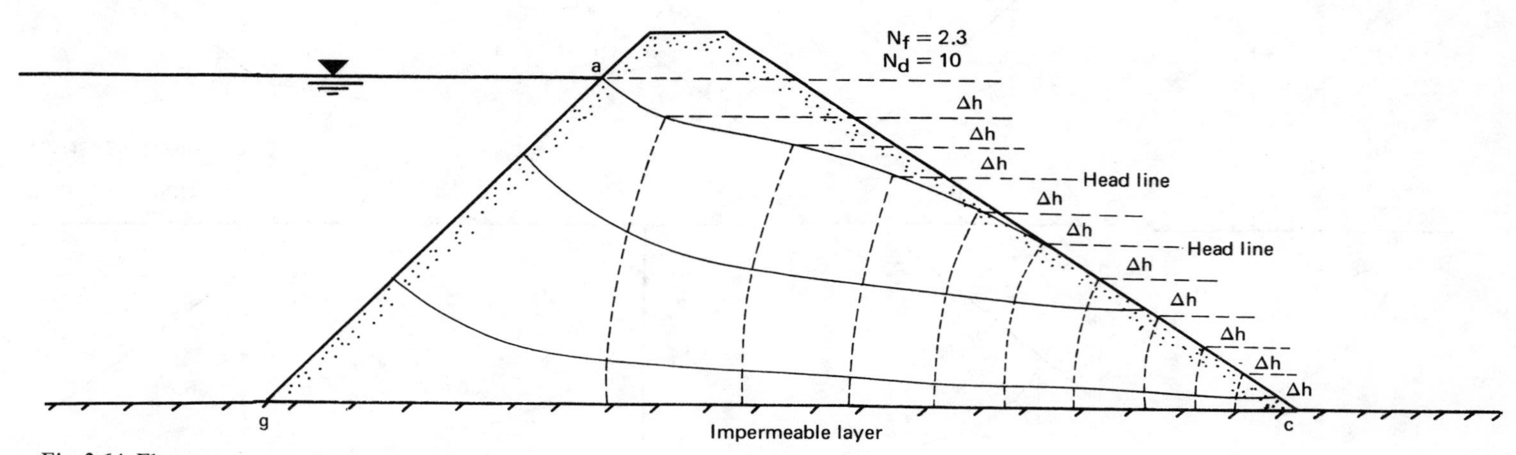

Fig. 2.64 Flow-net construction for an earth dam.

of the head lines and the phreatic lines are the points from which the equipotential lines should start.

5. Draw the flow net, keeping in mind that the equipotential lines and flow lines must intersect at right angles.
6. The rate of seepage through the earth dam can be calculated from the relation given in Eq. (2.118), $q = kh\,(N_f/N_d)$.

In Fig. 2.64, the number of flow channels, N_f, is equal to 2.3. The top two flow channels have square flow elements, and the bottom flow channel has elements with a width-to-length ratio of 0.3. Also, N_d in Fig. 2.64 is equal to 10.

If the dam section is anisotropic with respect to permeability, a transformed section should first be prepared in the manner outlined in Sec. 2.2.5. The flow net can then be drawn on the transformed section, and the rate of seepage obtained from Eq. (2.124).

If the phreatic line for the dam section is not known, a trial-and-error procedure will have to be adopted for the construction of flow nets (Cedergren, 1977). This technique is demonstrated in Fig. 2.65. The steps to obtain the flow net are as follows:

1. Draw the head lines on the cross section of the dam. Also draw the approximate zone of the phreatic line as shown in Fig. 2.65*a*.
2. Assume a trial saturation line (phreatic line) *ab* as shown in Fig. 2.65*b*. Draw an approximate flow net. Now check the number of flow channels between any two consecutive equipotential lines. If the flow net is correctly drawn, the number of flow channels between any two consecutive equipotential lines should be the same. If not, some adjustment has to be made by moving the phreatic line and the flow and equipotential lines. (The arrows in Fig. 2.65*b* show the direction of needed correction.)

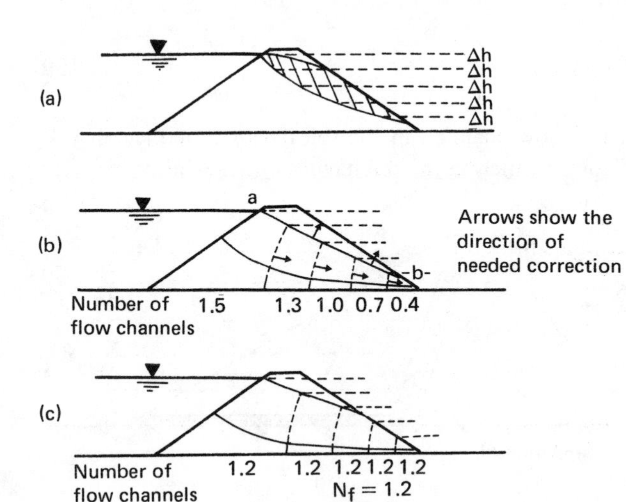

Fig. 2.65 Flow-net construction for an earth dam section with unknown phreatic line. *(Modified after H. R. Cedergren, "Seepage, Drainage and Flow Nets," 2d ed., Wiley, New York, 1977.)*

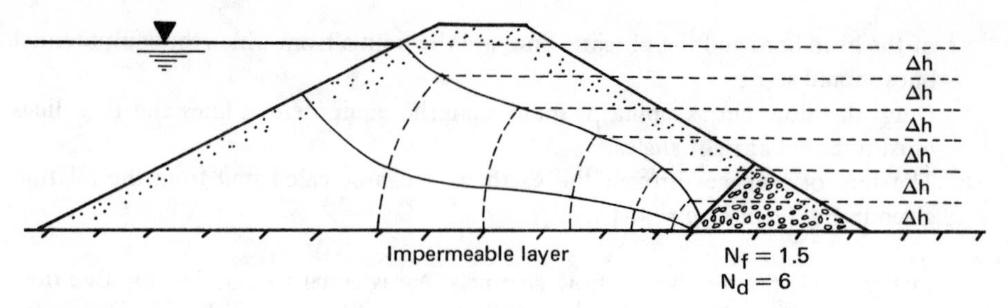

Fig. 2.66 Typical flow net for an earth dam with rock toe filter.

3. After a few trials, the final flow net can be obtained as shown in Fig. 2.65*c* ($N_f = 1.2$ and $N_d = 5$).

Figures 2.66 and 2.67 show some typical flow nets through earth dam sections.

A flow net for seepage through a zoned earth dam section is shown in Fig. 2.68. The soil for the upstream half of the dam has a permeability k_1, and the soil for the downstream half of the dam has a permeability $k_2 = 5k_1$. The phreatic line has to be plotted by trial and error. As shown in Fig. 2.37*b*, here the seepage is from a soil of low permeability (upstream half) to a soil of high permeability (downstream half). From Eq. (2.125),

$$\frac{k_1}{k_2} = \frac{b_2/l_2}{b_1/l_1}$$

If $b_1 = l_1$ and $k_2 = 5k_1$, $b_2/l_2 = 1/5$. For that reason, square flow elements have been plotted in the upstream half of the dam, and the flow elements in the downstream half have a width-to-length ratio of 1/5. The rate of seepage can be calculated by using the following equation:

$$q = k_1 \frac{h}{N_d} N_{f(1)} = k_2 \frac{h}{N_d} N_{f(2)} \tag{2.199}$$

where $N_{f(1)}$ is the number of full flow channels in the soil having a permeability k_1, and $N_{f(2)}$ is the number of full flow channels in the soil having a permeability k_2.

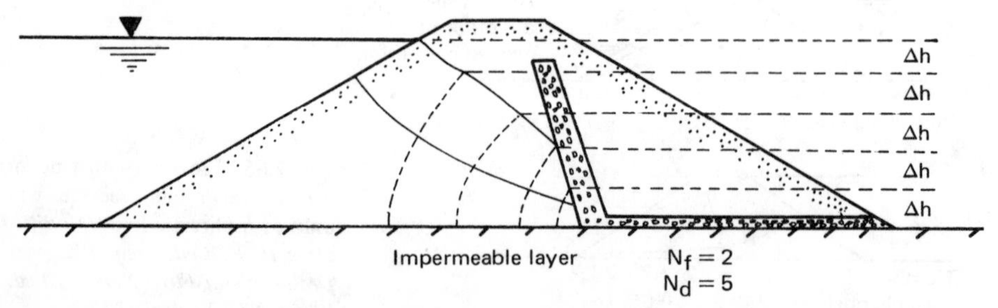

Fig. 2.67 Typical flow net for an earth dam with chimney drain.

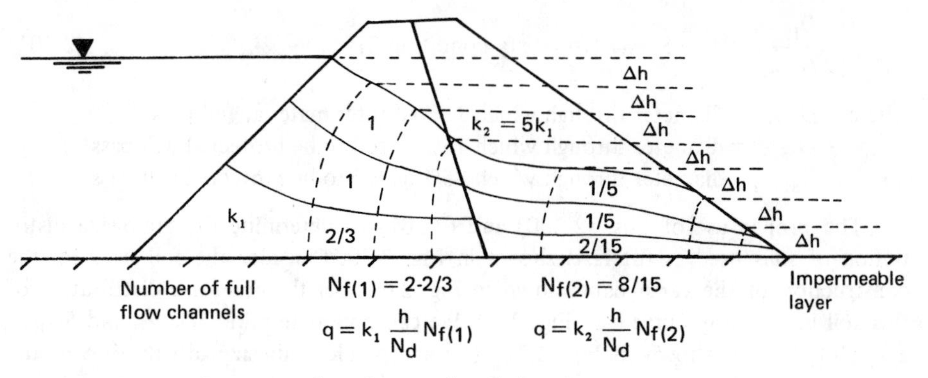

Fig. 2.68 Flow net for seepage through a zoned earth dam.

2.2.15 Filter Design

When seepage water flows from a soil with relatively fine grains into a coarser material, there is a danger that the fine soil particles may wash away into the coarse material. Over a period of time, this process may clog the void spaces in the coarser material. Such a situation can be prevented by the use of a filter or protective filter between the two soils. For example, consider the earth dam section shown in Fig. 2.66. If rockfills were only used at the toe of the dam, the seepage water would wash the fine soil grains into the toe and undermine the structure. Hence, for the safety of the structure, a filter should be placed between the fine soil and the rock toe (Fig. 2.69). For the proper selection of the filter material, two conditions should be kept in mind:

1. The size of the voids in the filter material should be small enough to hold the larger particles of the protected material in place.
2. The filter material should have a high permeability to prevent buildup of large seepage forces and hydrostatic pressures in the filters.

Based on the experimental investigation of protective filters, Bertram (1940) provided the following criteria to satisfy the above conditions:

$$\frac{D_{15(F)}}{D_{85(S)}} \leqslant 4 \text{ to } 5 \qquad \text{(to satisfy condition 1)} \qquad (2.200)$$

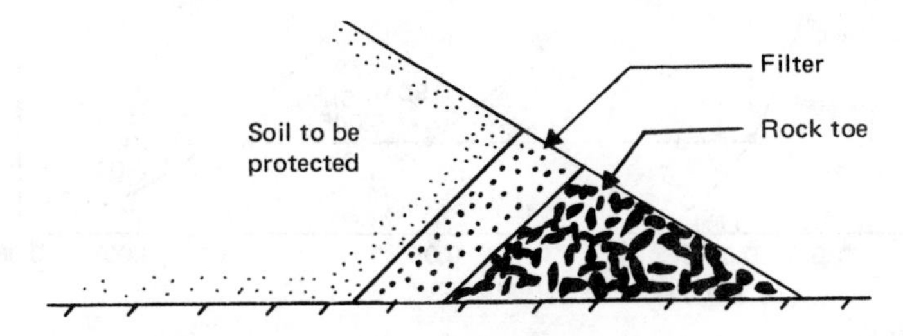

Fig. 2.69 Use of filter at the toe of an earth dam.

$$\frac{D_{15(F)}}{D_{15(S)}} \geqslant 4 \text{ to } 5 \qquad \text{(to satisfy condition 2)} \qquad (2.201)$$

where $D_{15(F)}$ = diameter through which 15% of filter material will pass
$D_{15(S)}$ = diameter through which 15% of soil to be protected will pass
$D_{85(S)}$ = diameter through which 85% of soil to be protected will pass

The proper use of Eqs. (2.200) and (2.201) to determine the grain-size distribution of soils used as filters is shown in Fig. 2.70. Consider the soil used for the construction of the earth dam shown in Fig. 2.69. Let the grain-size distribution of this soil be given by curve *a* in Fig. 2.70. We can now determine $5D_{85(S)}$ and $5D_{15(S)}$ and plot them as shown in Fig. 2.70. The acceptable grain-size distribution of the filter material will have to lie in the shaded zone.

The same principle can be adopted for determination of the size limits for the rock layer (Fig. 2.69) to protect the filter material from being washed away.

The U.S. Navy (1971) requires the following conditions for the design of filters.

1. For avoiding the movement of the particles of the protected soil:

$$\frac{D_{15(F)}}{D_{85(S)}} < 5$$

$$\frac{D_{50(F)}}{D_{50(S)}} < 25$$

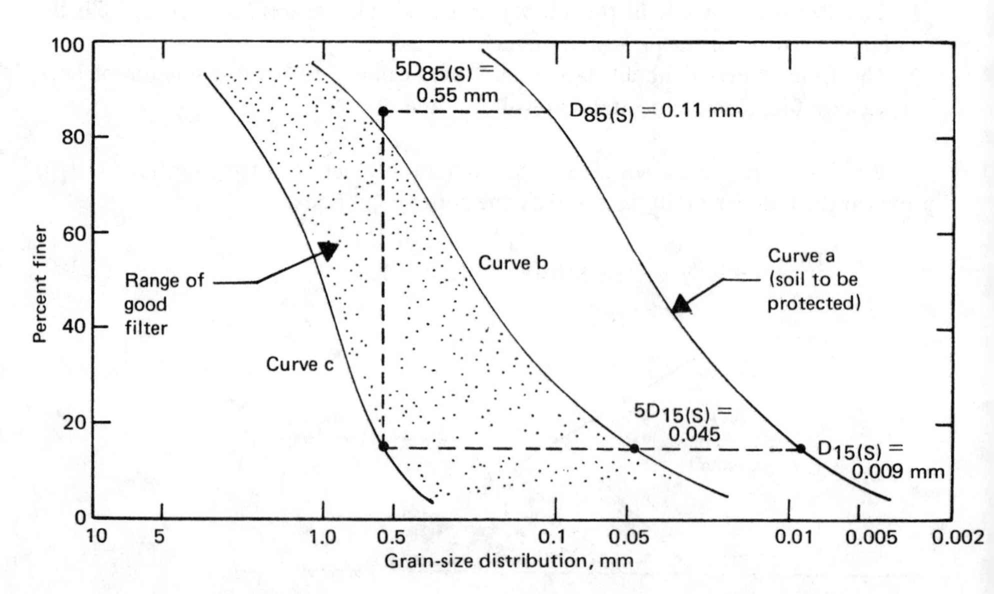

Fig. 2.70 Determination of grain-size distribution of filter using Eqs. (2.200) and (2.201).

$$\frac{D_{15(F)}}{D_{15(S)}} < 20$$

If the uniformity coefficient C_u of the protected soil is less than 1.5, $D_{15(F)}/D_{85(S)}$ may be increased to 6. Also, if C_u of the protected soil is greater than 4, $D_{15(F)}/D_{15(S)}$ may be increased to 40.

2. For avoiding buildup of large seepage force in the filter,

$$\frac{D_{15(F)}}{D_{15(S)}} > 4$$

3. The filter material should not have grain sizes greater than 3 in (76.2 mm). (This is to avoid segregation of particles in the filter.)
4. To avoid internal movement of fines in the filter, it should have no more than 5% passing a No. 200 sieve.
5. When perforated pipes are used for collecting seepage water, filters are also used around the pipes to protect the fine-grained soil from being washed into the pipes. To avoid the movement of the filter material into the drain-pipe perforations, the following additional conditions should be met:

$$\frac{D_{85(F)}}{\text{slot width}} > 1.2 \text{ to } 1.4$$

$$\frac{D_{85(F)}}{\text{hole diameter}} > 1.0 \text{ to } 1.2$$

Thanikachalam and Sakthivadivel (1974) analyzed experimental results for filters reported by Karpoff (1955), U.S. Corps of Engineers (1953), Leatherwood and Peterson (1954), Dayaprakash and Gupta (1972), and Belyashevskii et al. (1972). Based on this analysis, they recommended that when the soil to be protected is of a granular nature, the stable filter design criteria may be given by the following equations:

$$\frac{D_{60(S)}}{D_{10(S)}} = 0.4\,\frac{D_{10(F)}}{D_{10(S)}} - 2.0 \tag{2.202}$$

and

$$\frac{D_{60(F)}}{D_{10(F)}} = 0.941\,\frac{D_{10(F)}}{D_{10(S)}} - 5.65 \tag{2.203}$$

where $D_{60(S)}$ and $D_{10(S)}$ are, respectively, the diameters through which 60% and 10% of the soil to be protected is passing; and $D_{60(F)}$ and $D_{10(F)}$ are, respectively, the diameters through which 60% and 10% of the filter material is passing.

Cedergren (1960) constructed several flow nets, such as those shown in Fig. 2.71*a* and *b*, to study the condition of seepage into sloping filters placed at the downstream side of earth dams. Based on this work, he developed the chart given in Fig. 2.71*c* which allows us to determine the minimum thickness of filter material, W, required on the downstream side of an earth dam. (Note that in Fig. 2.71, k_F is the coefficient of permeability of the filter material, and k_S is the coefficient of permeability of the soil of the earth dam.)

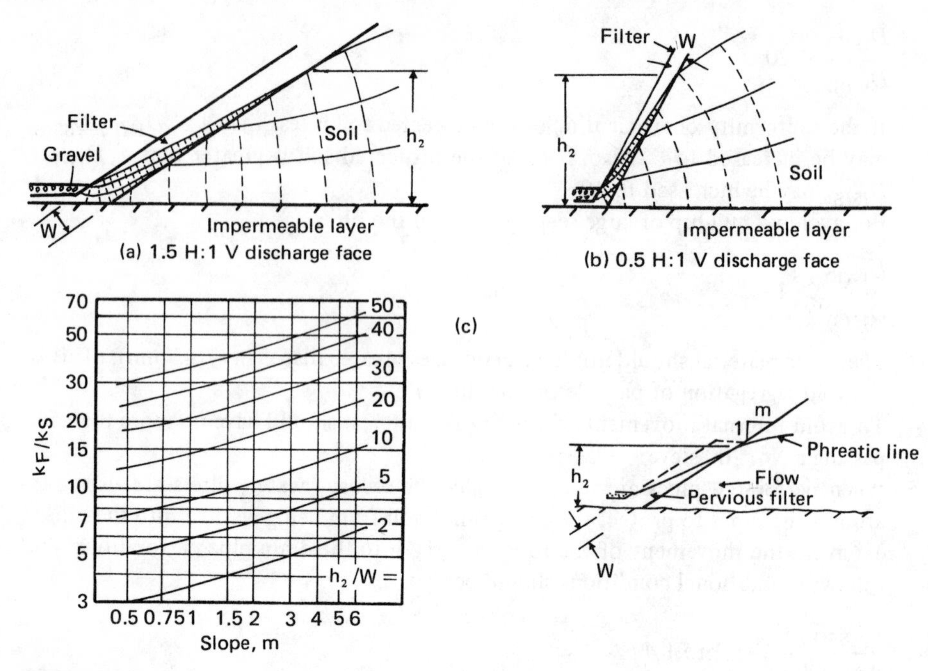

Fig. 2.71 Thickness of filter material on the downstream side of an earth dam. *(After H. R. Cedergren, Seepage Requirement of Filters and Pervious Bases,* J. Soil Mech. Found. Div., ASCE, *vol. 86, no. SM5 (part I), 1960.)*

PROBLEMS

2.1 The results of a constant-head permeability test in a fine sand are as follows: area of the soil specimen, 180 cm²; length of the specimen, 320 mm; constant head maintained, 460 mm; flow of water through the specimen, 200 ml in 5 min. Determine the coefficient of permeability.

2.2 The fine sand described in Prob. 2.1 was tested in a falling-head permeameter, and the results are as follows: area of the specimen, 90 cm²; length of the specimen, 320 mm; area of the standpipe, 5 cm²; head difference at the beginning of the test, 1000 mm. Calculate the head difference after 300 s from the start of the test (use the result of Prob. 2.1).

2.3 A silty sand specimen was subjected to a capillary permeability test. The results are as follows (refer to Fig. 2.10 for notations):

t, min	x^2, cm²	
2	40	
4	80	$h_1 = 48.75$ cm
6	120	
8	160	
		change from h_1 to h_2 at time $t = 10$ min
12	370	
14	545	$h_2 = 362.6$ cm
16	720	
18	900	

Given a soil porosity of 0.5 and $S_r = 0.9$, calculate the value of k.

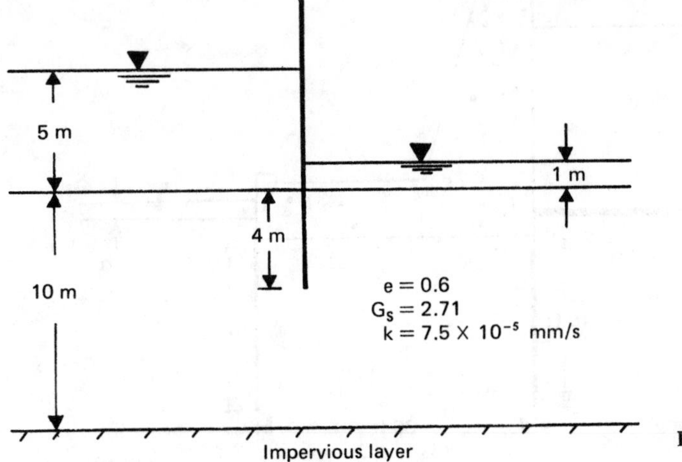

Fig. P2.1

2.4 A single row of sheet pile structure is shown in Fig. P2.1.

(*a*) Draw the flow net.

(*b*) Calculate the rate of seepage.

(*c*) Calculate the factor of safety against piping using Terzaghi's method [Eq. (2.179)] and then Harza's method.

2.5 For the single row of sheet piles shown in Fig. P2.1, calculate the hydraulic heads in the permeable layer using the numerical method shown in Sec. 2.2.8. From these results, draw the equipotential lines. Use $\Delta z = \Delta x = 2$ m.

2.6 A dam section is shown in Fig. P2.2. Given $k_x = 9 \times 10^{-5}$ mm/s and $k_z = 1 \times 10^{-5}$ mm/s, draw a flow net and calculate the rate of seepage.

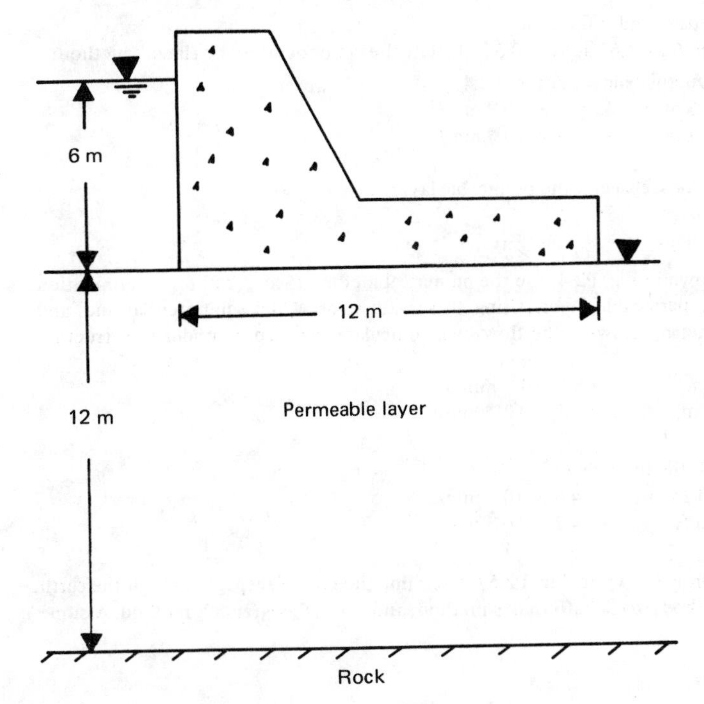

Fig. P2.2

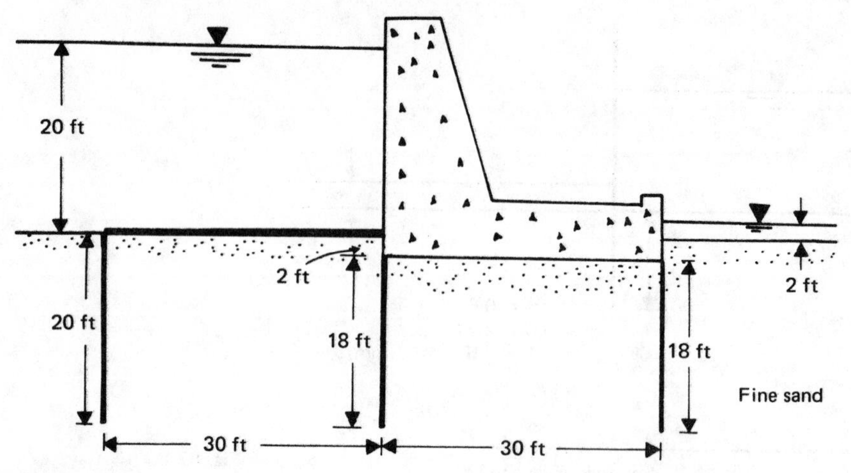

Fig. P2.3

2.7 For the dam section shown in Fig. P2.2, given $k = k_x = k_z = 5 \times 10^{-5}$ mm/s, calculate the hydraulic heads in the permeable soil layer. Use $\Delta x = \Delta z = 2m$ and the numerical method given in Sec. 2.2.8.

2.8 A dam section is shown in Fig. P2.3. Using Lane's method, calculate the weighted creep ratio. Is the dam safe against piping?

2.9 Refer to Fig. P2.3. Assume (1) that the sheet pile at the end of the upstream apron is not there, (2) that an impervious rock layer is located at a depth of 37 ft from the ground surface, and (3) that $k = 0.02$ mm/s.

(a) Draw a flow net.
(b) Calculate the seepage under the dam.
(c) Given, for the soil, $G_s = 2.65$ and $e = 0.5$, calculate the factor of safety by Harza's method.

2.10 For the sheet pile structure shown in Fig. P2.4,

$d = 2.5$m $H_1 = 3$ m $k_1 = 4 \times 10^{-3}$ mm/s
$d_1 = 5$ m $H_2 = 1$ m $k_2 = 2 \times 10^{-3}$ mm/s
$d_2 = 5$ m

(a) Draw a flow net for seepage in the permeable layer.
(b) Calculate the seepage.
(c) Find the exit gradient.

2.11 For the structure shown in Fig. P2.4, use the numerical method (Sec. 2.2.8) for determination of hydraulic heads in the permeable layer. Using these values, draw the equipotential lines and then complete the flow net by drawing the flow lines. Calculate the seepage under the structure and also the exit gradient.

$d = 3$ m $H_1 = 4$ m $k_1 = 6 \times 10^{-4}$ mm/s
$d_1 = 4$ m $H_2 = 1$ m $k_2 = 1.5 \times 10^{-4}$ mm/s
$d_2 = 4$ m

2.12 Redo Prob. 2.10 with the following

$d = 12$ ft $H_1 = 12$ ft $k_1 = 6 \times 10^{-5}$ mm/s
$d_1 = 18$ ft $H_2 = 3$ ft $k_2 = 2 \times 10^{-5}$ mm/s
$d_2 = 24$ ft

2.13 An earth dam section is shown in Fig. P2.5. Determine the rate of seepage through the earth dam using (a) Dupuit's method, (b) Schaffernak's method, and (c) L. Casagrande's method. Assume that $k = 10^{-5}$ mm/s.

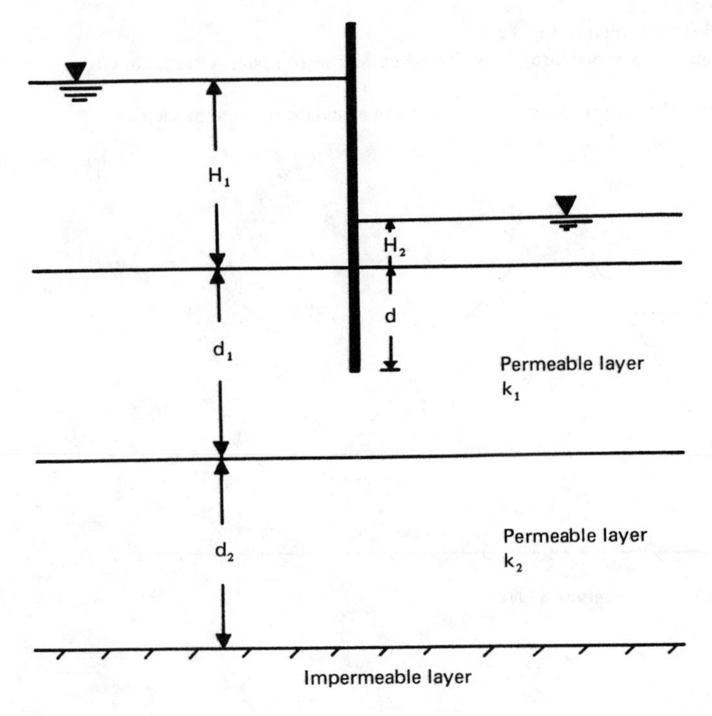

Fig. P2.4

2.14 Repeat Prob. 2.13 assuming that $k_x = 6 \times 10^{-5}$ mm/s and $k_z = 1 \times 10^{-5}$ mm/s.

2.15 For the earth dam section shown in Fig. P2.5, determine the rate of seepage through the dam using Pavlovsky's solution. Assume that $k = 4 \times 10^{-5}$ mm/s.

2.16 For the earth dam section shown in Fig. P2.5, determine the rate of seepage by using Pavlovsky's solution. Assume that $k_x = 4.6 \times 10^{-5}$ mm/s and $k_z = 1.15 \times 10^{-5}$ mm/s.

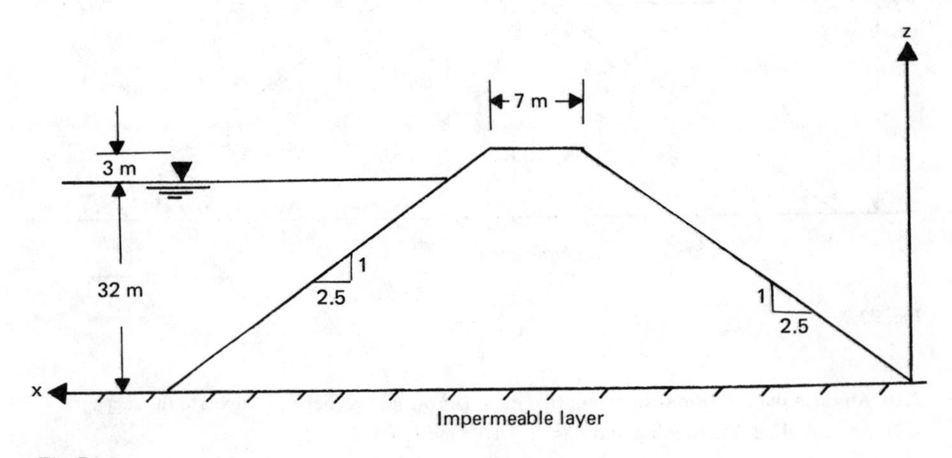

Fig. P2.5

2.17 An earth dam section is shown in Fig. P2.6.

(*a*) Using A. Casagrande's procedure, draw the phreatic line to scale. Assume that $k_x = k_z = 0.8 \times 10^{-6}$ mm/s.

(*b*) Using the results of part (*a*), draw the flow net and calculate the rate of seepage.

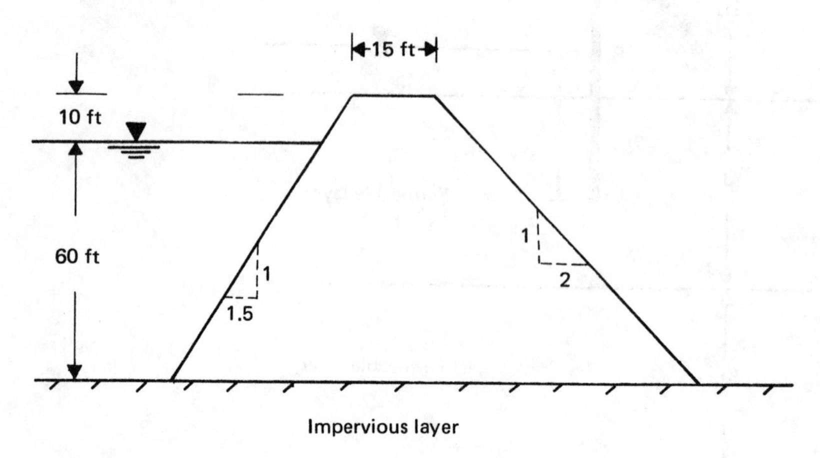

Fig. P2.6

2.18 Solve Prob. 2.17 assuming that $k_x = 3.2 \times 10^{-6}$ mm/s and $k_z = 0.8 \times 10^{-6}$ mm/s.

2.19 Solve Prob. 2.17 for the dam section shown in Fig. P2.7.

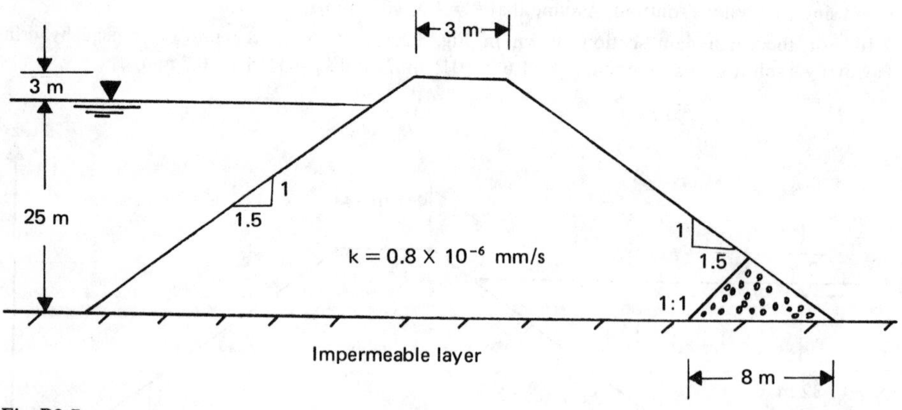

Fig. P2.7

2.20 An earth dam section is shown in Fig. P2.8. Draw the flow net and calculate the seepage.

2.21 Solve Prob. 2.20 assuming that $k_1 = 2 \times 10^{-6}$ mm/s and $k_2 = 2k_1$.

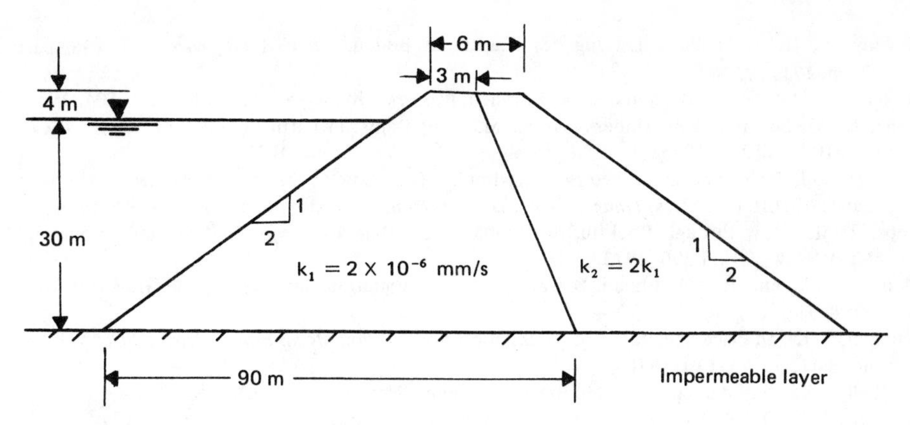

Fig. P2.8

REFERENCES

Belyashevskii, N. N., N. G. Bugai, I. I. Kalantyrenko, and S. L. Topchii, Behavior and Selection of the Composition of Graded Filters in the Presence of Fluctuating Flow, *Hydrotech. Construc.*, vol. 6, pp. 541–546, 1972.

Bertram, G. E., "An Experimental Investigation of Protective Filters," Harvard University Graduate School, Publication No. 267, 1940.

Carman, P. E., "Flow of Gases Through Porous Media," Academic, New York, 1956.

Casagrande, A., Seepage Through Dams, *in* "Contribution to Soil Mechanics 1925–1940," Boston Society of Civil Engineers, Boston, p. 295, 1937.

Casagrande, L., "Naeherungsmethoden zur Bestimmurg von Art und Menge der Sickerung durch geschuettete Daemme," Thesis, Technische Hochschule, Vienna, 1932.

Cedergren, H. R., Seepage Requirement of Filters and Pervious Bases, *J. Soil Mech. Found. Div., ASCE,* vol. 86, no. SM5, part I, pp. 15–33, 1960.

Cedergren, H. R., "Seepage, Drainage and Flow Nets," 2d ed., Wiley, New York, 1977.

Darcy, H., "Les Fontaines Publiques de la Ville de Dijon," Dalmont, Paris, 1856.

Dayaprakash, D. K., and R. C. Gupta, Laboratory Studies of Filter Materials Placed at Ramaganga Main Dam, *Indian Geotech. J.*, vol. 2, no. 3, pp. 203–219, 1972.

Dupuit, J., "Etudes theoriques et Practiques sur le Mouvement des eaux dans les Canaux Decouverts et a travers les Terrains Permeables," Dunod, Paris, 1863.

Forchheimer, P., Discussion of "The Bohio Dam" by George S. Morrison, *Trans. ASCE,* vol. 48, p. 302, 1902.

Gilboy, G., Mechanics of Hydraulic Fill Dams, *in* "Contribution to Soil Mechanics 1925–1940," Boston Society of Civil Engineers, Boston, 1934.

Gray, D. H., and J. K. Mitchell, Fundamental Aspects of Electro-osmosis in Soils, *J. Soil Mech. Found. Div., ASCE,* vol. 93, no. SM6, pp. 209–236, 1967.

Harr, M. E., "Groundwater and Seepage," McGraw-Hill, New York, 1962.

Harza, L. F., Uplift and Seepage under Dams in Sand, *Trans. ASCE,* vol. 100, 1935.

Hazen, A., Discussion of "Dams on Sand Foundation" by A. C. Koenig, *Trans. ASCE,* vol. 73, p. 199, 1911.

Helmholtz, H., *Wiedemanns Ann. Phys.*, vol. 7, p. 137, 1879.

Karpoff, K. P., The Use of Laboratory Tests to Develop Design Criteria for Protective Filters, *Proc. ASTM,* no. 55, pp. 1183–1193, 1955.

Kozeny, J., Ueber kapillare Leitung des Wassers in Boden, *Wien, Akad. Wiss.*, vol. 136, part 2a, p. 271, 1927.

Kozeny, J., Theorie und Berechnung der Brunnen, *Wasserkr. Wasserwritsch.*, vol. 28, p. 104, 1933.

Lane, E. W., Security from Under-Seepage: Masonry Dams on Earth Foundation, *Trans. ASCE*, vol. 100, p. 1235, 1935.

Leatherwood, F. N., and D. G. Peterson, Hydraulic Head Loss at the Interface between Uniform Sands of Different Sizes, *Trans. Amer. Geophys. Union*, vol. 35, no. 4, pp. 588–594, 1954.

Leps, T. M., Flow through Rockfill, *in* "Embankment-Dam Engineering—Casagrande Volume," Wiley, New York, p. 90, 1973.

Mansur, C. I., and R. I. Kaufman, Dewatering, *in* "Foundation Engineering," McGraw-Hill, New York, 1962.

Mitchell, J. K., In-Place Treatment of Foundation Soils, *J. Soil Mech. Found. Div., ASCE*, vol. 96, no. SM1, pp. 73–110, 1970.

Mitchell, J. K., "Fundamentals of Soil Behavior," Wiley, New York, 1976.

Mitchell, J. K., D. R. Hooper, and R. G. Campanella, Permeability of Compacted Clay, *J. Soil Mech. Found. Div., ASCE*, vol. 91, no. SM4, pp. 41–65, 1965.

Muskat, M., "The Flow of Homogeneous Fluids through Porous Media," McGraw-Hill, New York, 1937.

Olsen, H. W., "Hydraulic Flow through Saturated Clay," Sc.D. thesis, Massachusetts Institute of Technology, 1961.

Olsen, H. W., Hydraulic Flow through Saturated Clays, *Proc. 9th Nat. Conf. Clay and Clay Min.*, pp. 131–161, 1962.

Pavlovsky, N. N., "Seepage through Earth Dams," Inst. Gidrotekhniki i Melioratsii, Leningrad, 1931.

Schaffernak, F., Über die Standicherheit durchlaessiger geschuetteter Dämme, *Allgem. Bauzeitung*, 1917.

Schmid, G., Zur Elektrochemie Feinporiger Kapillarsystems, *Zh. Elektrochem.* vol. 54, p. 425, 1950; vol. 55, p. 684, 1951.

Smoluchowski, M., *in* L. Graetz (Ed.), "Handbuch der Elektrizital und Magnetismus," vol. 2, Barth, Leipzig, 1914.

Spangler, M. G., and R. L. Handy, "Soil Engineering," 3d ed., Intext Educational, New York, 1973.

Taylor, D. W., "Fundamentals of Soil Mechanics," Wiley, New York, 1948.

Terzaghi, K., "Theoretical Soil Mechanics," Wiley, New York, 1943.

Terzaghi, K., Der Grundbrunch on Stauwerken und Seine Verhutung, *Die Wasserkraft*, vol. 17, pp. 445–449, 1922. Reprinted in "From Theory to Practice in Soil Mechanics," Wiley, New York, pp. 146–148, 1960.

Thanikachalam, V., and R. Sakthivadivel, Rational Design Criteria for Protective Filters, *Can. Geotech. J.* vol. 11, no. 2, pp. 309–314, 1974.

U.S. Bureau of Reclamation, Department of the Interior, "Design of Small Dams," U.S. Government Printing Office, Washington, D.C., 1961.

U.S. Corps of Engineers, "Filter Experiments and Design Criteria," U.S. Waterways Experiment Station, Vicksburg, Miss., Tech. Memo. No. 3.360, 1953.

U.S. Department of the Navy, Naval Facilities Engineering Command, "Design Manual—Soil Mechanics, Foundations, and Earth Structures, NAVFAC DM-7, Washington, D.C, 1971.

CHAPTER

THREE

STRESSES IN SOIL MASS

This chapter deals with problems involving stresses induced by various types of loading. The expressions for stresses are obtained on the assumption that soil is a perfectly elastic material; problems related to plastic equilibrium are not treated in this text.

The chapter is divided into two major sections: (1) two-dimensional (plane strain) problems and (2) three-dimensional problems.

3.1 TWO-DIMENSIONAL PROBLEMS

3.1.1 Plane Strain State-of-Stress

In many soil mechanics problems, a type of state-of-stress that is encountered is the plane strain condition. Long retaining walls and strip foundations are two examples of where plane strain conditions are encountered. Referring to Fig. 3.1, for the strip footing the strain in the y direction at any point P in the soil mass is equal to zero. The normal stresses σ_y at all sections in the xz plane (i.e., normal to the y axis) are the same, and the shear stresses on these sections are zero. The normal and shear stresses on the plane normal to the x axis are equal to σ_x and τ_{xz}, respectively. Similarly, the normal and shear stresses on the plane normal to the z axis are σ_z and τ_{zx} ($= \tau_{xz}$), respectively. The relationship between the normal stresses can be expressed as

$$\sigma_y = \nu(\sigma_x + \sigma_y) \tag{3.1}$$

where ν is Poisson's ratio. Thus, these are essentially two-dimensional problems.

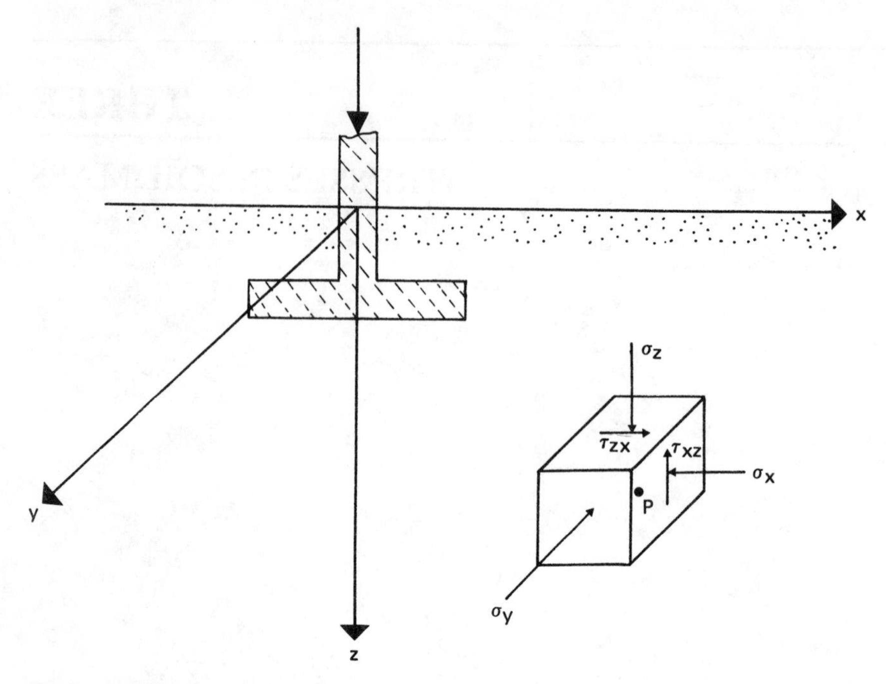

Fig. 3.1 Definition of plane strain state-of-stress.

3.1.2 Stresses on an Inclined Plane and Principal Stresses for Plane Strain Problems Using Mohr's Circle

If the stresses at a point in a soil mass—i.e., $\sigma_x, \sigma_y, \sigma_z, \tau_{xz}$ $(= \tau_{zx})$—are known, the normal stress σ and the shear stress τ on an inclined plane *BE* (Fig. 3.2) can be conveniently determined graphically by means of a Mohr's circle. The procedure for the construction of the Mohr's circle is explained below.

The sign convention for normal stress is positive for compression and negative for tension. The shear stress on a given plane is positive if it tends to produce a clockwise rotation about a point outside the soil element; it is negative if it tends to produce a counterclockwise rotation. This is shown in Fig. 3.3. Thus, referring to plane *AB* in Fig. 3.2, the normal stress is $+\sigma_x$ and the shear stress is $+\tau_{xz}$. Similarly, on the plane *AD* the stresses are $+\sigma_z$ and $-\tau_{xz}$. The stresses on the planes *AB* and *AD* can be plotted graphically with normal stresses along the abscissa and shear stresses along the ordinate. The points *B* and *D* in Fig. 3.4 refer to the stress conditions on the planes *AB* and *AD*, respectively. Now, if points *B* and *D* are joined by a straight line, the line will intersect the normal stress axis and O'. If a circle BP_1DP_3 is drawn with O' as the center and *OB* as the radius, it will be the Mohr's circle. The radius of the Mohr's circle is

$$O'B = \sqrt{O'G^2 + BG^2} = \sqrt{\left(\frac{\sigma_x - \sigma_z}{2}\right)^2 + \tau_{xz}^2} \tag{3.2}$$

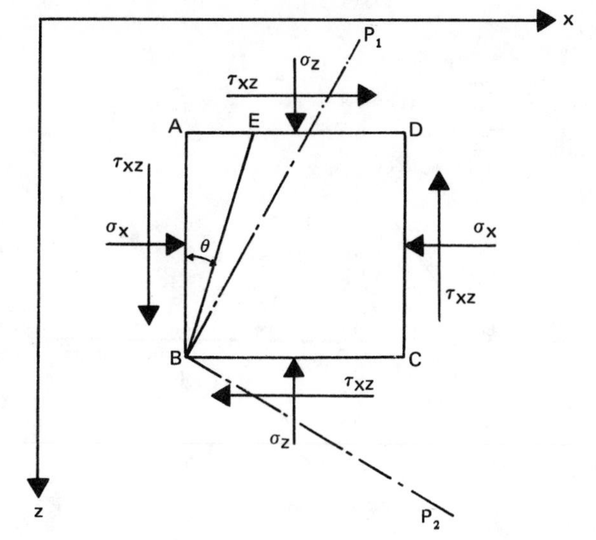

Fig. 3.2

Any radial line in a Mohr's circle represents a given plane, and the coordinates of the point of intersection of the radial line and the circumference of the Mohr's circle gives the stress condition on that plane. For example, let us find the stresses on the plane BE. If in Fig. 3.2 we start from the plane AB and move an angle θ in the clockwise direction, we reach the plane BE. In the Mohr's circle in Fig. 3.4, the radial line $O'B$ represents the plane AB. We will have to move an angle 2θ in the same clockwise direction to reach point F. Now, the radial line $O'F$ in Fig. 3.4 represents the plane BE of Fig. 3.2. The coordinates of the point F will give us the stresses on the plane BE.

Note that the ordinates of the points P_1 and P_3 are zero, which means that $O'P_1$

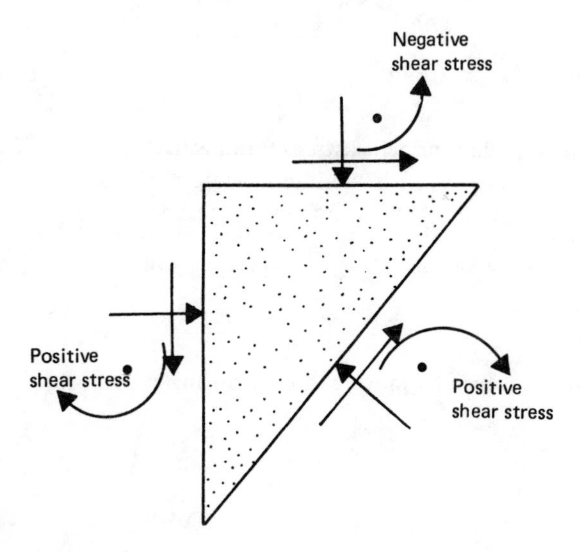

Fig. 3.3 Sign convention for shear stress used for the construction of Mohr's circle.

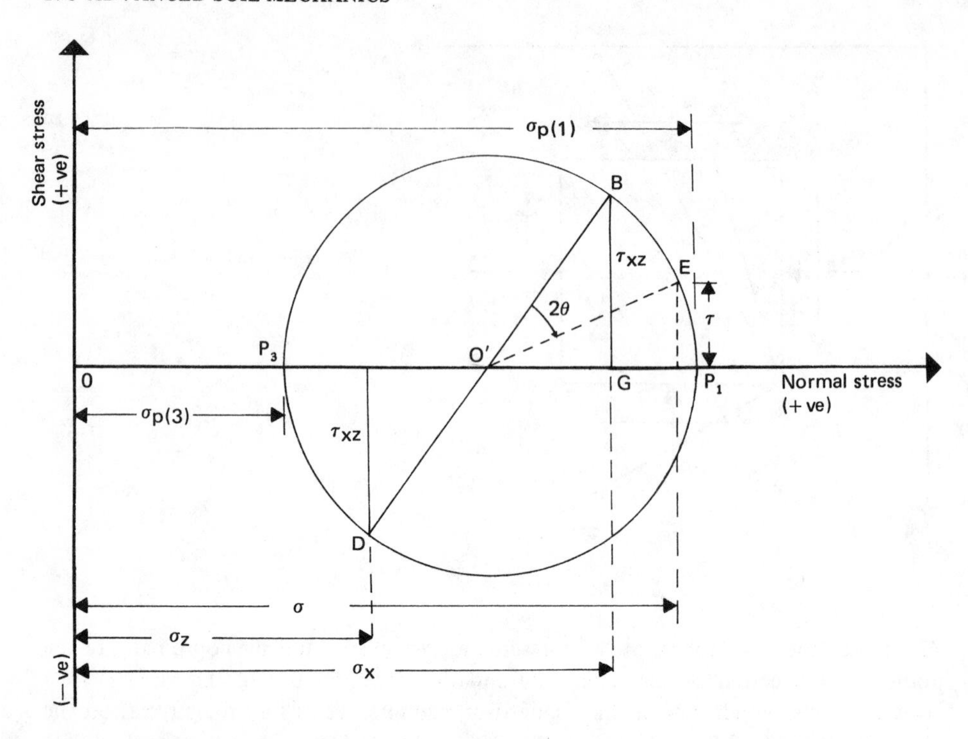

Fig. 3.4 Mohr's circle.

and $O'P_3$ represent the *major and minor principal planes* and that $\overline{OP_1} = \sigma_{P(1)}$ and $\overline{OP_3} = \sigma_{P(3)}$:

$$\sigma_{P(1)} = OP_1 = OO' + O'P_1 = \frac{\sigma_x + \sigma_z}{2} + \sqrt{\left(\frac{\sigma_x - \sigma_z}{2}\right)^2 + \tau_{xz}^2} \tag{3.3}$$

and $$\sigma_{P(3)} = OP_3 = OO' - O'P_3 = \frac{\sigma_x + \sigma_z}{2} - \sqrt{\left(\frac{\sigma_x - \sigma_z}{2}\right)^2 + \tau_{xz}^2} \tag{3.4}$$

where $\sigma_{P(1)}$ and $\sigma_{P(3)}$ are the major and minor principal stresses, respectively.

Note that $\sigma_y = \nu(\sigma_x + \sigma_z) = \nu(\sigma_{P(1)} + \sigma_{P(3)})$ is the intermediate principal stress. Also note that the principal plane $O'P_1$ in the Mohr's circle can be reached by moving clockwise from $O'B$ through an angle

$$BO'P_1 = \tan^{-1}\left(\frac{2\tau_{xz}}{\sigma_x - \sigma_z}\right)$$

The other principal plane $O'P_3$ can be reached by moving through an angle

$$180° + \tan^{-1}\left(\frac{2\tau_{xz}}{\sigma_x - \sigma_z}\right)$$

in the clockwise direction from $O'B$. So, in Fig. 3.2, if we move from plane AB through an angle

$$\frac{1}{2}\tan^{-1}\left(\frac{2\tau_{xz}}{\sigma_x-\sigma_z}\right)$$

we will reach the plane BP_1 on which the principal stress $\sigma_{P(1)}$ acts; similarly, moving clockwise from the plane AB through an angle

$$\frac{1}{2}\left[180^\circ+\tan^{-1}\left(\frac{2\tau_{xz}}{\sigma_x-\sigma_z}\right)\right]=90^\circ+\frac{1}{2}\tan^{-1}\left(\frac{2\tau_{xz}}{\sigma_x-\sigma_z}\right)$$

we reach the plane BP_2 on which the principal stress $\sigma_{P(3)}$ acts.

3.1.3 Stresses due to a Vertical Line Load on the Surface of a Semi-infinite Mass

Figure 3.5 shows the case where a line load of q per unit length is applied at the surface of a homogeneous, elastic, and isotropic soil mass. The stresses at a point P defined by r and θ can be determined by using the *stress function*

$$\phi=\frac{q}{\pi}r\theta\sin\theta \tag{3.5}$$

In the polar coordinate system, the expressions for the stresses are as follows (see any theory of elasticity text, e.g., Timoshenko and Goodier, 1970):

$$\sigma_r=\frac{1}{r}\frac{\partial\phi}{\partial r}+\frac{1}{r^2}\frac{\partial^2\phi}{\partial\theta^2} \tag{3.6}$$

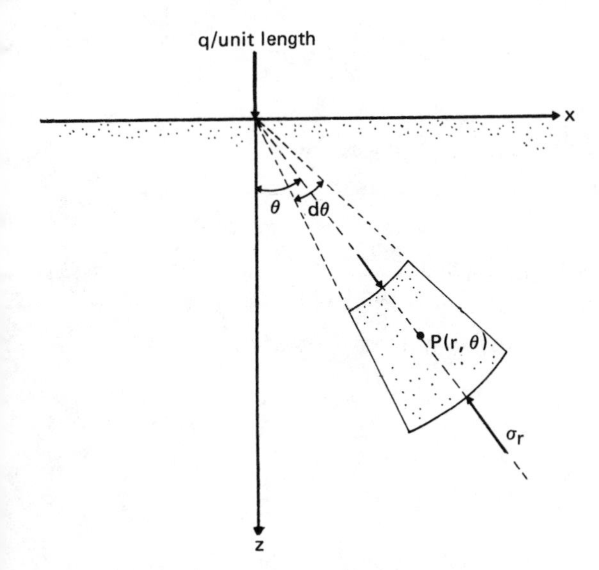

Fig. 3.5 Vertical line load on the surface of a semi-infinite mass.

$$\sigma_\theta = \frac{\partial^2 \phi}{\partial r^2} \tag{3.7}$$

and
$$\tau_{r\theta} = -\frac{\partial}{\partial r}\left(\frac{1}{r}\frac{\partial \phi}{\partial \theta}\right) \tag{3.8}$$

Substituting the values of ϕ in the above equations, we get

$$\sigma_r = \frac{1}{r}\left(\frac{q}{\pi}\theta \sin\theta\right) + \frac{1}{r^2}\left(\frac{q}{\pi} r\cos\theta + \frac{q}{\pi} r\cos\theta - \frac{q}{\pi} r\theta\sin\theta\right)$$

$$= \frac{2q}{\pi r}\cos\theta \tag{3.9}$$

Similarly,

$$\sigma_\theta = 0 \tag{3.10}$$

and
$$\tau_{r\theta} = 0 \tag{3.11}$$

The stress function assumed in Eq. (3.5) will satisfy the *compatibility equation*

$$\left(\frac{\partial^2}{\partial r^2} + \frac{1}{r}\frac{\partial}{\partial r} + \frac{1}{r^2}\frac{\partial^2}{\partial \theta^2}\right)\left(\frac{\partial^2 \phi}{\partial r^2} + \frac{1}{r}\frac{\partial \phi}{\partial r} + \frac{1}{r^2}\frac{\partial^2 \phi}{\partial \theta^2}\right) = 0 \tag{3.12}$$

Also, it can be seen that the stresses obtained in Eqs. (3.9) to (3.11) satisfy the boundary conditions. For $\theta = 90°$ and $r > 0$, $\sigma_r = 0$; and, at $r = 0$, σ_r is theoretically equal to infinity, which signifies that plastic flow will occur locally. Note that σ_r and σ_θ are the major and minor principal stresses at point P.

Using the above expressions for σ_r, σ_θ, and $\tau_{r\theta}$, we can derive the stresses in the rectangular coordinate system (Fig. 3.6):

$$\sigma_z = \sigma_r \cos^2\theta + \sigma_\theta \sin^2\theta - 2\tau_{r\theta}\sin\theta\cos\theta = \frac{2q}{\pi r}\cos^3\theta$$

$$= \frac{2q}{\pi\sqrt{x^2+z^2}}\left(\frac{z}{\sqrt{x^2+z^2}}\right)^3 = \frac{2qz^3}{\pi(x^2+z^2)^2} \tag{3.13}$$

Similarly,

$$\sigma_x = \sigma_r \sin^2\theta + \sigma_\theta \cos^2\theta + 2\tau_{r\theta}\sin\theta = \frac{2qx^2 z}{\pi(x^2+z^2)^2} \tag{3.14}$$

and
$$\tau_{xz} = -\sigma_\theta \sin\theta\cos\theta + \sigma_r \sin\theta\cos\theta + \tau_{r\theta}(\cos^2\theta - \sin^2\theta)$$

$$= \frac{2qxz^2}{\pi(x^2+z^2)^2} \tag{3.15}$$

For the plane strain case,

$$\sigma_y = \nu(\sigma_x + \sigma_z)$$

The values for σ_x, σ_z, and τ_{xz} in a nondimensional form are given in Table 3.1.

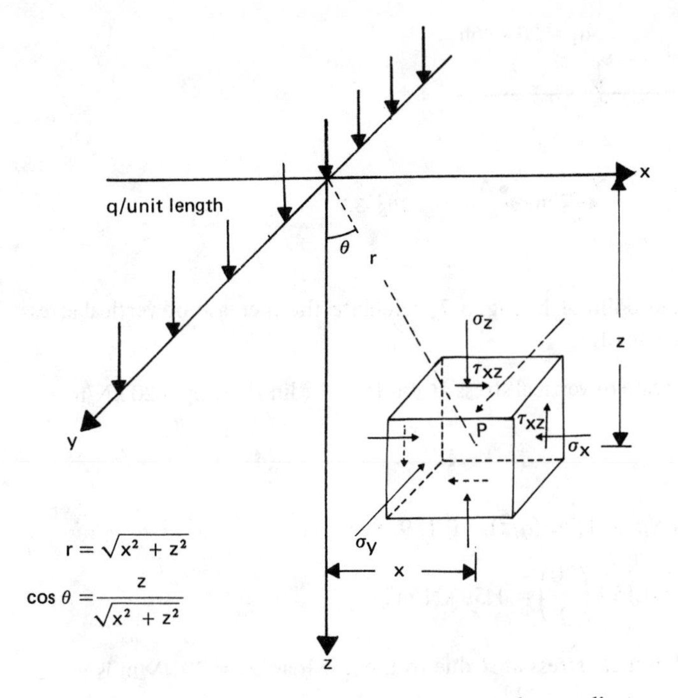

Fig. 3.6 Stresses due to a vertical line load in rectangular coordinates.

Table 3.1 Values of $\sigma_z/(q/z)$, $\sigma_x/(q/z)$, and $\tau_{xz}/(q/z)$ [Eqs. (3.13) to (3.15)]

x/z	$\sigma_z/(q/z)$	$\sigma_x/(q/z)$	$\tau_{xz}/(q/z)$
0	0.637	0	0
0.1	0.624	0.006	0.062
0.2	0.589	0.024	0.118
0.3	0.536	0.048	0.161
0.4	0.473	0.076	0.189
0.5	0.407	0.102	0.204
0.6	0.344	0.124	0.207
0.7	0.287	0.141	0.201
0.8	0.237	0.151	0.189
0.9	0.194	0.157	0.175
1.0	0.159	0.159	0.159
1.5	0.060	0.136	0.090
2.0	0.025	0.102	0.051
3.0	0.006	0.057	0.019

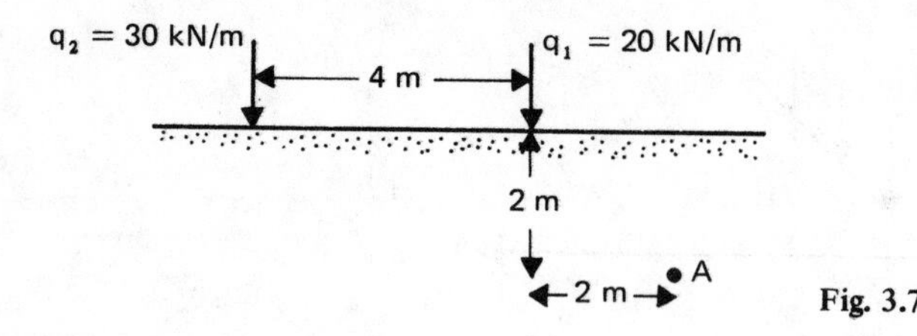

Fig. 3.7

Example 3.1 For the point A in Fig. 3.7, calculate the increase of vertical stress σ_z due to the two line loads.

SOLUTION The increase of vertical stress at A due to the line load $q_1 = 20$ kN/m is

$$\frac{x}{z} = \frac{2\,\text{m}}{2\,\text{m}} = 1$$

From Table 3.1, for $x/z = 1$, $\sigma_z/(q/z) = 0.159$. So

$$\sigma_{z(1)} = 0.159\left(\frac{q_1}{z}\right) = 0.159\left(\frac{20}{2}\right) = 1.59\ \text{kN/m}^2$$

The increase of vertical stress at A due to the line load $q_2 = 30$ kN/m is

$$\frac{x}{z} = \frac{6\,\text{m}}{2\,\text{m}} = 3$$

From Table 3.1, for $x/z = 3$, $\sigma_z/(q/z) = 0.006$. Thus,

$$\sigma_{z(2)} = 0.006\left(\frac{q_2}{z}\right) = 0.006\left(\frac{30}{2}\right) = 0.09\ \text{kN/m}^2$$

So, the total increase of vertical stress is

$$\sigma_z = \sigma_{z(1)} + \sigma_{z(2)} = 1.59 + 0.09 = \mathit{1.68\ kN/m^2}$$

3.1.4 Stresses due to a Horizontal Line Load on the Surface of a Semi-infinite Mass

The stresses due to a horizontal line load of q per unit length (Fig. 3.8) can be evaluated by a stress function of the form

$$\phi = \frac{q}{\pi} r\theta \cos\theta \tag{3.16}$$

Proceeding in a similar manner to that shown in Sec. 3.1.3 for the case of vertical line load, we obtain

$$\sigma_r = \frac{2q}{\pi r} \sin\theta \tag{3.17}$$

$$\sigma_\theta = 0 \tag{3.18}$$

$$\tau_{r\theta} = 0 \tag{3.19}$$

In the rectangular coordinate system,

$$\sigma_z = \frac{2q}{\pi} \frac{xz^2}{(x^2+z^2)^2} \tag{3.20}$$

$$\sigma_x = \frac{2q}{\pi} \frac{x^3}{(x^2+z^2)^2} \tag{3.21}$$

$$\tau_{xz} = \frac{2q}{\pi} \frac{x^2 z}{(x^2+z^2)^2} \tag{3.22}$$

For the plane strain case, $\sigma_y = \nu(\sigma_x + \sigma_z)$.

Some values of σ_x, σ_z, and τ_{xz} in a nondimensional form are given in Table 3.2.

3.1.5 Stresses due to a Line Load Inside a Semi-infinite Mass

Vertical line load. Melan (1932) gave the solution of stresses at a point P due to a vertical line load of q per unit length applied inside a semi-infinite mass (at point A, Fig. 3.9). The final equations are given below:

$$\sigma_z = \frac{q}{\pi}\left(\frac{1}{2(1-\nu)}\left\{\frac{(z-d)^3}{r_1^4} + \frac{(z+d)\,[(z+d)^2+2dz]}{r_2^4} - \frac{8dz(d+z)x^2}{r_2^6}\right\} + \frac{1-2\nu}{4(1-\nu)}\left(\frac{z-d}{r_1^2} + \frac{3z+d}{r_2^4} - \frac{4zx^2}{r_2^4}\right)\right) \tag{3.23}$$

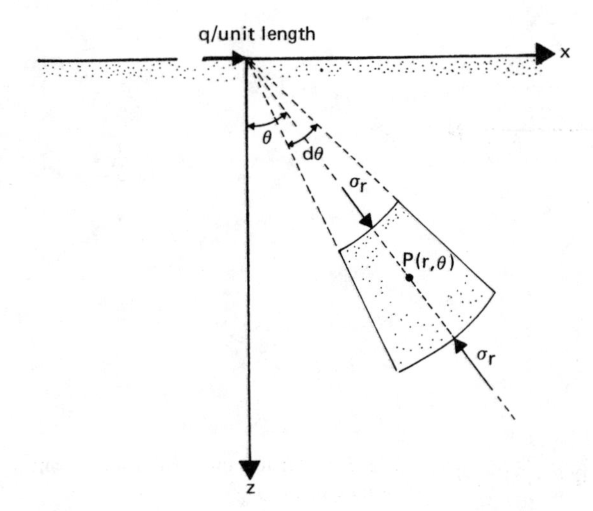

Fig. 3.8 Horizontal line load on the surface of a semi-infinite mass.

Table 3.2 Values of $\sigma_z/(q/z)$, $\sigma_x/(q/z)$, and $\tau_{xz}/(q/z)$ [Eqs. (3.20) to (3.22)]

x/z	$\sigma_z/(q/z)$	$\sigma_x/(q/z)$	$\tau_{xz}/(q/z)$
0	0	0	0
0.1	0.062	0.0006	0.006
0.2	0.118	0.0049	0.024
0.3	0.161	0.0145	0.048
0.4	0.189	0.0303	0.076
0.5	0.204	0.0509	0.102
0.6	0.207	0.0743	0.124
0.7	0.201	0.0984	0.141
0.8	0.189	0.1212	0.151
0.9	0.175	0.1417	0.157
1.0	0.159	0.1591	0.159
1.5	0.090	0.2034	0.136
2.0	0.051	0.2037	0.102
3.0	0.019	0.1719	0.057

$$\sigma_x = \frac{q}{\pi}\left\{\frac{1}{2(1-\nu)}\left[\frac{(z-d)x^2}{r_1^4} + \frac{(z+d)(x^2+2d^2)-2dx^2}{r_2^4} + \frac{8dz(d+z)x^2}{r_2^6}\right]\right.$$
$$\left. + \frac{1-2\nu}{4(1-\nu)}\left(\frac{d-z}{r_1^2} + \frac{z+3d}{r_2^2} + \frac{4zx^2}{r_2^4}\right)\right\} \tag{3.24}$$

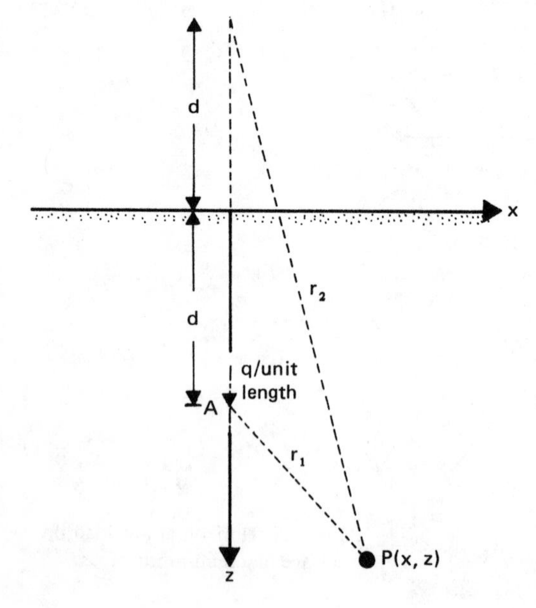

Fig. 3.9 Vertical line load inside a semi-infinite mass.

$$\tau_{xz} = \frac{qx}{\pi}\left\{\frac{1}{2(1-\nu)}\left[\frac{(z-d)^2}{r_1^4} + \frac{z^2 - 2dz - d^2}{r_2^4} + \frac{8dz(d+z)^2}{r_2^6}\right]\right.$$

$$\left. + \frac{1-2\nu}{4(1-\nu)}\left[\frac{1}{r_1^2} - \frac{1}{r_2^2} + \frac{4z(d+z)}{r_2^4}\right]\right\} \tag{3.25}$$

Horizontal line load. For a horizontal line load of intensity q per unit length (Fig. 3.10), Melan's solutions for stresses may be given as follows:

$$\sigma_z = \frac{qx}{\pi}\left\{\frac{1}{2(1-\nu)}\left[\frac{(z-d)^2}{r_1^4} - \frac{d^2 - z^2 + 6dz}{r_2^4} + \frac{8dz\,x^2}{r_2^6}\right]\right.$$

$$\left. - \frac{1-2\nu}{4(1-\nu)}\left[\frac{1}{r_1^2} - \frac{1}{r_2^2} - \frac{4z(d+z)}{r_2^4}\right]\right\} \tag{3.26}$$

$$\sigma_x = \frac{qx}{\pi}\left\{\frac{1}{2(1-\nu)}\left[\frac{x^2}{r_1^4} + \frac{x^2 + 8dz + 6d^2}{r_2^4} + \frac{8dz(d+z)^2}{r_2^6}\right]\right.$$

$$\left. + \frac{1-2\nu}{4(1-\nu)}\left[\frac{1}{r_1^2} + \frac{3}{r_2^2} - \frac{4z(d+z)}{r_2^4}\right]\right\} \tag{3.27}$$

$$\tau_{xz} = \frac{q}{\pi}\left\{\frac{1}{2(1-\nu)}\left[\frac{(z-d)x^2}{r_1^4} + \frac{(2dz + x^2)(d+z)}{r_2^4} - \frac{8dz(d+z)x^2}{r_2^6}\right]\right.$$

$$\left. + \frac{1-2\nu}{4(1-\nu)}\left[\frac{z-d}{r_1^2} + \frac{3z+d}{r_2^2} - \frac{4(d+z)^2}{r_2^4}\right]\right\} \tag{3.28}$$

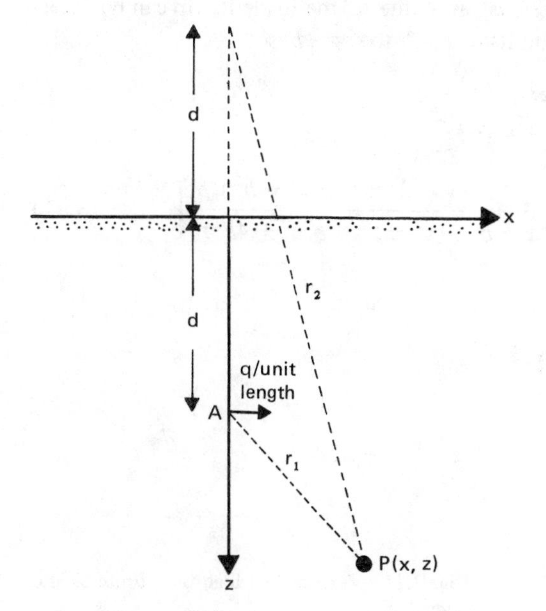

Fig. 3.10 Horizontal line load inside a semi-infinite mass.

3.1.6 Stresses due to a Vertical Line Load on an Elastic Soil Layer Underlain by a Rigid Rough Base

Figure 3.11 shows an elastic soil layer of thickness h and underlain by a rigid rough base. A line load q per unit length is acting on the surface of the soil layer. The vertical stress at a point can be expressed as (Poulos, 1966)

$$\sigma_z = \frac{q}{\pi h} I_z \tag{3.29}$$

where σ_z is the vertical stress at a point x, z, and I_z is the influence factor, which is a function of z/h, x/h, and ν. The values of I_z for various values of $z/h, x/h$, and ν are given in Table 3.3.

3.1.7 Uniform Vertical Loading on an Infinite Strip on the Surface of a Semi-infinite Mass

Figure 3.12 shows the case where a uniform vertical load of q per unit area is acting on a flexible infinite strip on the surface of a semi-infinite elastic mass. To obtain the stresses at a point $P(x, z)$, we can consider an elementary strip of width ds located at a distance s from the center line of the load. The load per unit length of this elementary strip is $q \cdot ds$, and it can be approximated as a line load.

The increase of vertical stress, σ_z, at P due to the elementary strip loading can be obtained by substituting $x - s$ for x and $q \cdot ds$ for q in Eq. (3.13), or

$$d\sigma_z = \frac{2q\,ds}{\pi} \frac{z^3}{[(x-s)^2 + z^2]^2} \tag{3.30}$$

The total increase of vertical stress, σ_z, at P due to the loaded strip can be determined by integrating Eq. (3.30) with limits of $s = b$ to $s = -b$; so,

$$\sigma_z = \int d\sigma_z = \frac{2q}{\pi}\int_{-b}^{+b} \frac{z^3}{[(x-s)^2 + z^2]^2}\, ds$$

$$= \frac{q}{\pi}\left[\tan^{-1}\frac{z}{x-b} - \tan^{-1}\frac{z}{x+b} - \frac{2bz(x^2 - z^2 - b^2)}{(x^2 + z^2 - b^2)^2 + 4b^2z^2}\right] \tag{3.31}$$

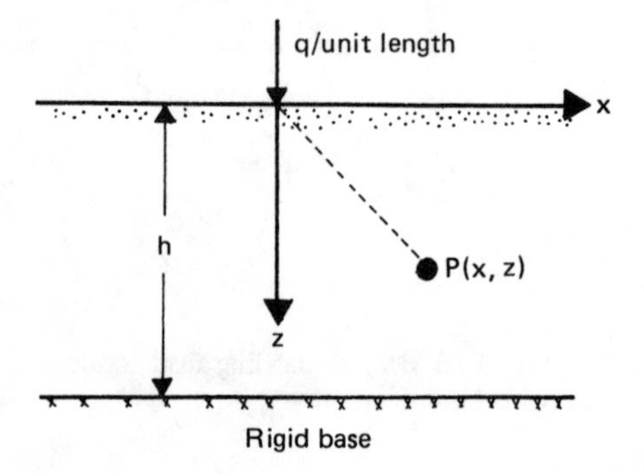

Fig. 3.11 Vertical line load on a finite elastic layer.

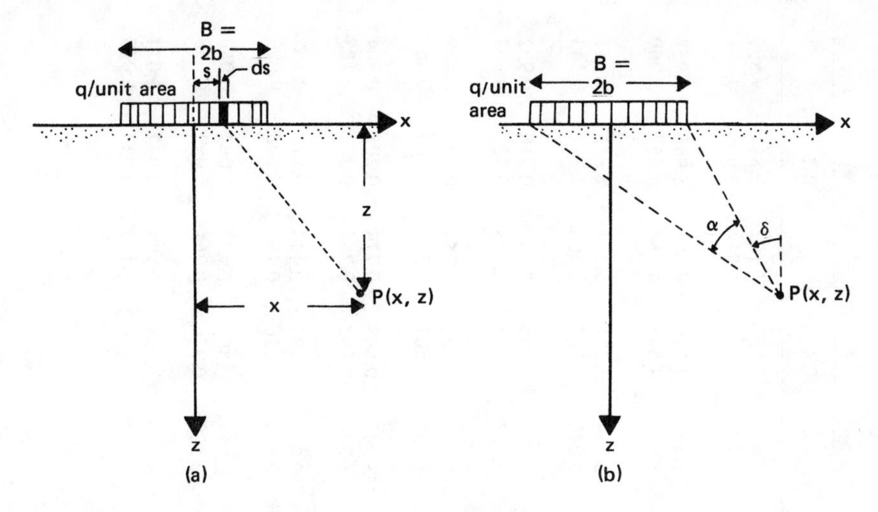

Fig. 3.12 Uniform vertical loading on an infinite strip.

In a similar manner, referring to Eqs. (3.14) and (3.15),

$$\sigma_x = \int d\sigma_x = \frac{2q}{\pi} \int_{-b}^{+b} \frac{(x-s)^2 z}{[(x-s)^2 + z^2]^2} \, ds$$

$$= \frac{q}{\pi} \left[\tan^{-1} \frac{z}{x-b} - \tan^{-1} \frac{z}{x+b} + \frac{2bz(x^2 - z^2 - b^2)}{(x^2 + z^2 - b^2)^2 + 4b^2 z^2} \right] \tag{3.32}$$

$$\tau_{xz} = \frac{2q}{\pi} \int_{-b}^{+b} \frac{(x-s)z^2}{[(x-s)^2 + z^2]^2} \, ds = \frac{4bqxz^2}{\pi (x^2 + z^2 - b^2)^2 + 4b^2 z^2} \tag{3.33}$$

Some nondimensional values of σ_z/q, σ_x/q, and τ_{xz}/q are given in Table 3.4.

The expressions for σ_z, σ_x, and τ_{xz} given in Eqs. (3.31) to (3.33) can be presented in a simplified form:

$$\sigma_z = \frac{q}{\pi} [\alpha + \sin \alpha \cos (\alpha + 2\delta)] \tag{3.34}$$

$$\sigma_x = \frac{q}{\pi} [\alpha - \sin \alpha \cos (\alpha + 2\delta)] \tag{3.35}$$

$$\tau_{xz} = \frac{q}{\pi} [\sin \alpha \sin (\alpha + 2\delta)] \tag{3.36}$$

where α and δ are the angles shown in Fig. 3.12*b*.

3.1.8 Uniform Horizontal Loading on an Infinite Strip on the Surface of a Semi-infinite Mass

If a uniform horizontal load is applied on an infinite strip of width $2b$ as shown in Fig. 3.13, the stresses at a point inside the semi-infinite mass can be determined by

Table 3.3 Influence values I_z for vertical stress σ_z due to a line load [Eq. (3.29)]*

	z/h													
x/h	1.0		0.9		0.8		0.7		0.6		0.4		0.2	
0	2.634	2.566	2.787	2.759	2.980	2.975	3.249	3.256	3.641	3.653	5.157	5.127	9.891	9.899
	2.539	2.580	2.787	2.865	3.025	3.113	3.312	3.391	3.704	3.772	5.201	5.234	9.905	9.911
0.1	2.573	2.503	2.713	2.682	2.885	2.877	3.118	3.122	3.443	3.452	4.516	4.505	5.946	5.952
	2.471	2.508	2.702	2.774	2.920	3.001	3.171	3.243	3.498	3.562	4.585	4.586	5.957	5.960
0.2	2.400	2.331	2.505	2.471	2.627	2.614	2.774	2.773	2.948	2.954	3.251	3.217	2.341	2.347
	2.291	2.312	2.477	2.528	2.642	2.703	2.810	2.876	2.990	3.040	3.283	3.305	2.351	2.343
0.3	2.144	2.075	2.203	2.162	2.261	2.240	2.311	2.302	2.335	2.333	2.099	2.081	0.918	0.922
	2.021	2.023	2.145	2.170	2.245	2.279	2.318	2.349	2.352	2.383	2.117	2.128	0.923	0.907
0.4	1.840	1.774	1.855	1.810	1.857	1.828	1.830	1.812	1.751	1.741	1.301	1.257	0.407	0.409
	1.711	1.689	1.773	1.766	1.810	1.815	1.808	1.815	1.744	1.755	1.307	1.307	0.409	0.389
0.5	1.525	1.462	1.504	1.455	1.465	1.429	1.391	1.365	1.265	1.247	0.803	0.774	0.205	0.204
	1.389	1.350	1.397	1.365	1.387	1.365	1.338	1.320	1.231	1.223	0.792	0.786	0.201	0.187
0.6	1.223	1.168	1.179	1.130	1.117	1.077	1.024	0.993	0.889	0.867	0.497	0.443	0.110	0.107
	1.092	1.039	1.061	1.008	1.021	0.978	0.951	0.916	0.837	0.814	0.476	0.466	0.105	0.099
0.7	0.954	0.906	0.898	0.850	0.827	0.785	0.733	0.698	0.611	0.584	0.308	0.270	0.062	0.056
	0.830	0.771	0.774	0.711	0.718	0.661	0.644	0.594	0.543	0.509	0.276	0.264	0.053	0.058
0.8	0.721	0.683	0.661	0.619	0.592	0.553	0.508	0.473	0.408	0.379	0.185	0.124	0.032	0.024
	0.615	0.553	0.547	0.478	0.485	0.421	0.416	0.360	0.335	0.294	0.150	0.137	0.023	0.035

0.9	0.536 0.447	0.507 0.390	0.479 0.375	0.443 0.309	0.417 0.314	0.380 0.251	0.345 0.254	0.311 0.195	0.267 0.192	0.237 0.149	0.108 0.070	0.066 0.057	0.015 0.005	0.006 0.020
1.0	0.357 0.294	0.338 0.241	0.306 0.224	0.279 0.162	0.254 0.167	0.224 0.106	0.199 0.117	0.169 0.062	0.144 0.075	0.116 0.032	0.045 0.010	−0.020 −0.005	−0.000 −0.007	−0.011 0.005
1.25	0.121 0.111	0.123 0.079	0.091 0.059	0.082 0.020	0.062 0.020	0.047 −0.023	0.035 −0.010	0.018 −0.055	0.013 −0.028	−0.005 −0.064	−0.016 −0.037	−0.057 −0.057	−0.014 −0.015	−0.024 −0.021
1.5	0.010 0.043	0.027 0.026	−0.005 0.011	0.003 −0.013	−0.018 −0.014	−0.016 −0.042	−0.028 −0.031	−0.031 −0.064	−0.033 −0.040	−0.040 −0.070	−0.035 −0.035	−0.097 −0.061	−0.019 −0.011	−0.026 −0.034
1.75	−0.030 0.024	−0.006 0.018	−0.036 0.005	−0.019 −0.006	−0.040 −0.010	−0.029 −0.027	−0.042 −0.021	−0.036 −0.046	−0.039 −0.026	−0.038 −0.049	−0.032 −0.021	−0.064 −0.047	−0.017 −0.005	−0.019 −0.034
2.0	−0.042 0.025	−0.014 0.024	−0.042 0.015	−0.019 0.010	−0.041 0.006	−0.022 −0.003	−0.039 −0.001	−0.026 −0.014	−0.035 −0.005	−0.026 −0.020	−0.025 −0.005	−0.131 −0.026	−0.013 0.002	−0.011 −0.025
2.5	−0.025 0.027	−0.003 0.034	−0.022 0.022	−0.004 0.027	−0.019 0.018	−0.004 0.020	−0.016 0.014	−0.004 0.001	−0.013 0.011	−0.004 0.009	−0.006 0.007	−0.042 −0.000	0.000 0.006	0.002 −0.006
3.0	−0.010 0.022	0.004 0.032	−0.007 0.019	0.004 0.028	−0.005 0.017	0.004 0.023	−0.003 0.014	0.004 0.015	−0.000 0.012	0.004 0.015	0.003 0.008	0.005 0.008	0.006 0.005	0.006 0.003
4.0	0.006 0.008	0.005 0.019	0.006 0.006	0.005 0.017	0.007 0.005	0.004 0.015	0.007 0.004	0.005 0.013	0.008 0.003	0.005 0.011	0.009 0.001	0.005 0.008	0.009 0.000	0.004 0.007
6.0	0.005 0.000	0.000 0.006	0.005 −0.000	0.000 0.005	0.005 −0.000	0.000 0.004	0.005 −0.001	0.000 0.004	0.005 −0.001	0.000 0.004	0.005 −0.001	0.000 0.003	0.005 −0.001	0.000 0.002
8.0	0.002 −0.001	−0.000 0.001	0.002 −0.001	−0.000 0.000	0.002 −0.001	−0.001 0.000	0.002 −0.001	−0.001 0.000	0.002 −0.001	−0.001 0.000	0.002 −0.001	−0.001 −0.000	0.002 −0.001	−0.001 −0.001

Source: H. G. Poulos and E. H. Davis, "Elastic Solutions for Soil and Rock Mechanics," Wiley, New York, 1974.

*The four values given under each entry are for different Poisson's ratios as follows: top left, $\nu = 0$; top right, $\nu = 0.2$; bottom left, $\nu = 0.4$; bottom right, $\nu = 0.5$.

Table 3.4 Values of σ_z/q, σ_x/q, and τ_{xz} for vertical strip loading [Eqs. (3.31) to (3.33)]

(1) x/b	(2) z/b	(3) σ_z/q	(4) σ_x/q	(5) τ_{xz}/q
0	0	1.000	1.000	0
	0.5	0.9594	0.4498	0
	1.0	0.8183	0.1817	0
	1.5	0.6678	0.0803	0
	2.0	0.5508	0.0410	0
	2.5	0.4617	0.0228	0
0.5	0	1.000	1.000	0
	0.25	0.9787	0.6214	0.0522
	0.5	0.9028	0.3920	0.1274
	1.0	0.7352	0.1863	0.1590
	1.5	0.6078	0.0994	0.1275
	2.0	0.5107	0.0542	0.0959
	2.5	0.4372	0.0334	0.0721
1.0	0.25	0.4996	0.4208	0.3134
	0.5	0.4969	0.3472	0.2996
	1.0	0.4797	0.2250	0.2546
	1.5	0.4480	0.1424	0.2037
	2.0	0.4095	0.0908	0.1592
	2.5	0.3701	0.0595	0.1243
1.5	0.25	0.0177	0.2079	0.0606
	0.5	0.0892	0.2850	0.1466
	1.0	0.2488	0.2137	0.2101
	1.5	0.2704	0.1807	0.2022
	2.0	0.2876	0.1268	0.1754
	2.5	0.2851	0.0892	0.1469
2.0	0.25	0.0027	0.0987	0.0164
	0.5	0.0194	0.1714	0.0552
	1.0	0.0776	0.2021	0.1305
	1.5	0.1458	0.1847	0.1568
	2.0	0.1847	0.1456	0.1567
	2.5	0.2045	0.1256	0.1442
2.5	0.5	0.0068	0.1104	0.0254
	1.0	0.0357	0.1615	0.0739
	1.5	0.0771	0.1645	0.1096
	2.0	0.1139	0.1447	0.1258
	2.5	0.1409	0.1205	0.1266

After L. Jurgenson, The Application of Theories of Elasticity and Plasticity to Foundation Problems, *Contribution to Soil Mechanics, 1925–1940,* Boston Society of Civil Engineers, Boston, 1934.

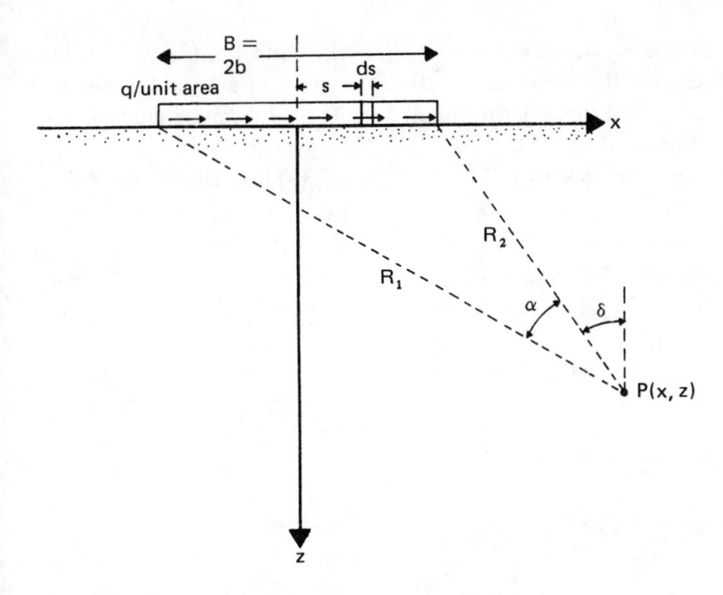

Fig. 3.13 Uniform horizontal loading on an infinite strip.

using a similar procedure of superposition as outlined in Sec. 3.1.7 for vertical loading. For an elementary strip of width ds, the load per unit length is $q \cdot ds$. Approximating this as a line load, we can substitute $q \cdot ds$ for q and $x - s$ for x in Eqs. (3.20) to (3.22). Thus,

$$\sigma_z = \int d\sigma_z = \frac{2q}{\pi} \int_{s=-b}^{s=+b} \frac{(x-s)z^2}{[(x-s)^2 + z^2]^2} ds$$

$$= \frac{4bqxz^2}{\pi [(x^2 + z^2 - b^2)^2 + 4b^2 z^2]} \tag{3.37}$$

$$\sigma_x = \int d\sigma_x = \frac{2q}{\pi} \int_{s=-b}^{s=+b} \frac{(x-s)^3 \, ds}{[(x-s)^2 + z^2]^2}$$

$$= \frac{q}{\pi} \left[2.303 \log \frac{(x+b)^2 + z^2}{(x-b)^2 + z^2} - \frac{4bxz^2}{(x^2 + z^2 - b^2)^2 + 4b^2 z^2} \right] \tag{3.38}$$

$$\tau_{xz} = \int d\tau_{xz} = \frac{2q}{\pi} \int_{s=-b}^{s=+b} \frac{(x-s)^2 z}{[(x-s)^2 + z^2]^2} ds$$

$$= \frac{q}{\pi} \left[\tan^{-1} \frac{z}{x-b} - \tan^{-1} \frac{z}{x+b} + \frac{2bz(x^2 - z^2 - b^2)}{(x^2 + z^2 - b^2)^2 + 4b^2 z^2} \right] \tag{3.39}$$

Note that Eq. (3.37) is in the same form as Eq. (3.33), and Eq. (3.39) is in the same form as Eq. (3.32). So the nondimensional values given in col. 5 of Table 3.4 can be used for Eq. (3.37), and those given in col. 4 of Table 3.4 can be used for Eq. (3.39). Some nondimensional values for σ_x [Eq. (3.38)] are given in Table 3.5.

The expressions for stresses given by Eqs. (3.37) to (3.39) may also be simplified as follows:

$$\sigma_z = \frac{q}{\pi} [\sin \alpha \sin (\alpha + 2\delta)] \tag{3.40}$$

$$\sigma_x = \frac{q}{\pi} \left[2.303 \log \frac{R_1^2}{R_2^2} - \sin \alpha \sin (\alpha + 2\delta)\right] \tag{3.41}$$

$$\tau_{xz} = \frac{q}{\pi} [\alpha - \sin \alpha \cos (\alpha + 2\delta)] \tag{3.42}$$

where R_1, R_2, α, and δ are as defined in Fig. 3.13.

3.1.9 Linearly Increasing Vertical Loading on an Infinite Strip on the Surface of a Semi-infinite Mass

Figure 3.14 shows a vertical loading on an infinite strip of width $2b$. The load increases from zero to q across the width. For an elementary strip of width ds, the load per unit length can be given as $(q/2b)s \cdot ds$. Approximating this as a line load, we can substitute $(q/2b)s \cdot ds$ for q and $x - s$ for x in Eqs. (3.13) to (3.15) to determine the stresses at a point (x, z) inside the semi-infinite mass. Thus,

$$\sigma_z = \int d\sigma_z = \left(\frac{1}{2b}\right)\left(\frac{2q}{\pi}\right) \int_{s=0}^{s=2b} \frac{z^3 s \, ds}{[(x-s)^2 + z^2]^2}$$

$$= \frac{q}{2\pi}\left(\frac{x}{b}\alpha - \sin 2\delta\right) \tag{3.43}$$

$$\sigma_x = \int d\sigma_x = \left(\frac{1}{2b}\right)\left(\frac{2q}{\pi}\right) \int_0^{2b} \frac{(x-s)^2 zs \, ds}{[(x-s)^2 + z^2]^2}$$

$$= \frac{q}{2\pi}\left(\frac{x}{b}\alpha - 2.303 \frac{z}{b} \log \frac{R_1^2}{R_2^2} + \sin 2\delta\right) \tag{3.44}$$

$$\tau_{xz} = \int d\tau_{xz} = \left(\frac{1}{2b}\right)\left(\frac{2q}{\pi}\right) \int_0^{2b} \frac{(x-s)z^2 \, ds}{[(x-s)^2 + z^2]^2}$$

$$= \frac{q}{2\pi}\left(1 + \cos 2\delta - \frac{z}{b}\alpha\right) \tag{3.45}$$

Nondimensional values of σ_z [Eq. (3.43)] are given in Table 3.6.

Table 3.5 Values of σ_x/q [Eq. (3.38)]

	z/b									
x/b	0	0.1	0.2	0.3	0.4	0.5	0.6	0.7	1.0	1.5
0	0	0	0	0	0	0	0	0	0	0
0.1	0.1287	0.1252	0.1180	0.1073	0.0946	0.0814	0.0687	0.0572	0.0317	0.0121
0.25	0.3253	0.3181	0.2982	0.2693	0.2357	0.2014	0.1692	0.2404	0.0780	0.0301
0.5	0.6995	0.6776	0.6195	0.5421	0.4608	0.3851	0.3188	0.2629	0.1475	0.0598
0.75	1.2390	1.1496	0.9655	0.7855	0.6379	0.5210	0.4283	0.3541	0.2058	0.0899
1.0	–	1.5908	1.1541	0.9037	0.7312	0.6024	0.5020	0.4217	0.2577	0.1215
1.25	1.3990	1.3091	1.1223	0.9384	0.7856	0.6623	0.5624	0.4804	0.3074	0.1548
1.5	1.0248	1.0011	0.9377	0.8517	0.7591	0.6697	0.5881	0.5157	0.3489	0.1874
1.75	0.8273	0.8170	0.7876	0.7437	0.6904	0.6328	0.5749	0.5190	0.3750	0.2162
2.0	0.6995	0.6939	0.6776	0.6521	0.6195	0.5821	0.5421	0.5012	0.3851	0.2386
2.5	0.5395	0.5372	0.5304	0.5194	0.5047	0.4869	0.4667	0.4446	0.3735	0.2627
3.0	0.4414	0.4402	0.4366	0.4303	0.4229	0.4132	0.4017	0.3889	0.3447	0.2658
4.0	0.3253	0.3248	0.3235	0.3212	0.3181	0.3143	0.3096	0.3042	0.2846	0.2443
5.0	0.2582	0.2580	0.2573	0.2562	0.2547	0.2527	0.2504	0.2477	0.2375	0.2151
6.0	0.2142	0.2141	0.2137	0.2131	0.2123	0.2112	0.2098	0.2083	0.2023	0.1888
	2.0	2.5	3.0	3.5	4.0	4.5	5.0	5.5	6.0	
0	0	0	0	0	0	0	0	0	0	
0.1	0.0051	0.0024	0.0013	0.0007	0.0004	0.0003	0.0002	0.00013	0.0001	
0.25	0.0129	0.0062	0.0033	0.00019	0.0012	0.0007	0.0005	0.00034	0.00025	
0.5	0.0269	0.0134	0.0073	0.0042	0.0026	0.0017	0.00114	0.00079	0.00057	
0.75	0.0429	0.0223	0.0124	0.0074	0.0046	0.0030	0.00205	0.00144	0.00104	
1.0	0.0615	0.0333	0.0191	0.0116	0.0074	0.0049	0.00335	0.00236	0.00171	
1.25	0.0825	0.0464	0.0275	0.0170	0.0110	0.0074	0.00510	0.00363	0.00265	
1.5	0.1049	0.0613	0.0373	0.0236	0.0155	0.0105	0.00736	0.00528	0.00387	
1.75	0.1271	0.0770	0.0483	0.0313	0.0209	0.0144	0.01013	0.00732	0.00541	
2.0	0.1475	0.0928	0.0598	0.0396	0.0269	0.0188	0.01339	0.00976	0.00727	
2.5	0.1788	0.1211	0.0826	0.0572	0.0403	0.0289	0.02112	0.01569	0.01185	
3.0	0.1962	0.1421	0.1024	0.0741	0.0541	0.0400	0.02993	0.02269	0.01742	
4.0	0.2014	0.1616	0.1276	0.0999	0.0780	0.0601	0.04789	0.03781	0.03006	
5.0	0.1888	0.1618	0.1362	0.1132	0.0934	0.0767	0.06285	0.05156	0.04239	
6.0	0.1712	0.1538	0.1352	0.1173	0.1008	0.0861	0.07320	0.06207	0.05259	

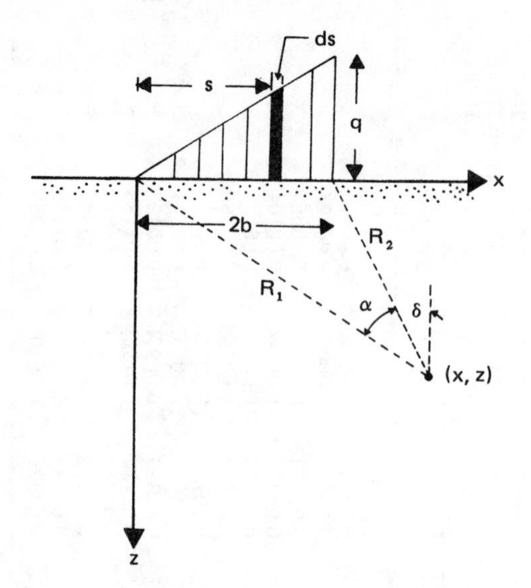

Fig. 3.14 Linearly increasing vertical loading on an infinite strip.

3.1.10 Vertical Stress in a Semi-infinite Mass due to Embankment Loading

In several practical cases, it is necessary to determine the increase of vertical stress in a soil mass due to embankment loading. This can be done by the method of superposition as shown in Fig. 3.15 and described below.

The stress at A due to the embankment loading as shown in Fig. 3.15a is equal to the stress at A due to the loading shown in Fig. 3.15b minus the stress at A due to the loading shown in Fig. 3.15c.

Referring to Eq. (3.43), the vertical stress at A due to the loading shown in Fig. 3.15b is

$$\frac{q + (b/a)q}{\pi}(\alpha_1 + \alpha_2)$$

Table 3.6 Values of σ_z/q [Eq. (3.43)]

	z/b								
x/b	0	0.5	1.0	1.5	2.0	2.5	3.0	4.0	5.0
−3	0	0.0003	0.0018	0.00054	0.0107	0.0170	0.0235	0.0347	0.0422
−2	0	0.0008	0.0053	0.0140	0.0249	0.0356	0.0448	0.0567	0.0616
−1	0	0.0041	0.0217	0.0447	0.0643	0.0777	0.0854	0.0894	0.0858
0	0	0.0748	0.1273	0.1528	0.1592	0.1553	0.1469	0.1273	0.1098
1	0.5	0.4797	0.4092	0.3341	0.2749	0.2309	0.1979	0.1735	0.1241
2	0.5	0.4220	0.3524	0.2952	0.2500	0.2148	0.1872	0.1476	0.1211
3	0	0.0152	0.0622	0.1010	0.1206	0.1268	0.1258	0.1154	0.1026
4	0	0.0019	0.0119	0.0285	0.0457	0.0596	0.0691	0.0775	0.0776
5	0	0.0005	0.0035	0.0097	0.0182	0.0274	0.0358	0.0482	0.0546

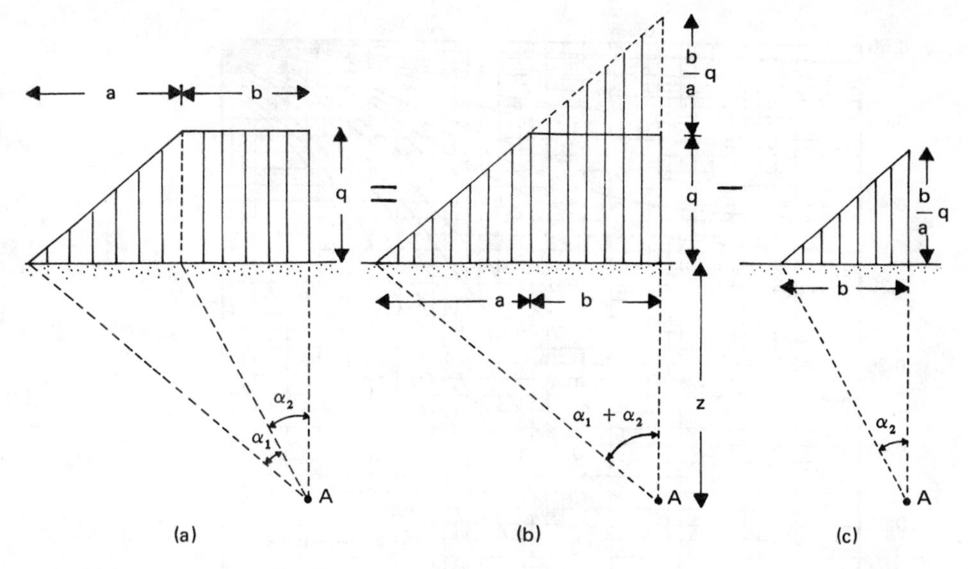

Fig. 3.15 Vertical stress due to embankment loading.

Note that in this case we have substituted $a + b$ for x, $(a + b)/2$ for b, and $\alpha_1 + \alpha_2$ for α in Eq. (3.43).

Similarly, the stress at A due to the loading shown in Fig. 3.15c is

$$\left(\frac{b}{a}q\right)\frac{1}{\pi}\alpha_2$$

Thus the stress at A due to embankment loading (Fig. 3.15a) is

$$\sigma_z = \frac{q}{\pi}\left[\left(\frac{a+b}{a}\right)(\alpha_1 + \alpha_2) - \frac{b}{a}\alpha_2\right]$$

or $$\sigma_z = Iq \tag{3.46}$$

where I is the influence factor,

$$I = \frac{1}{\pi}\left[\left(\frac{a+b}{a}\right)(\alpha_1 + \alpha_2) - \frac{b}{a}\alpha_2\right] = \frac{1}{\pi}f\left(\frac{a}{z}, \frac{b}{z}\right)$$

The values of the influence factor for various a/z and b/z are given in Fig. 3.16. A typical problem demonstrating the use of Fig. 3.16 is given in Example 3.2.

Example 3.2 A 10-ft high embankment is to be constructed as shown in Fig. 3.17. If the unit weight of compacted soil is 120 lb/ft³, calculate the vertical stress due solely to the embankment at A, B, and C.

SOLUTION $q = \gamma H = 120 \times 10 = 1200\ \text{lb/ft}^2$

Vertical stress at A: Using the method of superposition and referring to Fig. 3.18a,

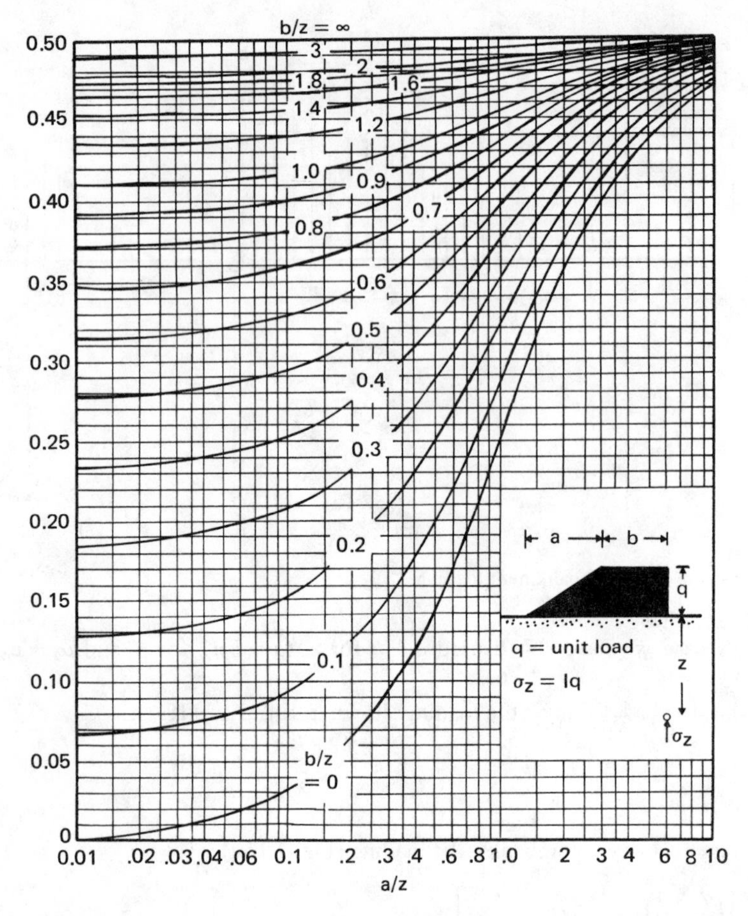

Fig. 3.16 Influence factor for embankment loading. *(After J. O. Osterberg, Influence Values for Vertical Stresses in Semi-infinite Mass Due to Embankment Loading,* Proc. 4th International Conference on Soil Mechanics and Foundation Engineering, *vol. 1, Butterworths, London, 1957.)*

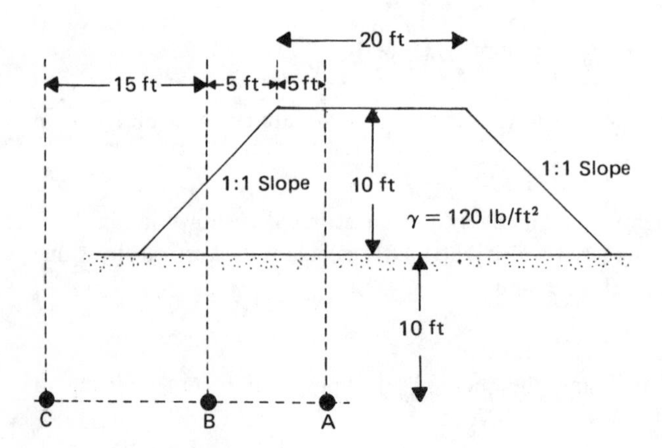

Fig. 3.17 (Not to scale.)

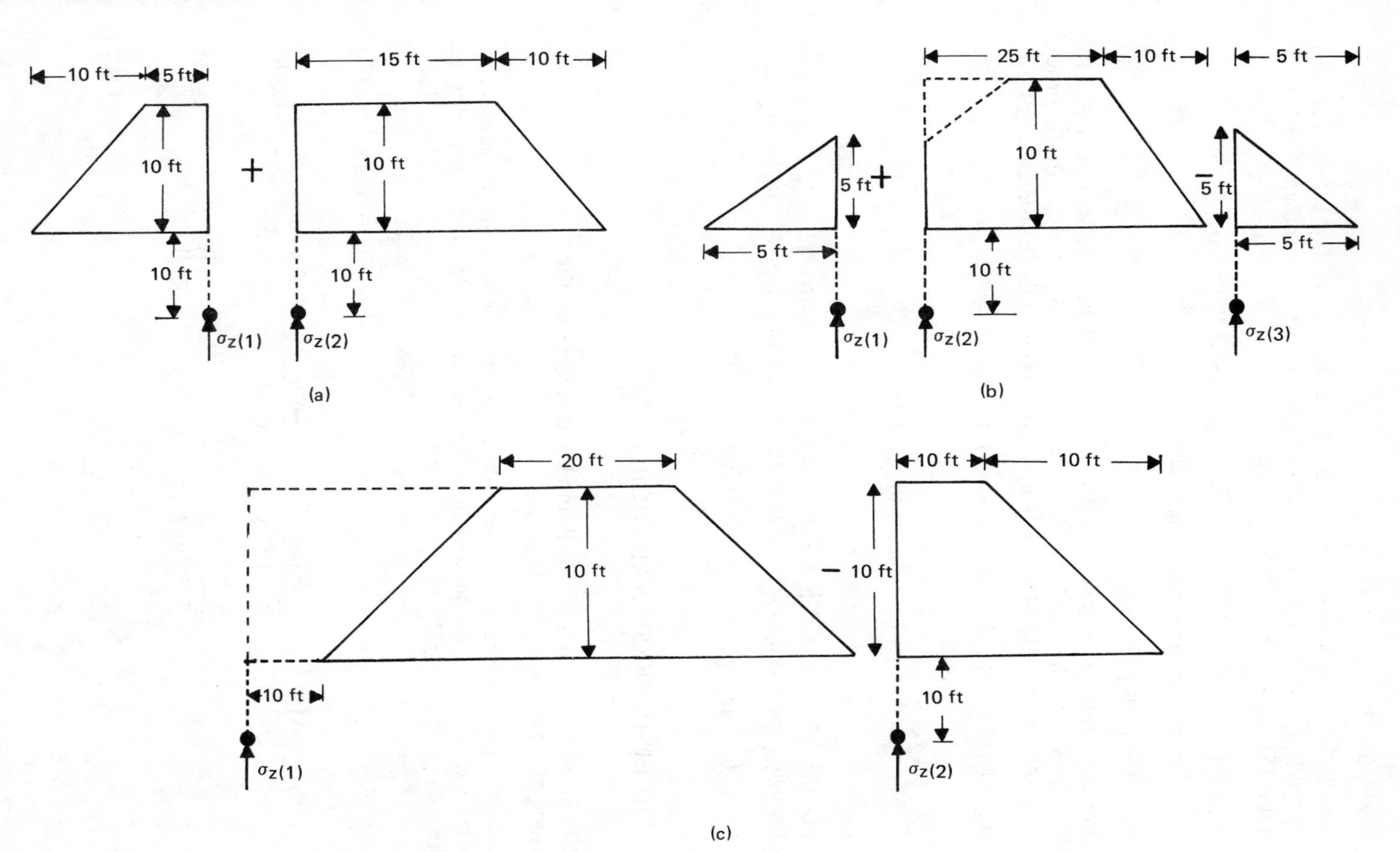

Fig. 3.18 (Not to scale.)

$$\sigma_{zA} = \sigma_{z(1)} + \sigma_{z(2)}$$

For the left-hand section, $b/z = 5/10 = 0.5$ and $a/z = 10/10 = 1$. From Fig. 3.16, $I_1 = 0.396$. For the right-hand section, $b/z = 15/10 = 1.5$ and $a/z = 10/10 = 1$. From Fig. 3.16, $I_2 = 0.477$. So,

$$\sigma_{z(A)} = (I_1 + I_2)q = (0.396 + 0.477)1200 = \mathit{1047.6\,lb/ft^2}$$

Vertical stress at B: Using Fig. 3.18*b*,

$$\sigma_{zB} = \sigma_{z(1)} + \sigma_{z(2)} - \sigma_{z(3)}$$

For the left-hand section, $b/z = 0/10 = 0$, $a/z = 5/10 = 0.5$. So, from Fig. 3.16, $I_1 = 0.14$. For the middle section, $b/z = 25/10 = 2.5$, $a/z = 10/10 = 1$. Hence, $I_2 = 0.493$. For the right-hand section, $I_3 = 0.14$ (same as the left-hand section). So,

$$\sigma_{zB} = I_1(120 \times 5) + I_2(120 \times 10) - I_3(120 \times 5) = 0.493(1200)$$
$$= \mathit{591.6\,lb/ft^2}$$

Vertical stress at C: Referring to Fig. 3.18*c*,

$$\sigma_{zC} = \sigma_{z(1)} - \sigma_{z(2)}$$

For the left-hand section, $b/z = 40/10 = 4$, $a/z = 10/10 = 1$. So $I_1 = 0.498$. For the right-hand section, $b/z = 10/10 = 1$, $a/z = 10/10 = 1$. So $I_2 = 0.456$. Hence,

$$\sigma_{zC} = (I_1 - I_2)q = (0.498 - 0.456)1200 = \mathit{50.4\,lb/ft^2}$$

3.2 THREE-DIMENSIONAL PROBLEMS

3.2.1 Stresses due to Vertical Point Load Acting on the Surface of a Semi-infinite Mass

Boussinesq (1883) solved the problem for stresses inside a semi-infinite mass due to a point load acting on the surface. In rectangular coordinates, the stresses may be expressed as follows (Fig. 3.19):

$$\sigma_z = \frac{3Qz^3}{2\pi R^5} \tag{3.47}$$

$$\sigma_x = \frac{3Q}{2\pi}\left\{\frac{x^2 z}{R^5} + \frac{1-2\nu}{3}\left[\frac{1}{R(R+z)} - \frac{(2R+z)x^2}{R^3(R+z)^2} - \frac{z}{R^3}\right]\right\} \tag{3.48}$$

$$\sigma_y = \frac{3Q}{2\pi}\left\{\frac{y^2 z}{R^5} + \frac{1-2\nu}{3}\left[\frac{1}{R(R+z)} - \frac{(2R+z)y^2}{R^3(R+z)^2} - \frac{z}{R^3}\right]\right\} \tag{3.49}$$

$$\tau_{xy} = \frac{3Q}{2\pi}\left[\frac{xyz}{R^5} - \frac{1-2\nu}{3}\frac{(2R+z)xy}{R^3(R+z)^2}\right] \tag{3.50}$$

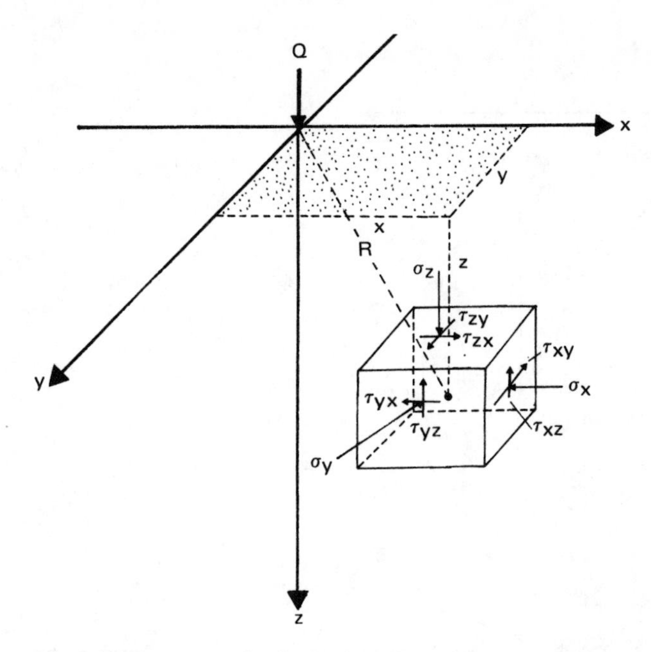

Fig. 3.19 Concentrated point load on the surface (rectangular coordinates).

$$\tau_{xz} = \frac{3Q}{2\pi} \frac{xz^2}{R^5} \tag{3.51}$$

$$\tau_{yz} = \frac{3Q}{2\pi} \frac{yz^2}{R^5} \tag{3.52}$$

where Q = point load
$r = \sqrt{x^2 + y^2}$
$R = \sqrt{z^2 + r^2}$
ν = Poisson's ratio

In cylindrical coordinates, the stresses may be expressed as follows (Fig. 3.20):

$$\sigma_z = \frac{3Qz^3}{2\pi R^5} \tag{3.53}$$

$$\sigma_r = \frac{Q}{2\pi} \left[\frac{3zr^2}{R^5} - \frac{1 - 2\nu}{R(R + z)} \right] \tag{3.54}$$

$$\sigma_\theta = \frac{Q}{2\pi} (1 - 2\nu) \left[\frac{1}{R(R + z)} - \frac{z}{R^3} \right] \tag{3.55}$$

$$\tau_{rz} = \frac{3Qrz^2}{2\pi R^5} \tag{3.56}$$

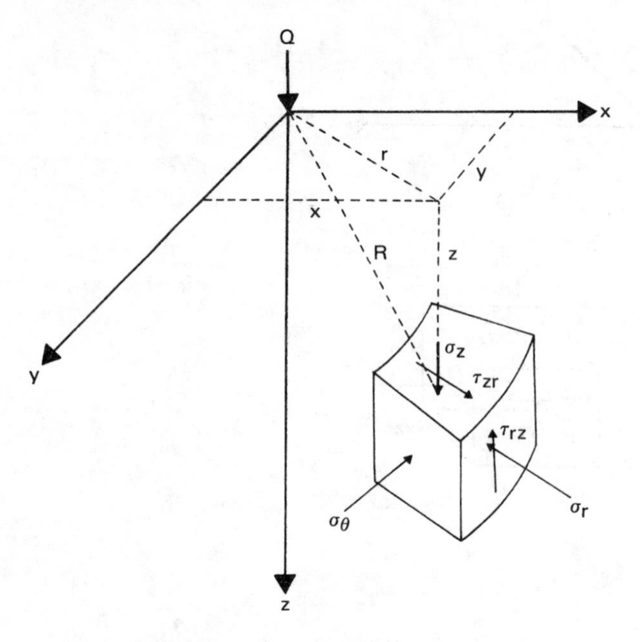

Fig. 3.20 Concentrated point load (vertical) on the surface (cylindrical coordinates).

3.2.2 Stresses due to Horizontal Point Loading on the Surface

Figure 3.21 shows a horizontal point load Q acting on the surface of a semi-infinite mass. The stresses at a point P due to this horizontal line load are as follows:

$$\sigma_z = \frac{3Qxz^2}{2\pi R^5} \tag{3.57}$$

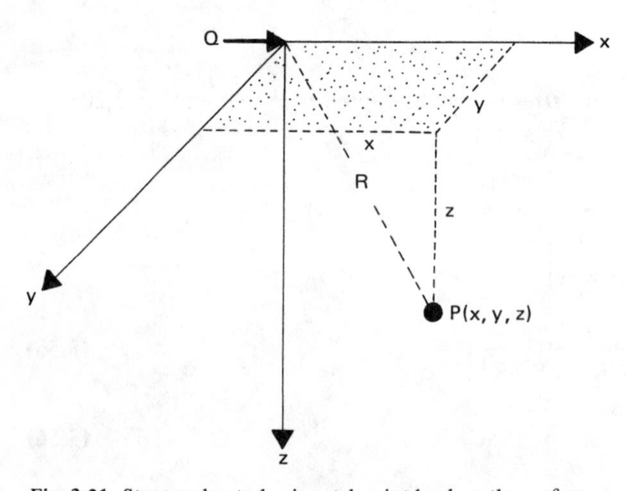

Fig. 3.21 Stresses due to horizontal point load on the surface.

$$\sigma_x = \frac{Q}{2\pi}\frac{x}{R^3}\left\{\frac{3x^2}{R^2} - (1-2\nu) + \frac{(1-2\nu)R^2}{(R+z)^2}\left[3 - \frac{x^2(3R+z)}{R^2(R+z)}\right]\right\} \tag{3.58}$$

$$\sigma_y = \frac{Q}{2\pi}\frac{x}{R^3}\left\{\frac{3y^2}{R^2} - (1-2\nu) + \frac{(1-2\nu)R^2}{(R+z)^2}\left[3 - \frac{y^2(3R+z)}{R^2(R+z)}\right]\right\} \tag{3.59}$$

$$\tau_{xy} = \frac{Q}{2\pi}\frac{y}{R^3}\left\{\frac{3x^2}{R^2} + \frac{(1-2\nu)R^2}{(R+z)^2}\left[1 - \frac{x^2(3R+z)}{R^2(R+z)}\right]\right\} \tag{3.60}$$

$$\tau_{xz} = \frac{3Q}{2\pi}\frac{x^2 z}{R^5} \tag{3.61}$$

$$\tau_{yz} = \frac{3Q}{2\pi}\frac{xyz}{R^5} \tag{3.62}$$

This is generally referred to as Cerutti's problem.

3.2.3 Stresses below a Circularly Loaded (Vertical) Flexible Area

Stresses below the center of the loaded area. Integration of the Boussinesq equation given in Sec. 3.2.1 can be adopted to obtain the stresses below the center of a circularly loaded flexible area. Figure 3.22 shows a circular area of radius b being subjected to a uniform load of q per unit area. Consider an elementary area dA. The load over the area is equal to $q \cdot dA$, and this can be treated as a point load. To determine the

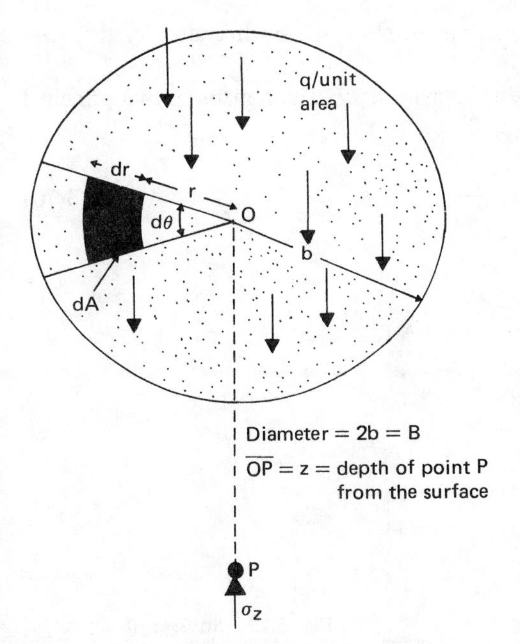

Fig. 3.22 Stresses below the center of a circularly loaded area.

vertical stress due to the elementary load at a point P, we can substitute $q \cdot dA$ for Q and $\sqrt{r^2 + z^2}$ for R in Eq. (3.47). Thus

$$d\sigma_z = \frac{(3q\,dA)z^3}{2\pi(r^2 + z^2)^{5/2}} \tag{3.63}$$

Since $dA = r\,d\theta\,dr$, the vertical stress at P due to the entire loaded area may now be obtained by substituting for dA in Eq. (3.63) and then integrating:

$$\sigma_z = \int_{\theta=0}^{\theta=2\pi}\int_{r=0}^{r=b} \frac{3q}{2\pi}\frac{z^3 r\,d\theta\,dr}{(r^2 + z^2)^{5/2}} = q\left[1 - \frac{z^3}{(b^2 + z^2)^{3/2}}\right] \tag{3.64}$$

Proceeding in a similar manner, we can also determine σ_r and σ_θ at point P as

$$\sigma_r = \sigma_\theta = \frac{q}{2}\left[1 + 2\nu - \frac{2(1+\nu)z}{(b^2 + z^2)^{1/2}} + \frac{z^3}{(b^2 + z^2)^{3/2}}\right] \tag{3.65}$$

Stresses at any point below the loaded area. A detailed tabulation of stresses below a uniformly loaded flexible circular area was given by Ahlvin and Ulery (1962). Referring to Fig. 3.23, the stresses at point P may be given by

$$\sigma_z = q(A' + B') \tag{3.66}$$

$$\sigma_r = q[2\nu A' + C + (1 - 2\nu)F] \tag{3.67}$$

$$\sigma_\theta = q[2\nu A' - D + (1 - 2\nu)E] \tag{3.68}$$

$$\tau_{rz} = \tau_{zr} = qG \tag{3.69}$$

where A', B', C, D, E, F, and G are functions of s/b and z/b; the values of these are given in Tables 3.7 to 3.13.

Note that σ_θ is a principal stress, due to symmetry. The remaining two principal stresses can be determined as

$$\sigma_P = \frac{(\sigma_z + \sigma_r) \pm \sqrt{(\sigma_z - \sigma_r)^2 + (2\tau_{rz})^2}}{2} \tag{3.70}$$

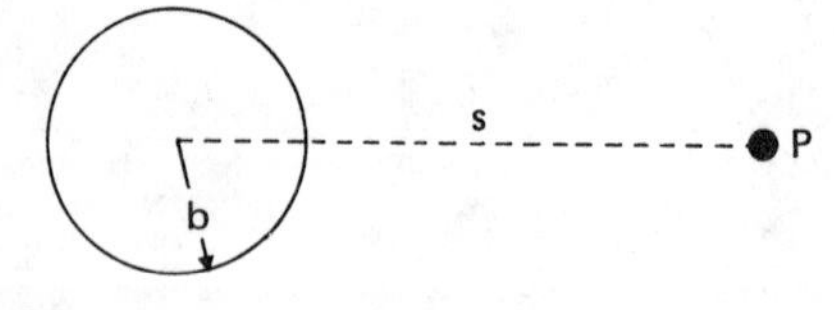

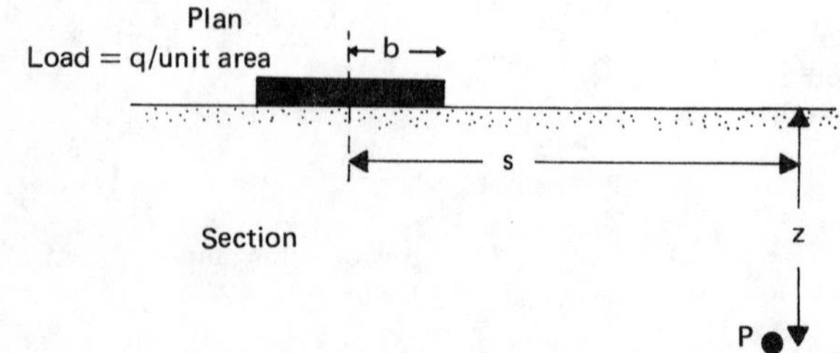

Fig. 3.23 Stresses at any point below a circularly loaded area.

Example 3.3 Refer to Fig. 3.23. Given that $q = 2000\,\text{lb/ft}^2$, $b = 8$ ft, and $\nu = 0.45$, determine the principal stresses at a point defined by $s = 12$ ft and $z = 16$ ft.

SOLUTION $s/B = 12\,\text{ft}/8\,\text{ft} = 1.5$; $z/B = 16\,\text{ft}/8\,\text{ft} = 2$. From Tables 3.7 to 3.13,

$A' = 0.06275$ $\quad E = 0.04078$
$B' = 0.06371$ $\quad F = 0.02197$
$C = -0.00782$ $\quad G = 0.07804$
$D = 0.05589$

So,

$$\sigma_z = q(A' + B') = 2000\,(0.06275 + 0.06371) = 252.92\,\text{lb/ft}^2$$

$$\sigma_\theta = q\,[2\nu A' - D + (1 - 2\nu)E]$$
$$= 2000\,\{2\,(0.45)\,(0.06275) - 0.05589 + [1 - (2)\,(0.45)]\,0.04078\}$$
$$= 9.33\,\text{lb/ft}^2$$

$$\sigma_r = q\,[2\nu A' + C + (1 - 2\nu)F]$$
$$= 2000\,[0.9\,(0.06275) - 0.00782 + 0.1\,(0.02197)] = 101.71\,\text{lb/ft}^2$$

$$\tau_{rz} = qG = (2000)\,(0.07804) = 156.08\,\text{lb/ft}^2$$

$$\sigma_\theta = 9.33\,\text{lb/ft}^2 = \sigma_2 \qquad \text{(intermediate principal stress)}$$

$$\sigma_P = \frac{(252.92 + 101.71) \pm \sqrt{(252.92 - 101.71)^2 + (2 \times 156.08)^2}}{2}$$

$$= \frac{354.63 \pm 346.85}{2}$$

$$\sigma_{P(1)} = 350.94\,lb/ft^2 \qquad \text{(major principal stress)}$$

$$\sigma_{P(3)} = 3.69\,lb/ft^2 \qquad \text{(minor principal stress)}$$

3.2.4 Vertical Stress below a Rectangular Loaded Area

The stress at a point P at a depth z below the corner of a uniformly loaded (vertical) flexible rectangular area (Fig. 3.24) can be determined by integration of Boussinesq's equations given in Sec. 3.2.1. The vertical load over the elementary area $dx \cdot dy$ may be treated as a point load of magnitude $q \cdot dx \cdot dy$. The vertical stress at P due to this elementary load can be evaluated with the aid of Eq. (3.47):

$$d\sigma_z = \frac{3q\,dx\,dy\,z^3}{2\pi(x^2 + y^2 + z^2)^{5/2}}$$

The total increase of vertical stress at P due to the entire loaded area may be determined by integration of the above equation with horizontal limits of $x = 0$ to $x = L$ and $y = 0$ to $y = B$. Newmark (1935) gave the results of the integration in the following form:

$$\sigma_z = qI_\sigma \tag{3.71}$$

Table 3.7 Function A'

z/b	s/b: 0	0.2	0.4	0.6	0.8	1	1.2	1.5	2
0	1.0	1.0	1.0	1.0	1.0	.5	0	0	0
0.1	.90050	.89748	.88679	.86126	.78797	.43015	.09645	.02787	.00856
0.2	.80388	.79824	.77884	.73483	.63014	.38269	.15433	.05251	.01680
0.3	.71265	.70518	.68316	.62690	.52081	.34375	.17964	.07199	.02440
0.4	.62861	.62015	.59241	.53767	.44329	.31048	.18709	.08593	.03118
0.5	.55279	.54403	.51622	.46448	.38390	.28156	.18556	.09499	.03701
0.6	.48550	.47691	.45078	.40427	.33676	.25588	.17952	.10010	
0.7	.42654	.41874	.39491	.35428	.29833	.21727	.17124	.10228	.04558
0.8	.37531	.36832	.34729	.31243	.26581	.21297	.16206	.10236	
0.9	.33104	.32492	.30669	.27707	.23832	.19488	.15253	.10094	
1	.29289	.28763	.27005	.24697	.21468	.17868	.14329	.09849	.05185
1.2	.23178	.22795	.21662	.19890	.17626	.15101	.12570	.09192	.05260
1.5	.16795	.16552	.15877	.14804	.13436	.11892	.10296	.08048	.05116
2	.10557	.10453	.10140	.09647	.09011	.08269	.07471	.06275	.04496
2.5	.07152	.07098	.06947	.06698	.06373	.05974	.05555	.04880	.03787
3	.05132	.05101	.05022	.04886	.04707	.04487	.04241	.03839	.03150
4	.02986	.02976	.02907	.02802	.02832	.02749	.02651	.02490	.02193
5	.01942	.01938				.01835			.01573
6	.01361					.01307			.01168
7	.01005					.00976			.00894
8	.00772					.00755			.00703
9	.00612					.00600			.00566
10								.00477	.00465

After R. G. Ahlvin and H. R. Ulery, Tabulated Values for Determining the Complete Pattern of Stresses, Strains and Deflections beneath a Uniform Load on a Homogeneous Half Space, *Highway Research Board, Bulletin 342,* 1962.

Table 3.8 Function B'

z/b	s/b: 0	0.2	0.4	0.6	0.8	1	1.2	1.5	2
0	0	0	0	0	0	0	0	0	0
0.1	.09852	.10140	.11138	.13424	.18796	.05388	−.07899	−.02672	−.00845
0.2	.18857	.19306	.20772	.23524	.25983	.08513	−.07759	−.04448	−.01593
0.3	.26362	.26787	.28018	.29483	.27257	.10757	−.04316	−.04999	−.02166
0.4	.32016	.32259	.32748	.32273	.26925	.12404	−.00766	−.04535	−.02522
0.5	.35777	.35752	.35323	.33106	.26236	.13591	.02165	−.03455	−.02651
0.6	.37831	.37531	.36308	.32822	.25411	.14440	.04457	−.02101	
0.7	.38487	.37962	.36072	.31929	.24638	.14986	.06209	−.00702	−.02329
0.8	.38091	.37408	.35133	.30699	.23779	.15292	.07530	.00614	
0.9	.36962	.36275	.33734	.29299	.22891	.15404	.08507	.01795	
1	.35355	.34553	.32075	.27819	.21978	.15355	.09210	.02814	−.01005

z/b	s/b 3	4	5	6	7	8	10	12	14
0	0	0	0	0	0	0	0	0	0
0.1	.00211	.00084	.00042						
0.2	.00419	.00167	.00083	.00048	.00030	.00020			
0.3	.00622	.00250							
0.4									
0.5	.01013	.00407	.00209	.00118	.0071	.00053	.00025	.00014	.00009
0.6									
0.7									
0.8									
0.9									
1	.01742	.00761	.00393	.00226	.00143	.00097	.00050	.00029	.00018
1.2	.01935	.00871	.00459	.00269	.00171	.00115			
1.5	.02142	.01013	.00548	.00325	.00210	.00141	.00073	.00043	.00027
2	.02221	.01160	.00659	.00399	.00264	.00180	.00094	.00056	.00036
2.5	.02143	.01221	.00732	.00463	.00308	.00214	.00115	.00068	.00043
3	.01980	.01220	.00770	.00505	.00346	.00242	.00132	.00079	.00051
4	.01592	.01109	.00768	.00536	.00384	.00282	.00160	.00099	.00065
5	.01249	.00949	.00708	.00527	.00394	.00298	.00179	.00113	.00075
6	.00983	.00795	.00628	.00492	.00384	.00299	.00188	.00124	.00084
7	.00784	.00661	.00548	.00445	.00360	.00291	.00193	.00130	.00091
8	.00635	.00554	.00472	.00398	.00332	.00276	.00189	.00134	.00094
9	.00520	.00466	.00409	.00353	.00301	.00256	.00184	.00133	.00096
10	.00438	.00397	.00352	.00326	.00273	.00241			

z/b	s/b 3	4	5	6	7	8	10	12	14
0	0	0	0	0	0	0	0	0	0
0.1	–.00210	–.00084	–.00042						
0.2	–.00412	–.00166	–.00083	–.00024	–.00015	–.00010			
0.3	–.00599	–.00245							
0.4									
0.5	–.00991	–.00388	–.00199	–.00116	–.00073	–.00049	–.00025	–.00014	–.00009
0.6									
0.7									
0.8									
0.9									
1	–.01115	–.00608	–.00344	–.00210	–.00135	–.00092	–.00048	–.00028	–.00018

Table 3.8 *(Continued)*

	s/b								
z/b	0	0.2	0.4	0.6	0.8	1	1.2	1.5	2
1.2	.31485	.30730	.28481	.24836	.20113	.14915	.10002	.04378	.00023
1.5	.25602	.25025	.23338	.20694	.17368	.13732	.10193	.05745	.01385
2	.17889	.18144	.16644	.15198	.13375	.11331	.09254	.06371	.02836
2.5	.12807	.12633	.12126	.11327	.10298	.09130	.07869	.06022	.03429
3	.09487	.09394	.09099	.08635	.08033	.07325	.06551	.05354	.03511
4	.05707	.05666	.05562	.05383	05145	.04773	.04532	.03995	.03066
5	.03772	.03760				.03384			.02474
6	.02666					.02468			.01968
7	.01980					.01868			.01577
8	.01526					.01459			.01279
9	.01212					.01170			.01054
10								.00924	.00879

After R. G. Ahlvin and H. R. Ulery, Tabulated Values for Determining the Complete Pattern of Stresses, Strains and Deflections beneath a Uniform Load on a Homogeneous Half Space, *Highway Research Board, Bulletin 342,* 1962.

Table 3.9 Function *C*

	s/b								
z/b	0	0.2	0.4	0.6	0.8	1	1.2	1.5	2
0	0	0	0	0	0	0	0	0	0
0.1	–.04926	–.05142	–.05903	–.07708	–.12108	.02247	.12007	.04475	.01536
0.2	–.09429	–.09755	–.10872	–.12977	–.14552	.02419	.14896	.07892	.02951
0.3	–.13181	–.13484	–.14415	–.15023	–.12990	.01988	.13394	.09816	.04148
0.4	–.16008	–.16188	–.16519	–.15985	–.11168	.01292	.11014	.10422	.05067
0.5	–.17889	–.17835	–.17497	–.15625	–.09833	.00483	.08730	.10125	.05690
0.6	–.18915	–.18664	–.17336	–.14934	–.08967	–.00304	.06731	.09313	
0.7	–.19244	–.18831	–.17393	–.14147	–.08409	–.01061	.05028	.08253	.06129
0.8	–.19046	–.18481	–.16784	–.13393	–.08066	–.01744	.03582	.07114	
0.9	–.18481	–.17841	–.16024	–.12664	–.07828	–.02337	.02359	.05993	
1	–.17678	–.17050	–.15188	–.11995	–.07634	–.02843	.01331	.04939	.05429
1.2	–.15742	–.15117	–.13467	–.10763	–.07289	–.03575	–.00245	.03107	.04552
1.5	–.12801	–.12277	–.11101	–.09145	–.06711	–.04124	–.01702	.01088	.03154
2	–.08944	–.08491	–.07976	–.06925	–.05560	–.04144	–.02687	–.00782	.01267
2.5	–.06403	–.06068	–.05839	–.05259	–.04522	–.03605	–.02800	–.01536	.00103
3	–.04744	–.04560	–.04339	–.04089	–.03642	–.03130	–.02587	–.01748	–.00528
4	–.02854	–.02737	–.02562	–.02585	–.02421	–.02112	–.01964	–.01586	–.00956
5	–.01886	–.01810				–.01568			–.00939
6	–.01333					–.01118			–.00819
7	–.00990					–.00902			–.00678
8	–.00763					–.00699			–.00552
9	–.00607					–.00423			–.00452
10								–.00381	–.00373

After R. G. Ahlvin and H. R. Ulery, Tabulated Values for Determining the Complete Pattern of Stresses, Strains and Deflections beneath a Uniform Load on a Homogeneous Half Space, *Highway Research Board, Bulletin 342,* 1962.

z/b	s/b = 3	4	5	6	7	8	10	12	14
1.2	–.00995	–.00632	–.00378	–.00236	–.00156	–.00107			
1.5	–.00669	–.00600	–.00401	–.00265	–.00181	–.00126	–.00068	–.00040	–.00026
2	.00028	–.00410	–.00371	–.00278	–.00202	–.00148	–.00084	–.00050	–.00033
2.5	.00661	–.00130	–.00271	–.00250	–.00201	–.00156	–.00094	–.00059	–.00039
3	.01112	.00157	–.00134	–.00192	–.00179	–.00151	–.00099	–.00065	–.00046
4	.01515	.00595	.00155	–.00029	–.00094	–.00109	–.00094	–.00068	–.00050
5	.01522	.00810	.00371	.00132	.00013	–.00043	–.00070	–.00061	–.00049
6	.01380	.00867	.00496	.00254	.00110	.00028	–.00037	–.00047	–.00045
7	.01204	.00842	.00547	.00332	.00185	.00093	–.00002	–.00029	–.00037
8	.01034	.00779	.00554	.00372	.00236	.00141	.00035	–.00008	–.00025
9	.00888	.00705	.00533	.00386	.00265	.00178	.00066	.00012	–.00012
10	.00764	.00631	.00501	.00382	.00281	.00199			

z/b	s/b = 3	4	5	6	7	8	10	12	14
0	0	0	0	0	0	0	0	0	0
0.1	.00403	.00164	.00082						
0.2	.00796	.00325	.00164	.00094	.00059	.00039			
0.3	.01169	.00483							
0.4									
0.5	.01824	.00778	.00399	.00231	.00146	.00098	.00050	.00029	.00018
0.6									
0.7									
1	.02726	.01333	.00726	.00433	.00278	.00188	.00098	.00057	.00036
1.2	.02791	.01467	.00824	.00501	.00324	.00221			
1.5	.02652	.01570	.00933	.00585	.00386	.00266	.00141	.00083	.00039
2	.02070	.01527	.01013	.00321	.00462	.00327	.00179	.00107	.00069
2.5	.01384	.01314	.00987	.00707	.00506	.00369	.00209	.00128	.00083
3	.00792	.01030	.00888	.00689	.00520	.00392	.00232	.00145	.00096
4	.00038	.00492	.00602	.00561	.00476	.00389	.00254	.00168	.00115
5	–.00293	–.00128	.00329	.00391	.00380	.00341	.00250	.00177	.00127
6	–.00405	–.00079	.00129	.00234	.00272	.00272	.00227	.00173	.00130
7	–.00417	–.00180	–.00004	.00113	.00174	.00200	.00193	.00161	.00128
8	–.00393	–.00225	–.00077	.00029	.00096	.00134	.00157	.00143	.00120
9	–.00353	–.00235	–.00118	–.00027	.00037	.00082	.00124	.00122	.00110
10	–.00314	–.00233	–.00137	–.00063	.00030	.00040			

Table 3.10 Function D

z/b	s/b: 0	0.2	0.4	0.6	0.8	1	1.2	1.5	2
0	0	0	0	0	0	0	0	0	0
0.1	.04926	.04998	.05235	.05716	.06687	.07635	.04108	.01803	.00691
0.2	.09429	.09552	.09900	.10546	.11431	.10932	.07139	.03444	.01359
0.3	.13181	.13305	.14051	.14062	.14267	.12745	.09078	.04817	.01982
0.4	.16008	.16070	.16229	.16288	.15756	.13696	.10248	.05887	.02545
0.5	.17889	.17917	.17826	.17481	.16403	.14074	.10894	.06670	.03039
0.6	.18915	.18867	.18573	.17887	.16489	.14137	.11186	.07212	
0.7	.19244	.19132	.18679	.17782	.16229	.13926	.11237	.07551	.03801
0.8	.19046	.18927	.18348	.17306	.15714	.13548	.11115	.07728	
0.9	.18481	.18349	.17709	.16635	.15063	.13067	.10866	.07788	
1	.17678	.17503	.16886	.15824	.14344	.12513	.10540	.07753	.04456
1.2	.15742	.15618	.15014	.14073	.12823	.11340	.09757	.07484	.04575
1.5	.12801	.12754	.12237	.11549	.10657	.09608	.08491	.06833	.04539
2	.08944	.09080	.08668	.08273	.07814	.07187	.06566	.05589	.04103
2,5	.06403	.06565	.06284	.06068	.05777	.05525	.05069	.04486	.03532
3	.04744	.04834	.04760	.04548	.04391	.04195	.03963	.03606	.02983
4	.02854	.02928	.02996	.02798	.02724	.02661	.02568	.02408	.02110
5	.01886	.01950				.01816			.01535
6	.01333					.01351			.01149
7	.00990					.00966			.00899
8	.00763					.00759			.00727
9	.00607					.00746			.00601
10								.00542	.00506

After R. G. Ahlvin and H. R. Ulery, Tabulated Values for Determining the Complete Pattern of Stresses, Strains and Deflections beneath a Uniform Load on a Homogeneous Half Space, *Highway Research Board, Bulletin 342*, 1962.

Table 3.11 Function E

z/b	s/b: 0	0.2	0.4	0.6	0.8	1	1.2	1.5	2
0	.5	.5	.5	.5	.5	.5	.34722	.22222	.12500
0.1	.45025	.449494	.44698	.44173	.43008	.39198	.30445	.20399	.11806
0.2	.40194	.400434	.39591	.38660	.36798	.32802	.26598	.18633	.11121
0.3	.35633	.35428	.33809	.33674	.31578	.28003	.23311	.16967	.10450
0.4	.31431	.31214	.30541	.29298	.27243	.24200	.20526	.15428	.09801
0.5	.27639	.27407	.26732	.25511	.23639	.21119	.18168	.14028	.09180
0.6	.24275	.24247	.23411	.22289	.20634	.18520	.16155	.12759	
0.7	.21327	.21112	.20535	.19525	.18093	.16356	.14421	.11620	.08027
0.8	.18765	.18550	.18049	.17190	.15977	.14523	.12928	.10602	
0.9	.16552	.16337	.15921	.15179	.14168	.12954	.11634	.09686	
1	.14645	.14483	.14610	.13472	.12618	.11611	.10510	.08865	.06552
1.2	.11589	.11435	.11201	.10741	.10140	.09431	.08657	.07476	.05728
1.5	.08398	.08356	.08159	.07885	.07517	.07088	.06611	.05871	.04703

z/b	s/b 3	4	5	6	7	8	10	12	14
0	0	0	0	0	0	0	0	0	0
0.1	.00193	.00080	.00041						
0.2	.00384	.00159	.00081	.00047	.00029	.00020			
0.3	.00927	.00238							
0.4									
0.5	.00921	.00390	.00200	.00116	.00073	.00049	.00025	.00015	.00009
0.6									
0.7									
0.8									
0.9									
1	.01611	.00725	.00382	.00224	.00142	.00096	.00050	.00029	.00018
1.2	.01796	.00835	.00446	.00264	.00169	.00114			
1.5	.01983	.00970	.00532	.00320	.00205	.00140	.00073	.00043	.00027
2	.02098	.01117	.00643	.00398	.00260	.00179	.00095	.00056	.00036
2.5	.02045	.01183	.00717	.00457	.00306	.00213	.00115	.00068	.00044
3	.01904	.01187	.00755	.00497	.00341	.00242	.00133	.00080	.00052
4	.01552	.01087	.00757	.00533	.00382	.00280	.00160	.00100	.00065
5	.01230	.00939	.00700	.00523	.00392	.00299	.00180	.00114	.00077
6	.00976	.00788	.00625	.00488	.00381	.00301	.00190	.00124	.00086
7	.00787	.00662	.00542	.00445	.00360	.00292	.00192	.00130	.00092
8	.00641	.00554	.00477	.00402	.00332	.00275	.00192	.00131	.00096
9	.00533	.00470	.00415	.00358	.00303	.00260	.00187	.00133	.00099
10	.00450	.00398	.00364	.00319	.00278	.00239			

z/b	s/b 3	4	5	6	7	8	10	12	14
0	.05556	.03125	.02000	.01389	.01020	.00781	.00500	.00347	.00255
0.1	.05362	.03045	.01959						
0.2	.05170	.02965	.01919	.01342	.00991	.00762			
0.3	.04979	.02886							
0.4									
0.5	.04608	.02727	.01800	.01272	.00946	.00734	.00475	.00332	.00246
0.6									
0.7									
0.8									
0.9									
1	.03736	.02352	.01602	.01157	.00874	.00683	.00450	.00318	.00237
1.2	.03425	.02208	.01527	.01113	.00847	.00664			
1.5	.03003	.02008	.01419	.01049	.00806	.00636	.00425	.00304	.00228

Table 3.11 *(Continued)*

	s/b								
z/b	0	0.2	0.4	0.6	0.8	1	1.2	1.5	2
2	.05279	.05105	.05146	.05034	.04850	.04675	.04442	.04078	.03454
2.5	.03576	.03426	.03489	.03435	.03360	.03211	.03150	.02953	.02599
3	.02566	.02519	.02470	.02491	.02444	.02389	.02330	.02216	.02007
4	.01493	.01452	.01495	.01526	.01446	.01418	.01395	.01356	.01281
5	.00971	.00927				.00929			.00873
6	.00680					.00632			.00629
7	.00503					.00493			.00466
8	.00386					.00377			.00354
9	.00306					.00227			.00275
10								.00210	.00220

After R. G. Ahlvin and H. R. Ulery, Tabulated Values for Determining the Complete Pattern of Stresses, Strains and Deflections beneath a Uniform Load on a Homogeneous Half Space, *Highway Research Board, Bulletin 342,* 1962.

Table 3.12 Function *F*

	s/b								
z/b	0	0.2	0.4	0.6	0.8	1	1.2	1.5	2
0	.5	.5	.5	.5	.5	0	–.34722	–.22222	–.12500
0.1	.45025	.44794	.43981	.41954	.35789	.03817	–.20800	–.17612	–.10950
0.2	.40194	.39781	.38294	.34823	.26215	.05466	–.11165	–.13381	–.09441
0.3	.35633	.35094	.34508	.29016	.20503	.06372	–.05346	–.09768	–.08010
0.4	.31431	.30801	.28681	.24469	.17086	.06848	–.01818	–.06835	–.06684
0.5	.27639	.26997	.24890	.20937	.14752	.07037	.00388	–.04529	–.05479
0.6	.24275	.23444	.21667	.18138	.13042	.07068	.01797	–.02749	
0.7	.21327	.20762	.18956	.15903	.11740	.06963	.02704	–.01392	–.03469
0.8	.18765	.18287	.16679	.14053	.10604	.06774	.03277	–.00365	
0.9	.16552	.16158	.14747	.12528	.09664	.06533	.03619	.00408	
1	.14645	.14280	.12395	.11225	.08850	.06256	.03819	.00984	–.01367
1.2	.11589	.11360	.10460	.09449	.07486	.05670	.03913	.01716	–.00452
1.5	.08398	.08196	.07719	.06918	.05919	.04804	.03686	.02177	.00413
2	.05279	.05348	.04994	.04614	.04162	.03593	.03029	.02197	.01043
2.5	.03576	.03673	.03459	.03263	.03014	.02762	.02406	.01927	.01188
3	.02566	.02586	.02255	.02395	.02263	.02097	.01911	.01623	.01144
4	.01493	.01536	.01412	.01259	.01386	.01331	.01256	.01134	.00912
5.	.00971	.01011				.00905			.00700
6	.00680					.00675			.00538
7	.00503					.00483			.00428
8	.00386					.00380			.00350
9	.00306					.00374			.00291
10								.00267	.00246

After R. G. Ahlvin and H. R. Ulery, Tabulated Values for Determining the Complete Pattern of Stresses, Strains and Deflections beneath a Uniform Load on a Homogeneous Half Space, *Highway Research Board, Bulletin 342,* 1962.

z/b	s/b 3	4	5	6	7	8	10	12	14
2	.02410	.01706	.01248	.00943	.00738	.00590	.00401	.00290	.00219
2.5	.01945	.01447	.01096	.00850	.00674	.00546	.00378	.00276	.00210
3	.01585	.01230	.00962	.00763	.00617	.00505	.00355	.00263	.00201
4	.01084	.00900	.00742	.00612	.00511	.00431	.00313	.00237	.00185
5	.00774	.00673	.00579	.00495	.00425	.00364	.00275	.00213	.00168
6	.00574	.00517	.00457	.00404	.00354	.00309	.00241	.00192	.00154
7	.00438	.00404	.00370	.00330	.00296	.00264	.00213	.00172	.00140
8	.00344	.00325	.00297	.00273	.00250	.00228	.00185	.00155	.00127
9	.00273	.00264	.00246	.00229	.00212	.00194	.00163	.00139	.00116
10	.00225	.00221	.00203	.00200	.00181	.00171			

z/b	s/b 3	4	5	6	7	8	10	12	14
0	–.05556	–.03125	–.02000	–.01389	–.01020	–.00781	–.00500	–.00347	–.00255
0.1	–.05151	–.02961	–.01917						
0.2	–.04750	–.02798	–.01835	–.01295	–.00961	–.00742			
0.3	–.04356	–.02636							
0.4									
0.5	–.03595	–.02320	–.01590	–.01154	–.00875	–.00681	–.00450	–.00318	–.00237
1	–.01994	–.01591	–.01209	–.00931	–.00731	–.00587	–.00400	–.00289	–.00219
1.2	–.01491	–.01337	–.01068	–.00844	–.00676	–.00550			
1.5	–.00879	–.00995	–.00870	–.00723	–.00596	–.00495	–.00353	–.00261	–.00201
2	–.00189	–.00546	–.00589	–.00544	–.00474	–.00410	–.00307	–.00233	–.00183
2.5	.00198	–.00226	–.00364	–.00386	–.00366	–.00332	–.00263	–.00208	–.00166
3	.00396	–.00010	–.00192	–.00258	–.00271	–.00263	–.00223	–.00183	–.00150
4	.00508	.00209	.00026	–.00076	–.00127	–.00148	–.00153	–.00137	–.00120
5	.00475	.00277	.00129	.00031	–.00030	–.00066	–.00096	–.00099	–.00093
6	.00409	.00278	.00170	.00088	.00030	–.00010	–.00053	–.00066	–.00070
7	.00346	.00258	.00178	.00114	.00064	.00027	–.00020	–.00041	–.00049
8	.00291	.00229	.00174	.00125	.00082	.00048	.00003	–.00020	–.00033
9	.00247	.00203	.00163	.00124	.00089	.00062	.00020	–.00005	–.00019
10	.00213	.00176	.00149	.00126	.00092	.00070			

Table 3.13 Function G

	s/b								
z/b	0	0.2	0.4	0.6	0.8	1	1.2	1.5	2
0	0	0	0	0	0	.31831	0	0	0
0.1	0	.00315	.00802	.01951	.06682	.31405	.05555	.00865	.00159
0.2	0	.01163	.02877	.06441	.16214	.30474	.13592	.03060	.00614
0.3	0	.02301	.05475	.11072	.21465	.29228	.18216	.05747	.01302
0.4	0	.03460	.07883	.14477	.23442	.27779	.20195	.08233	.02138
0.5	0	.04429	.09618	.16426	.23652	.26216	.20731	.10185	.03033
0.6	0	.04966	.10729	.17192	.22949	.24574	.20496	.11541	
0.7	0	.05484	.11256	.17126	.21772	.22924	.19840	.12373	.04718
0.8	0	.05590	.11225	.16534	.20381	.21295	.18953	.12855	
0.9	0	.05496	.10856	.15628	.18904	.19712	.17945	.28881	
1	0	.05266	.10274	.14566	.17419	.18198	.16884	.12745	.06434
1.2	0	.04585	.08831	.12323	.14615	.15408	.14755	.12038	.06967
1.5	0	.03483	.06688	.09293	.11071	.11904	.11830	.10477	.07075
2	0	.02102	.04069	.05721	.06948	.07738	.08067	.07804	.06275
2.5	0	.01293	.02534	.03611	.04484	.05119	.05509	.05668	.05117
3	0	.00840	.01638	.02376	.02994	.03485	.03843	.04124	.04039
4	0	.00382	.00772	.01149	.01480	.01764	.02004	.02271	.02475
5	0	.00214				.00992		.01343	.01551
6	0					.00602		.00845	.01014
7	0					.00396			.00687
8	0					.00270			.00481
9	0					.00177			.00347
10	0							.00199	.00258

After R. G. Ahlvin and H. R. Ulery, Tabulated Values for Determining the Complete Pattern of Stresses, Strains and Deflections beneath a Uniform Load on a Homogeneous Half Space, *Highway Research Board, Bulletin 342,* 1962.

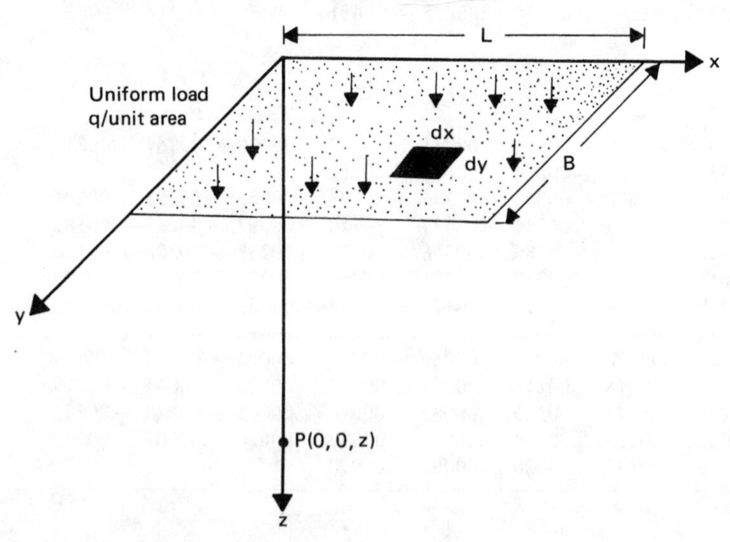

Fig. 3.24 Vertical stress below the corner of a uniformly loaded (normal) rectangular area.

z/b	s/b 3	4	5	6	7	8	10	12	14
0	0	0	0	0	0	0	0	0	0
0.1	.00023	.00007	.00003						
0.2	.00091	.00026	.00010	.00005	.00003	.00002			
0.3	.00201	.00059							
0.4									
0.5	.00528	.00158	.00063	.00030	.00016	.00009	.00004	.00002	.00001
0.6									
0.7									
0.8									
0.9									
1	.01646	.00555	.00233	.00113	.00062	.00036	.00015	.00007	.00004
1.2	.02077	.00743	.00320	.00159	.00087	.00051			
1.5	.02599	.01021	.00460	.00233	.00130	.00078	.00033	.00016	.00009
2	.03062	.01409	.00692	.00369	.00212	.00129	.00055	.00027	.00015
2.5	.03099	.01650	.00886	.00499	.00296	.00185	.00082	.00041	.00023
3	.02886	.01745	.01022	.00610	.00376	.00241	.00110	.00057	.00032
4	.02215	.01639	.01118	.00745	.00499	.00340	.00167	.00090	.00052
5	.01601	.01364	.01105	.00782	.00560	.00404	.00216	.00122	.00073
6	.01148	.01082	.00917	.00733	.00567	.00432	.00243	.00150	.00092
7	.00830	.00842	.00770	.00656	.00539	.00432	.00272	.00171	.00110
8	.00612	.00656	.00631	.00568	.00492	.00413	.00278	.00185	.00124
9	.00459	.00513	.00515	.00485	.00438	.00381	.00274	.00192	.00133
10	.00351	.00407	.00420	.00411	.00382	.00346			

$$I_\sigma = \frac{1}{4\pi}\left[\frac{2mn(m^2+n^2+1)^{1/2}}{m^2+n^2+m^2n^2+1}\,\frac{m^2+n^2+2}{m^2+n^2+1} + \tan^{-1}\frac{2mn(m^2+n^2+1)^{1/2}}{m^2+n^2-m^2n^2+1}\right] \tag{3.72}$$

where $m = B/z$ and $n = L/z$.

The values of I_σ for various values of m and n are given in a graphical form in Fig. 3.25. A similar plot of I_σ in a slightly different form was also given by Fadum (1948).

For equations concerning the determination of σ_x, σ_y, τ_{xz}, τ_{yz}, and τ_{xy}, the reader is referred to the works of Holl (1940) and Giroud (1970).

The use of Fig. 3.25 for determination of the vertical stress at any point below a rectangular loaded area is shown in Example 3.4.

Example 3.4 A distributed load of 50 kN/m^2 is acting on the flexible rectangular area 6 × 3 m as shown in Fig. 3.26. Determine the vertical stress at point A which is located at a depth of 3 m below the ground surface.

SOLUTION The total increase of stress at A may be evaluated by summing the stresses contributed by the four rectangular loaded areas shown in Fig. 3.27. Thus,

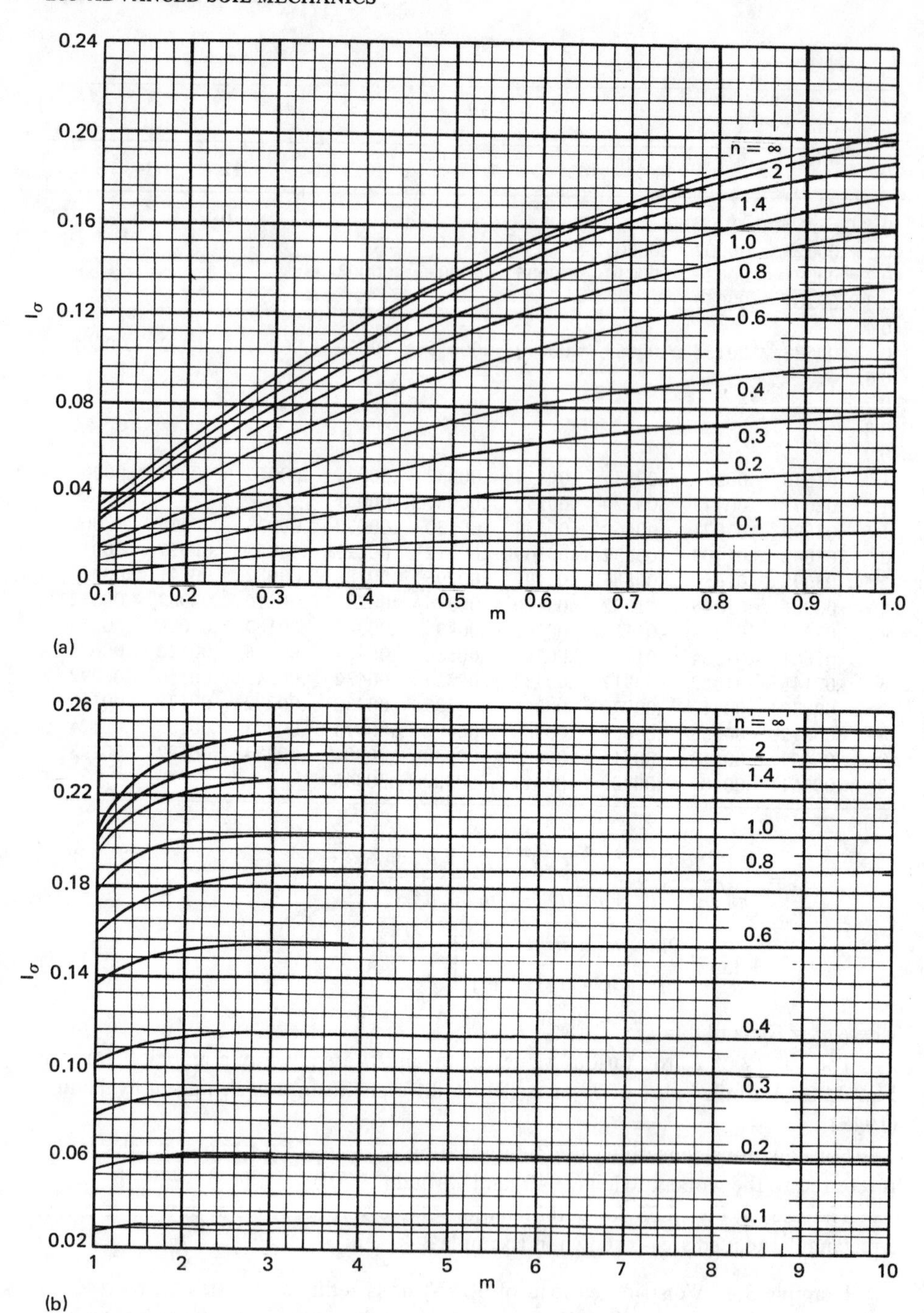

Fig. 3.25 Values of I_σ for determination of vertical stress below the corner of a flexible rectangular loaded area. (a) $m = 0.1$ to $1; n = 0.1$ to ∞. (b) $m = 1$ to $10; n = 0.1$ to ∞.

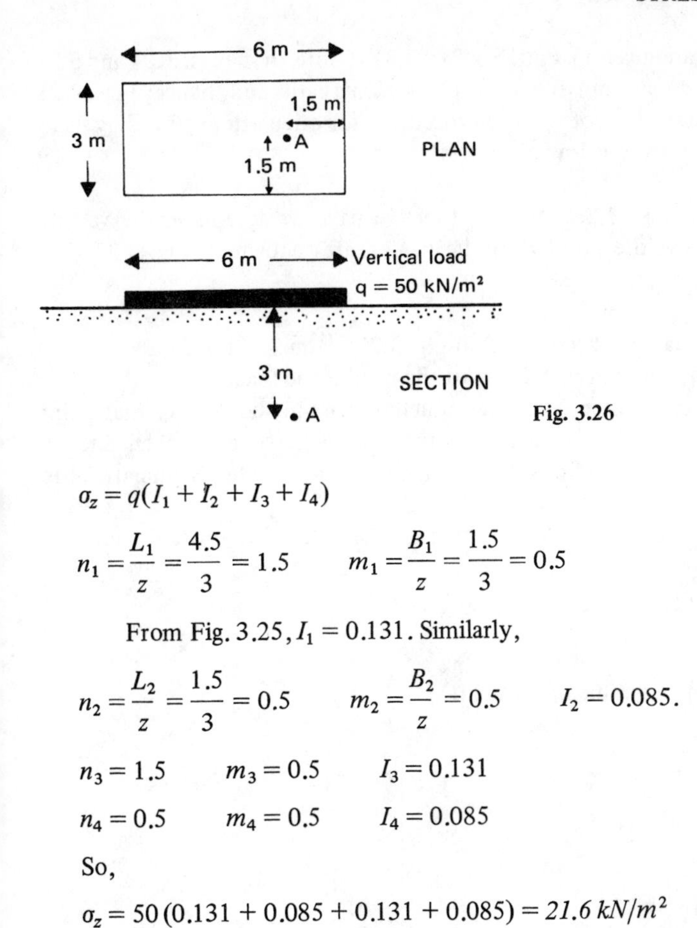

Fig. 3.26

$$\sigma_z = q(I_1 + I_2 + I_3 + I_4)$$

$$n_1 = \frac{L_1}{z} = \frac{4.5}{3} = 1.5 \qquad m_1 = \frac{B_1}{z} = \frac{1.5}{3} = 0.5$$

From Fig. 3.25, $I_1 = 0.131$. Similarly,

$$n_2 = \frac{L_2}{z} = \frac{1.5}{3} = 0.5 \qquad m_2 = \frac{B_2}{z} = 0.5 \qquad I_2 = 0.085.$$

$$n_3 = 1.5 \qquad m_3 = 0.5 \qquad I_3 = 0.131$$

$$n_4 = 0.5 \qquad m_4 = 0.5 \qquad I_4 = 0.085$$

So,

$$\sigma_z = 50\,(0.131 + 0.085 + 0.131 + 0.085) = \mathit{21.6\ kN/m^2}$$

3.2.5 Stresses due to any Type of Loaded Area

Newmark (1942) prepared several influence charts for determination of stresses at any point below any type of vertically loaded flexible area. These influence charts for

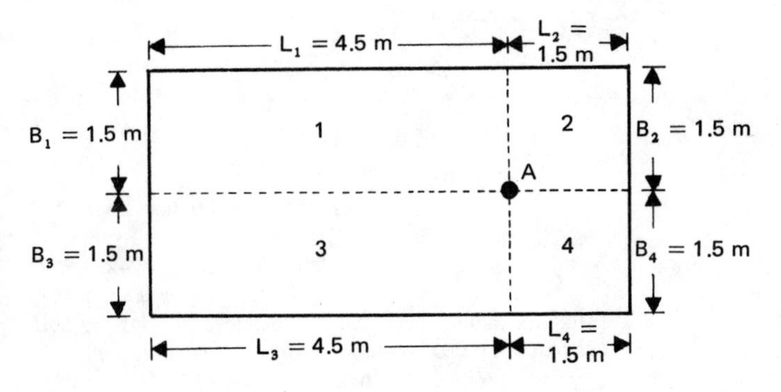

Fig. 3.27

σ_z, σ_x, σ_y, τ_{xz}, and τ_{xy} are given in Figs. 3.28 to 3.31. (Note that Figs. 3.29 and 3.31 are for $\nu = 0.5$; σ_z and τ_{xz} are not functions of Poisson's ratio and, hence, Figs. 3.28 and 3.30 are valid for all values of ν.) The procedures for calculating stresses by using these influence charts are given below.

Calculation of σ_z using Fig. 3.28. Assume that we have to determine the vertical stress at a depth z below the point P of the loaded area shown in Fig. 3.32. The following are the required steps:

1. Adopt a scale such that the distance AB in Fig. 3.28 is equal to the depth z.
2. Based on the scale adopted in step 1, replot the plan of the loaded area.
3. Place the plan plotted in step 2 on the influence chart in such a way that point P is located directly above the center of the chart (shown by broken lines in Fig. 3.28). Note that orientation of the positive x and y axes is immaterial in this case.
4. Count the number of blocks, N, of the influence chart which fall inside the plan.
5. Calculate σ_z as

$$\sigma_z = q(IV)(N) \tag{3.73}$$

where IV is the influence value of the chart.

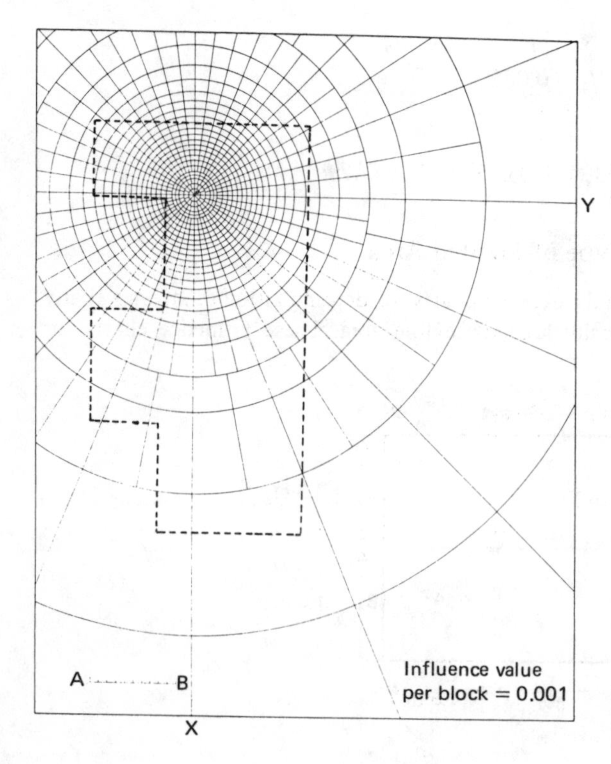

Fig. 3.28 Influence chart for vertical stress σ_z for all values of ν. *(After Newmark, 1942, taken from H. G. Poulos and E. H. Davis, "Elastic Solutions for Soil and Rock Mechanics," p. 78, Wiley, New York, 1974.)*

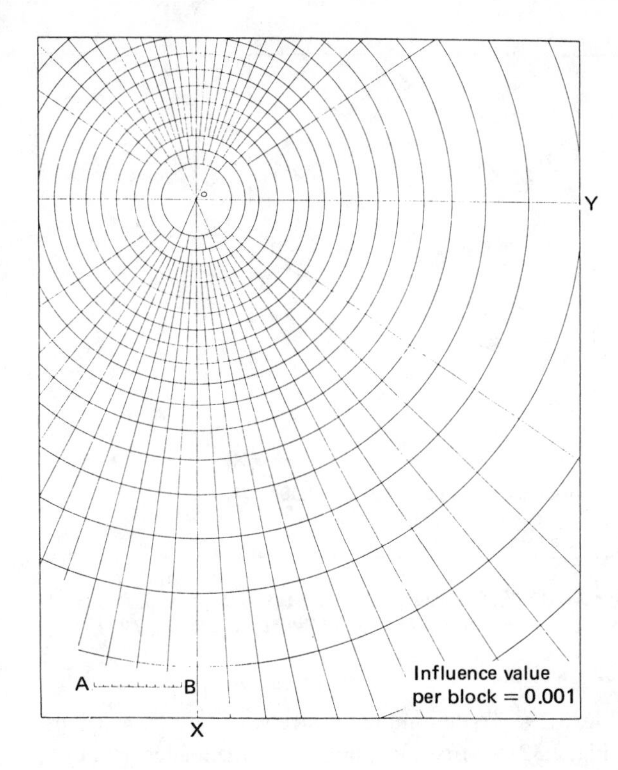

Fig. 3.29 Influence chart for σ_x and σ_y for $\nu = 0.5$. *(After Newmark, 1942, taken from H. G. Poulos and E. H. Davis, "Elastic Solutions for Soil and Rock Mechanics," p. 80, Wiley, New York, 1974.)*

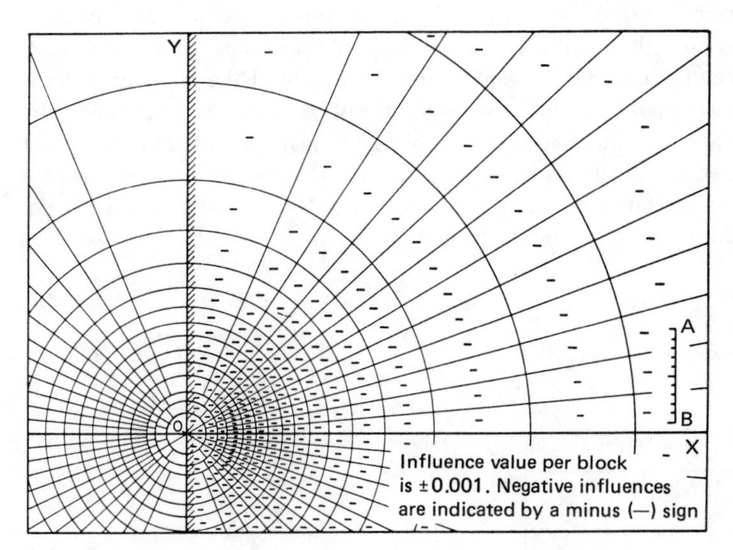

Fig. 3.30 Influence chart for τ_{xz} for all values of ν. *(After Newmark, 1942, taken from H. G. Poulos and E. H. Davis, "Elastic Solutions for Soil and Rock Mechanics," p. 81, Wiley, New York, 1974.)*

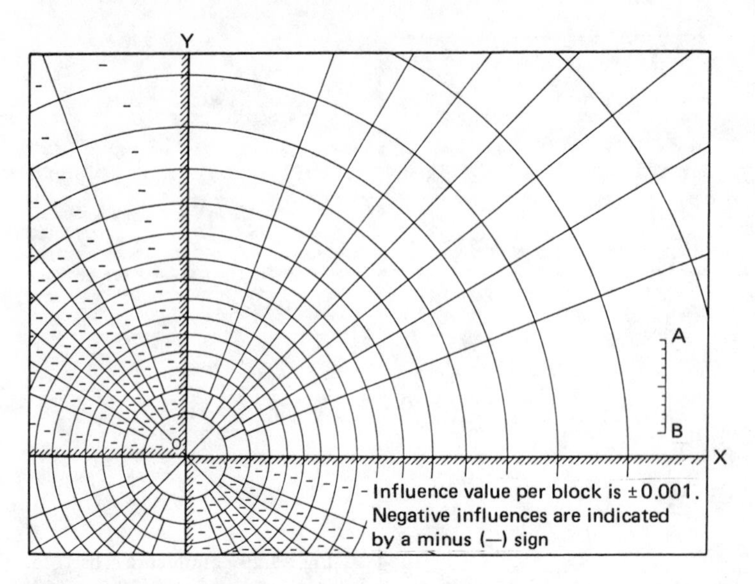

Fig. 3.31 Influence chart for τ_{xy} for $\nu = 0.5$. *(After Newmark, 1942, taken from H. G. Poulos and E. H. Davis, "Elastic Solutions for Soil and Rock Mechanics," p. 82, Wiley, New York, 1974.)*

Calculation of σ_x and σ_y using Fig. 3.29. To determine the stresses σ_x and σ_y at a depth z below the point P shown in Fig. 3.32, we first plot the plan of the loaded area using a scale $z = AB$ (given in Fig. 3.29). To find σ_x, the plan is placed over the influence chart in such a way that the point P is located directly above the center of the chart, and the positive x and y axes of the plan are parallel to and in the same direction as the positive x and y axes of the chart. The magnitude of σ_x can now be determined by using Eq. (3.73). For determination of σ_y, we place the plan over the influence chart such that P is located above the center of the chart and the positive x axis of the plan is parallel to and in the same direction as the positive y axis of the chart. Then Eq. (3.73) may be used for determination of the desired stress.

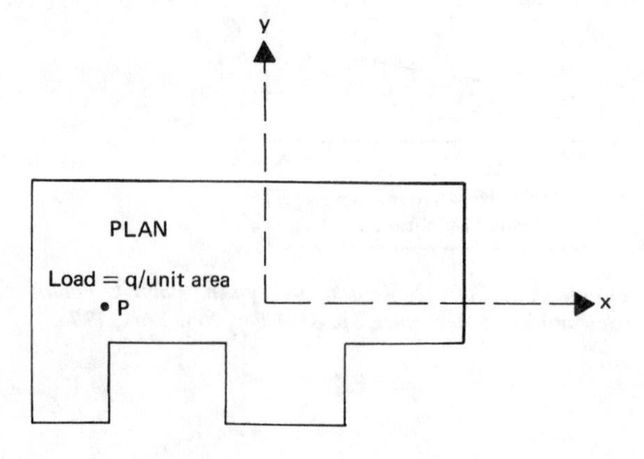

Fig. 3.32

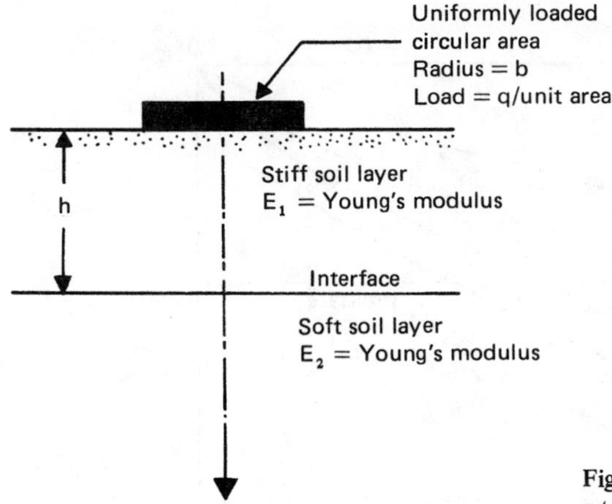

Fig. 3.33 Uniformly loaded circular area in a two-layered soil. (Note: $E_1 > E_2$.)

Calculation of τ_{xz} and τ_{xy} using Figs. 3.30 and 3.31. The basic procedure of replotting the plan is similar to that explained above, and the stresses can be calculated by using Eq. (3.73). However, care should be taken with the orientation of the positive x and y axes of the plan with respect to the positive x and y axes of the influence chart. The blocks in Figs. 3.30 and 3.31 shown as negative should be counted as negative. So, the net value of N to be used in the stress calculation is equal to $N_{\text{positive}} - N_{\text{negative}}$.

3.2.6 Stresses in Layered Medium

In the preceding sections, we discussed the stresses inside a homogeneous elastic medium due to various loading conditions. In actual cases of soil deposits it is possible to encounter layered soils, each with a different modulus of elasticity. A case of practical importance is that of a stiff soil layer on top of a softer layer, as shown in Fig. 3.33. For a given loading condition, the effect of the stiff layer will be to reduce the stress concentration in the lower layer. Burmister (1943) worked on such problems involving two- and three-layer flexible systems. This was later developed by Fox (1948), Burmister (1958), Jones (1962), and Peattie (1962).

The effect of the reduction of stress concentration due to the presence of a stiff top layer is demonstrated in Fig. 3.34. Consider a flexible circular area of radius b subjected to a loading of q per unit area at the surface of a two-layered system as shown in Fig. 3.33. E_1 and E_2 are the moduli of elasticity of the top and the bottom layer, respectively, with $E_1 > E_2$; and h is the thickness of the top layer. For $h = b$, the elasticity solution for the vertical stress σ_z at various depths below the center of the loaded area can be obtained from Fig. 3.34. The curves of σ_z/q against z/b for $E_1/E_2 = 1$ is the simple Boussinesq case, which is obtained by solving Eq. (3.64). However, for $E_1/E_2 > 1$, the value of σ_z/q for a given z/b decreases with the increase of E_1/E_2. It must be pointed out that in obtaining these results it is assumed that there is *no slippage at the interface.*

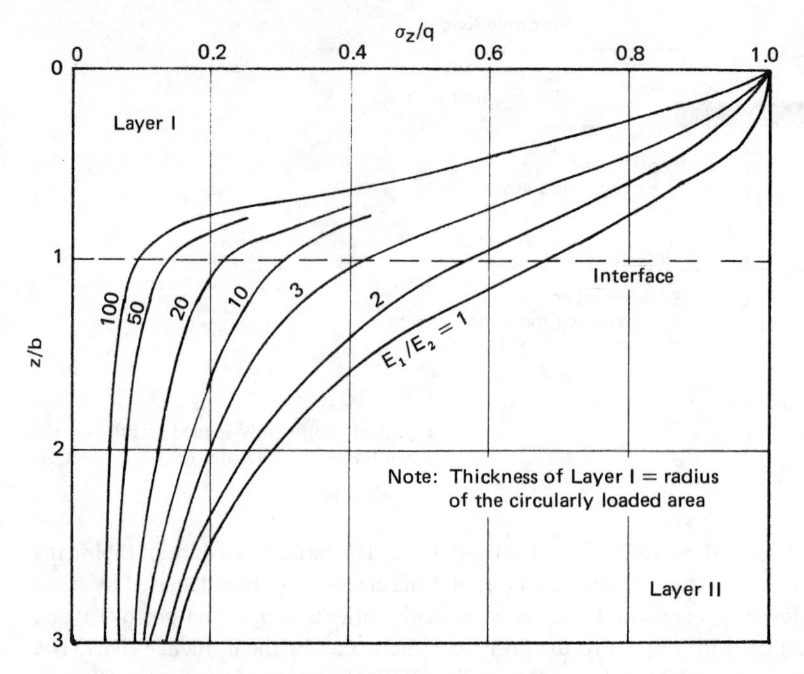

Fig. 3.34 Vertical stress below the center line of a uniformly loaded (vertical) circular area in a two-layered system. *(After D. M. Burmister, Evaluation of Pavement Systems of WASHO Road Testing Layered Systems Method*, Highway Research Board, Bulletin 177, *1958.)*

The study of the stresses in a flexible layered system is of importance in highway pavement design.

3.2.7 Vertical Stress at the Interface of a Three-Layer Flexible System

Peattie (1962) prepared a number of graphs for determination of the vertical stress σ_z at the interfaces of three-layer systems (Fig. 3.35) below the center of a uniformly loaded flexible circular area. These graphs are presented in Figs. 3.36 to 3.67. In the determination of these stresses, it is assumed that Poisson's ratio for all layers is 0.5. The following parameters have been used in the graphs:

$$K_1 = \frac{E_1}{E_2} \tag{3.74}$$

$$K_2 = \frac{E_2}{E_3} \tag{3.75}$$

$$A = \frac{b}{h_2} \tag{3.76}$$

$$H = \frac{h_1}{h_2} \tag{3.77}$$

For determination of the stresses σ_{z_1} and σ_{z_2} (vertical stresses at interfaces 1 and 2, respectively), we first obtain ZZ_1 and ZZ_2 from the graphs. The stresses can then be calculated from

$$\sigma_{z_1} = q(ZZ_1) \tag{3.78a}$$

and

$$\sigma_{z_2} = q(ZZ_2) \tag{3.78b}$$

Typical use of these graphs is shown in Example 3.5.

Example 3.5 A flexible circular area is subjected to a uniformly distributed load of 2000 lb/ft^2 as shown in Fig. 3.68. Determine the vertical stress σ_{z_1} at the interface of the stiff and medium-stiff clay.

SOLUTION

$$K_1 = \frac{E_1}{E_2} = \frac{1500}{1000} = 1.5 \qquad K_2 = \frac{E_2}{E_3} = \frac{1000}{250} = 4$$

$$A = \frac{b}{h_2} = \frac{2}{10} = 0.2 \qquad H = \frac{h_1}{h_2} = \frac{5}{10} = 0.5$$

Using the above parameters and the graphs for ZZ_1, the following table is prepared:

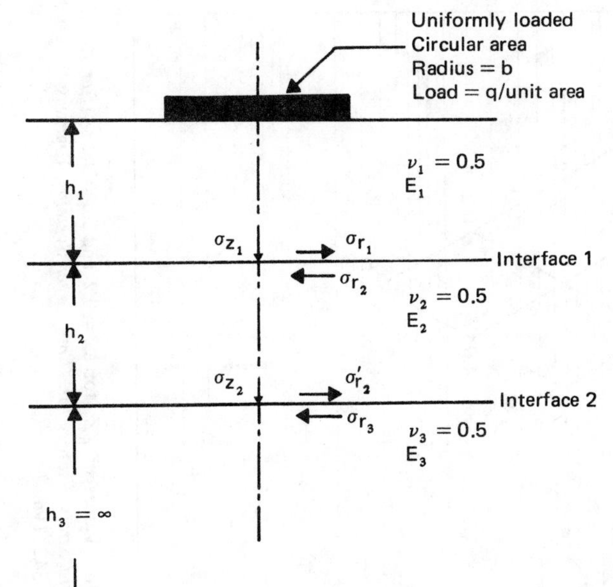

Fig. 3.35 Uniformly loaded circular area on a three-layered medium.

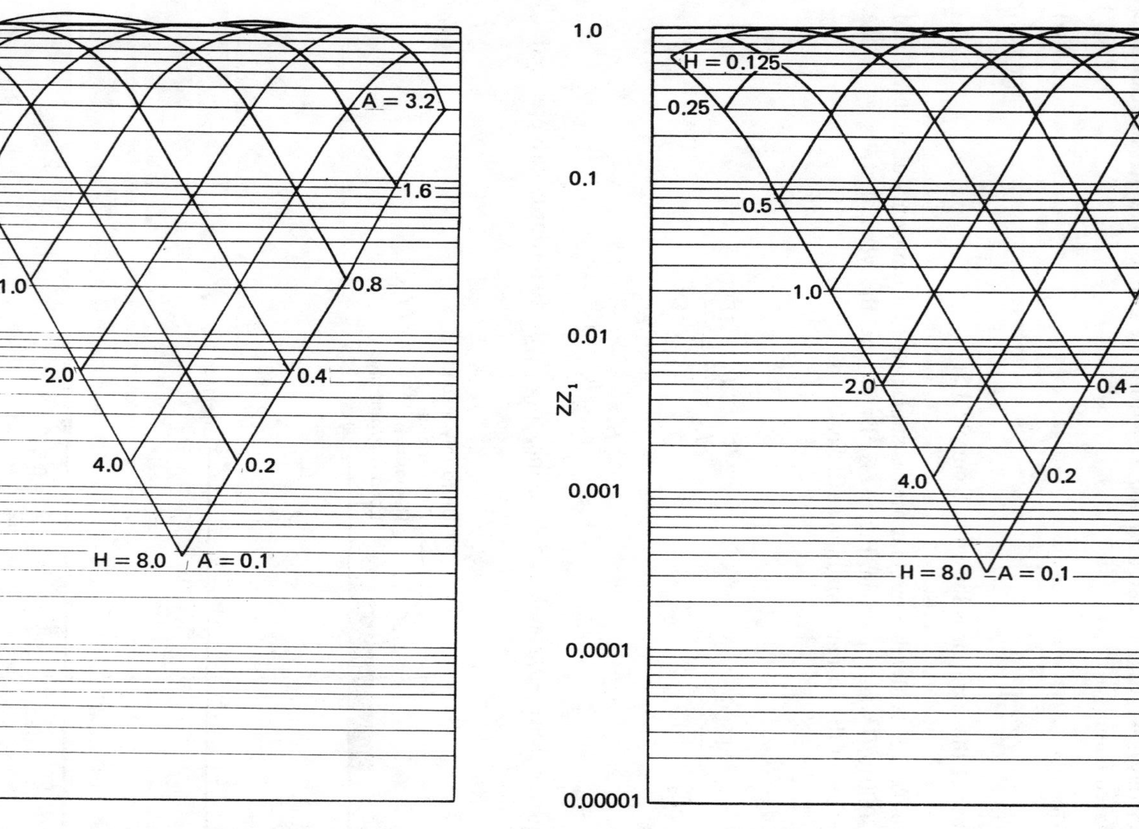

Fig. 3.36 Values of ZZ_1 for $K_1 = 0.2$ and $K_2 = 0.2$. *(After K. R. Peattie, Stress and Strain Factors for Three Layer Systems,* Highway Research Board, Bulletin 342, *1962.)*

Fig. 3.37 Values of ZZ_1 for $K_1 = 0.2$ and $K_2 = 2.0$. *(After K. R. Peattie, Stress and Strain Factors for Three Layer Systems*, Highway Research Board, Bulletin 342, *1962.)*

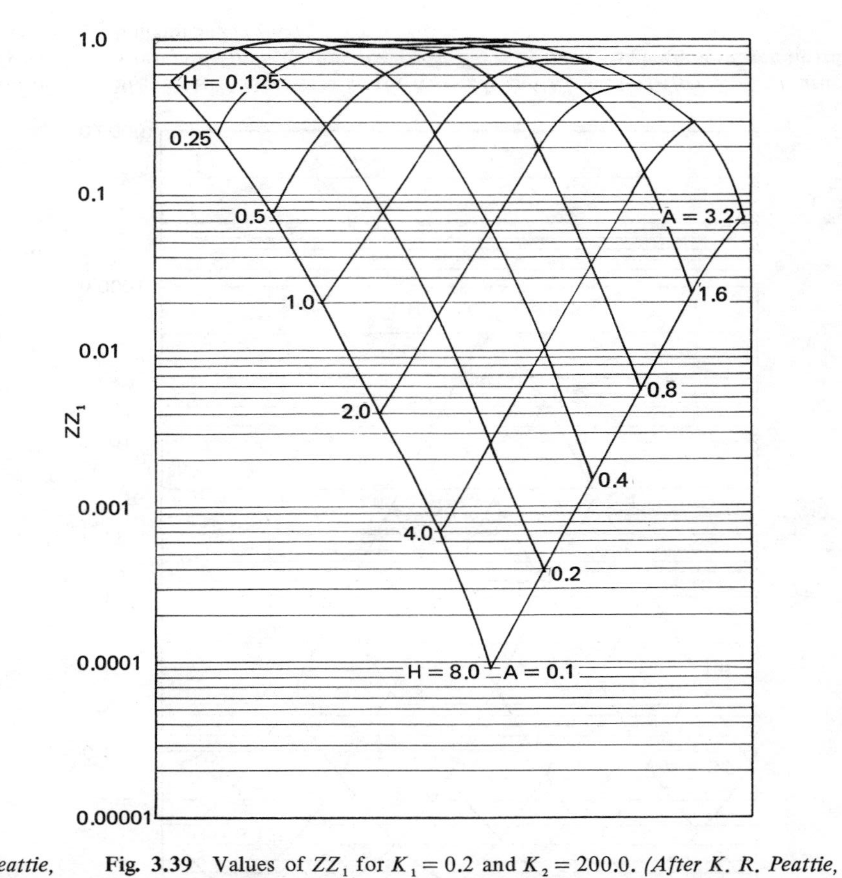

Fig. 3.38 Values of ZZ_1 for $K_1 = 0.2$ and $K_2 = 20.0$. *(After K. R. Peattie, Stress and Strain Factors for Three Layer Systems*, Highway Research Board, Bulletin 342, *1962.)*

Fig. 3.39 Values of ZZ_1 for $K_1 = 0.2$ and $K_2 = 200.0$. *(After K. R. Peattie, Stress and Strain Factors for Three Layer Systems*, Highway Research Board, Bulletin 342, *1962.)*

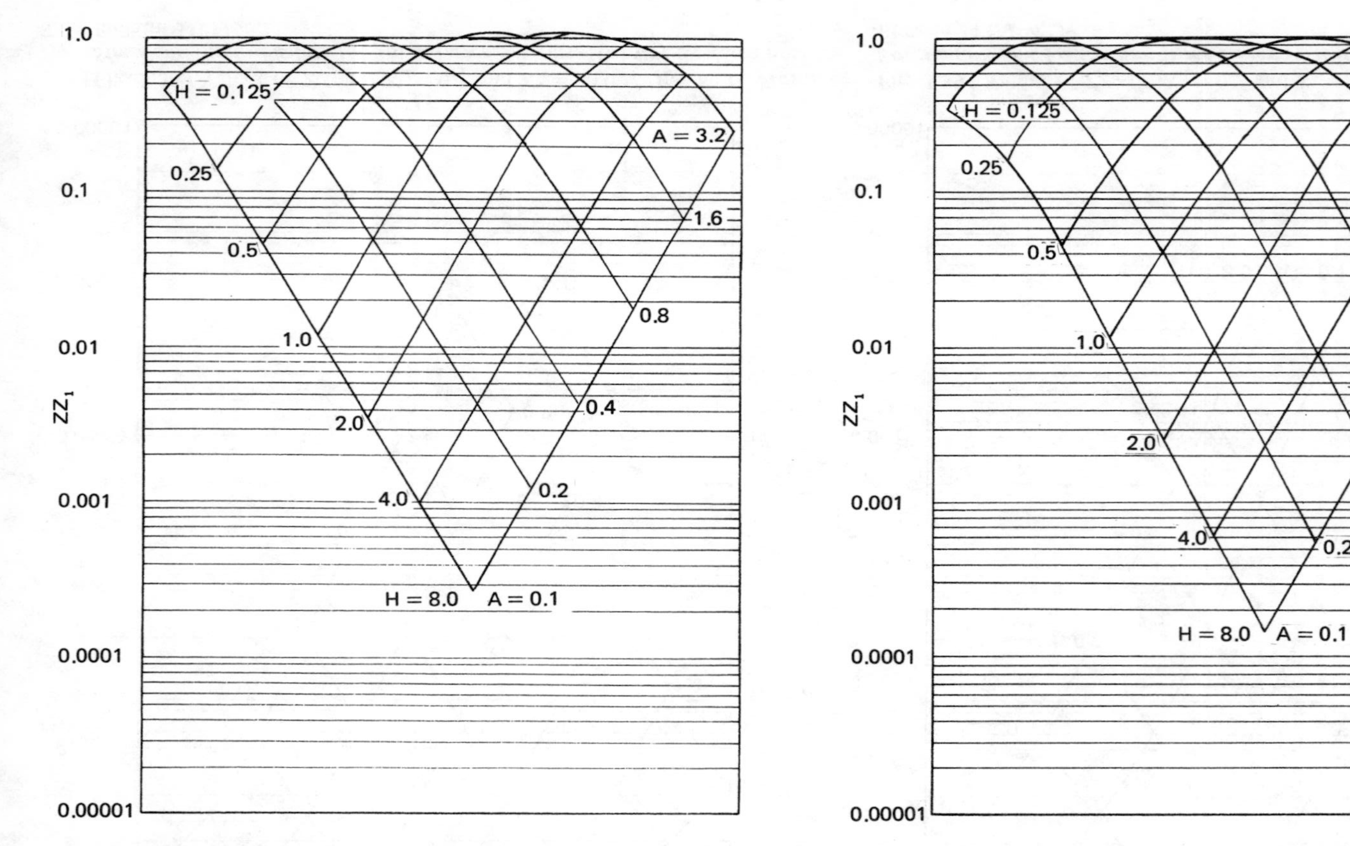

Fig. 3.40 Values of ZZ_1 for $K_1 = 2.0$ and $K_2 = 0.2$. *(After K. R. Peattie, Stress and Strain Factors for Three Layer Systems,* Highway Research Board, Bulletin 342, *1962.)*

Fig. 3.41 Values of ZZ_1 for $K_1 = 2.0$ and $K_2 = 2.0$. *(After K. R. Peattie, Stress and Strain Factors for Three Layer Systems,* Highway Research Board, Bulletin 342, *1962.)*

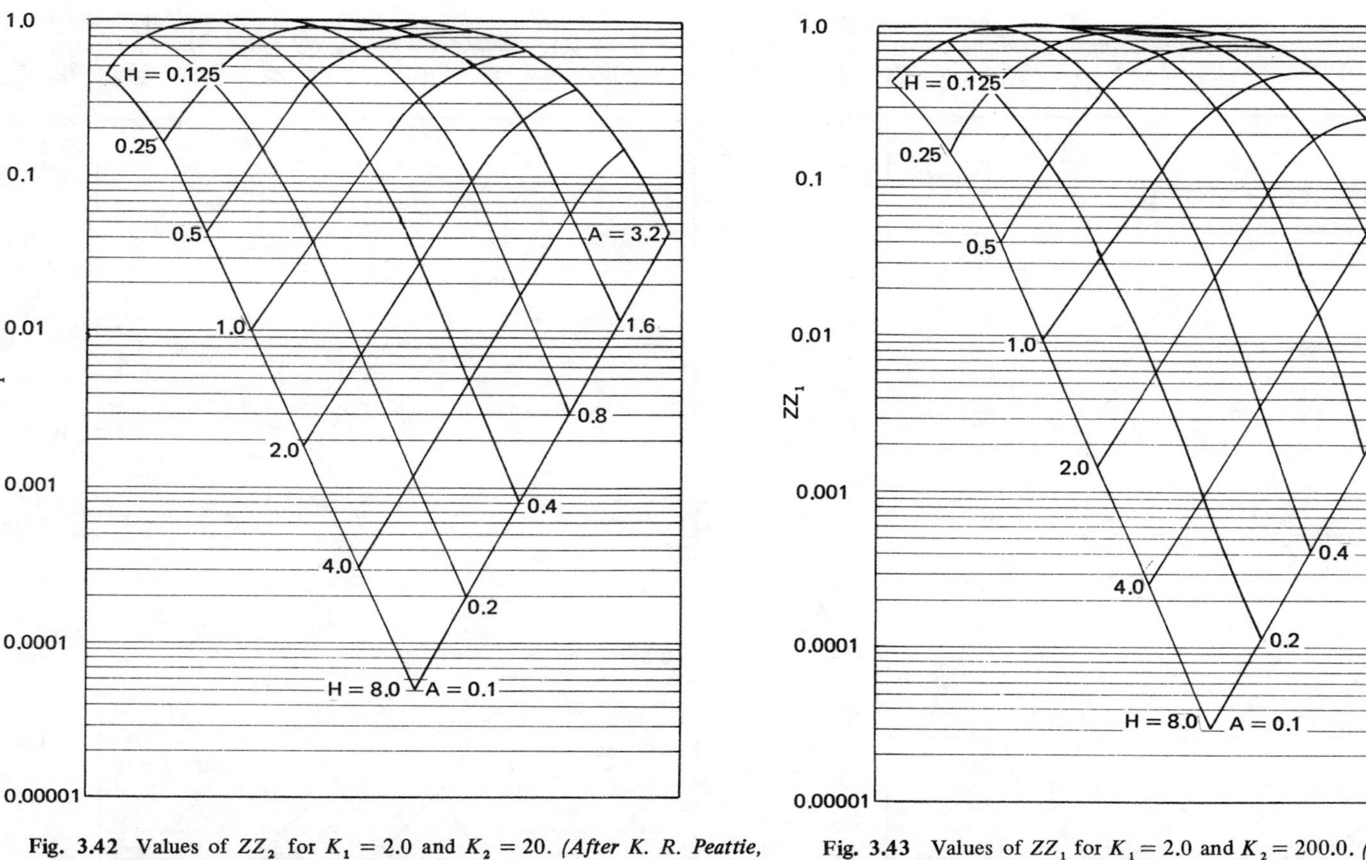

Fig. 3.42 Values of ZZ_2 for $K_1 = 2.0$ and $K_2 = 20$. *(After K. R. Peattie, Stress and Strain Factors for Three Layer Systems,* Highway Research Board, Bulletin 342, *1962.)*

Fig. 3.43 Values of ZZ_1 for $K_1 = 2.0$ and $K_2 = 200.0$. *(After K. R. Peattie, Stress and Strain Factors for Three Layer Systems,* Highway Research Board, Bulletin 342, *1962.)*

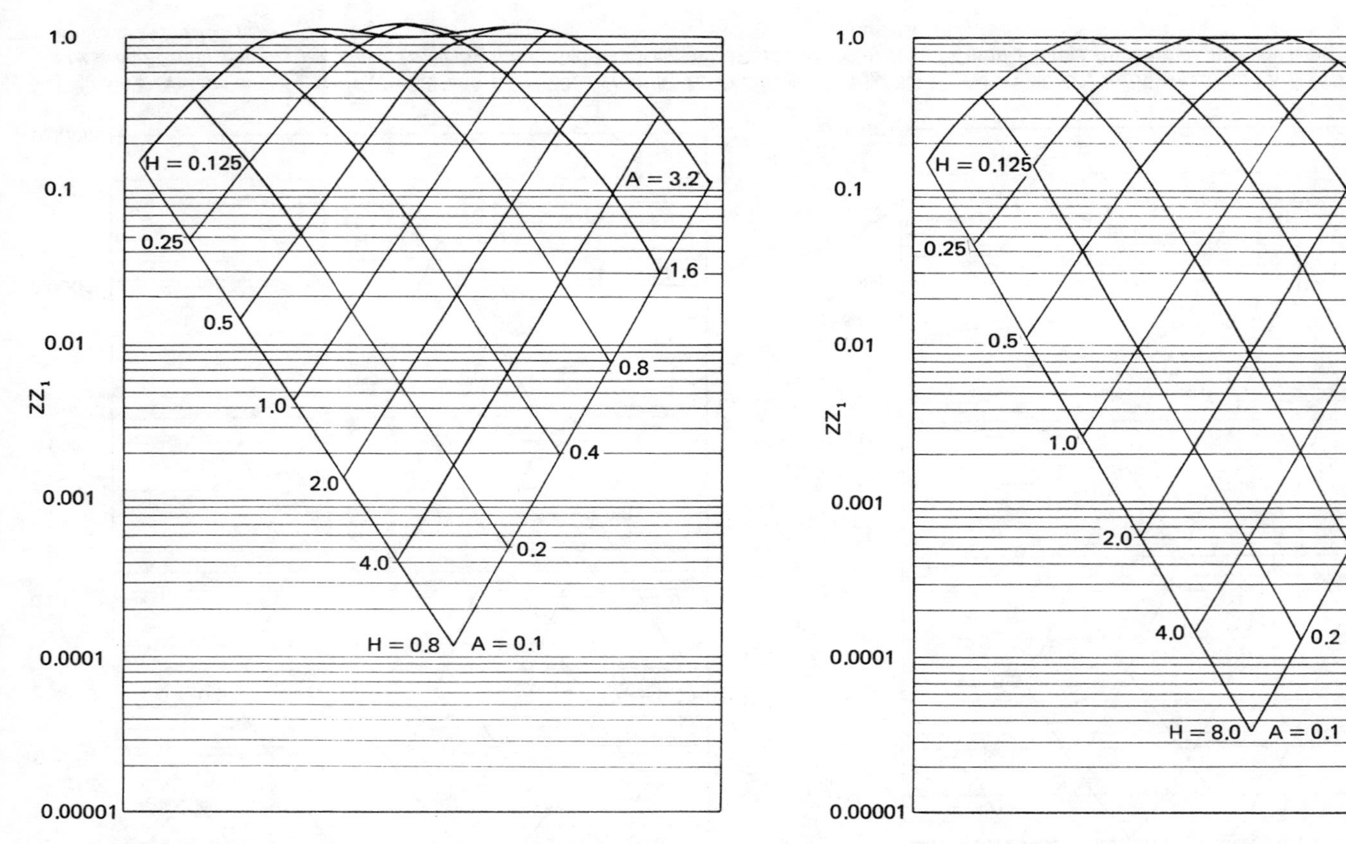

Fig. 3.44 Values of ZZ_1 for $K_1 = 20.0$ and $K_2 = 0.2$. *(After K. R. Peattie, Stress and Strain Factors for Three Layer Systems,* Highway Research Board, Bulletin 342, *1962.)*

Fig. 3.45 Values of ZZ_1 for $K_1 = 20$ and $K_2 = 2.0$. *(After K. R. Peattie, Stress and Strain Factors for Three Layer Systems,* Highway Research Board, Bulletin 342, *1962.)*

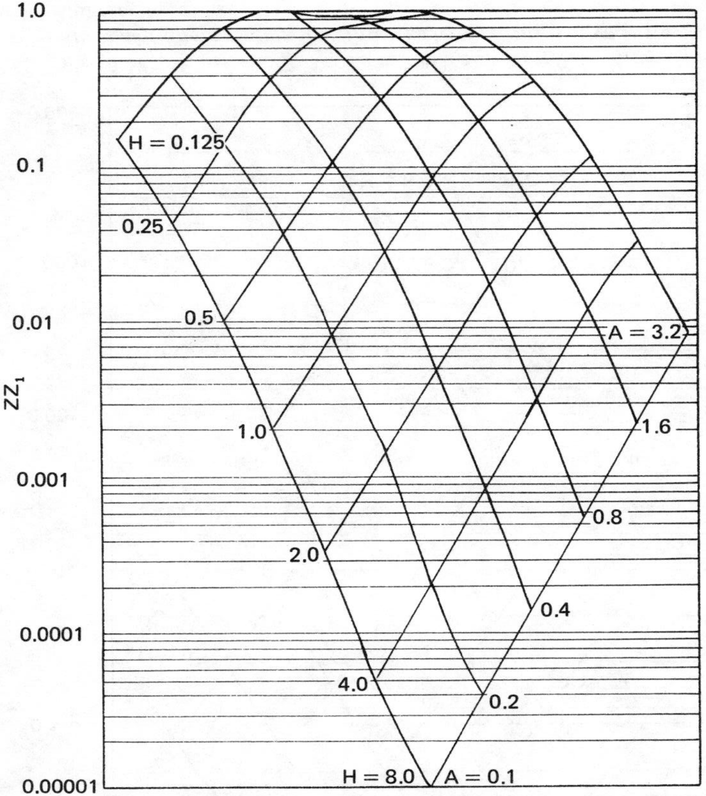

Fig. 3.46 Values of ZZ_1 for $K_1 = 20.0$ and $K_2 = 20.0$. *(After K. R. Peattie, Stress and Strain Factors for Three Layer Systems,* Highway Research Board, Bulletin 342, *1962.)*

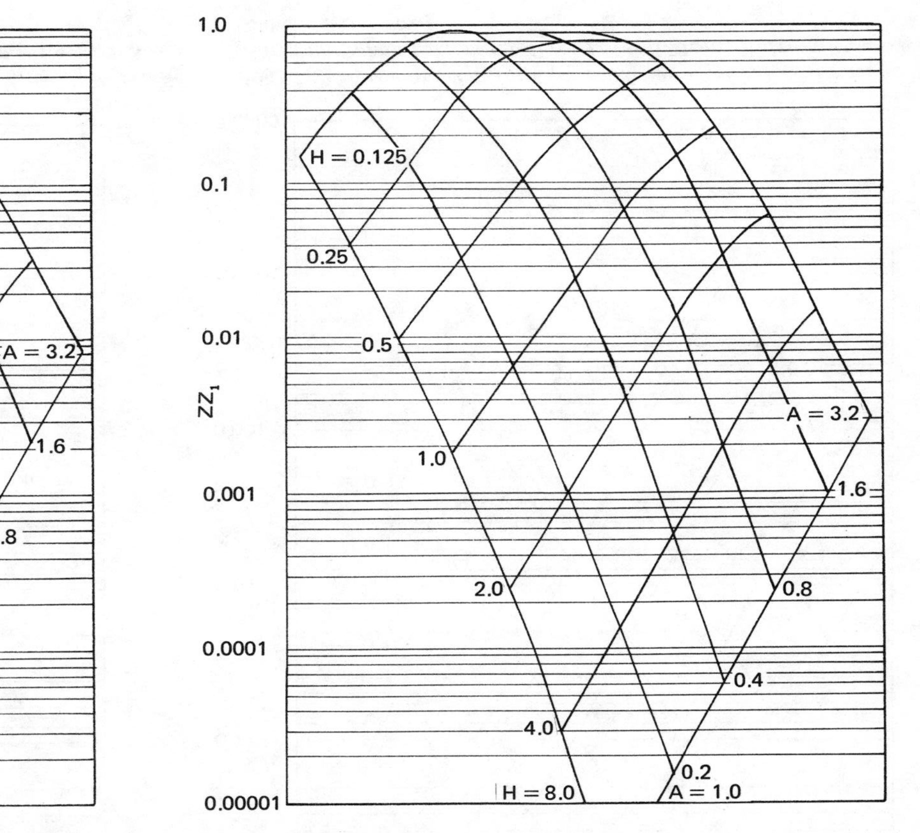

Fig. 3.47 Values of ZZ_1 for $K_1 = 20.0$ and $K_2 = 200.0$. *(After K. R. Peattie, Stress and Strain Factors for Three Layer Systems,* Highway Research Board, Bulletin 342, *1962.)*

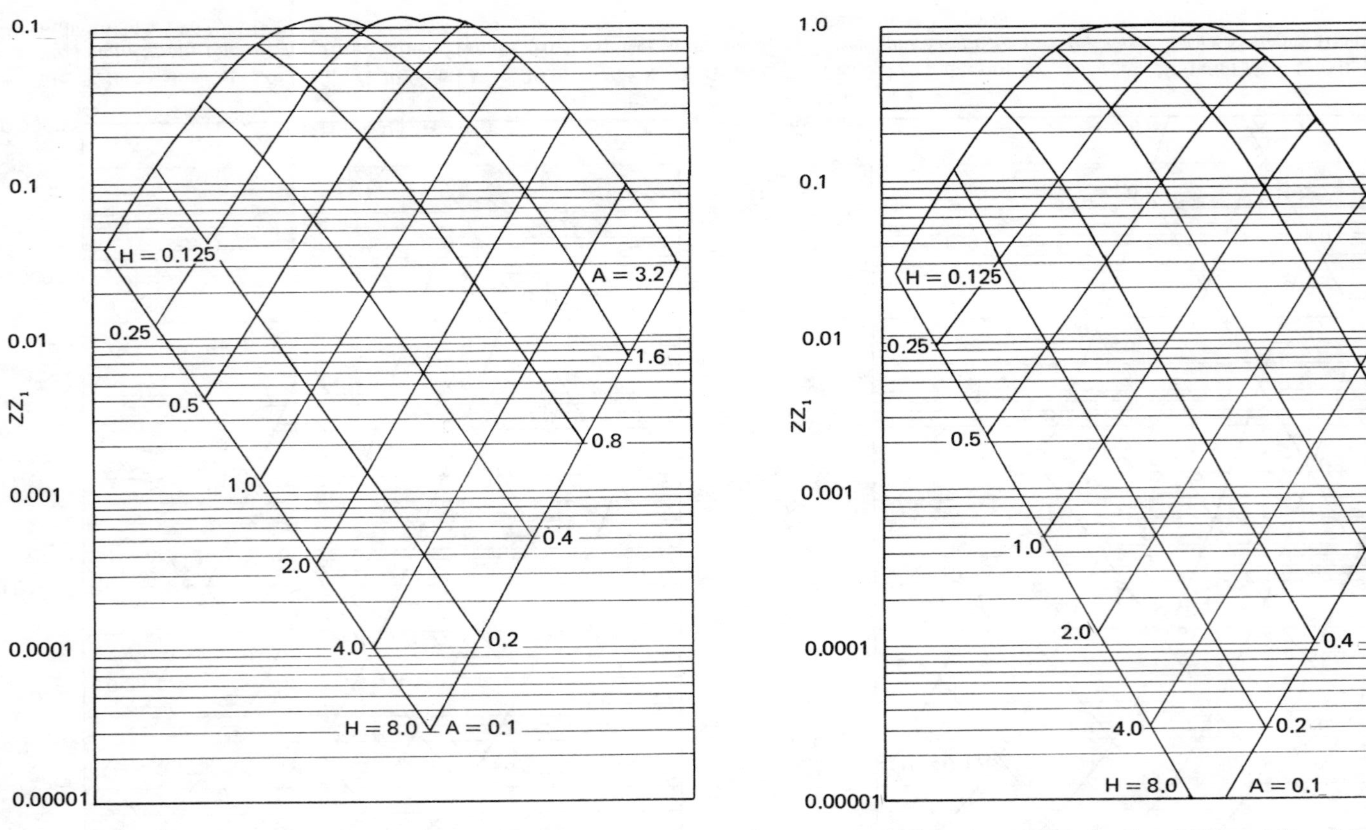

Fig. 3.48 Values of ZZ_1 for $K_1 = 200.0$ and $K_2 = 0.2$. *(After K. R. Peattie, Stress and Strain Factors for Three Layer Systems,* Highway Research Board, Bulletin 342, *1962.)*

Fig. 3.49 Values of ZZ_1 for $K_1 = 200.0$ and $K_2 = 2.0$. *(After K. R. Peattie, Stress and Strain Factors for Three Layer Systems,* Highway Research Board, Bulletin 342, *1962.)*

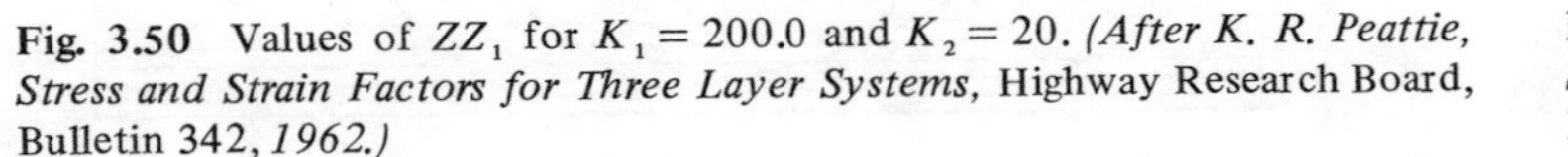

Fig. 3.50 Values of ZZ_1 for $K_1 = 200.0$ and $K_2 = 20$. *(After K. R. Peattie, Stress and Strain Factors for Three Layer Systems,* Highway Research Board, Bulletin 342, *1962.)*

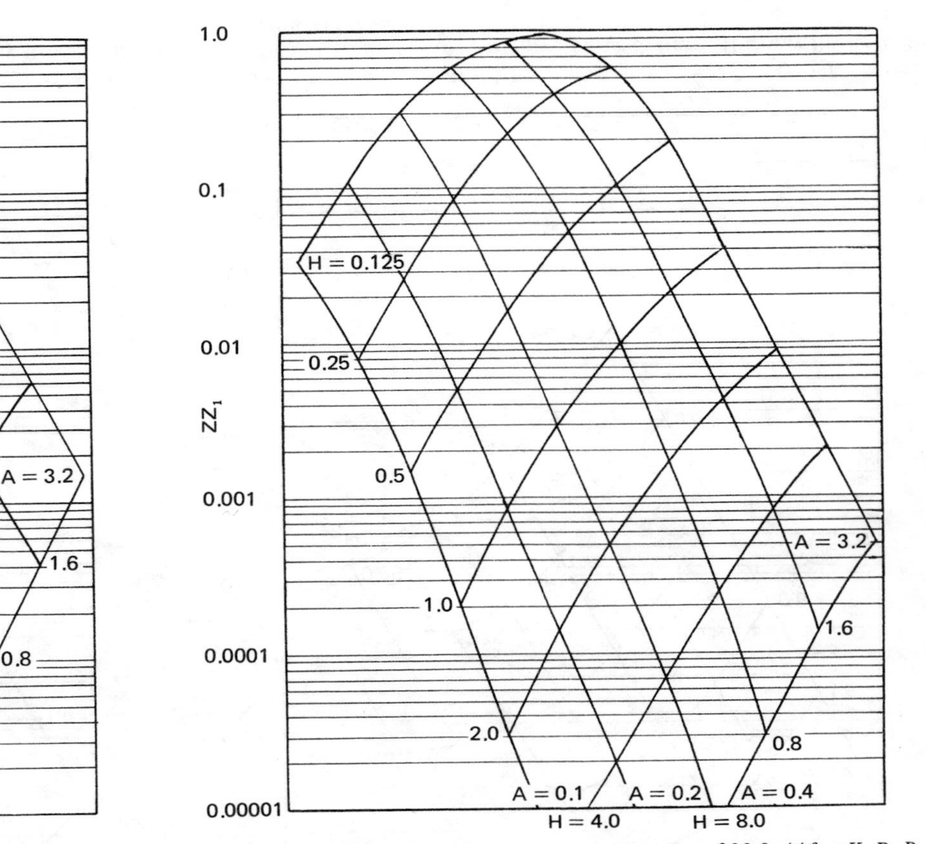

Fig. 3.51 Values of ZZ_1 for $K_1 = 200.0$ and $K_2 = 200.0$. *(After K. R. Peattie, Stress and Strain Factors for Three Layer Systems,* Highway Research Board, Bulletin 342, *1962.)*

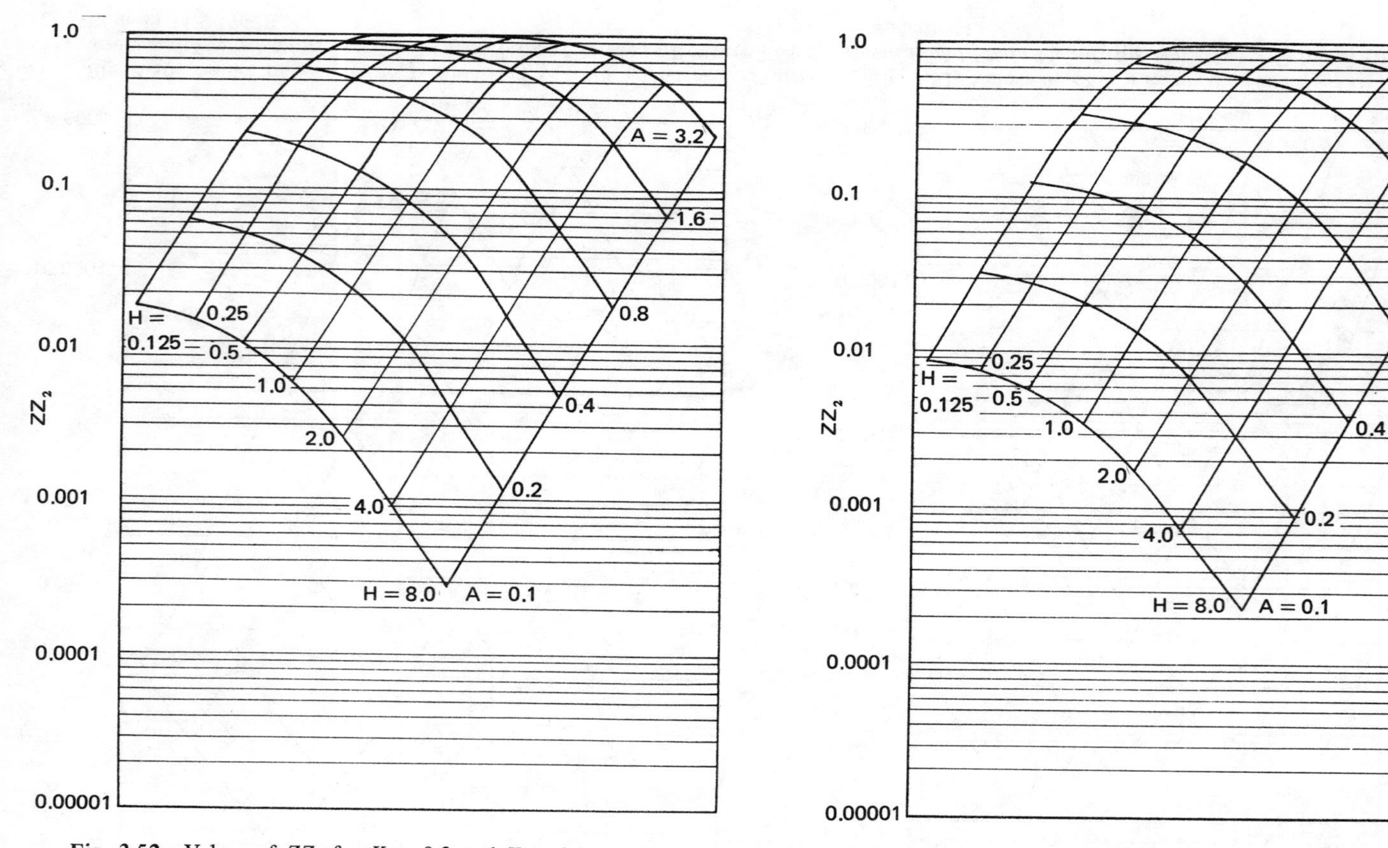

Fig. 3.52 Values of ZZ_2 for $K_1 = 0.2$ and $K_2 = 0.2$. *(After K. R. Peattie, Stress and Strain Factors for Three Layer Systems,* Highway Research Board, Bulletin 342, *1962.)*

Fig. 3.53 Values of ZZ_2 for $K_1 = 0.2$ and $K_2 = 2.0$. *(After K. R. Peattie, Stress and Strain Factors for Three Layer Systems,* Highway Research Board, Bulletin 342, *1962.)*

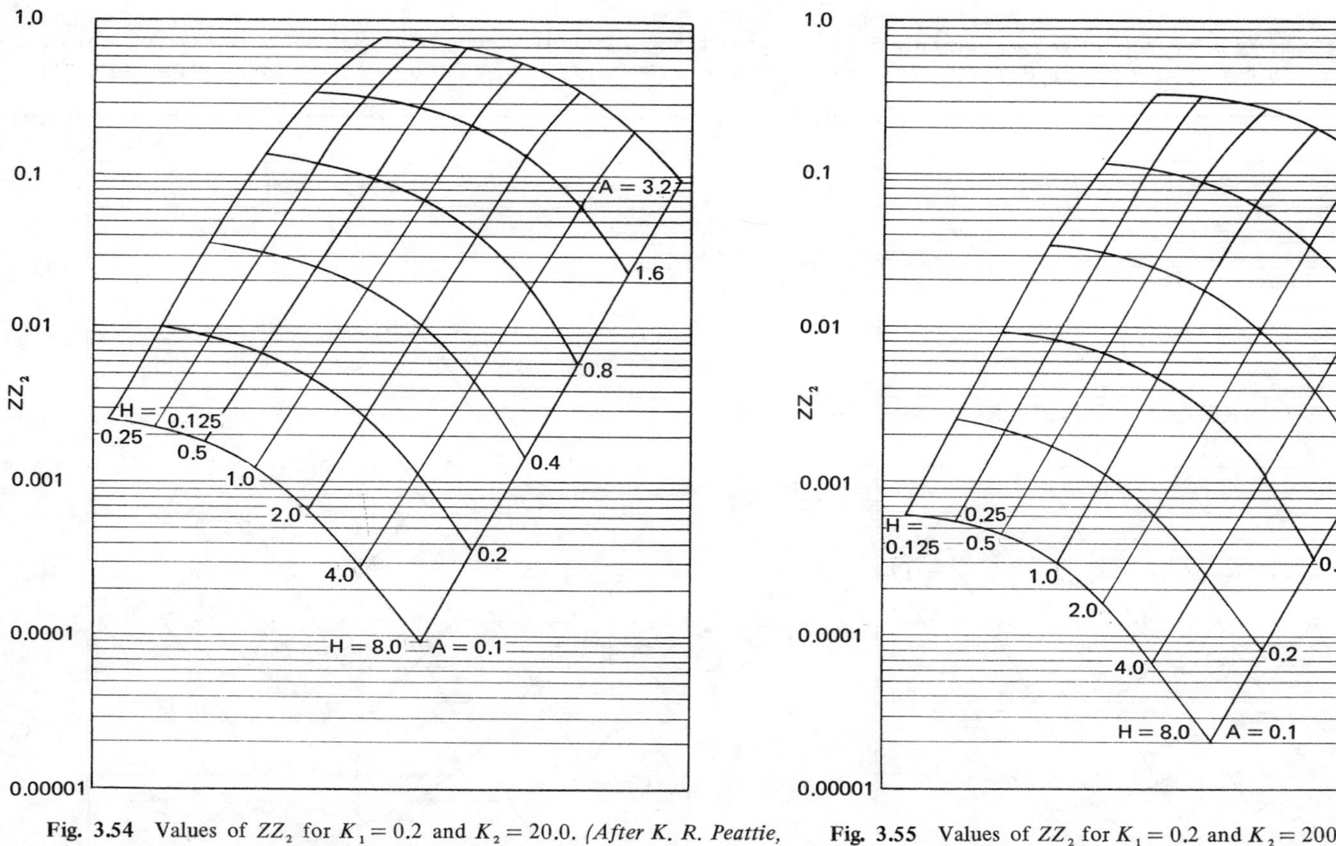

Fig. 3.54 Values of ZZ_2 for $K_1 = 0.2$ and $K_2 = 20.0$. *(After K. R. Peattie, Stress and Strain Factors for Three Layer Systems,* Highway Research Board, Bulletin 342, *1962.)*

Fig. 3.55 Values of ZZ_2 for $K_1 = 0.2$ and $K_2 = 200.0$. *(After K. R. Peattie, Stress and Strain Factors for Three Layer Systems,* Highway Research Board, Bulletin 342, *1962.)*

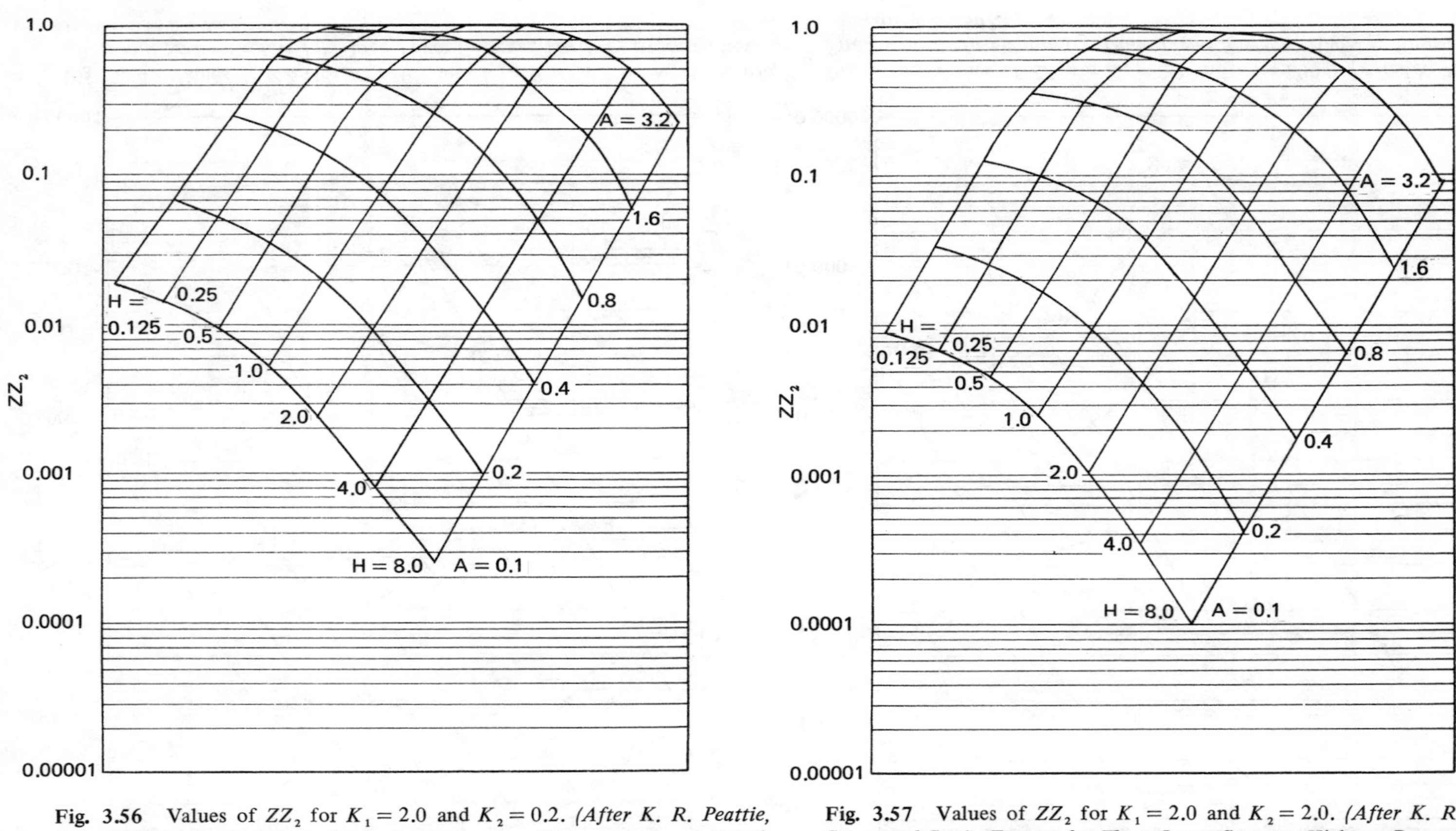

Fig. 3.56 Values of ZZ_2 for $K_1 = 2.0$ and $K_2 = 0.2$. *(After K. R. Peattie, Stress and Strain Factors for Three Layer Systems,* Highway Research Board, Bulletin 342, *1962.)*

Fig. 3.57 Values of ZZ_2 for $K_1 = 2.0$ and $K_2 = 2.0$. *(After K. R. Peattie, Stress and Strain Factors for Three Layer Systems,* Highway Research Board, Bulletin 342, *1962.)*

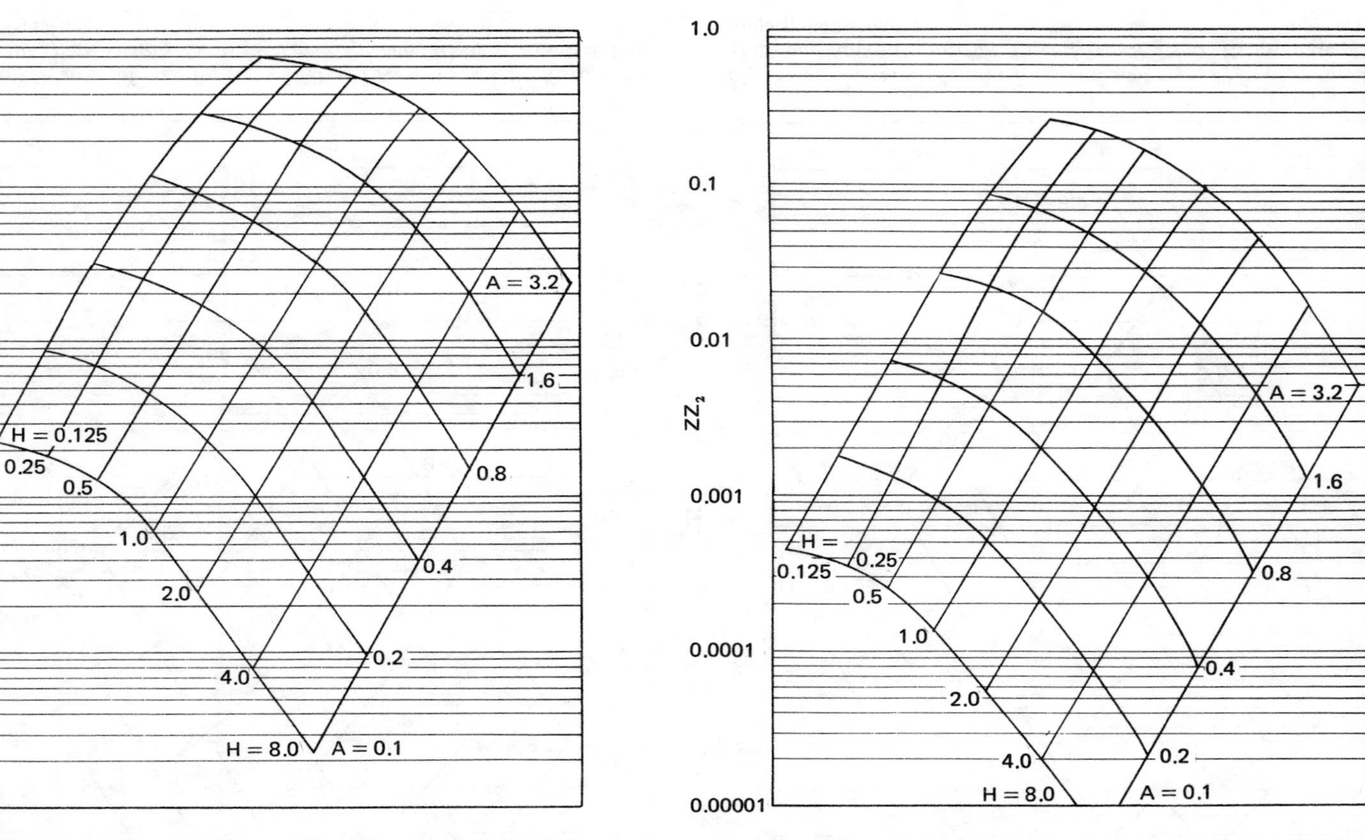

Fig. 3.58 Values of ZZ_2 for $K_1 = 2.0$ and $K_2 = 20.0$. *(After K. R. Peattie, Stress and Strain Factors for Three Layer Systems,* Highway Research Board, Bulletin 342, *1962.)*

Fig. 3.59 Values of ZZ_2 for $K_1 = 2.0$ and $K_2 = 200.0$. *(After K. R. Peattie, Stress and Strain Factors for Three Layer Systems,* Highway Research Board, Bulletin 342, *1962.)*

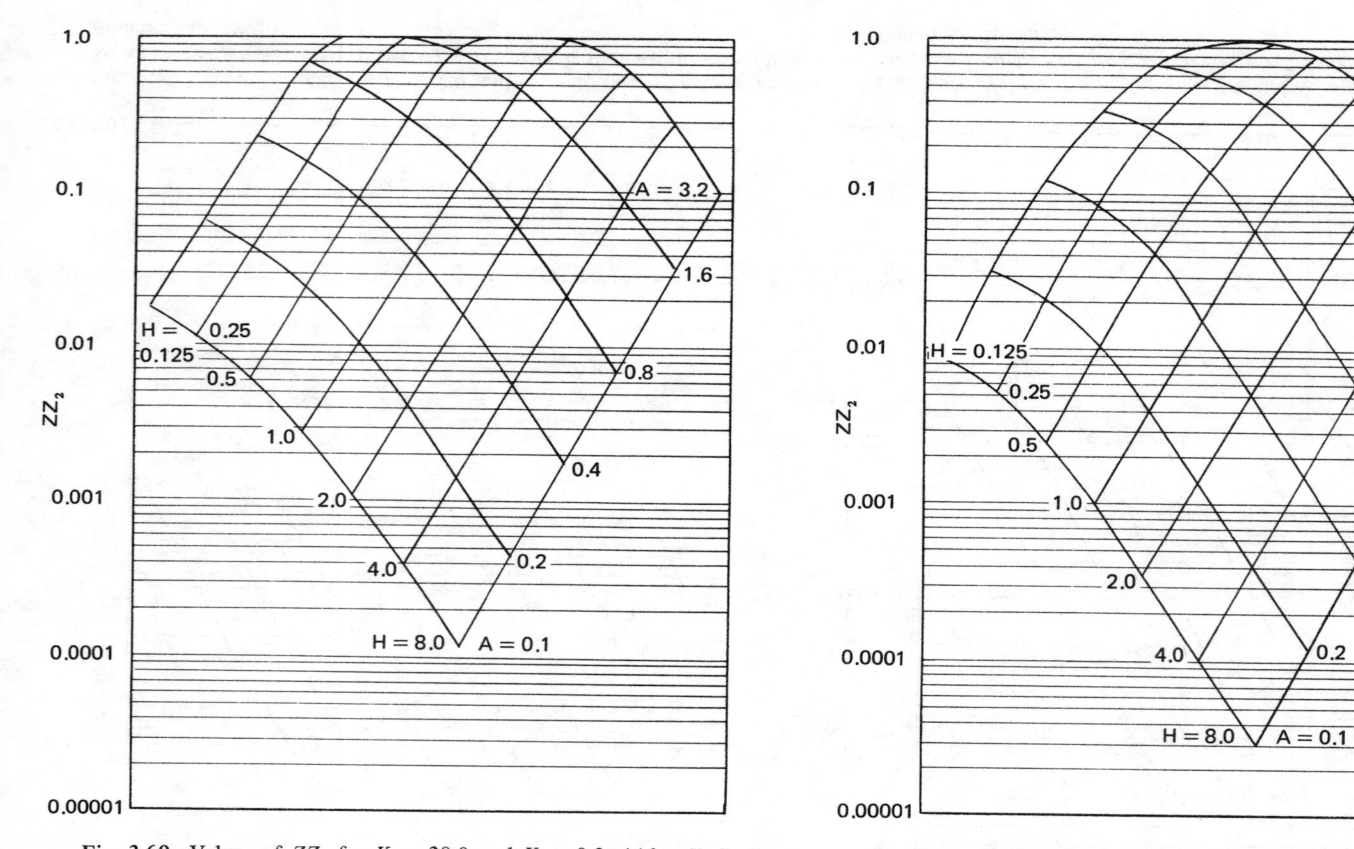

Fig. 3.60 Values of ZZ_2 for $K_1 = 20.0$ and $K_2 = 0.2$. *(After K. R. Peattie, Stress and Strain Factors for Three Layer Systems,* Highway Research Board, Bulletin 342, *1962.)*

Fig. 3.61 Values of ZZ_2 for $K_1 = 20.0$ and $K_2 = 20.0$. *(After K. R. Peattie, Stress and Strain Factors for Three Layer Systems,* Highway Research Board, Bulletin 342, *1962.)*

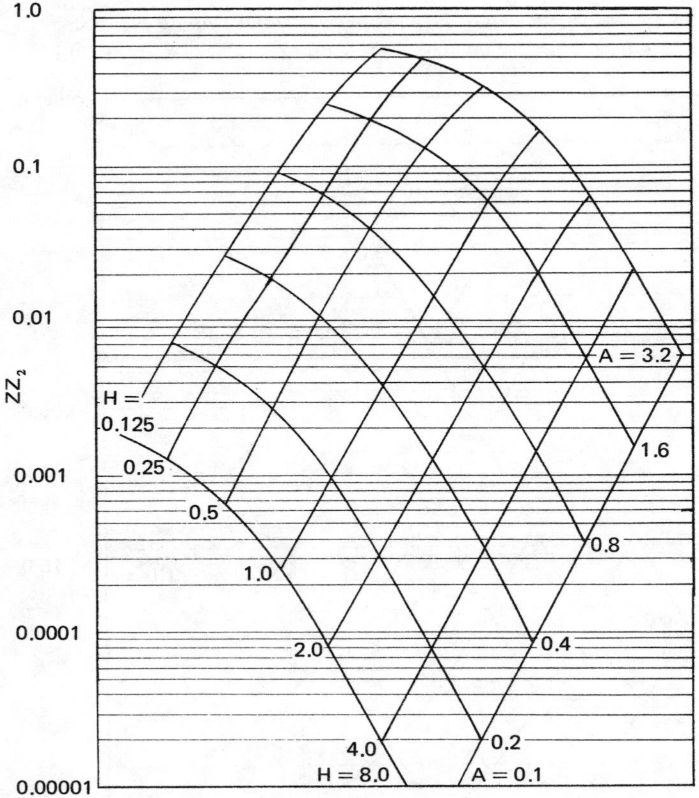

Fig. 3.62 Values of ZZ_2 for $K_1 = 20.0$ and $K_2 = 20.0$. *(After K. R. Peattie, Stress and Strain Factors for Three Layer Systems,* Highway Research Board, Bulletin 342, *1962.)*

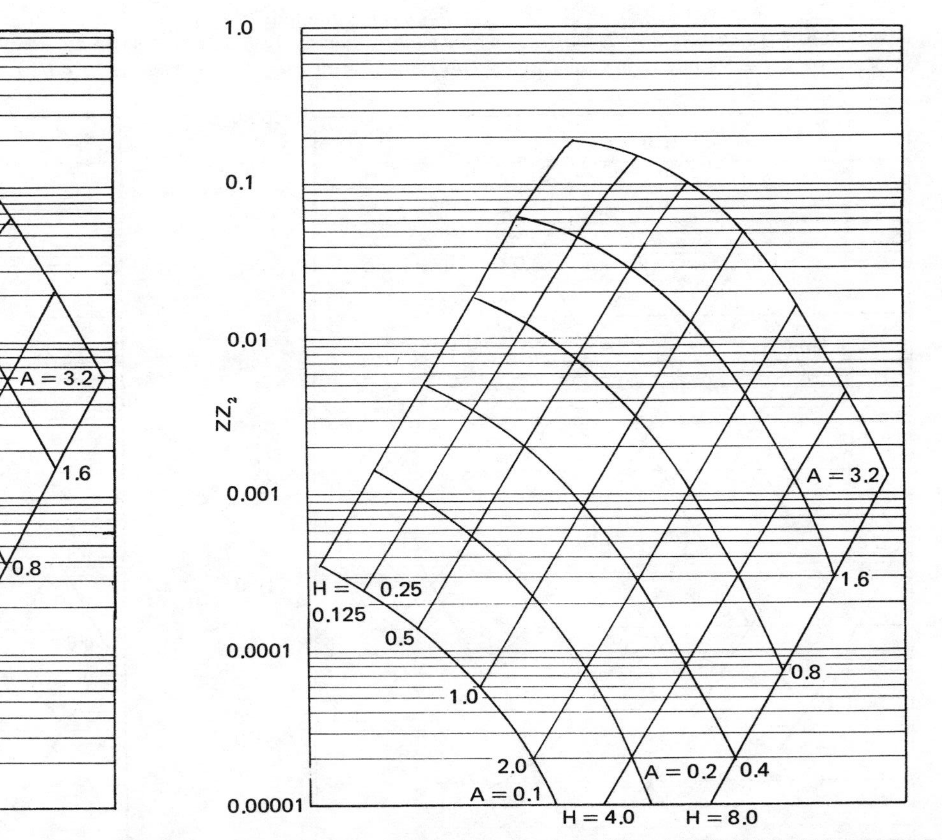

Fig. 3.63 Values of ZZ_2 for $K_1 = 20$ and $K_2 = 200.0$. *(After K. R. Peattie, Stress and Strain Factors for Three Layer Systems,* Highway Research Board, Bulletin 342, *1962.)*

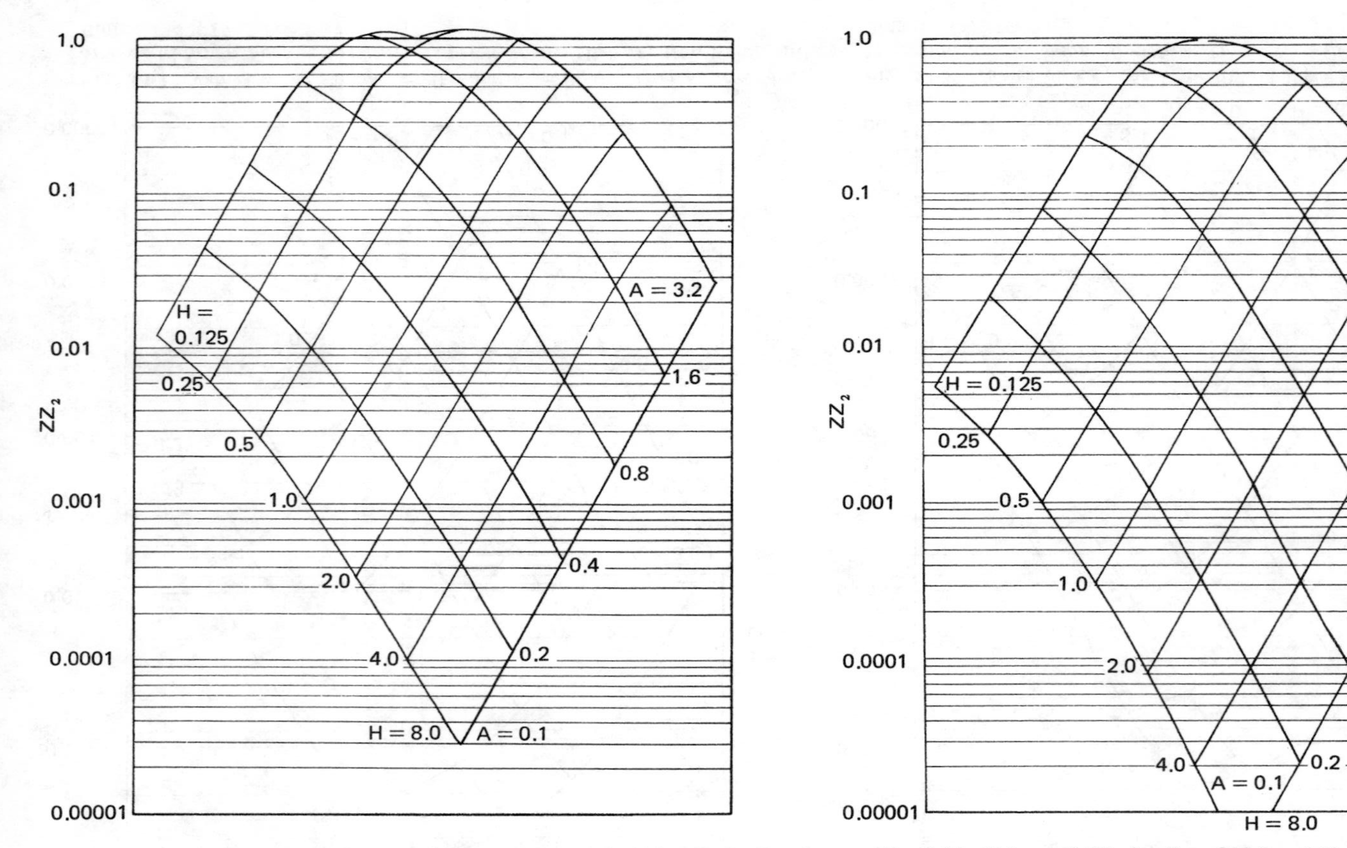

Fig. 3.64 Values of ZZ_2 for $K_1 = 200.0$ and $K_2 = 0.2$. *(After K. R. Peattie, Stress and Strain Factors for Three Layer Systems,* Highway Research Board, Bulletin 342, *1962.)*

Fig. 3.65 Values of ZZ_2 for $K_1 = 200.0$ and $K_2 = 2.0$. *(After K. R. Peattie, Stress and Strain Factors for Three Layer Systems,* Highway Research Board, Bulletin 342, *1962.)*

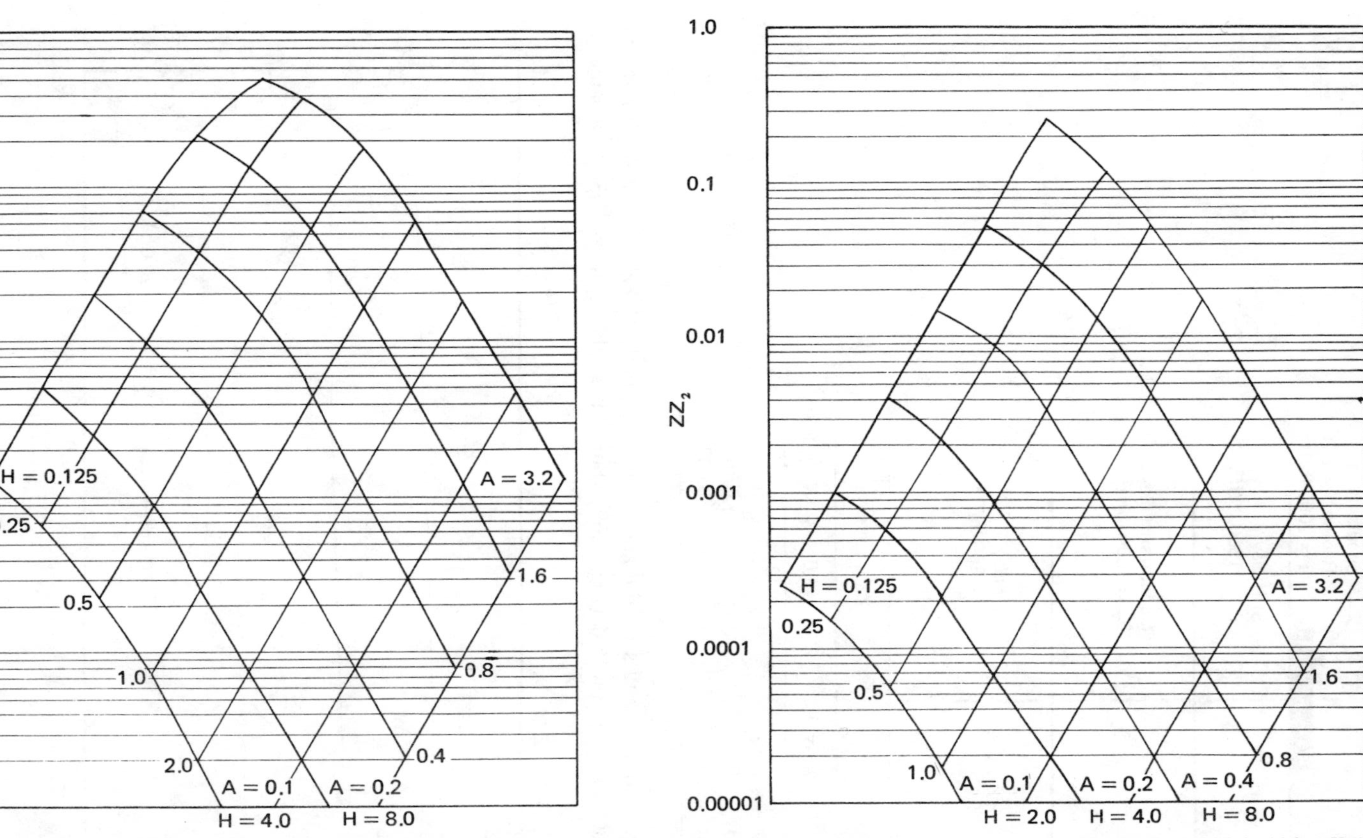

Fig. 3.66 Values of ZZ_2 for $K_1 = 200.0$ and $K_2 = 20.0$. *(After K. R. Peattie, Stress and Strain Factors for Three Layer Systems,* Highway Research Board, Bulletin 342, *1962.)*

Fig. 3.67 Values of ZZ_2 for $K_1 = 200.0$ and $K_2 = 200.0$. *(After K. R. Peattie, Stress and Strain Factors for Three Layer Systems,* Highway Research Board, Bulletin 342, *1962.)*

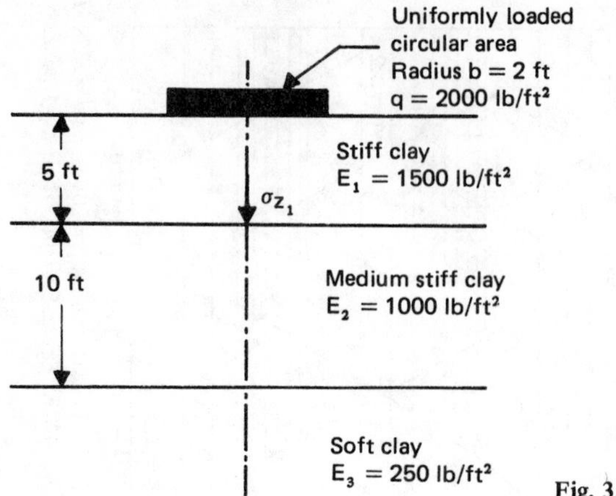

Fig. 3.68

K_1	ZZ_1		
	$K_2 = 0.2$	$K_2 = 2.0$	$K_2 = 20.0$
0.2	0.29	0.27	0.25
2.0	0.16	0.15	0.15
20.0	0.054	0.042	0.037

Based on the results of the above table, a graph of ZZ_1 against K_2 for various values of K_1 is plotted (Fig. 3.69). For this problem, $K_2 = 4$. So, the values of

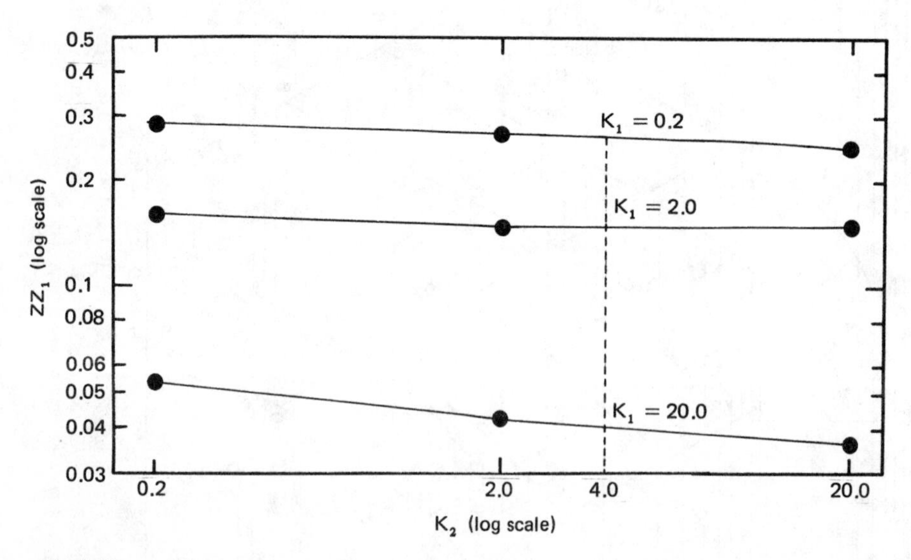

Fig. 3.69

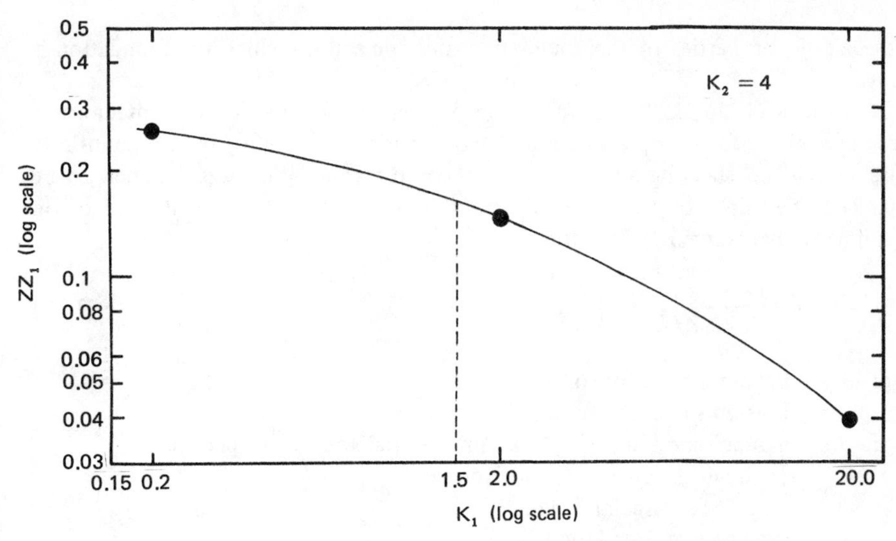

Fig. 3.70

ZZ_1 for $K_2 = 4$ and $K_1 = 0.2$, 2.0, and 20 are obtained from Fig. 3.69 and then plotted as in Fig. 3.70. From this graph, $ZZ_1 = 0.16$ for $K_1 = 1.5$. Thus,

$$\sigma_{z_1} = 2000\,(0.16) = \mathit{320\,lb/ft^2}$$

Example 3.6 For the problem given in Example 3.5, calculate the vertical stress at a depth of 5 ft below the center of the loaded area using Boussinesq's solution, Eq. (3.64), with $E = 1500\,\text{lb/in}^2$.

SOLUTION From Eq. (3.64),

$$\sigma_z = q\left[1 - \frac{z^3}{(b^2 + z^2)^{3/2}}\right] = q\left[1 - \frac{1}{[(b/z)^2 + 1]^{3/2}}\right]$$

$$\frac{b}{z} = \frac{2}{5} = 0.4$$

So,

$$\sigma_z = 2000\left[1 - \frac{1}{[(0.4)^2 + 1]^{3/2}}\right] = \mathit{400\,lb/ft^2}$$

3.2.8 Distribution of Contact Stress over Footings

In calculating vertical stress, we generally assume that the foundation of a structure is flexible. In practice this is not the case; no foundation is perfectly flexible, nor is it infinitely rigid. The actual nature of the distribution of contact stress will depend on the elastic properties of the foundation and the soil on which the foundation is resting.

Borowicka (1936, 1938) analyzed the problem of distribution of contact stress over uniformly loaded strip and circular rigid foundations resting on a semi-infinite elastic mass. The shearing stress at the base of the foundation was assumed to be zero. The analysis shows that the distribution of contact stress is dependent on a nondimensional factor K_r of the form

$$K_r = \frac{1}{6}\left(\frac{1-\nu_S^2}{1-\nu_F^2}\right)\left(\frac{E_F}{E_S}\right)\left(\frac{T}{b}\right)^3 \tag{3.79}$$

where ν_S = Poisson's ratio for soil
ν_F = Poisson's ratio for foundation material
E_F, E_S = Young's modulus of foundation material and soil, respectively
$b = \begin{cases}\text{half-width for strip foundation}\\ \text{radius for circular foundation}\end{cases}$
T = thickness of foundation

Figures 3.71*a* and *b* show the distribution of contact stress for circular and strip foundations. Note that $K_r = 0$ indicates a perfectly flexible foundation, and $K_r = \infty$ means a perfectly rigid foundation. This analysis indicates that, for a rigid strip ($K_r = \infty$) footing, $q_c/q \approx 0.67$ along the center line (q_c is the contact stress, and q is the load per unit area applied on the foundation); for rigid circular foundations, $q_c/q = 0.5$ at the center. However, in the case of rigid foundations, q_c approaches infinity at the edge.

The practical conditions of rigid and flexible footings resting on sand and clay soils are considered below.

Foundations on clay. When a flexible foundation resting on a saturated clay ($\phi = 0$) is loaded with a uniformly distributed load (q/unit area), it will deform and take a bowl shape (Fig. 3.72*a*). Maximum deflection will be at the center; however, the contact stress over the footing will be uniform (q per unit area).

A rigid foundation resting on the same clay will show a uniform settlement (Fig. 3.72*b*). The contact stress distribution will take a form such as that shown in Fig. 3.71, with only one exception: the stress at the edges of the footing cannot be infinity. Soil is not an infinitely elastic material; beyond a certain limiting stress $[q_{c(\max)}]$, plastic flow will begin.

Foundations on sand. For a flexible foundation resting on a cohesionless soil ($c = 0$), the distribution of contact pressure will be uniform (Fig. 3.73*a*). However, the edges of the foundation will undergo a larger settlement than the center. This occurs because the soil located at the edge of the foundation lacks lateral confining pressure and hence possesses less strength. The lower strength of the soil at the edge of the foundation will result in larger settlement.

A rigid foundation resting on a sand layer will settle uniformly. The contact pressure on the foundation will increase from zero at the edge to a maximum at the center, as shown in Fig. 3.73*b*.

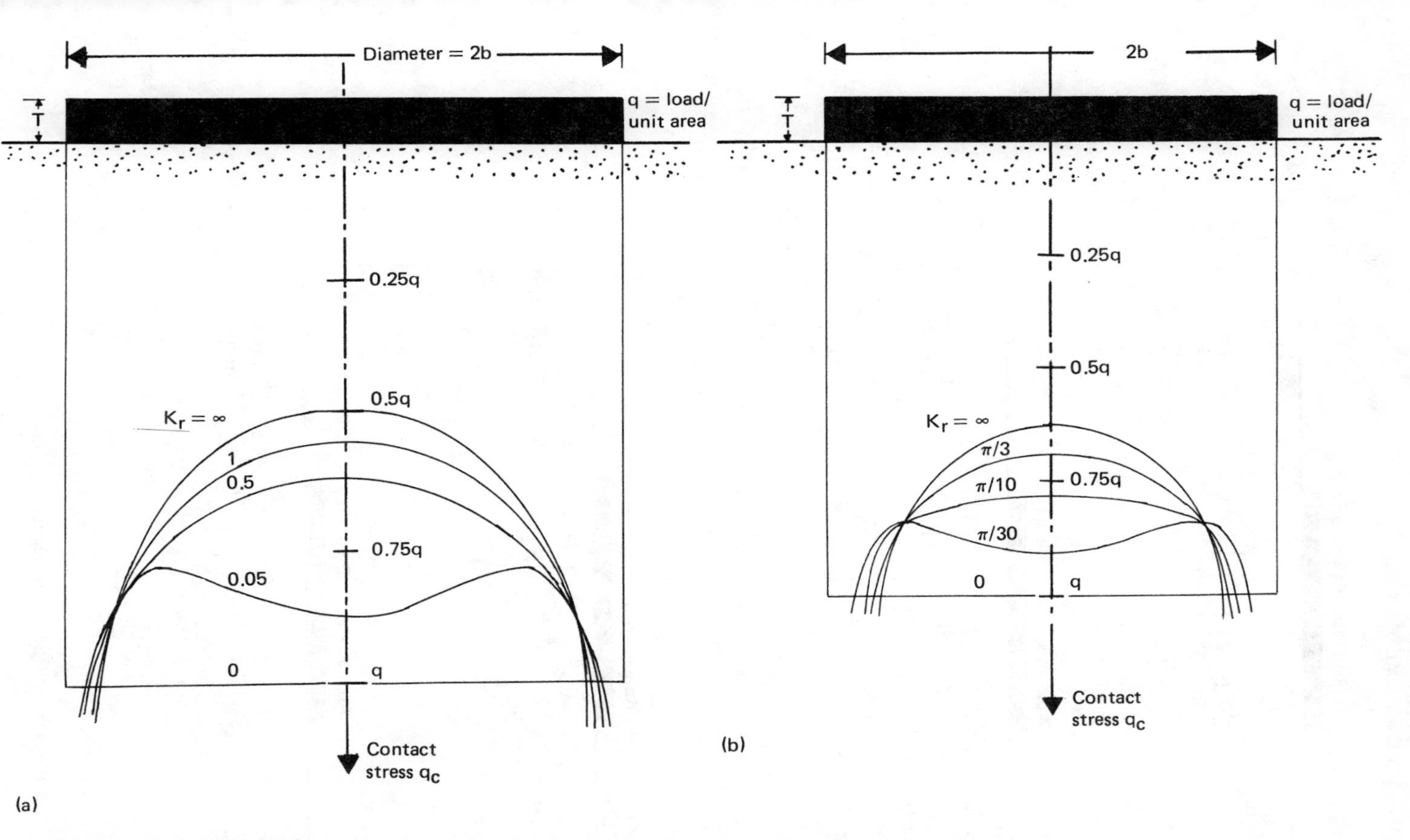

Fig. 3.71 Contact stress over rigid foundations resting on an elastic medium. (*a*) Circular foundation. (*b*) Strip foundation.

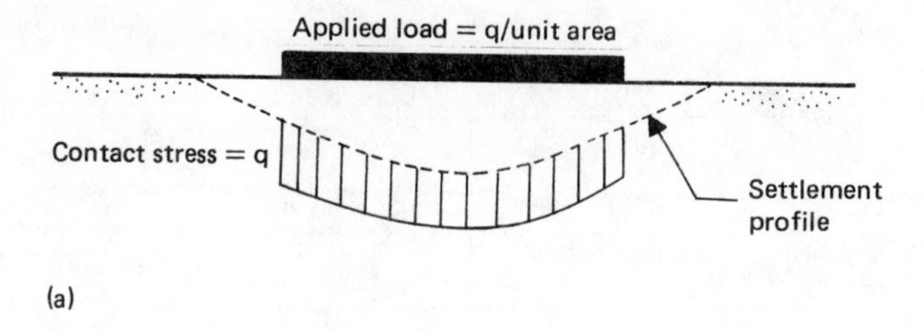

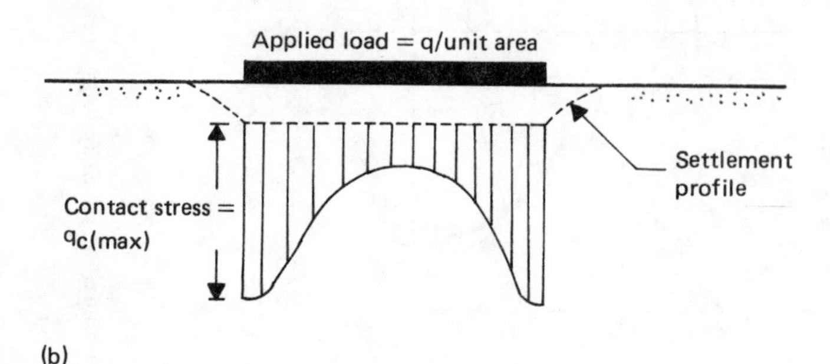

Fig. 3.72 Flexible (*a*) and rigid (*b*) foundations on clay.

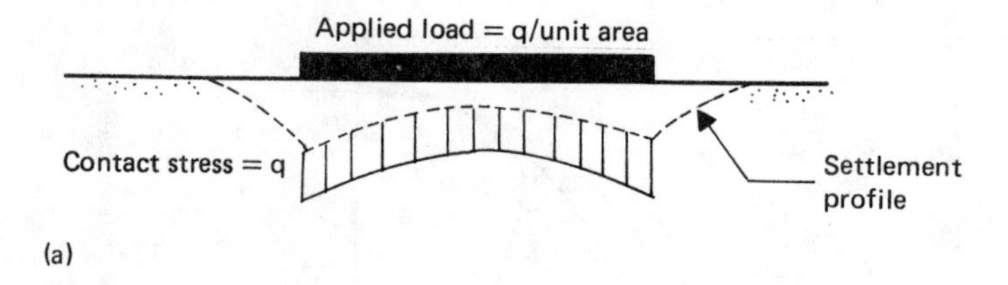

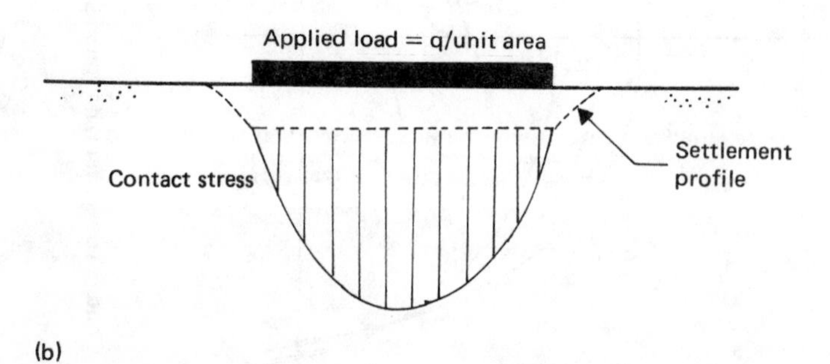

Fig. 3.73 Flexible (*a*) and rigid (*b*) foundations on sand.

3.2.9 Reliability of Stress Calculation by Using the Theory of Elasticity

Only a limited number of attempts have been made so far to compare theoretical results for stress distribution with the stresses observed under field conditions. The latter, of course, require elaborate field instrumentation. However, from the results available at present fairly good agreement is shown between theoretical considerations and field conditions especially in the case of vertical stress. In any case, a variation of about 20% to 30% between the theory and field conditions may be expected.

PROBLEMS

3.1 A line load of q per unit length is applied at the ground surface as shown in Fig. P3.1. Given: $q = 3000\,\text{lb/ft}^2$ and $\alpha = 0°$.

(*a*) Plot the variations of σ_z, σ_x, and τ_{xz} against x from $x = +20$ ft to $x = -20$ ft for $z = 8$ ft.

(*b*) Plot the variation of σ_z with z (from $z = 0$ ft to $z = 20$ ft) for $x = 0$.

(*c*) Plot the variation of σ_z with z (from $z = 0$ ft to $z = 20$ ft) for $x = 5$ ft.

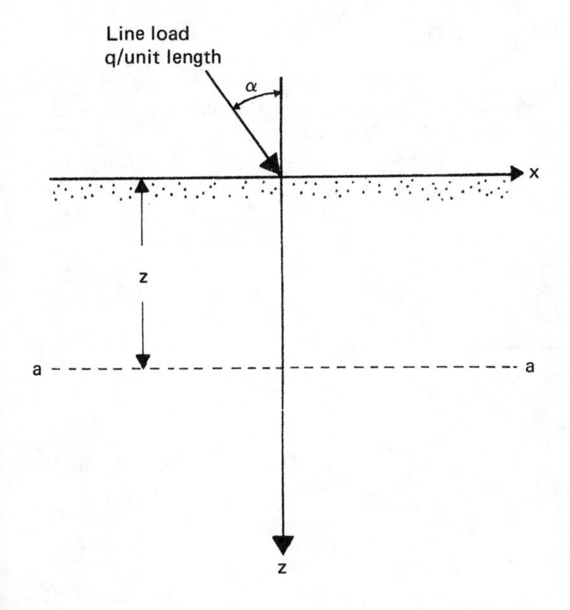

Fig. P3.1

3.2 Refer to Fig. P3.1. Assume that $q = 45$ kN/m and $\alpha = 90°$.

(*a*) If $z = 5$ m, plot the variation of σ_z, σ_x, and τ_{xz} against x for the range $x = \pm 10$ m.

(*b*) Plot the variation of σ_z with z for the range $z = 0$ m to $z = 10$ m (for $x = 0$ m).

(*c*) Plot the variation of σ_z with z for the range $z = 0$ m to $z = 10$ m (for $x = 5$ m).

3.3 Refer to Fig. P3.1. Given that $q = 3500$ lb/ft, $\alpha = 0°$, $\nu = 0.35$, and $z = 5$ ft, calculate the major, intermediate, and minor principal stresses at $x = 0, 5, 10, 15$, and 20 ft.

3.4 Refer to Fig. P3.1. Given that $q = 38$ kN/m, $\alpha = 90°$, $\nu = 0.3$, and $z = 1$ m, calculate the major, intermediate, and minor principal stresses at $x = 0, 0.5, 1$, and 1.5 m.

3.5 A line load of 3500 lb/ft is applied at the ground surface as shown in Fig. P3.2. Determine the major principal stresses at the grid points and draw the stress contours. What is the shape of these stress contours?

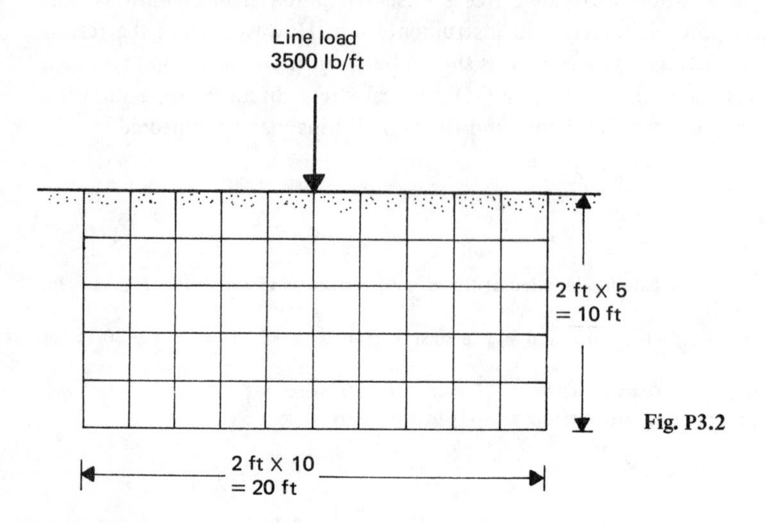

Fig. P3.2

3.6 Refer to Fig. P3.3. Given that $\alpha_1 = 90°$, $\alpha_2 = 90°$, $a = 5$ ft, $a_1 = 10$ ft, $a_2 = 10$ ft, $b = 5$ ft, $q_1 = 2500$ lb/ft, and $q_2 = 3500$ lb/ft, plot the variation of σ_z along MN.

3.7 Refer to Fig. P3.3. Assuming that $\alpha_1 = 30°$, $\alpha_2 = 45°$, $a = 2$ m, $a_1 = 3$ m, $a_2 = 5$ m, $b = 2$ m, $q_1 = 40$ kN/m, and $q_2 = 30$ kN/m, plot the variation of σ_z along MN.

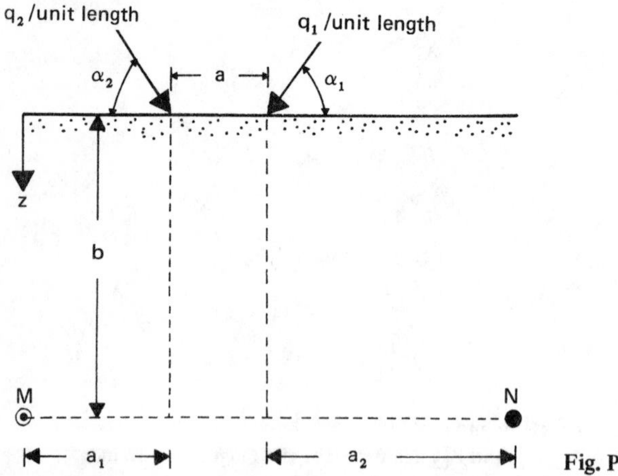

Fig. P3.3

3.8 Two line loads are applied on an elastic soil layer (sand) underlain by a rigid rough base as shown in Fig. P3.4. Given that $q_1 = 55$ kN/m, $q_2 = 35$ kN/m, $a = 3$ m, $h = 5$ m, $b = 2$ m, and $\nu = 0.2$, find the vertical stress σ_z at N.

3.9 Refer to Fig. P3.4. Given that $a = 10$ ft, $b = 6$ ft, $h = 10$ ft, and $\nu = 0.2$, if the stresses at points M and N are 280 lb/ft² and 315 lb/ft², respectively, determine the magnitude of the loads q_1 and q_2 in lb/ft.

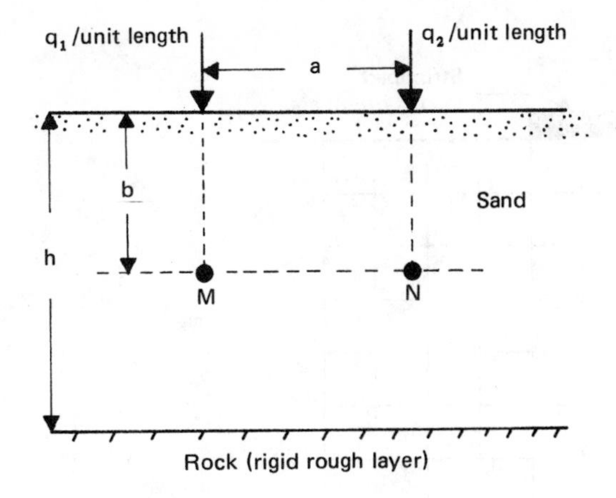

Fig. P3.4

3.10 Refer to Fig. 3.12*b* and Eqs. (3.34) to (3.36). For a uniformly distributed vertical load q on an infinite strip on the surface of a semi-infinite elastic layer, derive the following relations for the point P:

Maximum principal stress $\quad \sigma_1 = \dfrac{q}{\pi}(\alpha + \sin\alpha)$

Minimum principal stress $\quad \sigma_3 = \dfrac{q}{\pi}(\alpha - \sin\alpha)$

Maximum shear stress $\quad \tau_{max} = \dfrac{q}{\pi}\sin\alpha$

3.11 For the infinite strip load shown in Fig. P3.5, given $B = 4$ m, $q = 105$ kN/m², and $\nu = 0.3$, draw the variation of σ_x, σ_y, σ_z, $\sigma_{P(1)}$ (maximum principal stress), and $\sigma_{P(3)}$ (minimum principal stress) with x (from $x = +8$ m to $x = -8$ m) at $z = 3$ m.

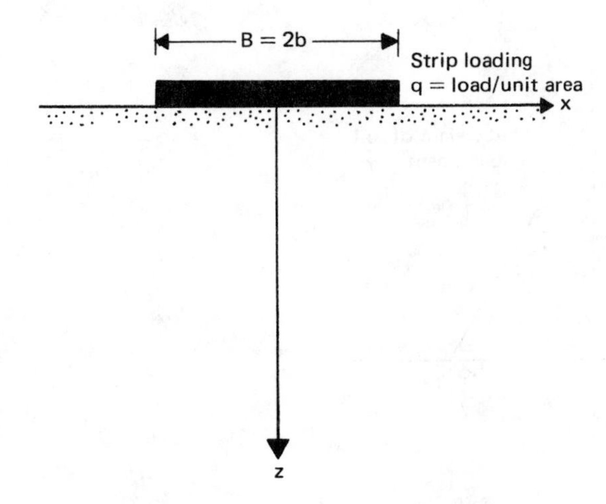

Fig. P3.5

3.12 (*a*) Determine the vertical stress σ_z due to the strip load shown in Fig. P3.6 at the grid points in terms of q.

(*b*) Plot the stress isobars for $\sigma_z/q = 0.9, 0.8, 0.7, 0.6, 0.5, 0.4, 0.3, 0.2$, and 0.1.

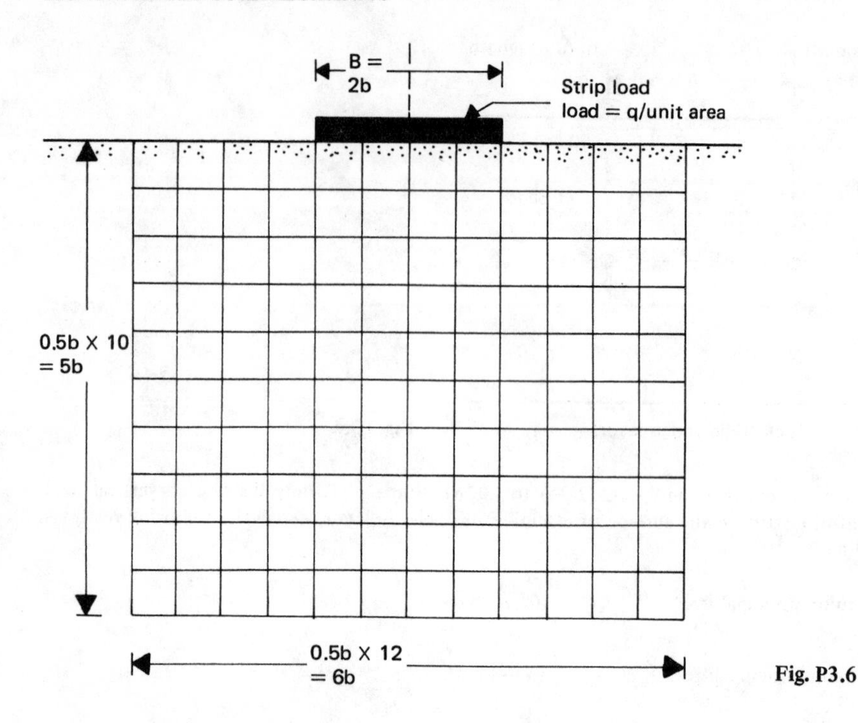

Fig. P3.6

3.13 An embankment is shown in Fig. P3.7. Given that $B = 5$ m, $H = 5$ m, $m = 1.5$, $z = 3$ m, $a = 3$ m, $b = 4$ m, and $\gamma = 18$ kN/m³, determine the stresses at A, B, C, D, and E.

3.14 Redo Prob. 3.13 for $B = 10$ ft, $H = 10$ ft, $m = 2$, $z = 10$ ft, $a = 8$ ft, $b = 8$ ft, and $\gamma = 110$ lb/ft³.

3.15 Refer to Fig. 3.23. Given that $\nu = 0.5$, $q = 192$ kN/m², $b = 3$ m, and $s = 0$, calculate the vertical stresses, σ_z, σ_r, and σ_θ at $z = 0$, 0.75, 1.5, 2.25, 3, 4.5, and 6 m, and plot the variations against z.

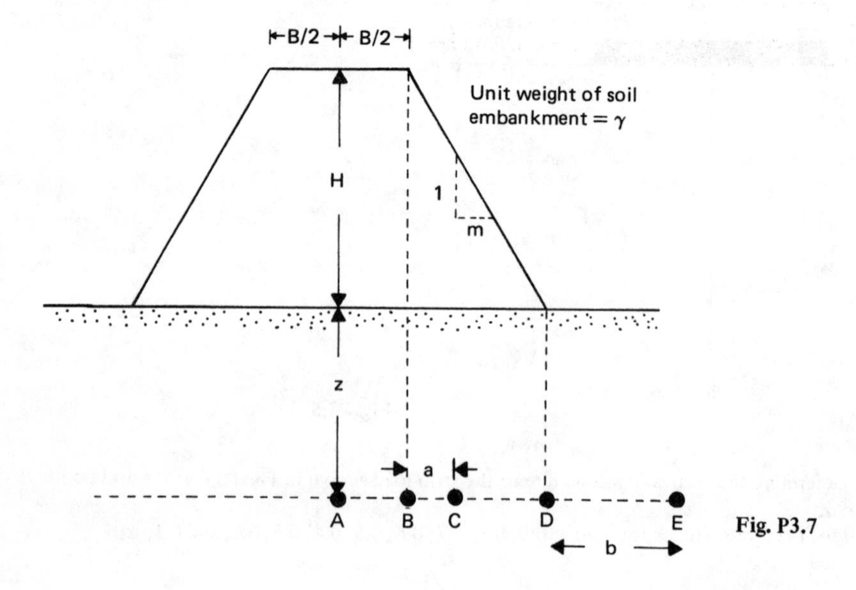

Fig. P3.7

3.16 Refer to Fig. 3.23. Given that $\nu = 0.35$, $q = 2800\,\text{lb/ft}^2$, $b = 5$ ft, and $s = 2.5$ ft, determine the principal stresses at $z = 2.5, 5$, and 10 ft.

3.17 Fig. P3.8 shows the plan of a loaded area on the surface of a clay layer. The uniformly distributed vertical loads on the area are also shown. Determine the vertical stress increase at A and B due to the loaded area. A and B are located at a depth of 3 m below the ground surface.

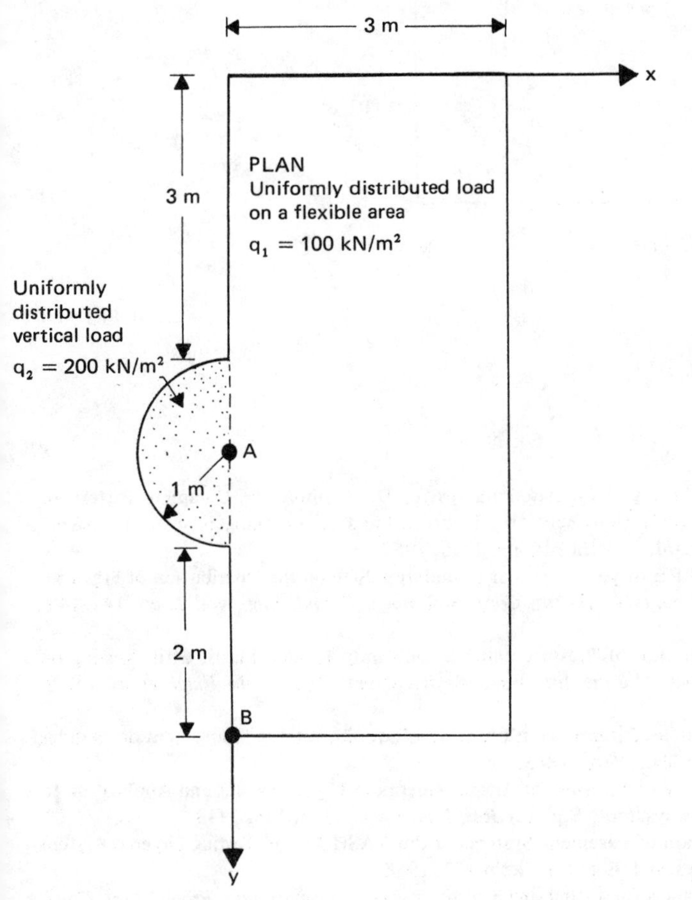

Fig. P3.8

3.18 The plan of a rectangular loaded area on the surface of a silty clay layer is shown in Fig. P3.7. The uniformly distributed vertical load on the rectangular area is $3500\,\text{lb/ft}^2$. Determine the vertical stresses due to the loaded area at A, B, C, D, and E. All points are located at a depth of 5 ft below the ground surface.

3.19 Solve Prob. 3.18 using Newmark's chart.

3.20 Solve Prob. 3.17 using Newmark's chart.

3.21 Determine σ_x and σ_y for points A, B, C, D, and E stated in Prob. 3.18 using Newmark's chart ($\nu = 0.5$).

3.22 Determine σ_x and σ_y for points A and B stated in Prob. 3.17 using Newmark's chart ($\nu = 0.5$).

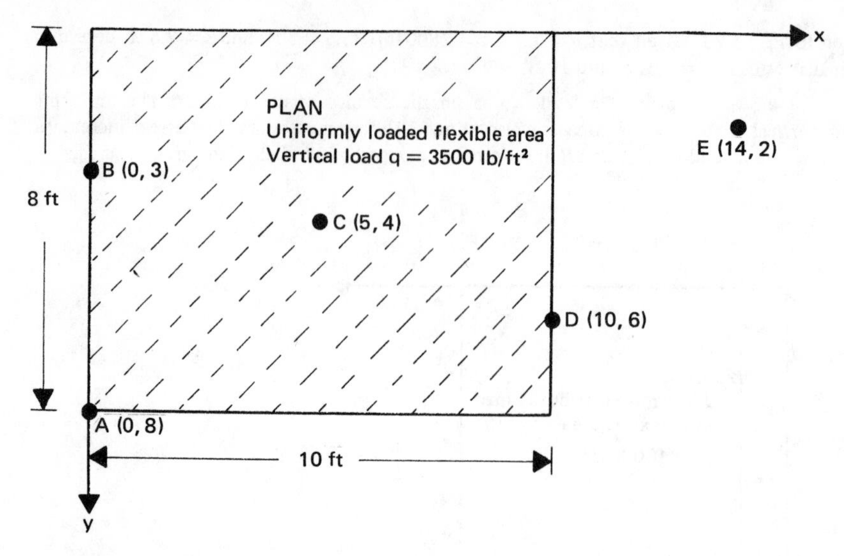

Fig. P3.9

REFERENCES

Ahlvin, R. G., and H. H. Ulery, "Tabulated Values for Determining the Complete Pattern of Stresses, Strains, and Deflections beneath a Uniform Load on a Homogeneous Half Space," Highway Research Record, Bulletin 342, pp. 1-13, 1962.

Borowicka, H., Influence of Rigidity of a Circular Foundation Slab on the Distribution of Pressures over the Contact Surface, *Proc. 1st Int. Conf. Soil Mech. Found. Eng.*, vol. 2, pp. 144–149, 1936.

Borowicka, H., The Distribution of Pressure under a Uniformly Loaded Elastic Strip Resting on Elastic-Isotropic Ground, *2d Cong. Int. Assoc. Bridge Struct. Eng., Berlin, Final report,* vol. 8, no. 3, 1938.

Boussinesq, J., "Application des Potentials a L'Etude de L'Equilibre et due Mouvement des Solides Elastiques," Gauthier-Villars, Paris, 1883.

Burmister, D. M., The Theory of Stresses and Displacements in Layer Systems and Application to Design of Airport Runways, *Proc. Highway Res. Board,* vol. 23, p. 126, 1943.

Burmister, D. M., "Evaluation of Pavement Systems of the WASHO Road Testing Layered System Methods," Highway Research Board, Bulletin 177, 1958.

Fadum, R. E., Influence Values for Estimating Stresses in Elastic Foundations, *Proc. 2d Int. Conf. Soil Mech. Found. Eng.*, vol. 3, pp. 77–84, 1948.

Fox, L., Computation of Traffic Stresses in a Simple Road Structure, *Proc., 2d Int. Conf. Soil Mech. Found. Eng.,* vol. 2, pp. 236–246, 1948.

Giroud, J. P., Stresses Under Linearly Loaded Rectangular Area, *J. Soil Mech. Found. Div., ASCE,* vol. 98, no. SM1, pp. 263–268, 1970.

Holl, D. L., Stress Transmission on Earths, *Proc. Highway Res. Board,* vol. 20, pp. 709–772, 1940.

Jones, A., "Tables of Stresses in Three-Layer Elastic Systems," Highway Research Board, Bulletin 342, pp. 176–214, 1962.

Jurgenson, L., The Application of Theories of Elasticity and Plasticity to Foundation Problems, *in* "Contribution to Soil Mechanics, 1925–1940," Boston Society of Civil Engineers, Boston, 1934.

Melan, E., Der Spanningzustand der durch eine Einzelkraft im Innern beanspruchten Halbschiebe, *Z. Angew, Math. Mech.* vol. 12, 1932.

Newmark, N. M., "Simplified Computation of Vertical Pressures in Elastic Foundations," University of Illinois Engineering Experiment Station, Circular 24, 1935.

Newmark, N. M., "Influence Charts for Computation of Stresses in Elastic Soils," University of Illinois Engineering Experiment Station, Bulletin No. 338, 1942.

Osterberg, J. O., Influence Values for Vertical Stresses in Semi-infinite Mass due to Embankment Loading, *Proc. 4th Int. Conf. Soil Mech. Found. Eng.*, vol. 1, p. 393, 1957.

Peattie, K. R., "Stresses and Strain Factors for Three-Layer Systems," Highway Research Board, Bulletin 342, pp. 215–253, 1962.

Poulos, H. G., "Stresses and Displacements in an Elastic Layer Underlain by a Rough Rigid Base." Civil Eng. Research Report No. R63, University of Sydney, Australia, 1966.

Poulos, H. G., and E. H. Davis, "Elastic Solutions for Soil and Rock Mechanics," Wiley, New York, 1974.

Timoshenko, S. P., and J. N. Goodier, "Theory of Elasticity" 3d ed., McGraw-Hill, New York, 1970.

CHAPTER

FOUR

PORE WATER PRESSURE DUE TO UNDRAINED LOADING

A knowledge of the increase of pore water pressure in soils due to various loading conditions without drainage is important in both theoretical and applied soil mechanics. If a load is applied very slowly on a soil such that sufficient time is allowed for pore water to drain out, there will be practically no increase of pore water pressure. However, when a soil is subjected to rapid loading and if the coefficient of permeability is small (e.g., as in the case of clay), there will be insufficient time for drainage of pore water. This will lead to an increase of the excess hydrostatic pressure. In this chapter, mathematical formulations for the excess pore water pressure for various types of undrained loading will be developed.

4.1 PORE WATER PRESSURE DEVELOPED DUE TO ISOTROPIC STRESS APPLICATION

Figure 4.1 shows an isotropic *saturated* soil element subjected to an isotropic stress increase of magnitude $\Delta\sigma$. If drainage from the soil is not allowed, the pore water pressure will increase by Δu.

The increase of pore water pressure will cause a change in volume of the pore fluid by an amount ΔV_p. This can be expressed as

$$\Delta V_p = n V_o C_p \, \Delta u \tag{4.1}$$

where n = porosity
C_p = compressibility of pore water
V_o = original volume of soil element.

The effective stress increase in all directions of the element is $\Delta\sigma' = \Delta\sigma - \Delta u$. The change in volume of the soil skeleton due to the effective stress increase can be given by

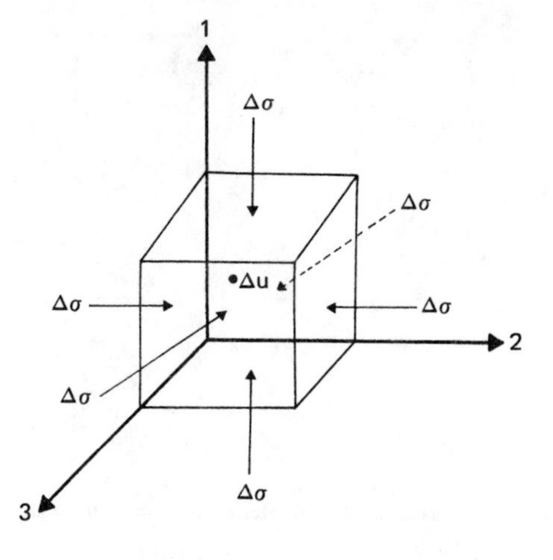

Fig. 4.1 Soil element under isotropic stress application.

$$\Delta V = 3C_c V_o \ \Delta\sigma' = 3C_c V_o(\Delta\sigma - \Delta u) \tag{4.2}$$

In Eq. (4.2), C_c is the compressibility of the soil skeleton obtained from laboratory compression results under uniaxial loading with zero excess pore water pressure, as shown in Fig. 4.2. It should be noted that compression–i.e., a reduction of volume–is taken as positive.

Since the change in volume of the pore fluid, ΔV_p, is equal to the change in the volume of the soil skeleton, ΔV, we obtain from Eqs. (4.1) and (4.2)

$$nV_o C_p \ \Delta u = 3C_c V_o(\Delta\sigma - \Delta u)$$

and hence

$$\frac{\Delta u}{\Delta\sigma} = B = \frac{1}{1 + n(C_p/3C_c)} \tag{4.3}$$

where B is the pore pressure parameter (Skempton, 1954).

The compressibility of pore water, C_p, is very small in comparison to C_c; and, for all practical purposes, the pore water pressure parameter B is equal to unity. So, for saturated soils, the increase of pore water pressure is equal to the increase of isotropic stress, i.e., $\Delta u = \Delta\sigma$.

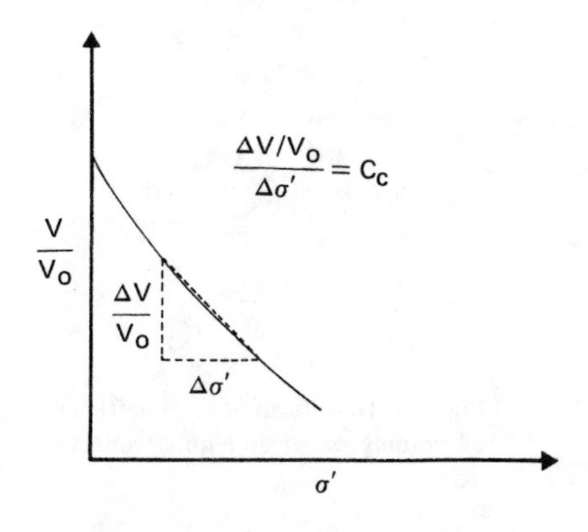

Fig. 4.2 Definition of C_c: volume change due to uniaxial stress application with zero excess pore water pressure. Note: V is the volume of the soil element at any value of σ'.)

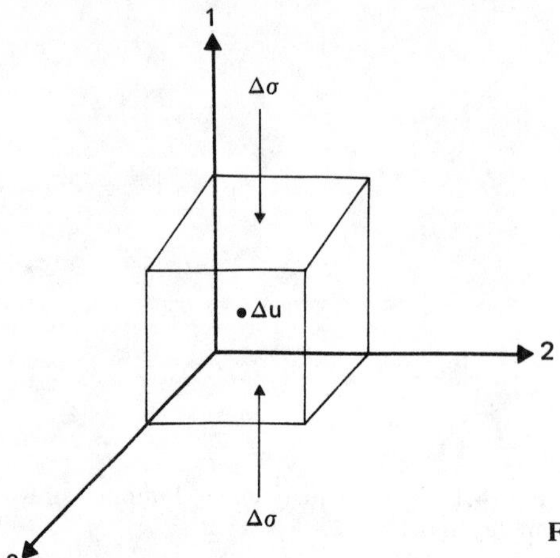

Fig. 4.3 Saturated soil element under uniaxial stress increment.

4.2 PORE WATER PRESSURE DUE TO UNIAXIAL LOADING

A saturated soil element under a uniaxial stress increment is shown in Fig. 4.3. Let the increase of pore water pressure be equal to Δu. As explained in the previous section, the change in the volume of the pore water is

$$\Delta V_p = nV_oC_p \, \Delta u$$

The increases of the effective stresses on the soil element in Fig. 4.3 are:

Direction 1: $\Delta\sigma' = \Delta\sigma - \Delta u$

Direction 2: $\Delta\sigma' = 0 - \Delta u = -\Delta u$

Direction 3: $\Delta\sigma' = 0 - \Delta u = -\Delta u$

This will result in a change in the volume of the soil skeleton, which may be written as

$$\Delta V = C_cV_o(\Delta\sigma - \Delta u) + C_eV_o(-\Delta u) + C_eV_o(-\Delta u) \tag{4.4}$$

where C_e is the coefficient of the volume expansibility (Fig. 4.4). Since $\Delta V_p = \Delta V$,

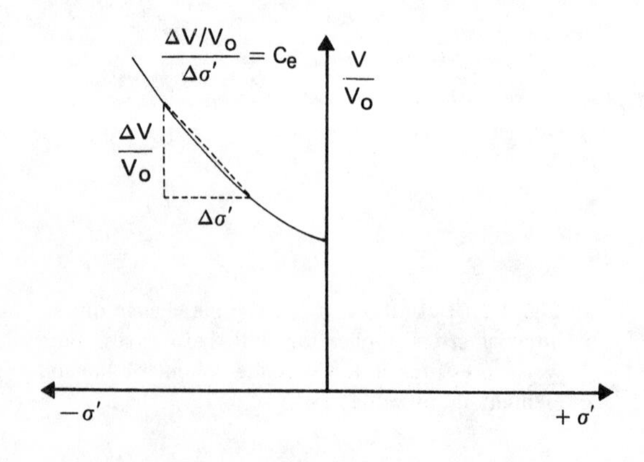

Fig. 4.4 Definition of C_e: coefficient of volume expansion under uniaxial loading.

Table 4.1 Typical values of A at failure

Type of soil	A
Clay with high sensitivity	$\frac{3}{4}$ to $1\frac{1}{2}$
Normally consolidated clay	$\frac{1}{2}$ to 1
Overconsolidated clay	$-\frac{1}{2}$ to 0
Compacted sandy clay	$\frac{1}{2}$ to $\frac{3}{4}$

$$nV_oC_p\,\Delta u = C_cV_o(\Delta\sigma - \Delta u) - 2C_eV_o\,\Delta u$$

or
$$\frac{\Delta u}{\Delta\sigma} = A = \frac{C_c}{nC_p + C_c + 2C_e} \tag{4.5}$$

where A is the pore pressure parameter (Skempton, 1954).

If we assume that the soil element is elastic, then $C_c = C_e$, or

$$A = \frac{1}{n(C_p/C_c) + 3} \tag{4.6}$$

Again, as pointed out previously, C_p is much smaller than C_c. So $C_p/C_c \approx 0$, which gives $A = \frac{1}{3}$. However, in reality, this is not the case—i.e., soil is not a perfectly elastic material—and the actual value of A varies widely. Some typical values of A at failure, determined from triaxial tests, are given in Table 4.1.

4.3 PORE WATER PRESSURE UNDER TRIAXIAL TEST CONDITIONS

A typical stress application on a soil element under triaxial test conditions is shown in Fig. 4.5*a* ($\Delta\sigma_1 > \Delta\sigma_3$). Δu is the increase of the pore water pressure without drainage. To develop a relation between Δu, $\Delta\sigma_1$, and $\Delta\sigma_3$, we can consider that the stress conditions shown in Fig. 4.5*a* are the sum of the stress conditions shown in Fig. 4.5*b* and *c*.

For the isotropic stress $\Delta\sigma_3$ as applied in Fig. 4.5*b*,

$$\Delta u_b = B\,\Delta\sigma_3 \tag{4.7}$$

[from Eq. (4.3)] and for a uniaxial stress $\Delta\sigma_1 - \Delta\sigma_3$ as applied in Fig. 4.5*c*,

$$\Delta u_a = A(\Delta\sigma_1 - \Delta\sigma_3) \tag{4.8}$$

[from Eq. (4.5)]. Now,

$$\Delta u = \Delta u_b + \Delta u_a = B\,\Delta\sigma_3 + A(\Delta\sigma_1 - \Delta\sigma_3) \tag{4.9}$$

For saturated soil, $B = 1$; so

$$\Delta u = \sigma_3 + A(\Delta\sigma_1 - \Delta\sigma_3) \tag{4.10}$$

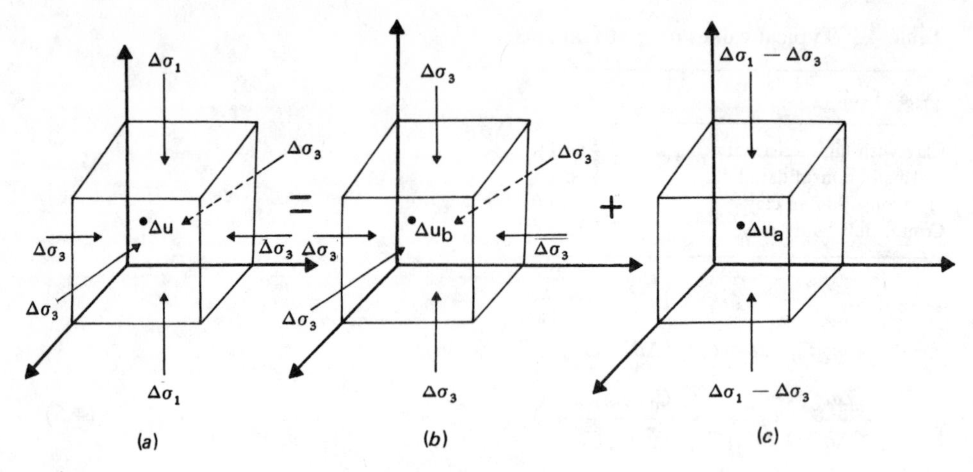

Fig. 4.5 Excess pore water pressure under undrained triaxial test conditions.

The pore water pressure parameters were first suggested by Skempton (1954) and, hence, are known as Skempton's parameters.

4.4 HENKEL'S MODIFICATION OF PORE WATER PRESSURE EQUATION

In several practical considerations in soil mechanics, the intermediate and minor principal stresses are not the same. To take the intermediate principal stress into consideration (Fig. 4.6), Henkel (1960) suggested a modification of Eq. (4.10):

$$\Delta u = \frac{\Delta\sigma_1 + \Delta\sigma_2 + \Delta\sigma_3}{3} + a\sqrt{(\Delta\sigma_1 - \Delta\sigma_2)^2 + (\Delta\sigma_2 - \Delta\sigma_3)^2 + (\Delta\sigma_3 - \Delta\sigma_1)^2} \tag{4.11}$$

or,

$$\Delta u = \Delta\sigma_{\text{oct}} + 3a\Delta\tau_{\text{oct}} \tag{4.12}$$

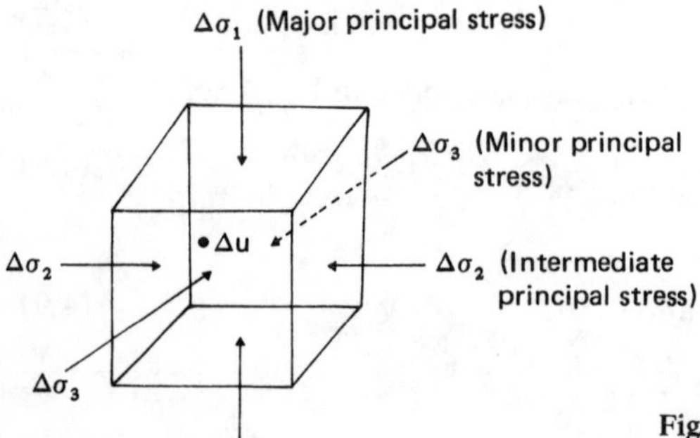

Fig. 4.6 Saturated soil element with major, intermediate, and minor principal stresses.

where a is Henkel's pore pressure diameter, and $\Delta\sigma_{oct}$ and $\Delta\tau_{oct}$ are the increases in the octahedral normal and shear stresses, respectively.

In triaxial compression tests, $\Delta\sigma_2 = \Delta\sigma_3$. For that condition,

$$\Delta u = \frac{\Delta\sigma_1 + 2\Delta\sigma_3}{3} + a\sqrt{2}\,(\Delta\sigma_1 - \Delta\sigma_3) \tag{4.13}$$

For uniaxial tests as in Fig. 4.5*c*, we can substitute $\Delta\sigma_1 - \Delta\sigma_3$ for $\Delta\sigma_1$ and zero for $\Delta\sigma_2$ and $\Delta\sigma_3$ in Eq. (4.11), which will yield

$$\Delta u = \frac{\Delta\sigma_1 - \Delta\sigma_3}{3} + a\sqrt{2}\,(\Delta\sigma_1 - \Delta\sigma_3)$$

or

$$\Delta u = (\tfrac{1}{3} + a\sqrt{2})\,(\Delta\sigma_1 - \Delta\sigma_3) \tag{4.14}$$

A comparison of Eqs. (4.8) and (4.14) gives

$$A = (\tfrac{1}{3} + a\sqrt{2})$$

or

$$a = \frac{1}{\sqrt{2}}\,(A - \tfrac{1}{3}) \tag{4.15}$$

The usefulness of this more fundamental definition of pore water pressure is that it enables us to predict the excess pore water pressure associated with loading conditions such as plane strain. This can be illustrated by deriving an expression for the excess pore water pressure developed in a saturated soil (undrained condition) below the center line of a flexible strip loading of uniform intensity, q (Fig. 4.7). The expressions for σ_x, σ_y, and σ_z for such loading are given in Chap. 3. Note that $\sigma_z > \sigma_y > \sigma_x$, and $\sigma_y = \nu(\sigma_x + \sigma_z)$. Substituting σ_z, σ_y, and σ_x for σ_1, σ_2, and σ_3 in Eq. (4.11),

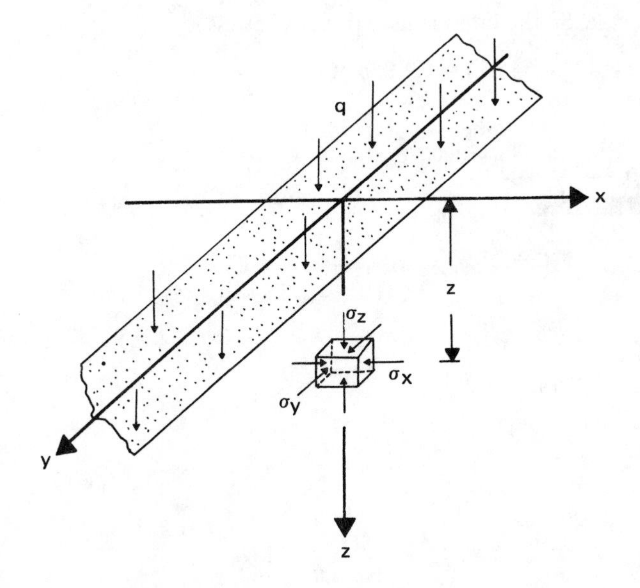

Fig. 4.7 Estimation of excess pore water pressure in a saturated soil below the center line of a flexible strip loading (undrained condition).

$$\Delta u = \frac{\sigma_z + \nu(\sigma_x + \sigma_z) + \sigma_x}{3} + \frac{1}{\sqrt{2}}\left(A - \frac{1}{3}\right)$$
$$\times \sqrt{[\sigma_z - \nu(\sigma_z + \sigma_x)]^2 + [\nu(\sigma_z + \sigma_x) - \sigma_x]^2 + (\sigma_x - \sigma_z)^2}$$

for $\nu = 0.5$, or,

$$\Delta u = \sigma_x + \left[\frac{\sqrt{3}}{2}\left(A - \frac{1}{3}\right) + \frac{1}{2}\right](\sigma_z - \sigma_x) \tag{4.16}$$

If a representative value of A can be determined from standard triaxial tests, Δu can be estimated.

Example 4.1 A uniform vertical load of 3000 lb/ft^2 is applied instantaneously over a very long strip, as shown in Fig. 4.8. Estimate the excess pore water pressure that will be developed due to the loading at A and B. Assume that $\nu = 0.45$ and that the representative value of the pore water pressure parameter A determined from standard triaxial tests for such loading is 0.6.

SOLUTION The values of σ_x, σ_z, and τ_{xz} at A and B can be determined from Table 3.4.

At A: $x/b = 0, z/b = 6/6 = 1$, and hence

1. $\sigma_z/q = 0.8183$, so $\sigma_z = 0.8183 \times 3000 = 2454.9$ lb/ft^2.
2. $\sigma_x/q = 0.1817$, so $\sigma_x = 545.1$ lb/ft^2.
3. $\tau_{xz}/q = 0$, so $\tau_{xz} = 0$.

Note that in this case σ_z and σ_x are the major (σ_1) and minor (σ_3) principal stresses, respectively.

This is a plane strain case. So the intermediate principal stress is

$$\sigma_2 = \nu(\sigma_1 + \sigma_3) = 0.45\,(2454.9 + 545.1) = 1350\ \text{lb/ft}^2$$

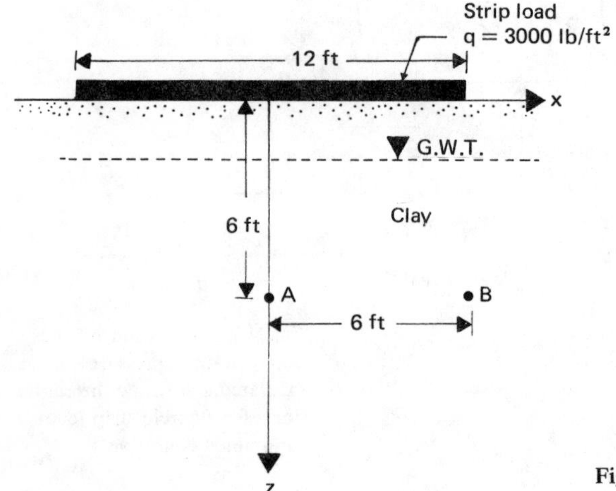

Fig. 4.8

From Eq. (4.15),

$$a = \frac{1}{\sqrt{2}}\left(A - \frac{1}{3}\right) = \frac{1}{\sqrt{2}}\left(0.6 - \frac{1}{3}\right) = 0.189$$

So

$$\Delta u = \frac{\sigma_1 + \sigma_2 + \sigma_3}{3} + a\sqrt{(\sigma_1 - \sigma_2)^2 + (\sigma_2 - \sigma_3)^2 + (\sigma_3 - \sigma_1)^2}$$

$$= \frac{2454.9 + 1350 + 545.1}{3}$$

$$+ 0.189\sqrt{(2454.9 - 1350)^2 + (1350 - 545.1)^2 + (545.1 - 2454.9)^2}$$

$$= \mathit{1893.9\,lb/ft^2}$$

At B: $x/b = 6/6 = 1, z/b = 6/6 = 1$, and hence

1. $\sigma_z/q = 0.4797$, so $\sigma_z = 0.4797 \times 3000 = 1439.1\ \text{lb/ft}^2$.
2. $\sigma_x/q = 0.2250$, so $\sigma_x = 0.2250 \times 3000 = 675\ \text{lb/ft}^2$.
3. $\tau_{xz}/q = 0.2546$, so $\tau_{xz} = 0.2546 \times 3000 = 763.8\ \text{lb/ft}^2$.

Calculation of the major and minor principal stresses is as follows:

$$\sigma_1, \sigma_3 = \frac{\sigma_z + \sigma_x}{2} \pm \sqrt{\left(\frac{\sigma_z - \sigma_x}{2}\right)^2 + \tau_{xz}^2}$$

$$= \frac{1439.1 + 675}{2} \pm \sqrt{\left(\frac{1439.1 - 675}{2}\right)^2 + 763.8^2}$$

Hence,

$$\sigma_1 = 1911.07\ \text{lb/ft}^2 \qquad \sigma_3 = 203.03\ \text{lb/ft}^2$$

$$\sigma_2 = 0.45\,(1911.07 + 203.03) = 951.3\ \text{lb/ft}^2$$

$$\Delta u = \frac{1911.07 + 203.03 + 951.3}{3}$$

$$+ 0.183\sqrt{(1911.07 - 951.3) + (951.3 - 203.03)^2 + (203.03 - 1911.07)^2}$$

$$= \mathit{1418.18\,lb/ft^2}$$

PROBLEMS

4.1 Derive an expression for the excess pore water pressure Δu that will be developed due to the undrained loading on a soil sample placed in an oedometer. $\Delta\sigma$ is the increase of vertical stress (Fig. P4.1).

4.2 A line load of $q = 60\ \text{kN/m}$ with $\alpha = 0$ is placed on a ground surface, as shown in Fig. P4.2.

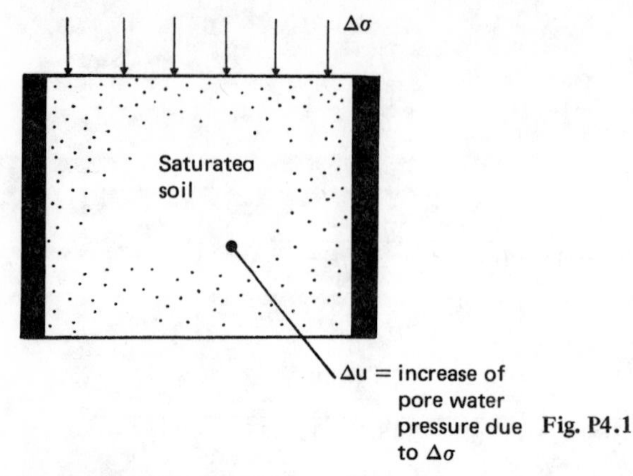

Fig. P4.1

Calculate the increase of pore water pressure at M immediately after application of the load for the cases given below.

(a) $z = 10\text{ m}; x = 0\text{ m}; \nu = 0.5; A = 0.45$.
(b) $z = 8\text{ m}; x = 0\text{ m}; \nu = 0.5; A = 0.5$.
(c) $z = 5\text{ m}; x = 5\text{ m}; \nu = 0.45; A = 0.65$.
(d) $z = 10\text{ m}; x = 2\text{ m}; \nu = 0.45; A = 0.6$.

4.3 Redo Prob. 4.2(a), (b), (c), and (d) with $q = 26$ kN/m and $\alpha = 90°$.

4.4 Redo Prob. 4.2(a), (b), (c), and (d) with $q = 60$ kN/m and $\alpha = 30°$.

4.5 Determine the increase of pore water at M due to the strip loading shown in Fig. P4.3. Assume $\nu = 0.5$ and $\alpha = 0$ for all the cases given below.

(a) $z = 8\text{ ft}; x = 0; A = 0.65$.
(b) $z = 12\text{ ft}; x = 0; A = 0.55$.
(c) $z = 8\text{ ft}; x = 8\text{ ft}; A = 0.5$.
(d) $z = 8\text{ ft}; x = 4\text{ ft}; A = 0.52$.

4.6 Redo Prob. 4.5(a), (b), (c), and (d) for $\alpha = 90°$.

4.7 Redo Prob. 4.5(a), (b), (c), and (d) for $\alpha = 45°$.

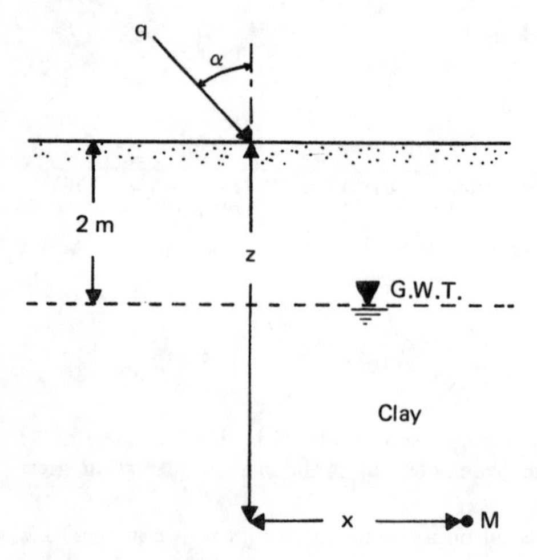

Fig. P4.2

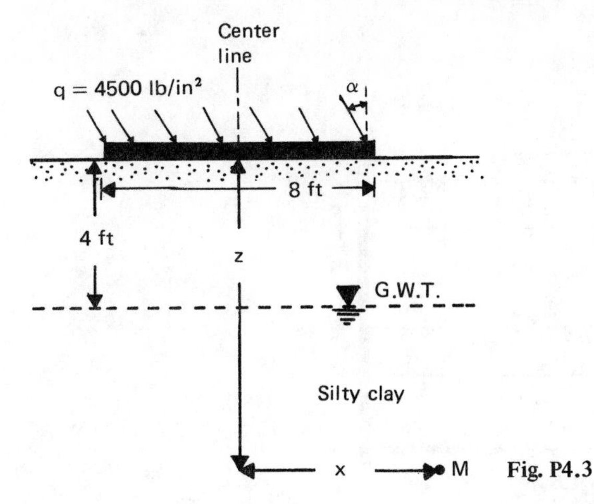

Fig. P4.3

4.8 The following are the results of a consolidated undrained triaxial test on a clay (Fig. P4.4):

σ_3, kN/m²	$\Delta\sigma$, kN/m²	Δu, kN/m²	Strain ϵ_1, %
69	27.0	8.91	0.2
69	40.0	15.2	0.4
69	58.0	26.68	0.8
69	63.0	30.87	1
69	70.0	37.1	1.2
69	77.0	46.2	1.6
69	83.0	53.2	2
69	87.0	57.42.	2.5
69	88.0	58.08	3
69	88.0	57.20	3.5
69	87.0	55.68	4
69	83.0	53.12	5

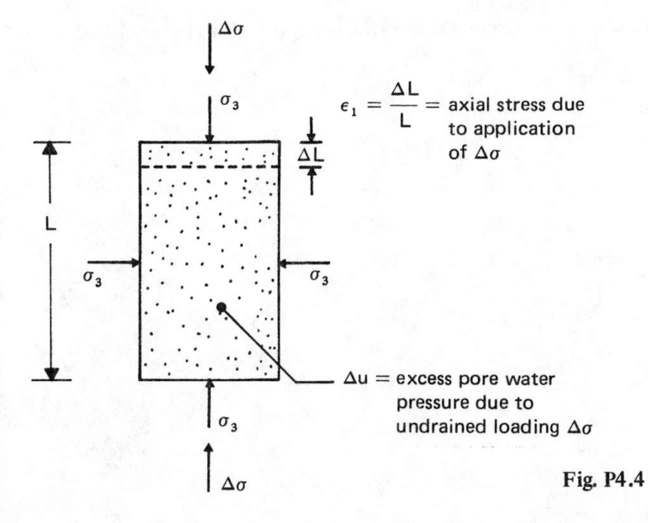

Fig. P4.4

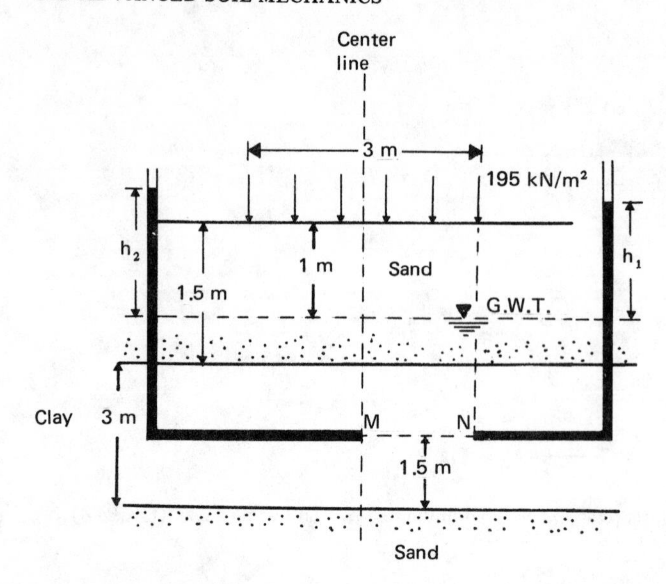

Fig. P4.5

Make the necessary calculations and plot the following:

(*a*) Variation of pore pressure parameter A against strain, in percent.

(*b*) Variation of Henkel's pore pressure parameter a against strain, in percent.

4.9 A surcharge of 195 kN/m² was applied over a circular area of diameter 3 m, as shown in Fig. P4.5. Estimate the height of water, h_1, that a piezometer would show immediately after the application of the surcharge. Assume that $A \approx 0.65$ and $\nu = 0.5$.

4.10 Redo Prob. 4.9 for point M; i.e., find h_2.

REFERENCES

Henkel, D. J., The Shear Strength of Saturated Remolded Clays, *Proc. Res. Conf. Shear Strength Cohesive Soils, ASCE,* pp. 533–554, 1960.

Skempton, A. W., The Pore Pressure Coefficients A and B, *Geotechnique,* vol. 4, pp. 143–147, 1954.

CHAPTER

FIVE

CONSOLIDATION

When a soil layer is subjected to a compressive stress, such as during the construction of a structure, it will exhibit a certain amount of compression. This compression is achieved through a number of ways, including rearrangement of the soil solids or extrusion of the pore air and/or water. According to Terzaghi (1943), "a decrease of water content of a saturated soil without replacement of the water by air is called a process of consolidation." When saturated clayey soils—which have a low coefficient of permeability—are subjected to a compressive stress due to a foundation loading, the pore water pressure will immediately increase; however, due to the *low permeability of the soil,* there will be a time lag between the application of load and the extrusion of the pore water and, thus, the settlement. This phenomenon is the subject of discussion of this chapter.

5.1 FUNDAMENTALS OF CONSOLIDATION

5.1.1 General Concepts of One-Dimensional Consolidation

To understand the basic concepts of consolidation, consider a clay layer of thickness H_t located below the groundwater level and between two highly permeable sand layers as shown in Fig. 5.1. If a surcharge of intensity $\Delta\sigma$ is applied at the ground surface over a very large area, the pore water pressure in the clay layer will increase. For a surcharge of *infinite extent,* the immediate increase of the pore water pressure, Δu, *at all depths* of the clay layer will be equal to the increase of the total stress, $\Delta\sigma$. Thus, immediately after the application of the surcharge,

$$\Delta u = \Delta\sigma$$

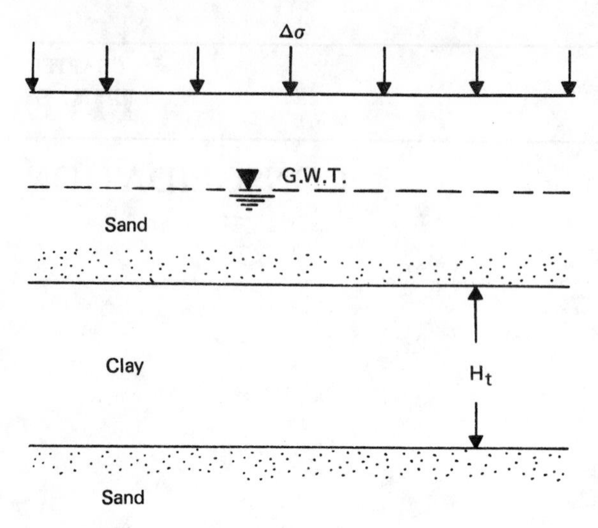

Fig. 5.1

Since the total stress is equal to the sum of the effective stress and the pore water pressure, at all depths of the clay layer the increase of effective stress due to the surcharge (immediately after application) will be equal to zero (i.e., $\Delta\sigma' = 0$, where $\Delta\sigma'$ is the increase of effective stress). In other words, at time $t = 0$, the entire stress increase at all depths of the clay is taken by the pore water pressure and none by the soil skeleton. This is shown in Fig. 5.2*a*. (It must be pointed out that, for loads applied over a limited area, it may not be true that the increase of the pore water pressure is equal to the increase of vertical stress at any depth at time $t = 0$; this fact is discussed in Secs. 6.3.2, 6.4, and 6.5.)

After application of the surcharge (i.e., at time $t > 0$), the water in the void spaces of the clay layer will be squeezed out and will flow toward both the highly permeable sand layers, thereby reducing the excess pore water pressure. This, in turn, will increase the effective stress by an equal amount since $\Delta\sigma' + \Delta u = \Delta\sigma$. Thus, at time $t > 0$,

$$\Delta\sigma' > 0$$

and $$\Delta u < \Delta\sigma$$

This fact is shown in Fig. 5.2*b*.

Theoretically, at time $t = \infty$, the excess pore water pressure at all depths of the clay layer will be dissipated by gradual drainage. Thus, at time $t = \infty$,

$$\Delta\sigma' = \Delta\sigma$$

and $$\Delta u = 0$$

This is shown in Fig. 5.2*c*.

This gradual process of increase of effective stress in the clay layer due to the surcharge will result in a settlement which is time-dependent and is referred to as the process of *consolidation*.

5.1.2 Theory of One-Dimensional Consolidation

The theory for the time rate of one-dimensional consolidation was first proposed by Terzaghi (1925). The underlying assumptions in the derivation of the mathematical equations are as follows:

1. The clay layer is homogeneous.
2. The clay layer is saturated.
3. The compression of the soil layer is due to the change in volume only, which, in turn, is due to the squeezing out of water from the void spaces.
4. Darcy's law is valid.
5. Deformation of soil occurs only in the direction of the load application.
6. The coefficient of consolidation C_v [Eq. (5.15)] is constant during the consolidation.

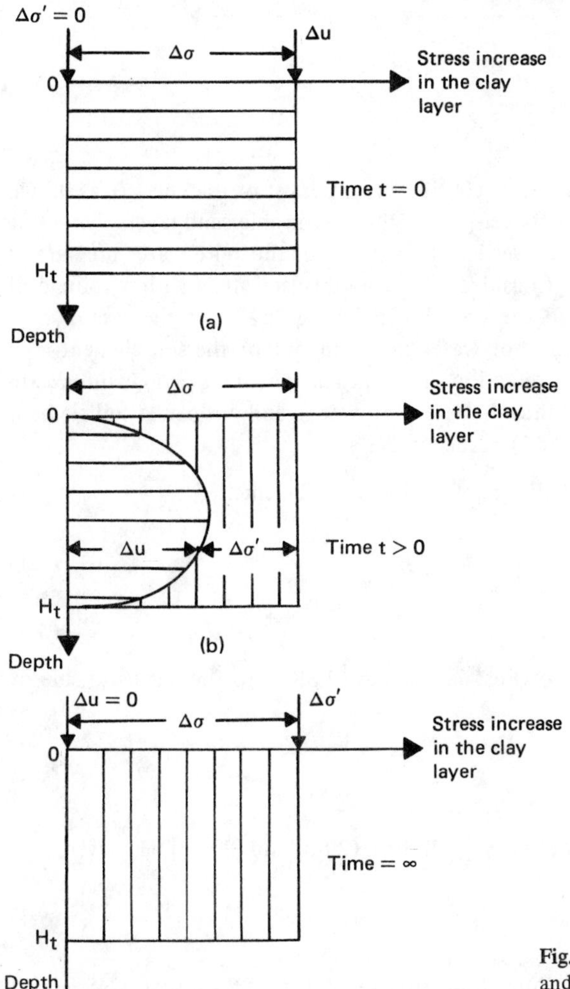

Fig. 5.2 Change of pore water pressure and effective stress in the clay layer shown in Fig. 5.1 due to the surcharge.

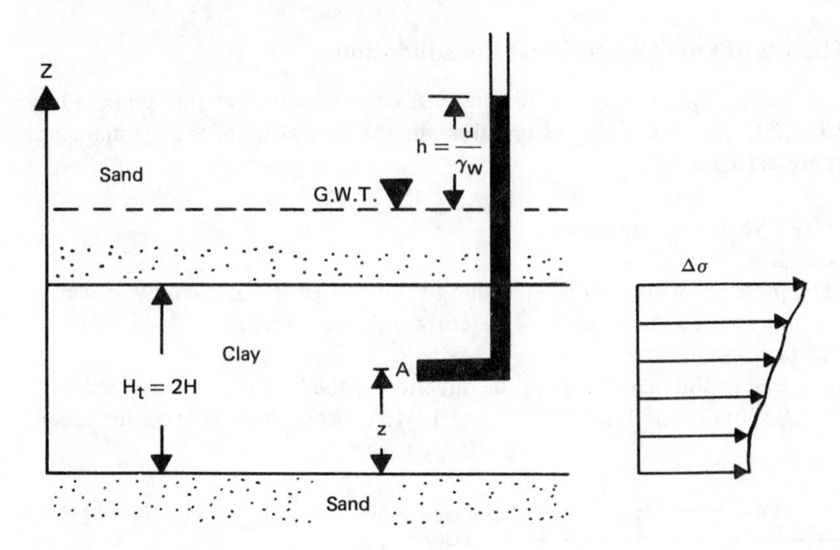

Fig. 5.3 Clay layer undergoing consolidation.

With the above assumptions, let us consider a clay layer of thickness H_t as shown in Fig. 5.3. The layer is located between two highly permeable sand layers. When the clay is subjected to an increase of vertical pressure, $\Delta\sigma$, the pore water pressure at any point A will increase by u. Consider an elementary soil mass with a volume of $dx \cdot dy \cdot dz$ at A; this is similar to the one shown in Fig. 2.27*b*. In the case of one-dimensional consolidation, the flow of water into and out of the soil element is in one direction only, i.e., in the z direction. This means that q_x, q_y, dq_x, and dq_y in Fig. 2.27*b* are equal to zero, and thus the rate of flow into and out of the soil element can be given by Eqs. (2.80) and (2.83), respectively. So,

$$(q_z + dq_z) - q_z = \text{rate of change of volume of soil element}$$

$$= \frac{\partial V}{\partial t} \tag{5.1}$$

where $\quad V = dx\, dy\, dz \tag{5.2}$

Substituting the right-hand sides of Eqs. (2.80) and (2.83) into the left-hand side of Eq. (5.1), we obtain

$$k \frac{\partial^2 h}{\partial z^2}\, dx\, dy\, dz = \frac{\partial V}{\partial t} \tag{5.3}$$

where k is the coefficient of permeability [k_z in Eqs. (2.80) and (2.83)]. However,

$$h = \frac{u}{\gamma_w} \tag{5.4}$$

where γ_w is the unit weight of water. Substitution of Eq. (5.4) into Eq. (5.3) and rearranging gives

$$\frac{k}{\gamma_w}\frac{\partial^2 u}{\partial z^2} = \frac{1}{dx\,dy\,dz}\frac{\partial V}{\partial t} \tag{5.5}$$

During consolidation the rate of change of volume is equal to the rate of change of the void volume. So,

$$\frac{\partial V}{\partial t} = \frac{\partial V_v}{\partial t} \tag{5.6}$$

where V_v is the volume of voids in the soil element. But

$$V_v = eV_s \tag{5.7}$$

where V_s is the volume of soil solids in the element, which is constant, and e is the void ratio. So,

$$\frac{\partial V}{\partial t} = V_s\frac{\partial e}{\partial t} = \frac{V}{1+e}\frac{\partial e}{\partial t} = \frac{dx\,dy\,dz}{1+e}\frac{\partial e}{\partial t} \tag{5.8}$$

Substituting the above relation into Eq. (5.5), we get

$$\frac{k}{\gamma_w}\frac{\partial^2 u}{\partial z^2} = \frac{1}{1+e}\frac{\partial e}{\partial t} \tag{5.9}$$

The change in void ratio, ∂e, is due to the increase of effective stress; assuming that these are linearly related, then

$$\partial e = -a_v\,\partial(\Delta\sigma') \tag{5.10}$$

where a_v is the coefficient of compressibility. Again, the increase of effective stress is due to the decrease of excess pore water pressure, ∂u. Hence,

$$\partial e = a_v\,\partial u \tag{5.11}$$

Combining Eqs. (5.9) and (5.11),

$$\frac{k}{\gamma_w}\frac{\partial^2 u}{\partial z^2} = \frac{a_v}{1+e}\frac{\partial u}{\partial t} = m_v\frac{\partial u}{\partial t} \tag{5.12}$$

where m_v = coefficient of volume compressibility $= \dfrac{a_v}{1+e}$ (5.13)

or
$$\frac{\partial u}{\partial t} = \frac{k}{\gamma_w m_v}\frac{\partial^2 u}{\partial z^2} = C_v\frac{\partial^2 u}{\partial z^2} \tag{5.14}$$

where C_v = coefficient of consolidation $= \dfrac{k}{\gamma_w m_v}$ (5.15)

Eq. (5.14) is the basic differential equation of Terzaghi's consolidation theory and can be solved with proper boundary conditions. To solve the equation, we assume u to be the product of two functions, i.e., the product of a function of z and a function of t, or

$$u = F(z)G(t) \tag{5.16}$$

So,
$$\frac{\partial u}{\partial t} = F(z)\frac{\partial}{\partial t}G(t) = F(z)G'(t) \tag{5.17}$$

and
$$\frac{\partial^2 u}{\partial z^2} = \frac{\partial^2}{\partial z^2}F(z)G(t) = F''(z)G(t) \tag{5.18}$$

From Eqs. (5.14), (5.17), and (5.18),

$$F(z)G'(t) = C_v F''(z)G(t)$$

or
$$\frac{F''(z)}{F(z)} = \frac{G'(t)}{C_v G(t)} \tag{5.19}$$

The right-hand side of Eq. (5.19) is a function of z only and is independent of t; the left-hand side of the equation is a function of t only and is independent of z. Therefore, they must be equal to a constant, say $-B^2$. So,

$$F''(z) = -B^2 F(z) \tag{5.20}$$

A solution to Eq. (5.20) can be given by

$$F(z) = A_1 \cos Bz + A_2 \sin Bz \tag{5.21}$$

where A_1 and A_2 are constants.

Again, the right-hand side of Eq. (5.19) may be written as

$$G'(t) = -B^2 C_v G(t) \tag{5.22}$$

The solution to Eq. (5.22) is given by

$$G(t) = A_3 \exp(-B^2 C_v t) \tag{5.23}$$

where A_3 is a constant. Combining Eqs. (5.16), (5.21), and (5.23),

$$u = (A_1 \cos Bz + A_2 \sin Bz)A_3 \exp(-B^2 C_v t)$$

$$= (A_4 \cos Bz + A_5 \sin Bz) \exp(-B^2 C_v t) \tag{5.24}$$

where $A_4 = A_1 A_3$ and $A_5 = A_2 A_3$.

The constants in Eq. (5.24) can be evaluated from the boundary conditions, which are as follows:

1. At time $t = 0, u = u_i$ (initial excess pore water pressure at any depth).
2. $u = 0$ at $z = 0$.
3. $u = 0$ at $z = H_t = 2H$.

Note that H is the length of the longest drainage path. In this case, which is a two-way drainage condition (top *and* bottom of the clay layer), H is equal to half the total thickness of the clay layer, H_t.

The second boundary condition dictates that $A_4 = 0$, and from the third boundary condition we get

$$A_5 \sin 2BH = 0 \qquad \text{or} \qquad 2BH = n\pi$$

where n is an integer. From the above, a general solution of Eq. (5.24) can be given in the form

$$u = \sum_{n=1}^{n=\infty} A_n \sin \frac{n\pi z}{2H} \exp\left(\frac{-n^2\pi^2 T_v}{4}\right) \tag{5.25}$$

where T_v is the nondimensional time factor and is equal to $C_v t/H^2$.

To satisfy the first boundary condition, we must have the coefficients of A_n such that

$$u_i = \sum_{n=1}^{n=\infty} A_n \sin \frac{n\pi z}{2H} \tag{5.26}$$

Equation (5.26) is a Fourier sine series, and A_n can be given by

$$A_n = \frac{1}{H}\int_0^{2H} u_i \sin \frac{n\pi z}{2H}\, dz \tag{5.27}$$

Combining Eqs. (5.25) and (5.27),

$$u = \sum_{n=1}^{n=\infty}\left(\frac{1}{H}\int_0^{2H} u_i \sin \frac{n\pi z}{2H}\, dz\right) \sin \frac{n\pi z}{2H} \exp\left(\frac{-n^2\pi^2 T_v}{4}\right) \tag{5.28}$$

So far we have not made any assumptions regarding the variation of u_i with the depth of the clay layer. Several possible types of variation for u_i are considered below.

Constant u_i with depth. If u_i is constant with depth—i.e., if $u_i = u_o$ (Fig. 5.4)—then, referring to Eq. (5.28),

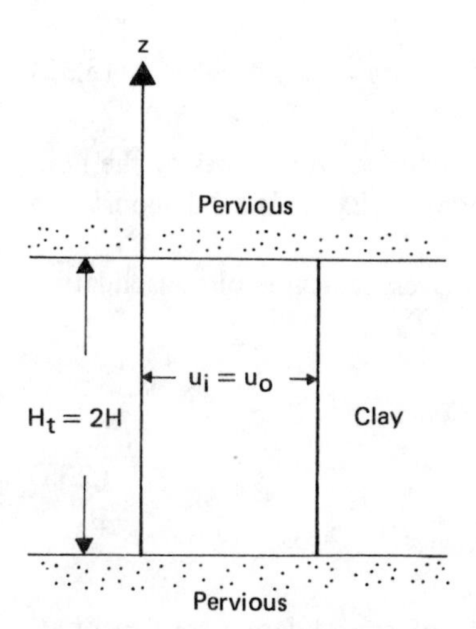

Fig. 5.4 Initial excess pore water pressure—constant with depth (double drainage).

$$\frac{1}{H}\int_0^{2H} \underset{=u_o}{\underset{\uparrow}{u_i}} \sin\frac{n\pi z}{2H}\,dz = \frac{2u_o}{n\pi}(1-\cos n\pi)$$

So,
$$u = \sum_{n=1}^{n=\infty} \frac{2u_o}{n\pi}(1-\cos n\pi)\sin\frac{n\pi z}{2H}\exp\left(\frac{-n^2\pi^2 T_v}{4}\right) \tag{5.29}$$

Note that the term $1-\cos n\pi$ in the above equation is zero for cases when n is even; therefore, u is also zero. For the nonzero terms, it is convenient to substitute $n = 2m+1$, where m is an integer. So Eq. (5.29) will now read

$$u = \sum_{m=0}^{m=\infty} \frac{2u_o}{(2m+1)\pi}[1-\cos(2m+1)\pi]\sin\frac{(2m+1)\pi z}{2H}$$

$$\times \exp\left[\frac{-(2m+1)^2\pi^2 T_v}{4}\right]$$

or
$$u = \sum_{m=0}^{m=\infty} \frac{2u_o}{M}\sin\frac{Mz}{H}\exp(-M^2 T_v) \tag{5.30}$$

where $M = (2m+1)\pi/2$. At a given time, the degree of consolidation at any depth z is defined as

$$U_z = \frac{\text{excess pore water pressure dissipated}}{\text{initial excess pore water pressure}}$$

$$= \frac{u_i - u}{u_i} = 1 - \frac{u}{u_i} = \frac{\sigma'}{u_i} = \frac{\sigma'}{u_o} \tag{5.31}$$

where σ' is the increase of effective stress at a depth z due to consolidation. From Eqs. (5.30) and (5.31),

$$U_z = 1 - \sum_{m=0}^{m=\infty} \frac{2}{M}\sin\frac{Mz}{H}\exp(-M^2 T_v) \tag{5.32}$$

Figure 5.5 shows the variation of U_z with depth for various values of the non-dimensional time factor, T_v; these curves are called isocrones. Example 5.1 demonstrates the procedure for calculation of U_z using Eq. (5.32).

In most cases, however, we need to obtain the average degree of consolidation for the entire layer. This is given by

$$U_{av} = \frac{(1/H_t)\int_0^{H_t} u_i\,dz - (1/H_t)\int_0^{H_t} u\,dz}{(1/H_t)\int_0^{H_t} u_i\,dz} \tag{5.33}$$

The average degree of consolidation is also the ratio of consolidation settlement at

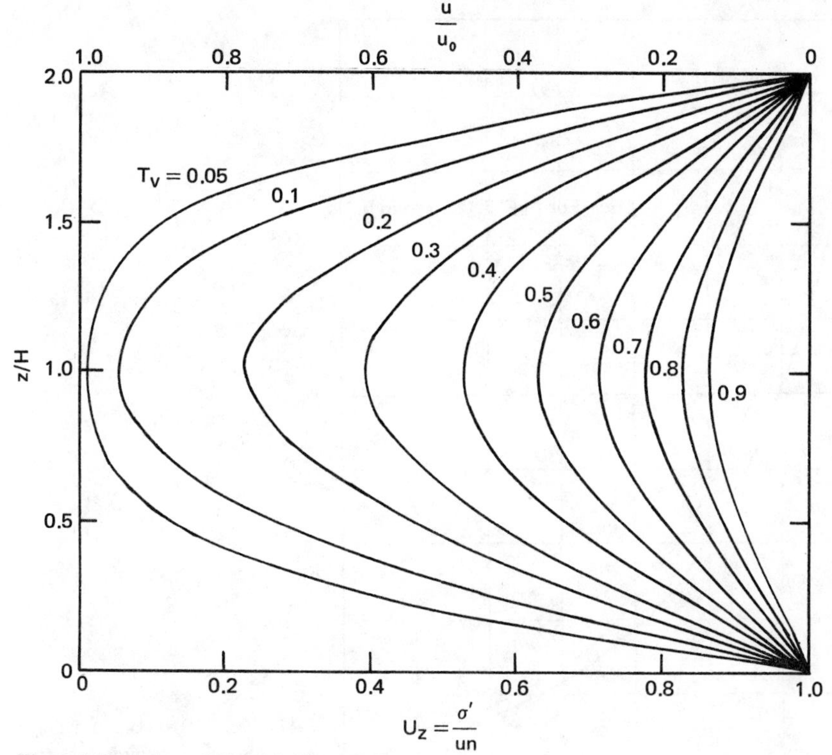

Fig. 5.5 Variation of U_z with z/H and T_v.

any time to maximum consolidation settlement. Note, in this case, that $H_t = 2H$ and $u_i = u_o$.

Combining Eqs. (5.30) and (5.33),

$$U_{av} = 1 - \sum_{m=0}^{m=\infty} \frac{2}{M^2} \exp(-M^2 T_v) \tag{5.34}$$

Figure 5.6 gives the variation of U_{av} vs. T_v (also see Table 5.1).

Terzaghi suggested the following equations for U_{av} to approximate the values obtained from Eq. (5.34):

$$\text{For } U_{av} = 0 \text{ to } 53\%: \quad T_v = \frac{\pi}{4}\left(\frac{U\%}{100}\right)^2 \tag{5.35}$$

$$\text{For } U_{av} = 53 \text{ to } 100\%: \quad T_v = 1.781 - 0.933\,[\log(100 - U\%)] \tag{5.36}$$

Sivaram and Swamee (1977) gave the following equation for U_{av} varying from 0 to 100%:

$$\frac{U_{av}\%}{100} = \frac{(4T_v/\pi)^{0.5}}{[1 + (4T_v/\pi)^{2.8}]^{0.179}} \tag{5.37}$$

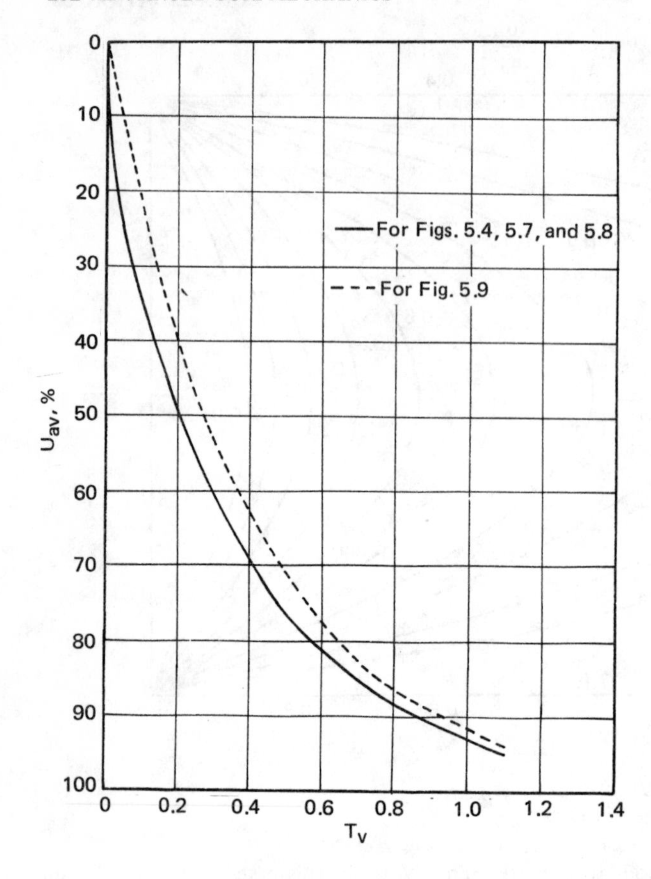

Fig. 5.6 Variation of average degree of consolidation (for conditions given in Figs. 5.4, 5.7, 5.8, and 5.9).

or
$$T_v = \frac{(\pi/4)\,(U_{av}\%/100)^2}{[1-(U_{av}\%/100)^{5.6}]^{0.357}} \tag{5.38}$$

Equations (5.37) and (5.38) give an error in T_v of less than 1% for $0\% < U_{av} < 90\%$ and less than 3% for $90\% < U_{av} < 100\%$.

Table 5.1 Variation of T_v with U_{av} [Eq. (5.34)]

U_{av}, %	T_v	U_{av}, %	T_v
0	0	60	0.287
10	0.008	65	0.342
20	0.031	70	0.403
30	0.071	75	0.478
35	0.096	80	0.567
40	0.126	85	0.684
45	0.159	90	0.848
50	0.197	95	1.127
55	0.238	100	∞

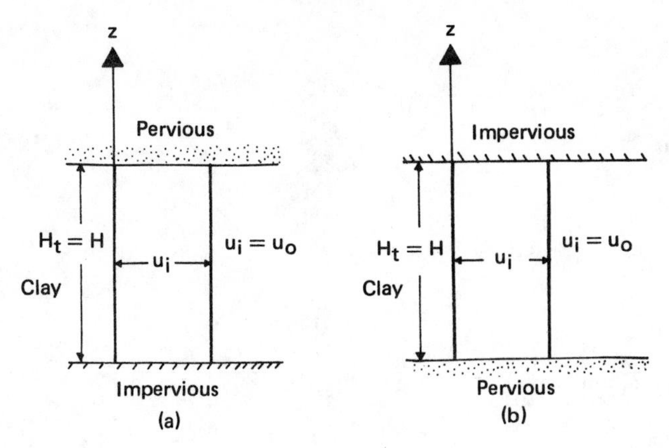

Fig. 5.7 Initial excess pore pressure distribution – one-way drainage, u_i constant with depth.

It must be pointed out that, if we have a situation of one-way drainage as shown in Fig. 5.7*a* and *b*, Eq. (5.34) would still be valid. Note, however, that the length of the drainage path is equal to the total thickness of the clay layer.

Linear variation of u_i. The linear variation of the initial excess pore water pressure, as shown in Fig. 5.8, may be written as

$$u_i = u_1 - u_2 \frac{H - z}{H} \tag{5.39}$$

Substitution of the above relation for u_i into Eq. (5.28) yields

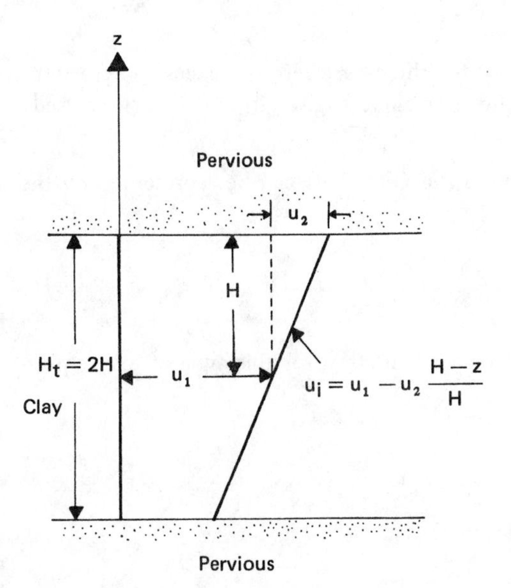

Fig. 5.8 Linearly varying initial excess pore water pressure distribution – two-way drainage.

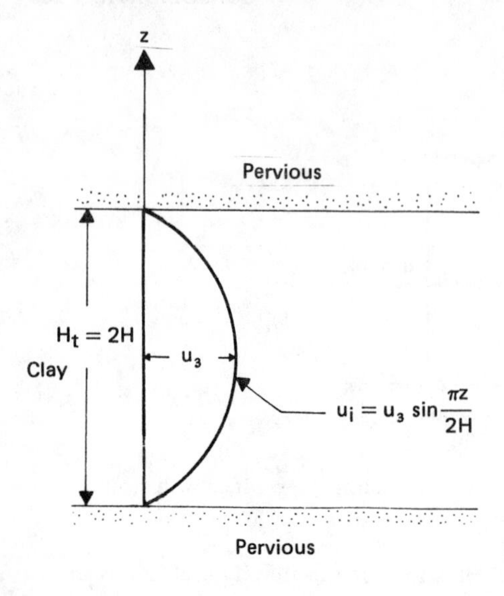

Fig. 5.9 Sinusoidal initial excess pore water pressure distribution – two-way drainage.

$$u = \sum_{n=1}^{n=\infty} \left[\frac{1}{H} \int_0^{2H} \left(u_1 - u_2 \frac{H-z}{H} \right) \sin \frac{n\pi z}{2H} \, dz \right] \sin \frac{n\pi z}{2H} \times \exp \left(\frac{-n^2 \pi^2 T_v}{4} \right) \tag{5.40}$$

The average degree of consolidation can be obtained by solving Eqs. (5.40) and (5.33):

$$U_{av} = 1 - \sum_{m=0}^{m=\infty} \frac{2}{M^2} \exp(-M^2 T_v)$$

This is identical to Eq. (5.34), which was for the case where the excess pore water pressure is constant with depth, and so the same curve as given in Fig. 5.6 can be used.

Sinusoidal variation of u_i. Sinusoidal variation (Fig. 5.9) can be represented by the equation

$$u_i = u_3 \sin \frac{\pi z}{2H} \tag{5.41}$$

The solution for the average degree of consolidation for this type of excess pore water pressure distribution is of the form

$$U_{av} = 1 - \exp \left(\frac{-\pi^2 T_v}{4} \right) \tag{5.42}$$

The variation of U_{av} for various values of T_v is given in Fig. 5.6.

5.1.3 Relations of U_{av} and T_v for Other Forms of Initial Excess Pore Water Pressure Distribution

Using the basic equation [Eq. (5.28)] for excess pore water pressure and with proper boundary conditions, relations for U_{av} and T_v for various other types of initial excess pore water pressure distribution can be obtained. Figures 5.10 and 5.11 present some of these cases.

Example 5.1 Consider the case of an initial excess hydrostatic pore water that is constant with depth, i.e., $u_i = u_o$ (Fig. 5.12). For $T_v = 0.3$, determine the degree of consolidation at a depth $H/3$ measured from the top of the layer.

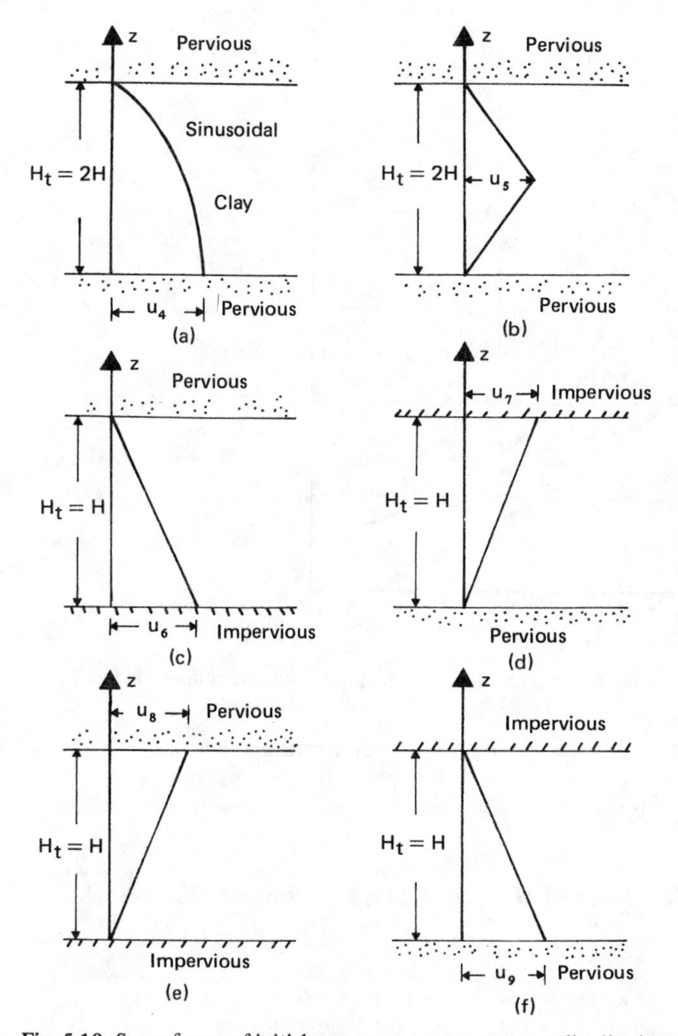

Fig. 5.10 Some forms of initial excess pore water pressure distribution.

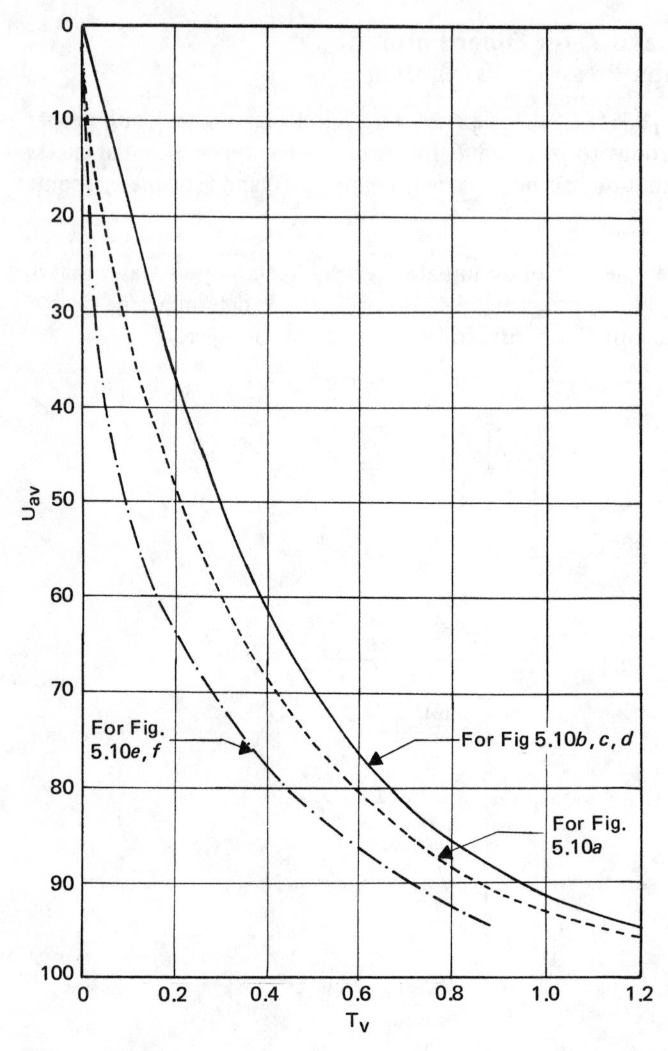

Fig. 5.11 Variation of U_{av} with T_v for initial excess pore water pressure diagrams shown in Fig. 5.10.

SOLUTION From Eq. (5.32), for constant pore water pressure increase,

$$U_z = 1 - \sum_{m=0}^{m=\infty} \frac{2}{M} \sin \frac{Mz}{H} \exp(-M^2 T_v)$$

Here $z = H/3$, or $z/H = 1/3$, and $M = (2m + 1)\pi/2$. We can now make a table to calculate U_z:

1.	z/H	1/3	1/3	1/3
2.	T_v	0.3	0.3	0.3
3.	m	0	1	2

4.	M	$\pi/2$	$3\pi/2$	$5\pi/2$	
5.	Mz/H	$\pi/6$	$\pi/2$	$5\pi/6$	
6.	$2/M$	1.273	0.4244	0.2546	
7.	$\exp(-M^2 T_v)$	0.4770	0.00128	≈ 0	
8.	$\sin(Mz/H)$	0.5	1.0	0.5	
9.	$(2/M)[\exp(-M^2 T_v)\sin(Mz/H)]$	0.3036	0.0005	≈ 0	$\Sigma = 0.3041$

Using the value of 0.3041 calculated in step 9, the degree of consolidation at depth $H/3$ is

$$U_{(H/3)} = 1 - 0.3041 = 0.6959 = \mathit{69.59\%}$$

Note that in the above table we need not go beyond $m = 2$, since the expression in step 9 is negligible for $m \geqslant 3$.

Example 5.2 Due to certain loading conditions, the excess pore water pressure in a clay layer (drained at top and bottom) increased in the manner shown in Fig. 5.13. For a time factor $T_v = 0.3$, calculate the average degree of consolidation.

SOLUTION The excess pore water pressure diagram shown in Fig. 5.13 can be expressed as the difference of two diagrams, as shown in Fig. 5.14*b* and *c*. The excess pore water pressure diagram in Fig. 5.14*b* shows a case where u_i varies linearly with depth. Figure 5.14*c* can be approximated as a sinusoidal variation.

The area of the diagram in Fig. 5.14*b* is

$$A_1 = 20(\tfrac{1}{2})(300 + 100) = 4000 \text{ lb/ft}$$

The area of the diagram in Fig. 5.14*c* is

$$A_2 = \sum_{z=0}^{z=20} 40 \sin \frac{\pi z}{2H}\, dz = \int_0^{20} 40 \sin \frac{\pi z}{20}\, dz$$

$$= (40)\left(\frac{20}{\pi}\right)\left(-\cos \frac{\pi z}{20}\right)_0^{20} = \frac{800}{\pi}(2) = \frac{1600}{\pi} = 509.29 \text{ lb/ft}$$

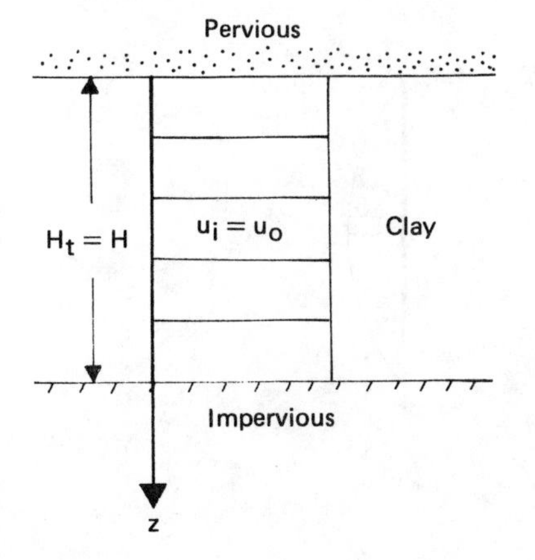

Fig. 5.12

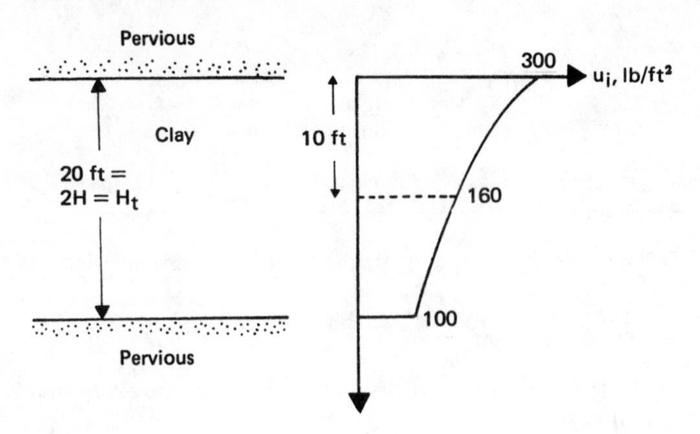

Fig. 5.13

The average degree of consolidation can now be calculated as follows:

$$\underbrace{U_{av(T_v=0.3)}}_{\text{For Fig. 5.14}a} = \frac{\overbrace{U_{av(T_v=0.3)}}^{\text{For Fig. 5.14}b} A_1 - \overbrace{U_{av(T_v=0.3)}}^{\text{For Fig. 5.14}c} A_2}{\underbrace{A_1 - A_2}_{\text{Net area of Fig. 5.14}a}}$$

From Fig. 5.6, for $T_v = 0.3$, $U_{av} = 61\%$ for area A_1; $U_{av} = 52.3\%$ for area A_2. So

$$U_{av} = \frac{61(4000) - (509.29)52.3}{4000 - 509.29} = \frac{271{,}364.3}{3{,}490.71} = 62.3\%$$

Example 5.3 A uniform surcharge of $q = 2000\ \text{lb/ft}^2$ is applied on the ground surface as shown in Fig. 5.15*a*.

(*a*) Determine the initial excess pore water pressure distribution in the clay layer.

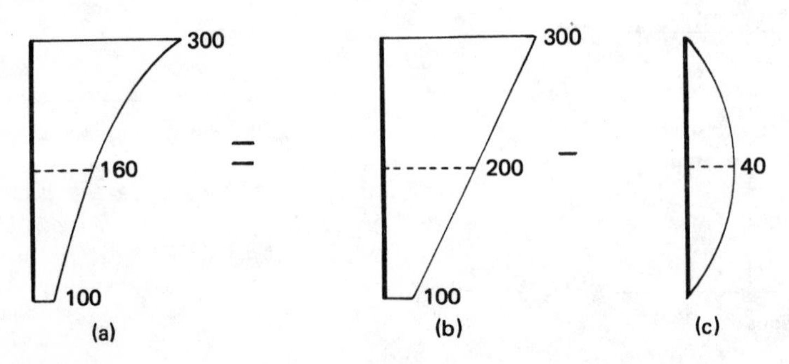

Fig. 5.14

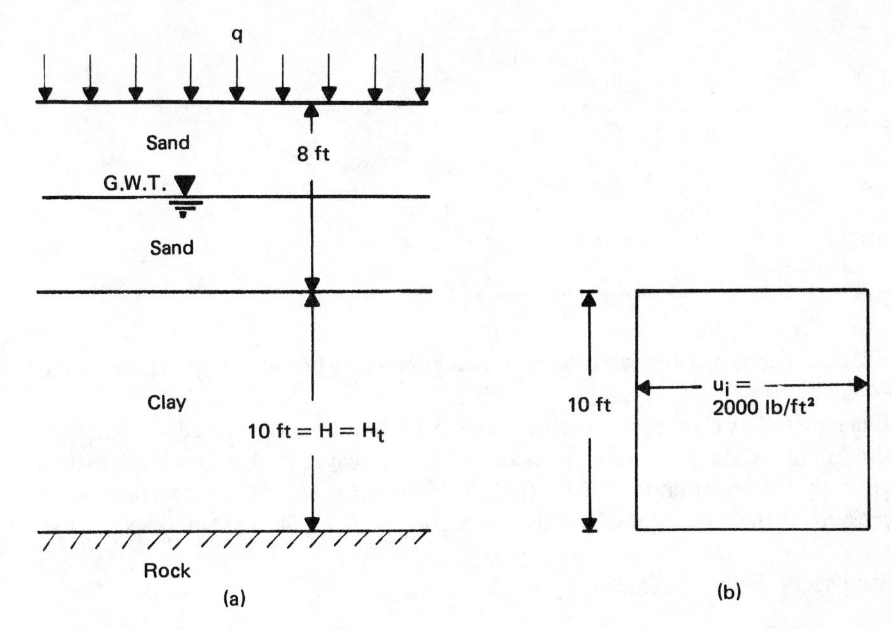

Fig. 5.15

(*b*) Plot the distribution of the excess pore water pressure with depth in the clay layer at a time for which $T_v = 0.5$.

SOLUTION *Part (a):* The initial excess pore water pressure will be 2000 lb/ft^2 and will be the same throughout the clay layer (Fig. 5.15*b*; refer to Prob. 4.1).

Part (b): From Eq. (5.31), $U_z = 1 - u/u_i$, or $u = u_i(1 - U_z)$. For $T_v = 0.5$, the values of U_z can be obtained from the top half of Fig. 5.5 as shown in Fig. 5.16*a*, and then the following table can be prepared:

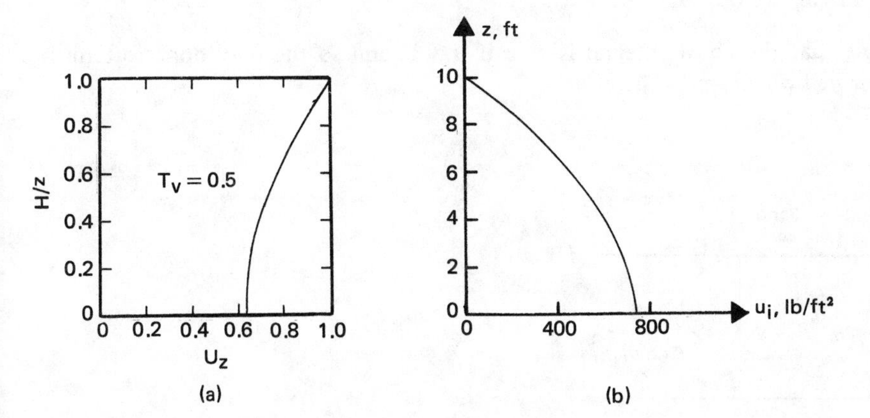

Fig. 5.16

z/H	z, ft	U_z	$u = u_i(1 - U_z)$, lb/ft²
0	0	0.63	740
0.2	2	0.65	700
0.4	4	0.71	580
0.6	6	0.78	440
0.8	8	0.89	220
1.0	10	1	0

Figure 5.16*b* shows the variation of excess pore water pressure with depth.

Example 5.4 A clay layer is shown in Fig. 5.17. Due to a certain loading condition, the initial excess pore water pressure in the clay layer is of a sinusoidal nature, given by the equation $u_i = 50 \sin(\pi z/2H)$ kN/m². Calculate the excess pore water pressure at the midheight of the clay layer for $T_v = 0.2, 0.4, 0.6$, and 0.8.

SOLUTION From Eq. (5.28),

$$u = \sum_{n=1}^{n=\infty} \underbrace{\left(\frac{1}{H}\int_0^{2H} u_i \sin\frac{n\pi z}{2H}\,dz\right)}_{A} \left(\sin\frac{n\pi z}{2H}\right) \exp\left(\frac{-n^2\pi^2 T_v}{4}\right)$$

Let us evaluate the term A:

$$A = \frac{1}{H}\int_0^{2H} u_i \sin\frac{n\pi z}{2H}\,dz$$

or

$$A = \frac{1}{H}\int_0^{2H} 50 \sin\frac{\pi z}{2H} \sin\frac{n\pi z}{2H}\,dz$$

Note that the above integral is zero if $n \neq 1$, and so the only nonzero term is obtained when $n = 1$. Therefore,

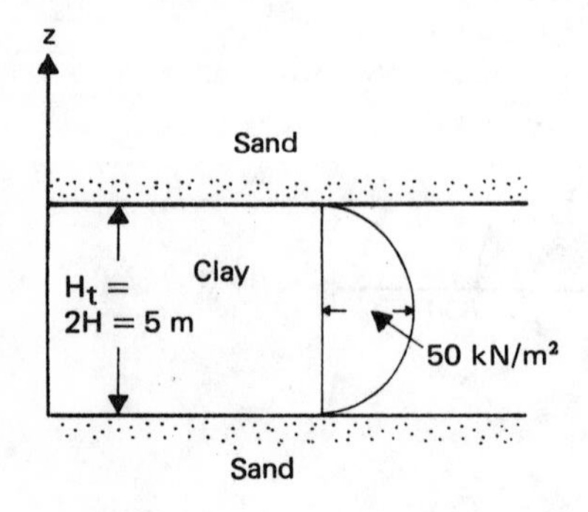

Fig. 5.17

$$A = \frac{50}{H}\int_0^{2H} \sin^2 \frac{\pi z}{2H}\,dz = \frac{50}{H}H = 50$$

Since only for $n = 1$ is A not zero,

$$u = 50 \sin \frac{\pi z}{2H} \exp\left(\frac{-\pi^2 T_v}{4}\right)$$

At the midheight of the clay layer, $z = H$, and so

$$u = 50 \sin \frac{\pi}{2} \exp\left(\frac{-\pi^2 T_v}{4}\right) = 50 \exp\left(\frac{-\pi^2 T_v}{4}\right)$$

The values of the excess pore water pressure are tabulated below:

T_v	$u = 50 \exp\left(\frac{-\pi^2 T_v}{4}\right)$, kN/m²
0.2	30.52
0.4	18.64
0.6	11.38
0.8	6.95

5.1.4 Numerical Solution for One-Dimensional Consolidation

The principles of finite-difference solutions were introduced in Sec. 2.2.8. In this section, we will consider the finite-difference solution for one-dimensional consolidation, starting from the basic differential equation of Terzaghi's consolidation theory:

$$\frac{\partial u}{\partial t} = C_v \frac{\partial^2 u}{\partial z^2} \tag{5.14}$$

Let u_R, t_R, and z_R be any arbitrary reference excess pore water pressure, time, and distance, respectively. From these we can define the following nondimensional terms:

$$\text{Nondimensional excess pore water pressure:} \quad \bar{u} = \frac{u}{u_R} \tag{5.43}$$

$$\text{Nondimensional time:} \quad \bar{t} = \frac{t}{t_R} \tag{5.44}$$

$$\text{Nondimensional depth:} \quad \bar{z} = \frac{z}{z_R} \tag{5.45}$$

From Eqs. (5.43), (5.44), and the left-hand side of Eq. (5.14),

$$\frac{\partial u}{\partial t} = \frac{u_R}{t_R} \frac{\partial \bar{u}}{\partial \bar{t}} \tag{5.46}$$

Similarly, from Eqs. (5.43), (5.45), and the right-hand side of Eq. (5.14),

$$C_v \frac{\partial^2 u}{\partial z^2} = C_v \frac{u_R}{z_R^2} \frac{\partial^2 \bar{u}}{\partial \bar{z}^2} \tag{5.47}$$

From Eqs. (5.46) and (5.47),

$$\frac{u_R}{t_R} \frac{\partial \bar{u}}{\partial \bar{t}} = C_v \frac{u_R}{z_R^2} \frac{\partial^2 \bar{u}}{\partial \bar{z}^2}$$

or

$$\frac{1}{t_R} \frac{\partial \bar{u}}{\partial \bar{t}} = \frac{C_v}{z_R^2} \frac{\partial^2 \bar{u}}{\partial \bar{z}^2} \tag{5.48}$$

If we adopt the reference time in such a way that $t_R = z_R^2/C_v$, then Eq. (5.48) will be of the form

$$\frac{\partial \bar{u}}{\partial \bar{t}} = \frac{\partial^2 \bar{u}}{\partial \bar{z}^2} \tag{5.49}$$

The left-hand side of Eq. (5.49) can be written as

$$\frac{\partial \bar{u}}{\partial \bar{t}} = \frac{1}{\Delta \bar{t}} (\bar{u}_{0,\bar{t}+\Delta \bar{t}} - \bar{u}_{0,\bar{t}}) \tag{5.50}$$

where $\bar{u}_{0,\bar{t}}$ and $\bar{u}_{0,\bar{t}+\Delta \bar{t}}$ are the nondimensional pore water pressures at point O (Fig. 5.18*a*) at nondimensional times t and $t + \Delta t$. Again, similar to Eq. (2.144),

$$\frac{\partial^2 \bar{u}}{\partial \bar{z}^2} = \frac{1}{(\Delta \bar{z})^2} (\bar{u}_{1,\bar{t}} + \bar{u}_{3,\bar{t}} - 2\bar{u}_{0,\bar{t}}) \tag{5.51}$$

Equating the right sides of Eqs. (5.50) and (5.51),

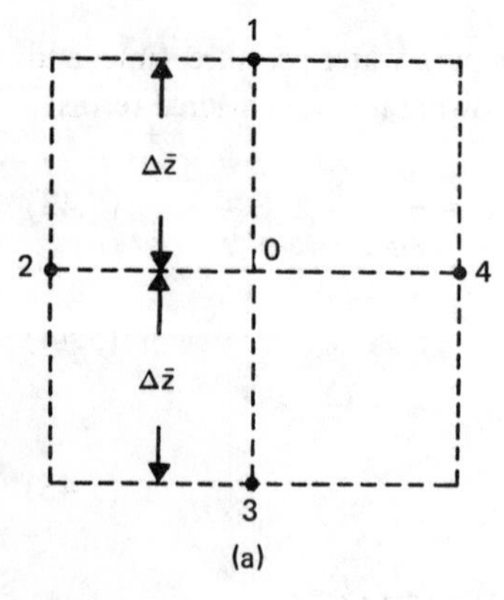

Fig. 5.18

$$\frac{1}{\Delta \bar{t}}(\bar{u}_{0,\bar{t}+\Delta\bar{t}} - \bar{u}_{0,\bar{t}}) = \frac{1}{(\Delta\bar{z})^2}(\bar{u}_{1,\bar{t}} + \bar{u}_{3,\bar{t}} - 2\bar{u}_{0,\bar{t}})$$

or

$$\bar{u}_{0,\bar{t}+\Delta\bar{t}} = \frac{\Delta\bar{t}}{(\Delta\bar{z})^2}(\bar{u}_{1,\bar{t}} + \bar{u}_{3,\bar{t}} - 2\bar{u}_{0,\bar{t}}) + \bar{u}_{0,\bar{t}} \tag{5.52}$$

For Eq. (5.52) to converge, $\Delta\bar{t}$ and $\Delta\bar{z}$ must be chosen such that $\Delta\bar{t}/(\Delta\bar{z})^2$ is less than 0.5.

When solving for pore water pressure at the interface of a clay layer and an impervious layer, Eq. (5.52) can be used. However, we need to take point 3 as the mirror image of point 1 (Fig. 5.18*b*); thus $\bar{u}_{1,\bar{t}} = \bar{u}_{3,\bar{t}}$. So Eq. (5.52) becomes

$$\bar{u}_{0,\bar{t}+\Delta\bar{t}} = \frac{\Delta\bar{t}}{(\Delta\bar{z})^2}(2\bar{u}_{1,\bar{t}} - 2\bar{u}_{0,\bar{t}}) + \bar{u}_{0,\bar{t}} \tag{5.53}$$

Consolidation in a layered soil. It is not always possible to develop a closed-form solution for consolidation in layered soils. There are several variables involved, such as different coefficients of permeability, the thickness of layers, and different values of coefficient of consolidation. Figure 5.19 shows the nature of the degree of consolidation of a two-layered soil.

In view of the above, numerical solutions provide a better approach. If we are involved with the calculation of excess pore water pressure at the interface of two different types (i.e., different values of C_v) of clayey soils, Eq. (5.52) will have to be modified to some extent. Referring to Fig. 5.20, this can be achieved as follows (Scott, 1963): from Eq. (5.14),

$$\underset{\substack{\uparrow \\ \text{change in volume}}}{\frac{k}{C_v}\frac{\partial u}{\partial t}} = \underset{\substack{\uparrow \\ \text{difference between the rate of flow}}}{k\frac{\partial^2 u}{\partial z^2}}$$

Based on the derivations of Eq. (2.163*a*)

$$k\frac{\partial^2 u}{\partial z^2} = \frac{1}{2}\left[\frac{k_1}{(\Delta z)^2} + \frac{k_2}{(\Delta z)^2}\right]\left(\frac{2k_1}{k_1+k_2}u_{1,t} + \frac{2k_2}{k_1+k_2}u_{3,t} - 2u_{0,t}\right) \tag{5.54}$$

where k_1 and k_2 are the coefficients of permeability in layers 1 and 2, respectively. $u_{0,t}$, $u_{1,t}$, and $u_{3,t}$ are the excess pore water pressures at time t for points 0, 1, and 3, respectively.

Also, the average volume change for the element at the boundary is

$$\frac{k}{C_v}\frac{\partial u}{\partial t} = \frac{1}{2}\left(\frac{k_1}{C_{v_1}} + \frac{k_2}{C_{v_2}}\right)\frac{1}{\Delta t}(u_{0,t+\Delta t} - u_{0,t}) \tag{5.55}$$

where $u_{0,t}$ and $u_{0,t+\Delta t}$ are the excess pore water pressures at point 0 at times t and $t + \Delta t$, respectively. Equating the right-hand sides of Eqs. (5.54) and (5.55), we get

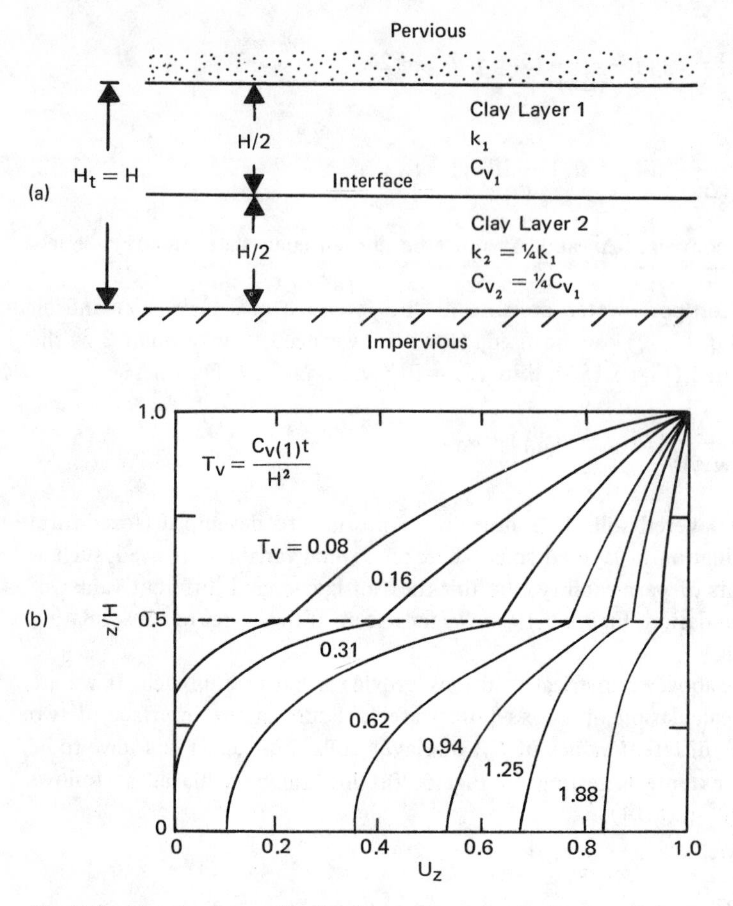

Fig. 5.19 Degree of consolidation in two-layered soil. *(Figure 5.19b after U. Luscher, Discussion,* J. Soil Mech. Found. Div., ASCE, *vol. 91, no. SM1, 1965.)*

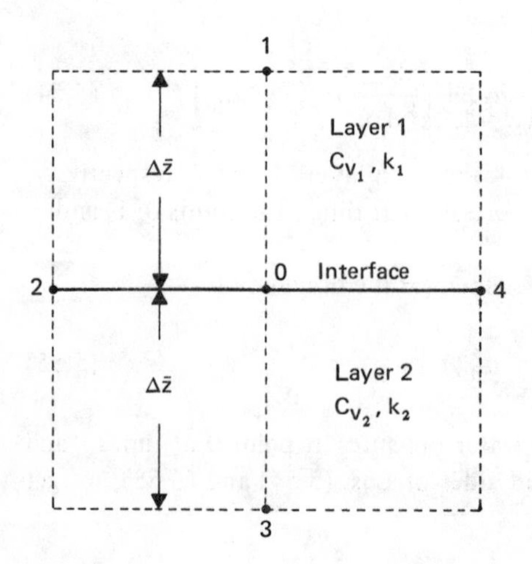

Fig. 5.20

$$\left(\frac{k_1}{C_{v_1}} + \frac{k_2}{C_{v_2}}\right)\frac{1}{\Delta t}(u_{0,t+\Delta t} - u_{0,t})$$

$$= \frac{1}{(\Delta z)^2}(k_1 + k_2)\left(\frac{2k_1}{k_1 + k_2}u_{1,t} + \frac{2k_2}{k_1 + k_2}u_{3,t} - 2u_{0,t}\right)$$

or
$$u_{0,t+\Delta t} = \frac{\Delta t}{(\Delta z)^2}\frac{k_1 + k_2}{k_1/C_{v_1} + k_2/C_{v_2}} \times \left(\frac{2k_1}{k_1 + k_2}u_{1,t} + \frac{2k_2}{k_1 + k_2}u_{3,t} - 2u_{0,t}\right) + u_{0,t}$$

or
$$u_{0,t+\Delta t} = \frac{\Delta t\, C_{v_1}}{(\Delta z)^2}\frac{1 + k_2/k_1}{1 + (k_2/k_1)(C_{v_1}/C_{v_2})} \times \left(\frac{2k_1}{k_1 + k_2}u_{1,t} + \frac{2k_2}{k_1 + k_2}u_{3,t} - 2u_{0,t}\right) + u_{0,t} \tag{5.56}$$

Assuming $1/t_R = C_{v_1}/z_R^2$ and combining Eqs. (5.43) to (5.45) and (5.56), we get

$$u_{0,\bar{t}+\Delta\bar{t}} = \frac{1 + k_2/k_1}{1 + (k_2/k_1)(C_{v_1}/C_{v_2})}\frac{\Delta\bar{t}}{(\Delta\bar{z})^2} \times \left(\frac{2k_1}{k_1 + k_2}\bar{u}_{1,\bar{t}} + \frac{2k_2}{k_1 + k_2}\bar{u}_{3,\bar{t}} - 2\bar{u}_{0,\bar{t}}\right) + u_{0,\bar{t}} \tag{5.57}$$

Example 5.5 A uniform surcharge of $q = 150\,\text{kN/m}^2$ is applied at the ground surface of the soil profile shown in Fig. 5.21. Using the numerical method, determine the distribution of excess pore water pressure for the clay layers after 10 days of load application.

SOLUTION Since this is a uniform surcharge, the excess pore water pressure immediately after the load application will be $150\,\text{kN/m}^2$ throughout the clay layers. However, due to the drainage conditions, the excess pore water pressures at the top of the layer 1 and bottom of layer 2 will immediately become zero. Now, let $z_R = 8\,\text{m}$ and $u_R = 1.5\,\text{kN/m}^2$. So $\bar{z} = (8\,\text{m})/(8\,\text{m}) = 1$ and $\bar{u} = (150\,\text{kN/m}^2)/(1.5\,\text{kN/m}^2) = 100$. Fig. 5.22 shows the distribution of $\bar{u}$ at time $t = 0$; note that $\Delta\bar{z} = 2/8 = 0.25$. Now,

$$t_R = \frac{z_R^2}{C_v} \qquad \bar{t} = \frac{t}{t_R} \qquad \frac{\Delta t}{\Delta\bar{t}} = \frac{z_R^2}{C_v} \qquad \text{or} \qquad \Delta\bar{t} = \frac{C_v\,\Delta t}{z_R^2}$$

Let $\Delta t = 5$ days for both layers. So, for layer 1,

$$\Delta\bar{t}_{(1)} = \frac{C_{v_1}\Delta t}{z_R^2} = \frac{0.26(5)}{8^2} = 0.0203 \qquad \frac{\Delta\bar{t}_{(1)}}{(\Delta\bar{z})^2} = \frac{0.0203}{0.25^2} = 0.325 \qquad (<0.5)$$

For layer 2,

$$\Delta\bar{t}_{(2)} = \frac{C_{v_2}\Delta t}{z_R^2} = \frac{0.38(5)}{8^2} = 0.0297 \qquad \frac{\Delta\bar{t}_{(2)}}{(\Delta\bar{z})^2} = \frac{0.0297}{0.25^2} = 0.475 \qquad (<0.5)$$

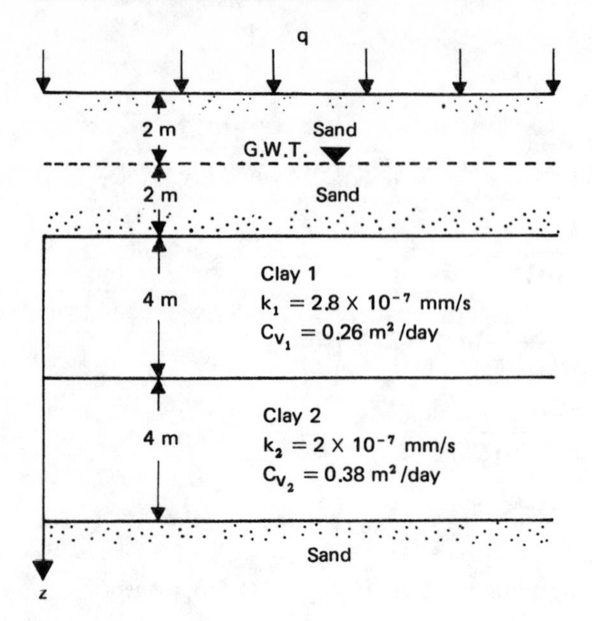

Fig. 5.21

$\bar{u}$ *at* $t = 5$ *days:* At $\bar{z} = 0$

$$\bar{u}_{0,\bar{t}+\Delta\bar{t}} = 0$$

At $\bar{z} = 0.25$,

$$\bar{u}_{0,\bar{t}+\Delta\bar{t}} = \frac{\Delta\bar{t}_{(1)}}{(\Delta\bar{z})^2}\,(\bar{u}_{1,\bar{t}} + \bar{u}_{3,\bar{t}} - 2\bar{u}_{0,\bar{t}}) + \bar{u}_{0,\bar{t}} \tag{5.52}$$

$$= 0.325\,[0 + 100 - 2(100)] + 100 = 67.5$$

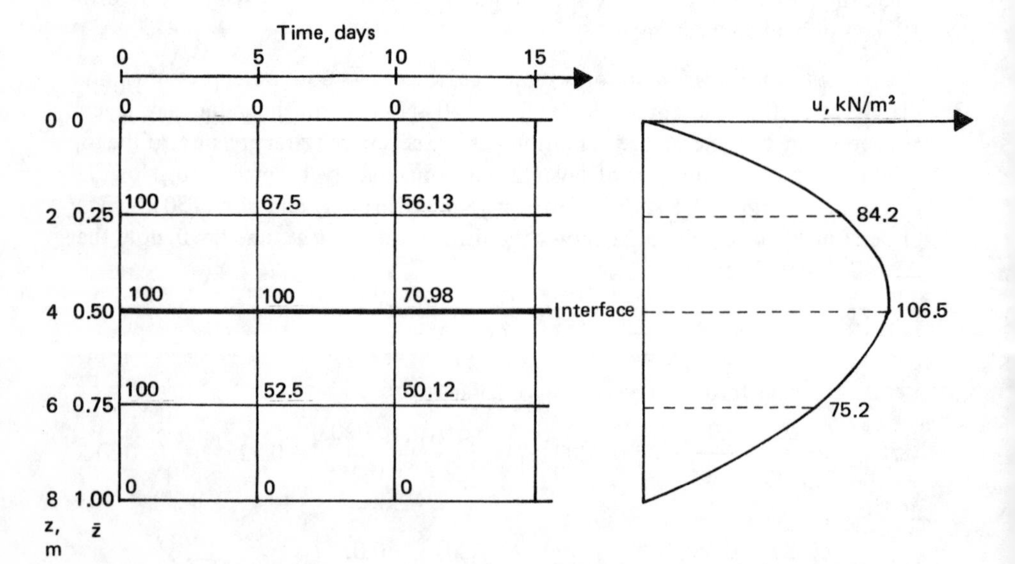

Fig. 5.22

At $\bar{z} = 0.5$ [note: this is the boundary of two layers, so we will use Eq. (5.57)],

$$\bar{u}_{0,\bar{t}+\Delta\bar{t}} = \frac{1 + k_2/k_1}{1 + (k_2/k_1)(C_{v_2}/C_{v_1})} \frac{\Delta\bar{t}_{(1)}}{(\Delta\bar{z})^2}$$

$$\times \left(\frac{2k_1}{k_1 + k_2} \bar{u}_{1,\bar{t}} + \frac{2k_2}{k_1 + k_2} \bar{u}_{3,\bar{t}} - 2\bar{u}_{0,\bar{t}} \right) + \bar{u}_{0,\bar{t}}$$

$$= \frac{1 + 2/2.8}{1 + (2 \times 0.26)/(2.8 \times 0.38)} (0.325)$$

$$\times \left[\frac{2 \times 2.8}{2 + 2.8} (100) + \frac{2 \times 2}{2 + 2.8} (100) - 2(100) \right] + 100$$

or

$$\bar{u}_{0,\bar{t}+\Delta\bar{t}} = (1.152)(0.325)(116.67 + 83.33 - 200) + 100 = 100$$

At $\bar{z} = 0.75$,

$$\bar{u}_{0,\bar{t}+\Delta\bar{t}} = \frac{\Delta\bar{t}_{(2)}}{(\Delta\bar{z})^2} (\bar{u}_{1,\bar{t}} + \bar{u}_{3,\bar{t}} - 2u_{0,\bar{t}}) + \bar{u}_{0,\bar{t}}$$

$$= 0.475\ [100 + 0 - 2(100)] + 100 = 52.5$$

At $\bar{z} = 1.0$,

$$\bar{u}_{0,\bar{t}+\Delta\bar{t}} = 0$$

$\bar{u}$ at $t = 10$ days: At $\bar{z} = 0$

$$\bar{u}_{0,\bar{t}+\Delta\bar{t}} = 0$$

At $\bar{z} = 0.25$,

$$\bar{u}_{0,\bar{t}+\Delta\bar{t}} = 0.325\ [0 + 100 - 2(67.5)] + 67.5 = 56.13$$

At $\bar{z} = 0.5$,

$$\bar{u}_{0,\bar{t}+\Delta\bar{t}} = (1.152)(0.325) \left[\frac{2 \times 2.8}{2 + 2.8} (67.5) + \frac{2 \times 2}{2 + 2.8} (52.5) - 2(100) \right] + 100$$

$$= (1.152)(0.325)(78.75 + 43.75 - 200) + 100 = 70.98$$

At $\bar{z} = 0.75$,

$$\bar{u}_{0,\bar{t}+\Delta\bar{t}} = 0.475\ [100 + 0 - 2(52.5)] + 52.5 = 50.12$$

At $\bar{z} = 1.0$,

$$\bar{u}_{0,\bar{t}+\Delta\bar{t}} = 0$$

The variation of the nondimensional excess pore water pressure is shown in Fig. 5.22. Knowing $\bar{u} = (\bar{u})(u_R) = \bar{u}(1.5)\,\text{kN/m}^2$, we can plot the variation of u with depth.

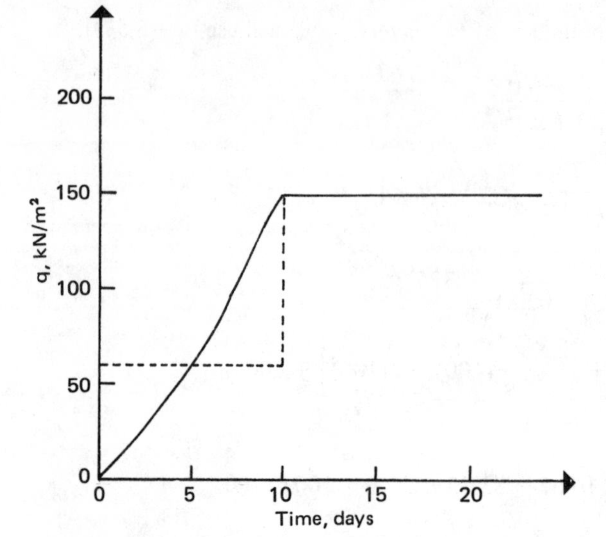

Fig. 5.23

Example 5.6 For Example 5.5, assume that the surcharge q is applied gradually. The relationship between time and q is shown in Fig. 5.23. Using the numerical method, determine the distribution of excess pore water pressure after 15 days from the start of loading.

SOLUTION As before, $z_R = 8\,\text{m}$, $u_R = 1.5\,\text{kN/m}^2$. For $\Delta t = 5$ days,

$$\frac{\Delta \bar{T}_{(1)}}{(\Delta \bar{z})^2} = 0.325 \qquad \frac{\Delta \bar{T}_{(2)}}{(\Delta \bar{z})^2} = 0.475$$

The continuous loading can be divided into step loads such as $60\,\text{kN/m}^2$ from 0 to 10 days and an added $90\,\text{kN/m}^2$ from the tenth day on. This is shown by dashed lines in Fig. 5.23.

$\bar{u}$ at $t = 0$ days:

$\bar{z} = 0 \qquad \bar{u} = 0$

$\bar{z} = 0.25 \qquad \bar{u} = 60/1.5 = 40$

$\bar{z} = 0.5 \qquad \bar{u} = 40$

$\bar{z} = 0.75 \qquad \bar{u} = 40$

$\bar{z} = 1 \qquad \bar{u} = 0$

$\bar{u}$ at $t = 5$ days: At $\bar{z} = 0$,

$\bar{u} = 0$

At $\bar{z} = 0.25$, from Eq. (5.52),

$$\bar{u}_{0,\bar{t}+\Delta\bar{t}} = 0.325\ [0 + 40 - 2(40)] + 40 = 27$$

At $\bar{z} = 0.5$, from Eq. (5.57),

$$\bar{u}_{0,\bar{t}+\Delta\bar{t}} = (1.532)(0.325)\left[\frac{2 \times 2.8}{2+2.8}(40) + \frac{2 \times 2}{2+2.8}(40) - 2(40)\right] + 40 = 40$$

At $\bar{z} = 0.75$, from Eq. (5.52),

$\bar{u}_{0,\bar{t}+\Delta\bar{t}} = 0.475\ [40 + 0 - 2(40)] + 40 = 21$

At $\bar{z} = 1$,

$\bar{u}_{0,\bar{t}+\Delta\bar{t}} = 0$

$\bar{u}$ at t = 10 days: At $\bar{z} = 0$,

$\bar{u} = 0$

At $\bar{z} = 0.25$, from Eq. (5.52),

$\bar{u}_{0,\bar{t}+\Delta\bar{t}} = 0.325\ [0 + 40 - 2(27)] + 27 = 22.45$

At this point, a new load of 90 kN/m² is added. So $\bar{u}$ will increase by an amount $90/1.5 = 60$. So, the new $\bar{u}_{0,\bar{t}+\Delta\bar{t}}$ is $60 + 22.45 = 82.45$. At $\bar{z} = 0.5$, from Eq. (5.57),

$$\bar{u}_{0,\bar{t}+\Delta\bar{t}} = (1.152)(0.325)\left[\frac{2 \times 2.8}{2+2.8}(27) + \frac{2 \times 2}{2+2.8}(21) - 2(40)\right] + 40 = 28.4$$

New $\bar{u}_{0,\bar{t}+\Delta\bar{t}} = 28.4 + 60 = 88.4$

At $\bar{z} = 0.75$, from Eq. (5.52),

$\bar{u}_{0,\bar{t}+\Delta\bar{t}} = 0.475\ [40 + 0 - 2(21)] + 21 = 20.05$

New $\bar{u}_{0,\bar{t}+\Delta\bar{t}} = 60 + 20.05 = 80.05$

At $\bar{z} = 1$,

$\bar{u} = 0$

$\bar{u}$ at t = 15 days: At $\bar{z} = 0$,

$\bar{u} = 0$

At $\bar{z} = 0.25$,

$\bar{u}_{0,\bar{t}+\Delta\bar{t}} = 0.325\ [0 + 88.4 - 2(82.45)] + 82.45 = \mathit{57.6}$

At $\bar{z} = 0.5$,

$$\bar{u}_{0,\bar{t}+\Delta\bar{t}} = (1.152)(0.325) \times \left[\frac{2 \times 2.8}{2+2.8}(82.45) + \frac{2 \times 2}{2+2.8}(80.05) - 2(88.4)\right] + 88.4 = \mathit{83.2}$$

At $\bar{z} = 0.75$,

$\bar{u}_{0,\bar{t}+\Delta\bar{t}} = 0.475\ [88.4 + 0 - 2(80.05)] + 80.05 = \mathit{46.0}$

At $\bar{z} = 1$,

$\bar{u} = 0$

The distribution of excess pore water pressure is shown in Fig. 5.24.

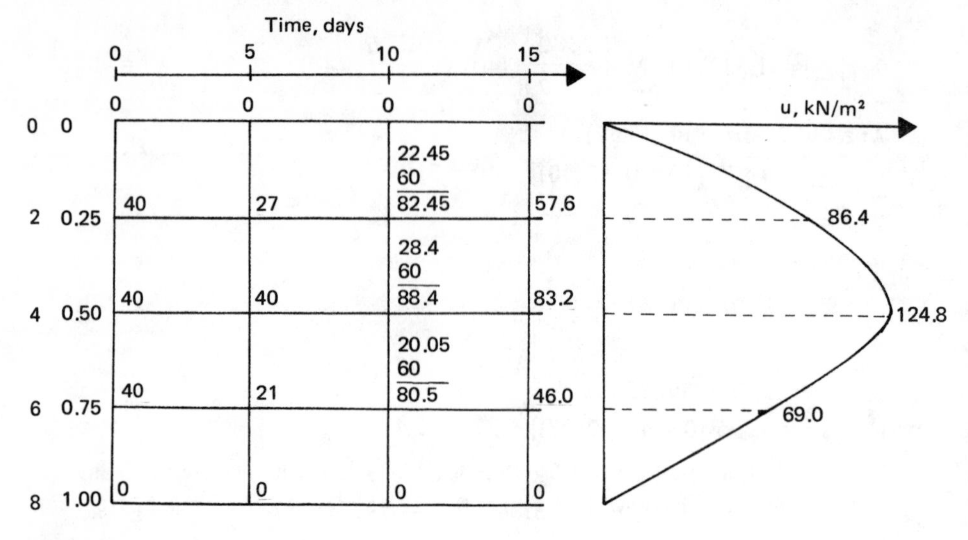

Fig. 5.24

5.1.5 Degree of Consolidation under Time-Dependent Loading

Olson (1977) presented a mathematical solution for one-dimensional consolidation due to a single ramp load. Olson's solution can be explained with the help of Fig. 5.25, in which a clay layer is drained at the top and at the bottom (H is the drainage distance). A uniformly distributed load q is applied at the ground surface. Note that q is a function of time, as shown in Fig. 5.25*b*.

The expression for the excess pore water pressure for the case where $u_i = u_o$ is given in Eq. (5.30) as

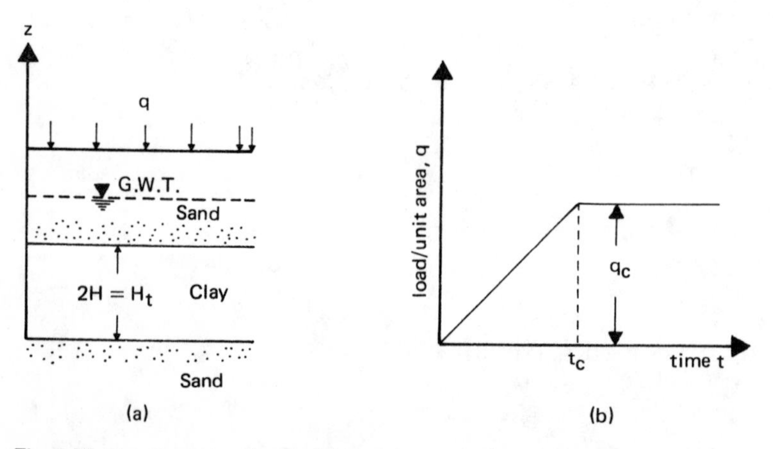

Fig. 5.25 One-dimensional consolidation due to single ramp load.

$$u = \sum_{m=0}^{m=\infty} \frac{2u_o}{M} \sin \frac{Mz}{H} \exp(-M^2 T_v)$$

where $T_v = C_v t/H^2$.

As stated above, the applied load is a function of time:

$$q = f(t_a) \tag{5.58}$$

where t_a is the time of application of any load.

For a differential load dq applied at time t_a, the instantaneous pore pressure increase will be $du_i = dq$. At time t, the remaining excess pore water pressure du at a depth z can be given by the expression

$$du = \sum_{m=0}^{m=\infty} \frac{2du_i}{M} \sin \frac{Mz}{H} \exp\left[\frac{-M^2 C_v(t-t_a)}{H^2}\right]$$

$$= \sum_{m=0}^{m=\infty} \frac{2dq}{M} \sin \frac{Mz}{H} \exp\left[\frac{-M^2 C_v(t-t_a)}{H^2}\right] \tag{5.59}$$

The average degree of consolidation can be defined as

$$U_{\text{av}} = \frac{\alpha q_c - (1/H_t)\int_0^{H_t} u\,dz}{q_c} = \frac{\text{settlement at time } t}{\text{settlement at time } t = \infty} \tag{5.60}$$

where αq_c is the total load per unit area applied at the time of the analysis. The settlement at time $t = \infty$ is, of course, the ultimate settlement. Note that the term q_c in the denominator of Eq. (5.60) is equal to the instantaneous excess pore water pressure ($u_i = q_c$) that might have been generated throughout the clay layer had the stress q_c been applied instantaneously.

Proper integration of Eqs. (5.59) and (5.60) gives the following:

For $T_v \leqslant T_c$:

$$u = \sum_{m=0}^{m=\infty} \frac{2q_c}{M^3 T_c} \sin \frac{Mz}{H} \left[1 - \exp(-M^2 T_v)\right] \tag{5.61}$$

and

$$U_{\text{av}} = \frac{T_v}{T_c}\left\{1 - \frac{2}{T_v}\sum_{m=0}^{m=\infty} \frac{1}{M^4}\left[1 - \exp(-M^2 T_v)\right]\right\} \tag{5.62}$$

For $T_v \geqslant T_c$:

$$u = \sum_{m=0}^{m=\infty} \frac{2q_c}{M^3 T_c} \left[\exp(M^2 T_c) - 1\right] \sin \frac{Mz}{H} \exp(-M^2 T_v) \tag{5.63}$$

and

$$U_{\text{av}} = 1 - \frac{2}{T_c}\sum_{m=0}^{m=\infty} \frac{1}{M^4}\left[\exp(M^2 T_c) - 1\right] \exp(-M^2 T_c) \tag{5.64}$$

where $T_c = \dfrac{C_v t_c}{H^2}$ (5.65)

Fig. 5.26 shows the plot of U_{av} against T_v for various values of T_c.

Example 5.7 Based on one-dimensional consolidation test results on a clay, the coefficient of consolidation for a given pressure range was obtained as 8×10^{-3} mm²/s. In the field, there is a 2-m thick layer of the same clay, as shown in Fig. 5.27. Based on the assumption that a uniform surcharge of 70 kN/m² was to be applied instantaneously, the total consolidation settlement was estimated to be 150 mm. However, during the construction, the loading was gradual; the resulting surcharge can be approximated as shown in Fig. 5.27*b*. Estimate the settlement at $t = 30$ and 120 days after the beginning of the construction.

SOLUTION

$$T_c = \frac{C_v t_c}{H^2} \tag{5.65}$$

Now, $t_c = 60$ days $= 60 \times 24 \times 60 \times 60$ s; also, $H_t = 2\text{m} = 2H$ (two-way drainage), and so $H = 1\text{m} = 1000$ mm. Hence,

$$T_c = \frac{(8 \times 10^{-3})(60 \times 24 \times 60 \times 60)}{(1000)^2} = 0.0414$$

At $t = 30$ days,

$$T_v = \frac{C_v t}{H^2} = \frac{(8 \times 10^{-3})(30 \times 24 \times 60 \times 60)}{(1000)^2} = 0.0207$$

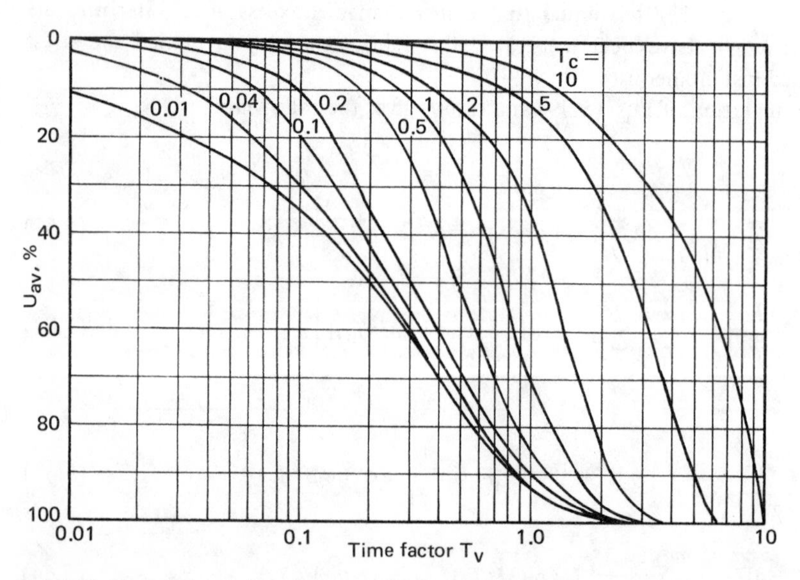

Fig. 5.26 Plot of U_{av} against time factor T_v for single ramp load. *(After R. E. Olsen, Consolidation under Time Dependent Loading,* J. Geotech. Eng. Div., ASCE, *vol. 103, no. GT1, 1977.)*

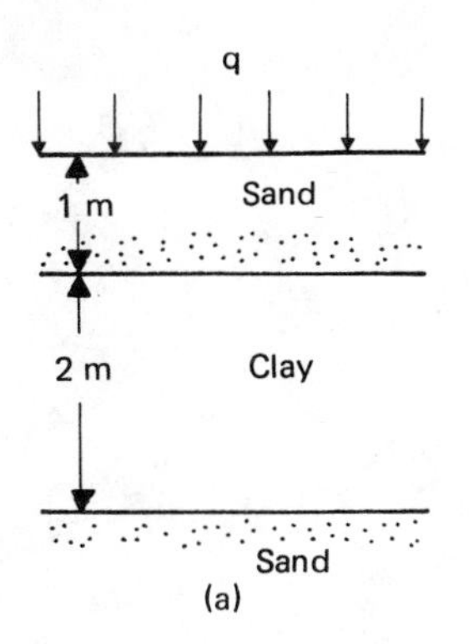

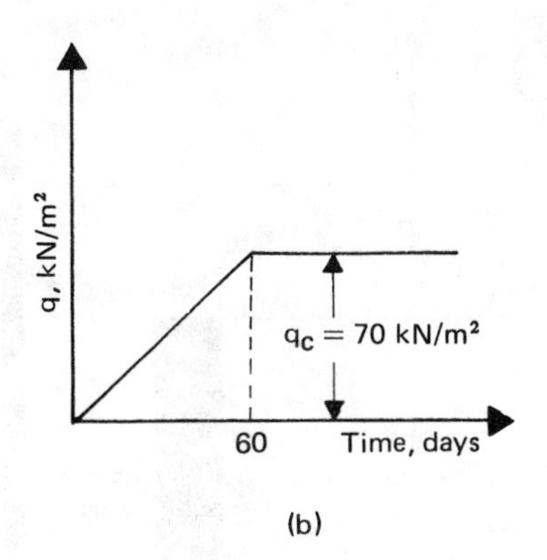

Fig. 5.27

From Fig. 5.26, for $T_v = 0.0207$ and $T_c = 0.0414$, $U_{av} \approx 5\%$. So,

Settlement = (0.05) (150) = *7.5 mm*

At t = 120 days,

$$T_v = \frac{(8 \times 10^{-3})(120 \times 24 \times 60 \times 60)}{(1000)^2} = 0.083$$

From Fig. 5.26, for $T_v = 0.083$ and $T_c = 0.0414$, $U_{av} \approx 27\%$. So,

Settlement = (0.27) (150) = *40.5 mm*

5.1.6 Standard One-Dimensional Consolidation Test and Interpretation

The standard one-dimensional consolidation test is usually carried out on saturated specimens about 1 in (25.4 mm) thick and 2.5 in (63.5 mm) in diameter (Fig. 5.28). The soil sample is kept inside a metal ring, with a porous stone at the top and another at the bottom. The load P on the sample is applied through a lever arm, and the compression of the specimen is measured by a micrometer dial gauge. The load is usually doubled every 24 hours. The specimen is kept under water throughout the test. (For detailed test procedures, see ASTM test designation D-2435.)

For each load increment, the sample deformation and the corresponding time t is plotted on semilogarithmic graph paper. Figure 5.29 shows a typical deformation vs. log t graph. The graph consists of three distinct parts:

1. Upper curved portion (stage I). This is mainly the result of precompression of the specimen.
2. A straight-line portion (stage II). This is referred to as primary consolidation. At the end of the primary consolidation, the excess pore water pressure generated by the incremental loading is dissipated to a large extent.
3. A lower straight-line portion (stage III). This is called secondary consolidation. During this stage, the specimen undergoes small deformation with time. In fact, there must be immeasurably small excess pore water pressure in the specimen during secondary consolidation.

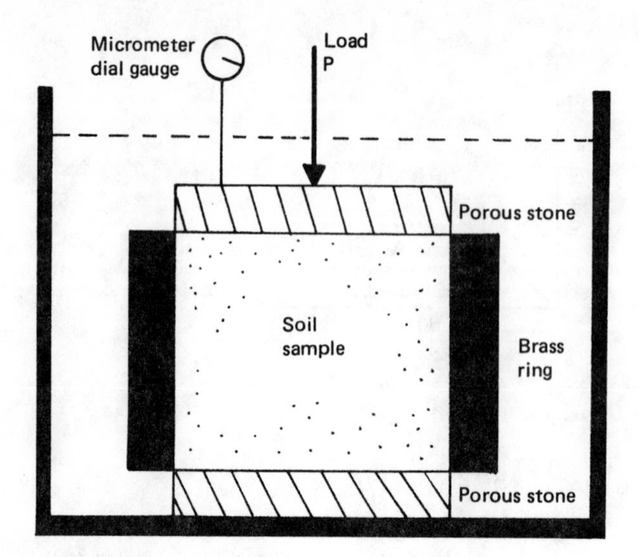

Fig. 5.28 Consolidometer.

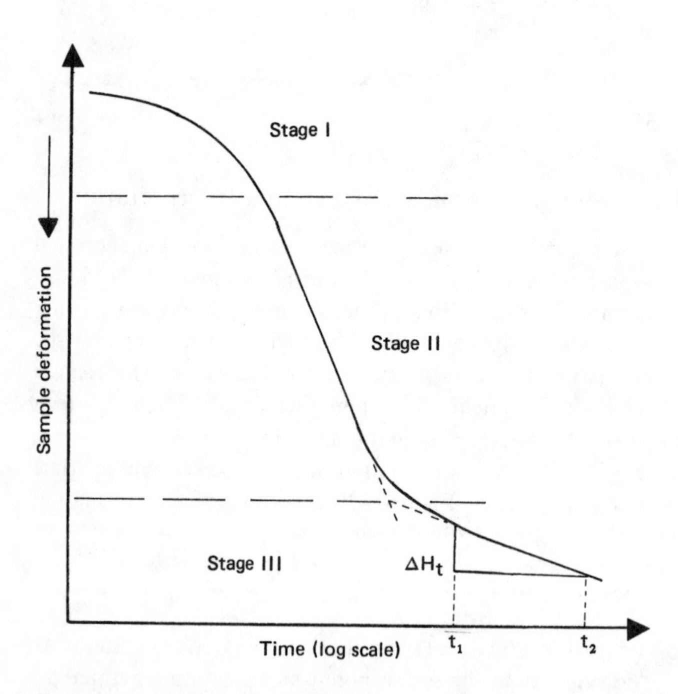

Fig. 5.29 Typical sample deformation vs. log-of-time plot for a given load increment.

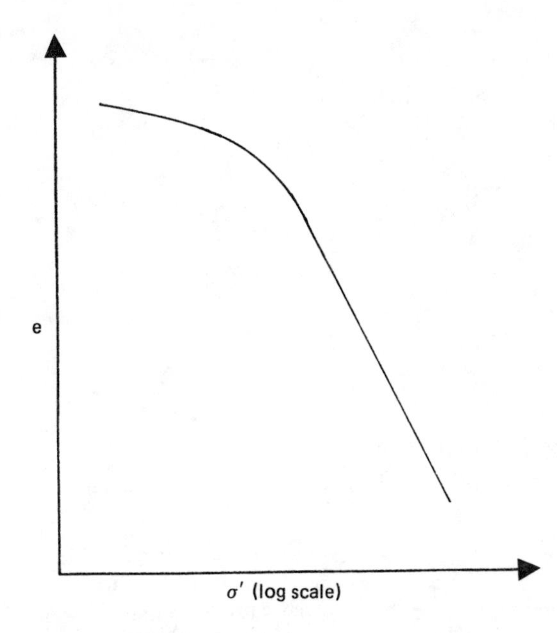

Fig. 5.30 Typical e vs. log σ' plot.

Note that at the end of the test for each incremental loading the stress on the specimen is the effective stress, σ'. Once the specific gravity of the soil solids, the initial specimen dimensions, and the specimen deformation at the end of each load have been determined, the corresponding void ratio can be calculated. A typical void ratio vs. effective pressure relationship plotted on semilogarithmic graph paper is shown in Fig. 5.30.

Preconsolidation pressure. In the typical e vs. log σ' plot shown in Fig. 5.30, it can be seen that the upper part is curved; however, at higher pressures, e and log σ' bear a linear relationship. The upper part is curved because when the soil specimen was obtained from the field, it was subjected to a certain maximum effective pressure. During the process of soil exploration, the pressure is released. In the laboratory, when the soil sample is loaded, it will show relatively small decrease of void ratio with load up to the maximum effective stress to which the soil was subjected in the past. This is represented by the upper curved portion in Fig. 5.30. If the effective stress on the soil sample is increased further, the decrease of void ratio with stress level will be larger. This is represented by the straight-line portion in the e vs. log σ' plot. The effect can also be demonstrated in the laboratory by unloading and reloading a soil sample, as shown in Fig. 5.31. In this figure, *cd* is the void ratio–effective stress relation as the sample is unloaded, and *dfgh* is the reloading branch. At d, the sample is being subjected to a lower effective stress than the maximum stress σ'_1 to which the soil was ever subjected. So *df* will show a flatter curved portion. Beyond point f, the void ratio will decrease at a larger rate with effective stress, and *gh* will have the same slope as *bc*.

Based on the above explanation, we can now define the two conditions of a soil:

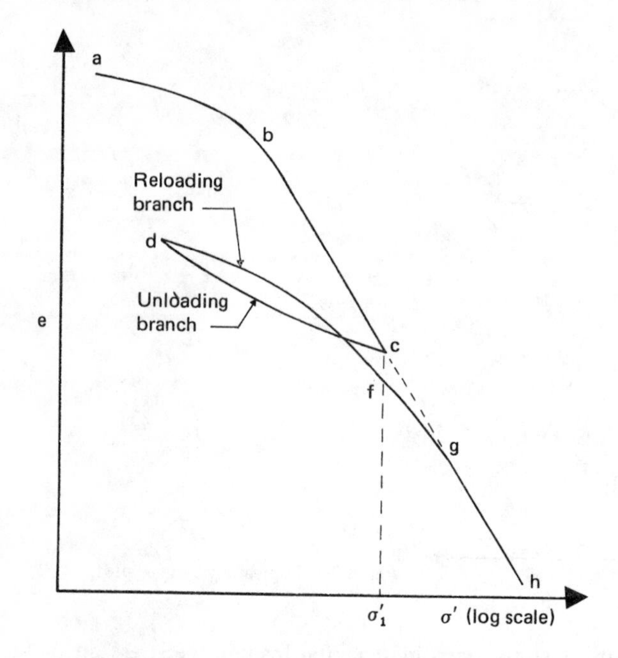

Fig. 5.31 Plot of void ratio vs. effective pressure showing unloading and reloading branches.

1. *Normally consolidated.* A soil is called normally consolidated if the present effective overburden pressure is the maximum to which the soil has ever been subjected, i.e., $\sigma'_{\text{present}} \geqslant \sigma'_{\text{past maximum}}$.
2. *Overconsolidated.* A soil is called overconsolidated if the present effective overburden pressure is less than the maximum to which the soil was ever subjected in the past, i.e., $\sigma'_{\text{present}} < \sigma'_{\text{past maximum}}$.

In Fig. 5.31, the branches *ab*, *cd*, and *df* are the overconsolidated state of a soil, and the branches *bc* and *fh* are the normally consolidated state of a soil.

In the natural condition in the field, a soil may be either normally consolidated or overconsolidated. A soil in the field may become overconsolidated through several mechanisms, some of which are listed in Table 5.2.

The preconsolidation pressure from a e vs. log σ' plot is generally determined by a graphical procedure suggested by Casagrande (1936), as shown in Fig. 5.32. The steps are as follows:

1. Visually determine the point P (on the upper curved portion of the e vs. log σ' plot) that has the maximum curvature.
2. Draw a horizontal line PQ.
3. Draw a tangent PR at P.
4. Draw the line PS bisecting the angle QPR.
6. Produce the straight-line portion of the e vs. log σ' plot backward to intersect PS at T.
6. The effective pressure corresponding to point T is the preconsolidation pressure σ'_c.

Another method for the determination of σ'_c is given in Burmister (1951).

Table 5.2 Mechanisms causing overconsolidation

Mechanisms	Remarks and references
Changes in total stress due to:	
Removal of overburden pressure	
Past structures	
Glaciation	
Changes in pore water pressure due to change in water table elevation:	Kenny (1964) gives sea level changes
Artesian pressures	Common in glaciated areas
Deep pumping	Common in many cities
Desiccation due to drying	Many have occureed during deposition
Desiccation due to plant life	Many have occurred during deposition
Changes in soil structure due to secondary compression (aging)*	Raju (1965); Leonards and Ramiah (1960); Leonards and Altschaeffl (1964); Bjerrum (1967, 1972)
Environmental changes such as pH, temperature, and salt concentration	Lambe (1958)
Chemical alteration due to "weathering," precipitation of cementing agents, ion exchange	Bjerrum (1967)
Change of strain rate on loading	Lowe (1974)

After W. F. Brumund, E. Jonas, and C. C. Ladd, Estimating In Situ Maximum Past (Preconsolidation) Pressure of Saturated Clays from Results of Laboratory Consolidation Tests, *Transportation Research Board, Special Report 163*, 1976.

*See also Sec. 5.1.9.

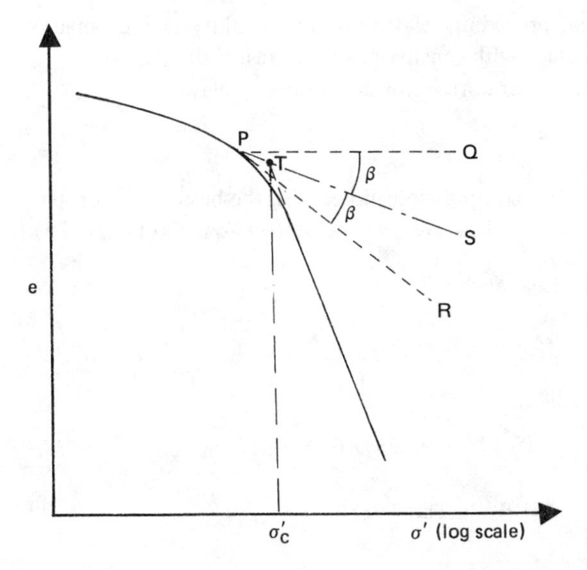

Fig. 5.32 Graphical procedure for determination of preconsolidation pressure.

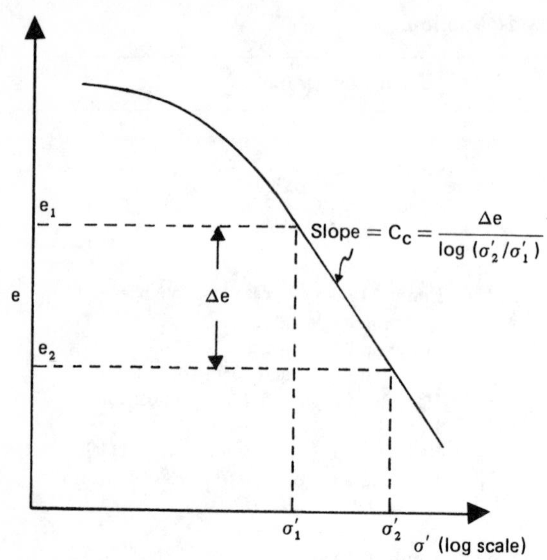

Fig. 5.33 Compression index C_c.

Compression index. The slope of the e vs. log σ' plot for normally consolidated soil is referred to as the compression index C_c. From Fig. 5.33,

$$C_c = \frac{e_1 - e_2}{\log \sigma_2' - \log \sigma_1'} = \frac{\Delta e}{\log(\sigma_2'/\sigma_1)} \tag{5.66}$$

For normally consolidated clays, Terzaghi and Peck (1967) gave a correlation for the compression index as

$$C_c = 0.009(LL - 10) \tag{5.67}$$

where LL is the liquid limit. The preceding relation has a reliability in the range of ±30% and should not be used for clays with sensitivity ratios greater than 4.

Terzaghi and Peck also gave a similar correlation for remolded clays:

$$C_c = 0.007(LL - 10)$$

Several other correlations for the compression index with the basic index properties of soils have been made, and some of these are given below (see Azzouz et al., 1976):

$$C_c = 0.01 w_N \qquad \text{(for Chicago clays)} \tag{5.68}$$

$$C_c = 0.0046(LL - 9) \qquad \text{(for Brazilian clays)} \tag{5.69}$$

$$C_c = 1.21 + 1.055(e_o - 1.87) \qquad \text{(for Motley clays from Sao Paulo city)} \tag{5.70}$$

$$C_c = 0.208 e_o + 0.0083 \qquad \text{(for Chicago clays)} \tag{5.71}$$

$$C_c = 0.0115 w_N \qquad \text{(for organic soil, peats, etc.)} \tag{5.72}$$

where w_N is the natural moisture content (%), and e_o is the in situ void ratio.

Nacci et al. (1975) tested some natural deep-ocean soil samples from the North Atlantic. The calcite content varied from 10 to 80%. Based on their results, the following equation has also been proposed:

$$C_c = 0.02 + 0.014(PI) \tag{5.73}$$

where PI is the plasticity index.

Effect of sample disturbance on the e vs. log σ' curve. Soil samples obtained from the field are somewhat disturbed. When consolidation tests are conducted on these samples, we obtain e vs. log σ' plots that are slightly different from those in the field. This is demonstrated in Fig. 5.34.

Curve I in Fig. 5.34*a* shows the nature of the e vs. log σ' variation that an undisturbed normally consolidated clay (present effective overburden pressure σ'_o; void ratio e_o) in the field would exhibit. This is called the *virgin compression curve*. A laboratory consolidation test on a carefully recovered sample would result in an e vs. log σ' plot such as curve II. If the same soil is completely remolded and then tested in a consolidometer, the resulting void ratio-pressure plot will be like curve III. The virgin compression curve (curve I) and the laboratory e vs. log σ' curve obtained from a carefully recovered sample (curve II) intersect at a void ratio of about $0.4e_o$ (Terzaghi and Peck, 1967).

Curve I in Fig. 5.34*b* shows the nature of the field consolidation curve of an overconsolidated clay. Note that the present effective overburden pressure is σ'_o and

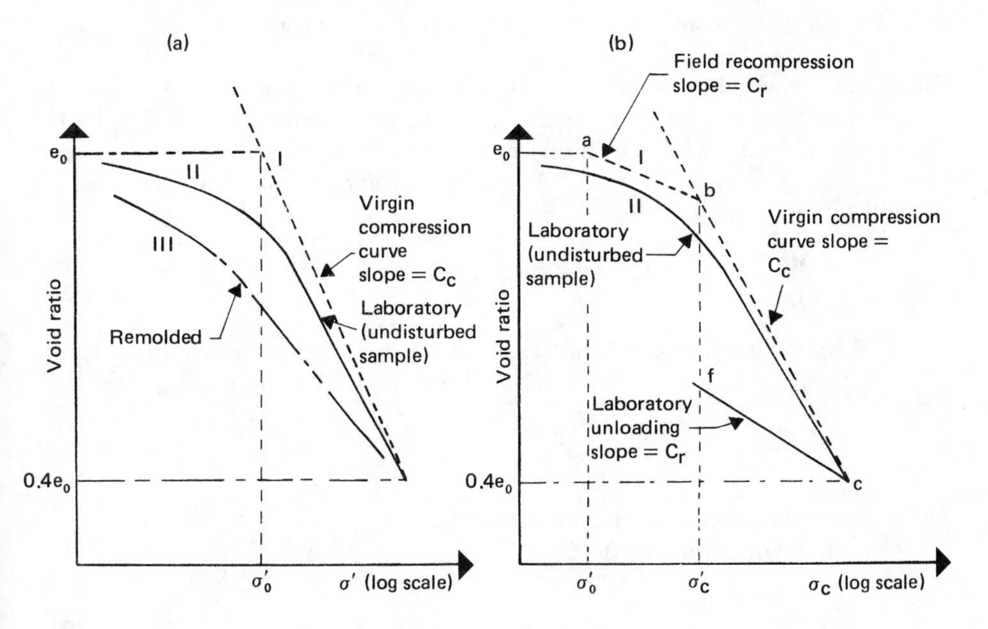

Fig. 5.34 Effect of sample disturbance on e vs. log σ' curve.

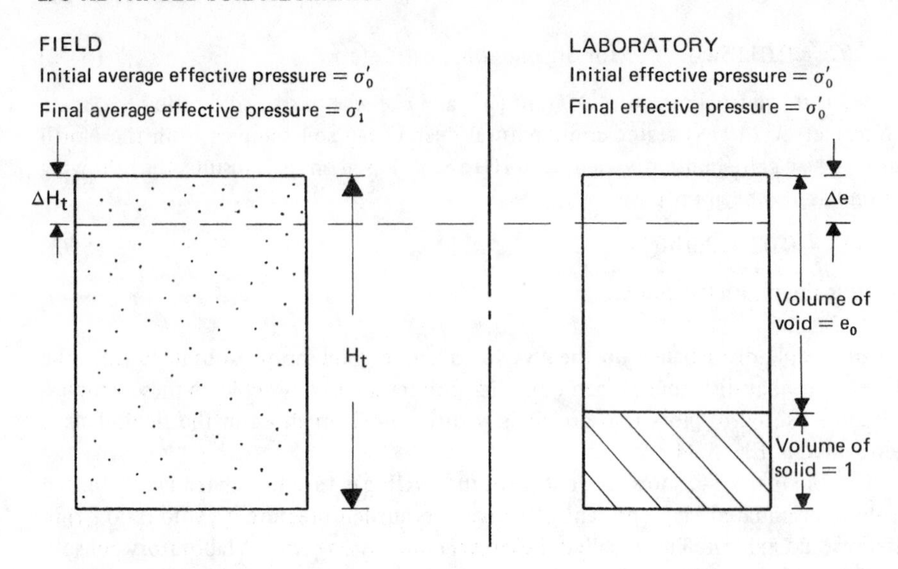

Fig. 5.35 Calculation of one-dimensional consolidation settlement.

the corresponding void ratio is e_o. σ_c' is the preconsolidation pressure, and *bc* is a part of the virgin compression curve. Curve II is the corresponding laboratory consolidation curve. After careful testing, Schmertmann (1953) concluded that the field recompression branch (*ab* in Fig. 5.34*b*) has approximately the same slope as the laboratory unloading branch, *cf*. The slope of the laboratory unloading branch is referred to as C_r. The range of C_r is approximately from one-fifth to one-tenth of C_c.

Calculation of one-dimensional consolidation settlement. The basic principle of one-dimensional consolidation settlement calculation is demonstrated in Fig. 5.35. If a clay layer of total thickness H_t is subjected to an increase of average effective overburden pressure from σ_o' to σ_1', it will undergo a consolidation settlement of ΔH_t. Hence the strain can be given by

$$\epsilon = \frac{\Delta H_t}{H_t} \tag{5.74}$$

where ϵ is strain. Again, if an undisturbed laboratory specimen is subjected to the same effective stress increase, the void ratio will decrease by Δe. Thus, the strain is equal to

$$\epsilon = \frac{\Delta e}{1 + e_o} \tag{5.75}$$

where e_o is the void ratio at an effective stress of σ_o'.

Thus, from Eqs. (5.74) and (5.75),

$$\Delta H_t = \frac{\Delta e\, H_t}{1 + e_o} \tag{5.76}$$

For a normally consolidated clay in the field (Fig. 5.36*a*),

$$\Delta e = C_c \log \frac{\sigma_1'}{\sigma_o'} = C_c \log \frac{\sigma_o' + \Delta\sigma}{\sigma_o'} \tag{5.77}$$

For an overconsolidated clay, (1) if $\sigma_1' < \sigma_c'$ (i.e., overconsolidation pressure) (Fig. 5.36*b*),

$$\Delta e = C_r \log \frac{\sigma_1'}{\sigma_o'} = C_r \log \frac{\sigma_o' + \Delta\sigma}{\sigma_o'} \tag{5.78}$$

and (2) if $\sigma_o' < \sigma_c' < \sigma_1'$ (Fig. 5.36*c*),

$$\Delta e = \Delta e_1 + \Delta e_2 = C_r \log \frac{\sigma_c'}{\sigma_o'} + C_c \log \frac{\sigma_o' + \Delta\sigma}{\sigma_c'} \tag{5.79}$$

The procedure for calculation of one-dimensional consolidation settlement is described in more detail in Chap. 6.

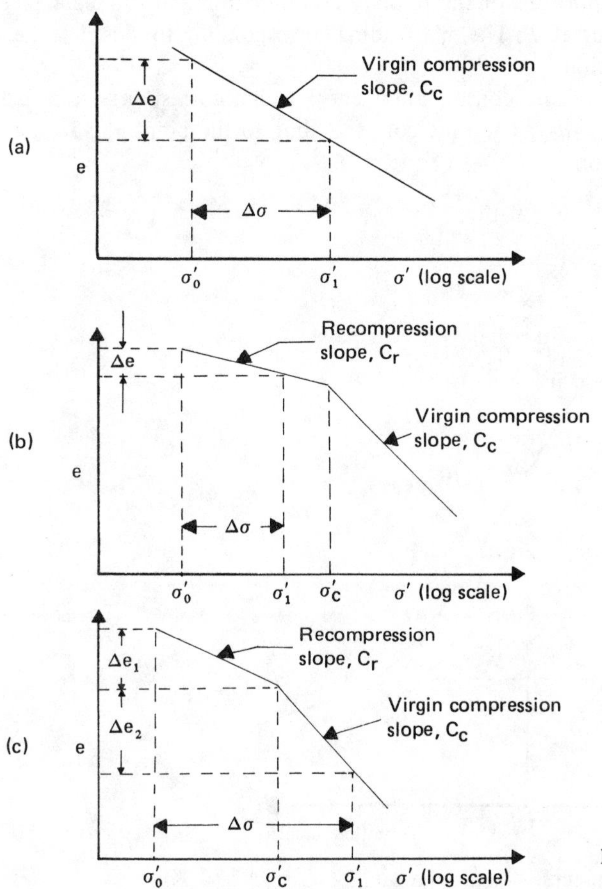

Fig. 5.36 Calculation of Δe [Eqs. (5.77) to (5.79)].

Calculation of coefficient of consolidation from laboratory test results. For a given load increment, the coefficient of consolidation C_v can be determined from laboratory observations of *time vs. dial reading.* Two graphical procedures are commonly used for this: the *logarithm-of-time method* proposed by Casagrande and Fadum (1940), and the *square-root-of-time method* proposed by Taylor (1942). There are also two other useful methods, which were proposed by Su (1958) and Sivaram and Swamee (1977). Each of these four methods is described below.

Logarithm-of-time method

1. Plot the dial readings for sample deformation for a given load increment against time on semilog graph paper as shown in Fig. 5.37.
2. Plot two points, P and Q on the upper portion of the consolidation curve which correspond to time t_1 and t_2, respectively. Note that $t_2 = 4t_1$.
3. The difference of dial readings between P and Q is equal to x. Locate point R, which is at a distance x above point P.
4. Draw the horizontal line RS. The dial reading corresponding to this line is d_0, which corresponds to 0% consolidation.
5. Project the straight-line portions of the primary consolidation and the secondary consolidation to intersect at T. The dial reading corresponding to T is d_{100}, i.e., 100% primary consolidation.
6. Determine the point V on the consolidation curve which corresponds to a dial reading of $(d_0 + d_{100})/2 = d_{50}$. The time corresponding to the point V is t_{50}, i.e., time for 50% consolidation.

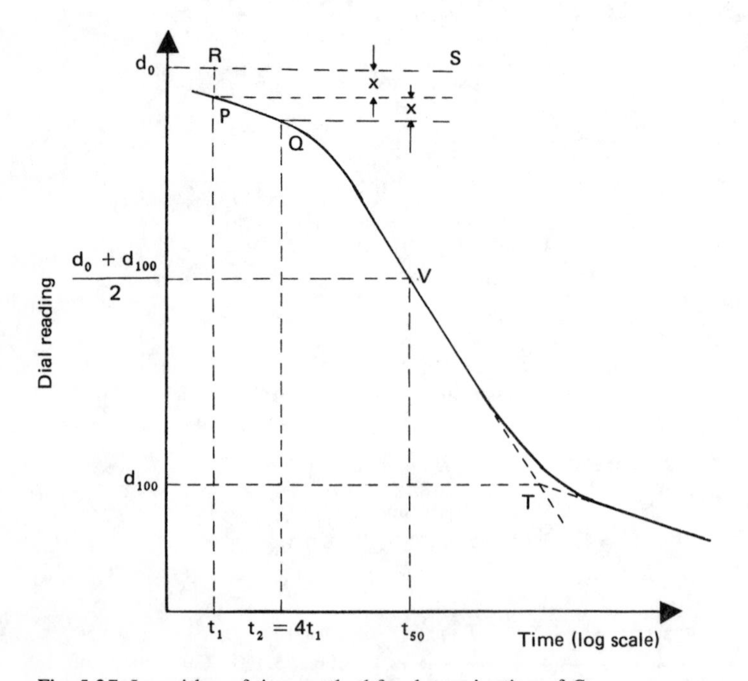

Fig. 5.37 Logarithm-of-time method for determination of C_v.

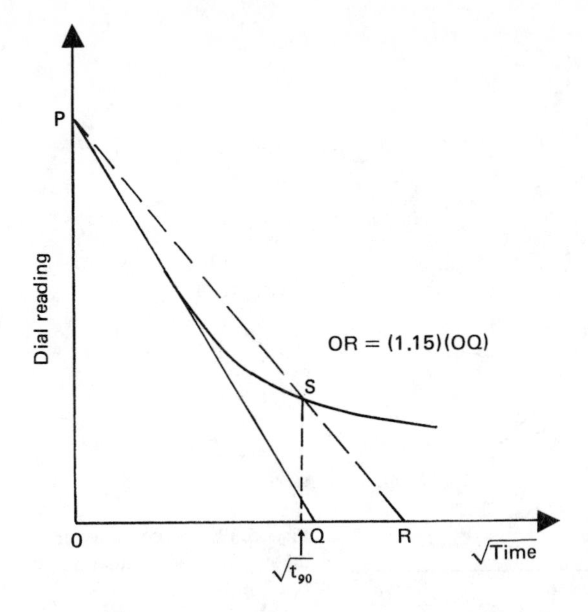

Fig. 5.38 Square-root-of-time method for determination of C_v.

7. Determine C_v from the equation $T_v = C_v t / H^2$. The value of T_v for $U_{av} = 50\%$ is 0.197 (Table 5.1). So,

$$C_v = \frac{0.197 H^2}{t_{50}} \tag{5.80}$$

Square-root-of-time method

1. Plot the dial reading and the corresponding *square-root-of-time* $\sqrt{t}$ as shown in Fig. 5.38.
2. Draw the tangent PQ to the early portion of the plot.
3. Draw a line PR such that $OR = (1.15)\,(OQ)$.
4. The abscissa of the point S (i.e., the intersection of PR and the consolidation curve) will give $\sqrt{t_{90}}$ (i.e., the square-root-of-time for 90% consolidation).
5. The value of T_v for $U_{av} = 90\%$ is 0.848. So,

$$C_v = \frac{0.848 H^2}{t_{50}} \tag{5.81}$$

Su's maximum slope method

1. Plot the dial reading against time on semilog graph paper as shown in Fig. 5.39.
2. Determine d_0 in the same manner as in the case of the logarithm-of-time method (steps 2 through 4).
3. Draw a tangent PQ to the steepest part of the consolidation curve.
4. Find h, which is the slope of the tangent PQ.
5. Find d_u as

$$d_u = d_0 + \frac{h}{0.688} U_{av} \tag{5.82}$$

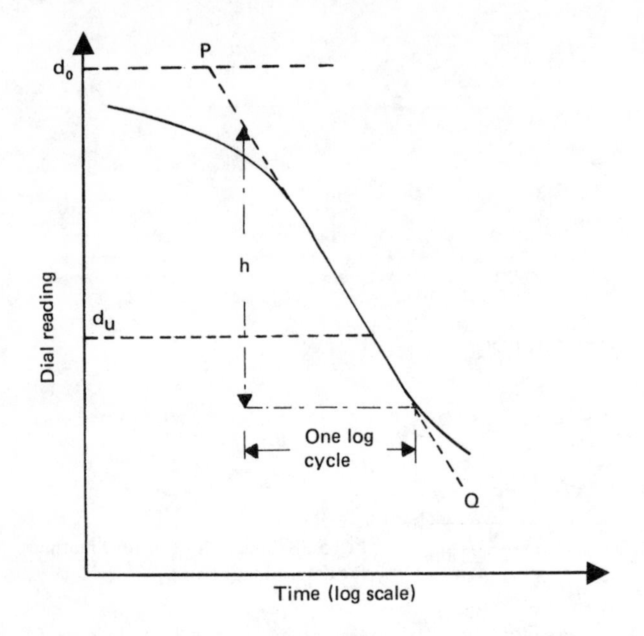

Fig. 5.39 Maximum slope method for determination of C_v.

where d_u is the dial reading corresponding to any given average degree of consolidation, U_{av}.

6. The time corresponding to the dial reading d_u can now be determined, and

$$C_v = \frac{T_v H^2}{t} \tag{5.83}$$

Su's method is more applicable for consolidation curves that do not exhibit the typical S-shape (see Sec. 5.1.8).

Sivaram and Swamee's computational method

1. Note two dial readings, d_1 and d_2, and their corresponding times, t_1 and t_2, from the early phase of consolidation. ("Early phase" means that the degree of consolidation should be less than 53%.)
2. Note a dial reading, d_3, at time t_3 after considerable settlement has taken place.
3. Determine d_0 as

$$d_0 = \frac{d_1 - d_2\sqrt{t_1/t_2}}{1 - \sqrt{t_1/t_2}} \tag{5.84}$$

4. Determine d_{100} as

$$d_{100} = d_0 - \frac{d_0 - d_3}{\{1 - [(d_0 - d_3)(\sqrt{t_2} - \sqrt{t_1})/(d_1 - d_2)\sqrt{t_3}]^{5.6}\}^{0.179}} \tag{5.85}$$

5. Determine C_v as

$$C_v = \frac{\pi}{4}\left(\frac{d_1 - d_2}{d_0 - d_{100}} \frac{H}{\sqrt{t_2} - \sqrt{t_1}}\right)^2 \tag{5.86}$$

where H is the length of the maximum drainage path.

Example 5.8 The results of an oedometer test on a normally consolidated clay are given below (two-way drainage):

σ', lb/ft²	e
1000	1.01
2000	0.90

The time for 50% consolidation for the load increment from 1000 to 2000 lb/ft² was 12 min, and the average thickness of the sample was 0.95 in. Determine the coefficient of permeability and the compression index.

SOLUTION

$$T_v = \frac{C_v t}{H^2}$$

For $U_{av} = 50\%$, $T_v = 0.197$. Hence,

$$0.197 = \frac{C_v(12)}{(0.95/2)^2} \qquad C_v = 0.0037 \text{ in}^2/\text{min}$$

$$C_v = \frac{k}{m_v \gamma_w} = \frac{k}{[\Delta e/\Delta\sigma(1 + e_{av})]\,\gamma_w}$$

For the given data, $\Delta e = 1.01 - 0.90 = 0.11$; $\Delta\sigma = 2000 - 1000 = 1000 \text{ lb/ft}^2 = 6.944 \text{ lb/in}^2$; $\gamma_w = 62.4/1728 \text{ lb/in}^3$; and $e_{av} = (1.01 + 0.9)/2 = 0.955$. So,

$$k = C_v \frac{\Delta e}{\Delta\sigma(1 + e_{av})}\gamma_w = (0.0037)\left[\frac{0.11}{6.944(1 + 0.955)}\right]\left(\frac{62.4}{1728}\right)$$

$$= 1.08 \times 10^{-6} \text{ in/min}$$

$$\text{Compression index} = C_c = \frac{\Delta e}{\log(\sigma_2'/\sigma_1')} = \frac{1.01 - 0.9}{\log(2000/1000)} = 0.365$$

5.1.7 Secondary Consolidation

We pointed out previously that clays continue to settle under sustained loading at the end of primary consolidation, and this is due to the continued readjustment of clay particles. Several investigations have been carried out for qualitative and quantitative

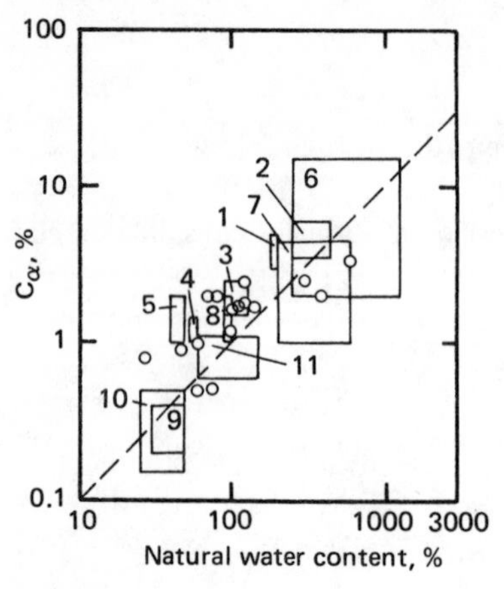

Fig. 5.40 Coefficient of secondary consolidation for natural soil deposits. *(After G. Mesri, Coefficient of Secondary Compression,* J. Soil Mech. Found. Div., ASCE, *vol. 99, no. SM1, 1973.)*

evaluation of secondary consolidation. The magnitude of secondary consolidation is often defined by (Fig. 5.29)

$$C_\alpha = \frac{\Delta H_t / H_t}{\log t_2 - \log t_1} \tag{5.87}$$

where C_α is the coefficient of secondary consolidation.

Mesri (1973) published an extensive list of the works of various investigators in this area. Figure 5.40 details the general range of the coefficient of secondary consolidation observed in a number of clayey soils. Secondary compression is high in plastic clays and organic soils. Table 5.3 provides a classification of soil based on secondary compressibility.

The proportion of secondary to primary consolidation depends on factors such as sample thickness and load increment ratio, $\Delta\sigma/\sigma'$ ($\Delta\sigma$ is the stress increment, and σ' is the effective stress on the sample before the application of the load increment).

Table 5.3 Classification of soil based on secondary compressibility

C_α	Secondary compressibility
<0.002	Very low
0.004	Low
0.008	Medium
0.016	High
0.032	Very high
0.064	Extremely high

After G. Mesri, Coefficient of Secondary Compression, *J. Soil Mech. Found. Div., ASCE,* vol. 99, no. SMI, 1973.

For similar load increment ratios, the proportion of secondary to primary compression increases with the decrease of sample thickness (Fig. 5.41). Also, the ratio of secondary to primary compression increases with decrease of $\Delta\sigma/\sigma'$.

In order to study the effect of remolding and preloading on secondary compression, Mesri (1973) conducted a series of one-dimensional consolidation tests on an organic Paulding clay. Figure 5.42 shows the results in the form of a plot of $\Delta e/(\Delta \log t)$ vs. consolidation pressure. For these tests, each specimen was loaded to a final pressure with load increment ratios of 1 and with only sufficient time allowed for excess pore water pressure dissipation. Under the final pressure, secondary compression was observed for a period of 6 months. The following conclusions can be drawn from the results of these tests:

1. For sedimented (undisturbed) soils, $\Delta e/(\Delta \log t)$ decreases with the increase of the final consolidation pressure.
2. Remolding of clays creates a more dispersed fabric. This results in a decrease of the coefficient of secondary consolidation at lower consolidation pressures as compared to that for undisturbed samples. However, it increases with consolidation pressure to a maximum value and then decreases, finally merging with the values for normally consolidated undisturbed samples.
3. Precompressed clays show a smaller value of coefficient of secondary consolidation. The degree of reduction appears to be a function of the degree of precompression.

Procedures for calculation of settlement due to secondary consolidation are given in Chap. 6.

5.1.8 Some Comments on Standard One-Dimensional Consolidation Test

The standard one-dimensional consolidation test procedure was described in Sec. 5.1.6, and it was pointed out that every 24 hours the load is generally doubled (i.e., $\Delta\sigma/\sigma' = 1$). Some questions may arise regarding the test procedure, such as:

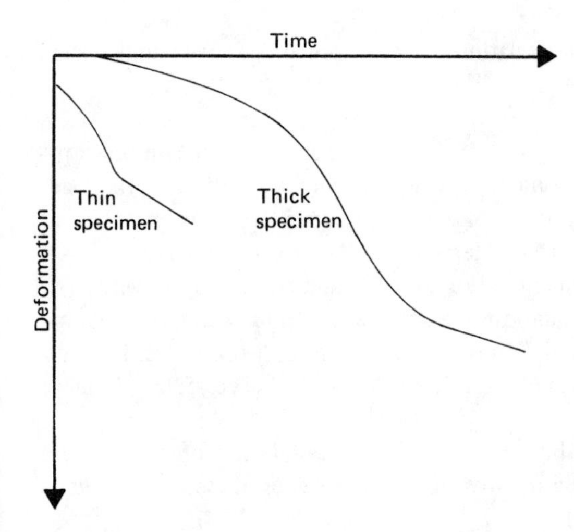

Fig. 5.41 The effect of similar load increment ratio, $\Delta\sigma/\sigma'$, on sample thickness.

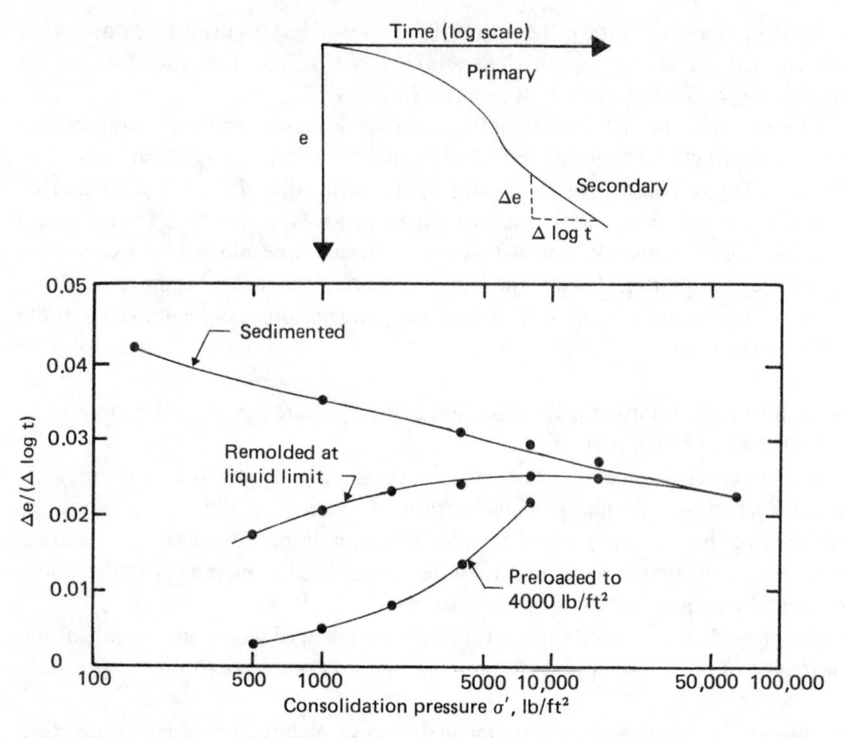

Fig. 5.42 Coefficient of secondary compression for organic Paulding clay. (Note: $1\ \text{lb/ft}^2 = 47.9\ \text{N/m}^3$.) *(Redrawn after G. Mesri, Coefficient of Secondary Compression,* J. Soil Mech. Found. Div., ASCE, *vol. 99, no. SM1, 1973.)*

1. What will happen if the load increment ratio $\Delta\sigma/\sigma'$ is not doubled?
2. What will happen if a given load on the soil specimen is kept for a duration of other than 24 hours?

In this section, we will discuss the deviations in e vs. $\log \sigma'$ observed under such conditions.

Effect of load-increment ratio σ/σ'. Striking changes in the shape of the compression-time curves for one-dimensional consolidation tests are generally noticed if the load-increment ratio is reduced to a substantially low value. Figure 5.43 shows the shape of dial reading vs. time curves for undisturbed Mexico City clay. Curve I is for $\Delta\sigma/\sigma' = 1$. Curves II and III are for load-increment ratios of 0.25 and 0.22, respectively. The position of the end of primary consolidation—i.e., zero excess pore water pressure due to incremental loading in curves II and III—is somewhat difficult to resolve. In determination of the coefficient of consolidation C_v, Su's method (Sec. 5.1.6) is more applicable for these curves.

The load-increment ratio has a high influence on consolidation of clay. Figure 5.44 shows the nature of the e vs. $\log \sigma'$ curve for various values of $\Delta\sigma/\sigma'$. If $\Delta\sigma/\sigma'$ is

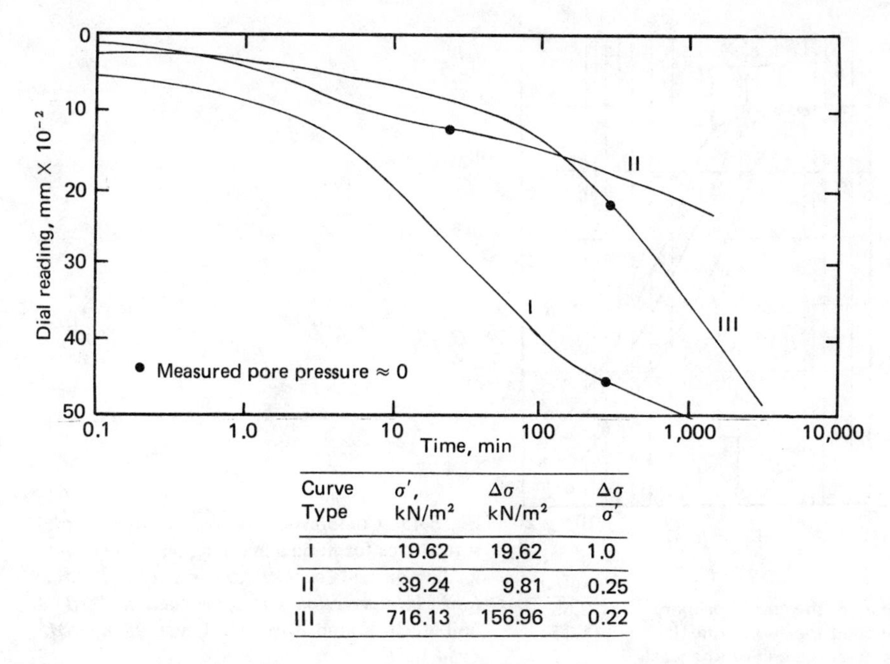

Curve Type	σ', kN/m²	$\Delta\sigma$ kN/m²	$\frac{\Delta\sigma}{\sigma'}$
I	19.62	19.62	1.0
II	39.24	9.81	0.25
III	716.13	156.96	0.22

Fig. 5.43 Effect of $\Delta\sigma/\sigma'$ on consolidation curves for Mexico City clay. *(Redrawn after G. A. Leonards and A. A. Altschaeffl, Compressibility of Clay,* J. Soil Mech. Found. Div., ASCE, *vol. 90, no. SM5, 1964.)*

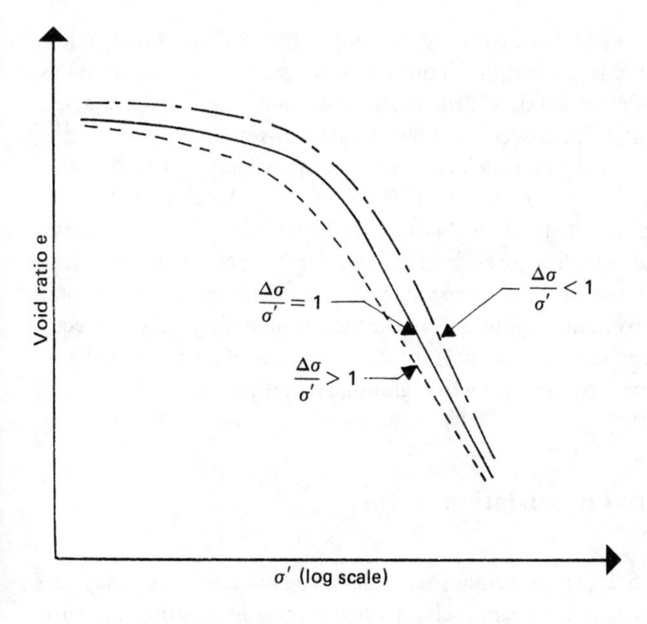

Fig. 5.44 Effect of load-increment ratio on e vs. $\log \sigma'$ curve.

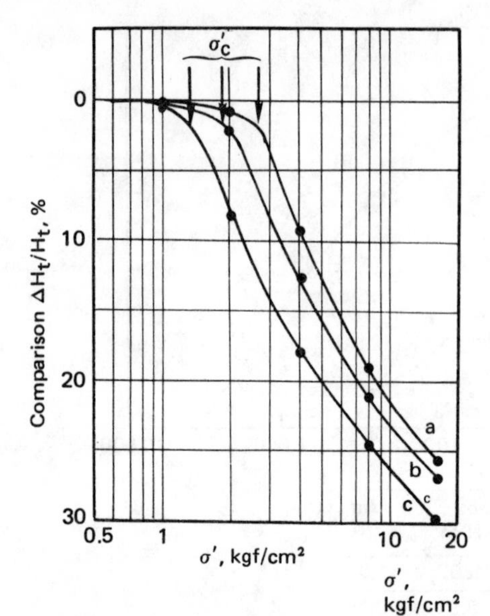

		σ', kgf/cm²
Curve a:	at the end of primary	2.65
Curve b:	at the end of one day	1.85
Curve c:	at the end of one week	1.33

Fig. 5.45 Compression vs. log of effective pressure curves for normal and long-term incremental loading on Leda clay. *(Redrawn after C. B. Crawford, Interpretation of Consolidation Tests,* J. Soil Mech. Found. Div., ASCE, *vol. 90, no. SM5, 1964.)*

small, the ability of individual clay particles to readjust to their positions of equilibrium is small, which results in a smaller compression as compared to that for larger $\Delta\sigma/\sigma'$.

Effect of load duration. In conventional testing, in which the soil specimen is left under a given load for about a day, a certain amount of secondary consolidation takes place before the next load increment is added. If the specimen is left under a given load for more than a day, additional secondary consolidation settlement will occur. This additional amount of secondary consolidation will have an effect on the e vs. log σ' plot, as shown in Fig. 5.45. Curve a is based on the results at the end of primary consolidation. Curve b is based on standard 24-hour load increment duration. Curve c refers to the condition for which a given load is kept for 1 week before the next load increment is applied. The strain for a given value of σ' is calculated from the total deformation that the specimen has undergone before the next load increment is applied.

Another point that must be made is that, for each of these curves, the preconsolidation pressure determined by Casagrande's graphical method is different and varies from 260 to 130 kN/m².

5.1.9 Effect of Secondary Consolidation on the Preconsolidation Pressure

It was pointed out in Table 5.2 that continued secondary consolidation in a clay soil will change its structure and will have some effect on the preconsolidation pressure σ'_c. This is also demonstrated in Fig. 5.45 for Leda clay. The principle of this mechanism can be explained by the highly idealized diagram given in Fig. 5.46.

A clay that has recently been deposited and comes to equilibrium by its own weight can be called a "young, normally consolidated clay." If such a clay, with an effective overburden pressure of σ_o' at an equilibrium void ratio of e_o, is now removed from the ground and tested in a consolidometer, it will show an e vs. log σ' curve like that marked as curve a in Fig. 5.46. Note that the preconsolidation pressure for curve a is σ_o'.

On the other hand, if the same clay is allowed to remain undisturbed for 10,000 years, for example, under the same effective overburden pressure σ_o', there will be creep or secondary consolidation. This will reduce the void ratio to e_1. The clay may now be called an "aged, normally consolidated clay." If this clay at a void ratio of e_1 and effective overburden pressure of σ_o' is removed and tested in a consolidometer, the e vs. log σ' curve will be like curve b in Fig. 5.46. The preconsolidation pressure, when determined by standard procedure, will be σ_1'. Now, $\sigma_c' = \sigma_1' > \sigma_o'$. This is sometimes referred to as a quasi-preconsolidation effect.

It was also mentioned in Sec. 5.1.7 that the effect of secondary consolidation is

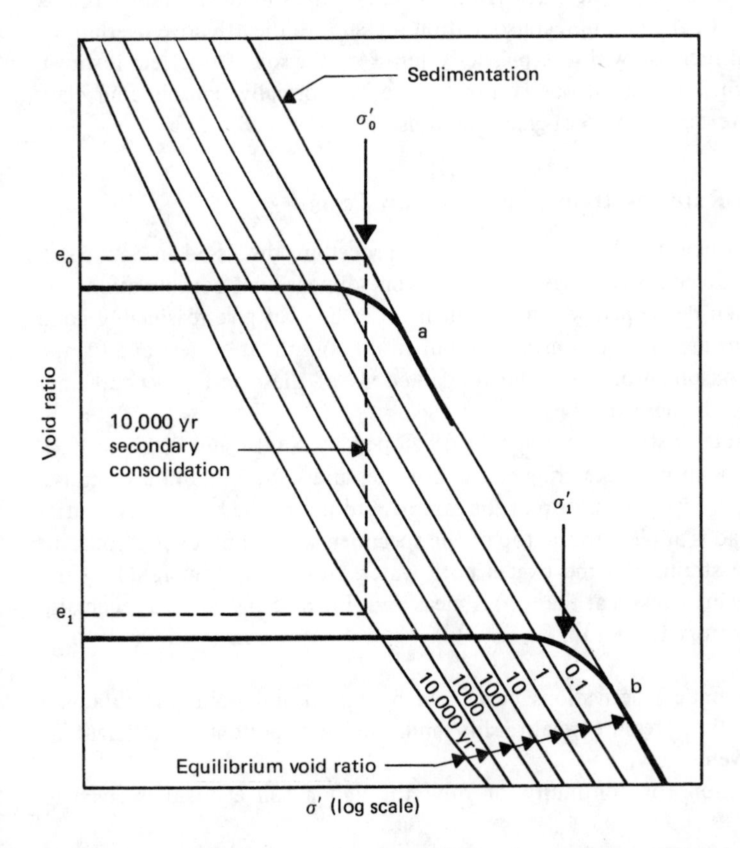

Fig. 5.46 Geological history and compressibility of normally consolidated clays. Curve a: Young, normally consolidated clay, $\sigma_0' = \sigma_c'$. Curve b: Aged, normally consolidated clay, $\sigma_1' = \sigma_c'$. *(Redrawn after L. Bjerrum, Embankments on Soft Ground,* Proc. Specialty Conference on Performance of Earth and Earth Supported Structures, ASCE, *vol. 2, 1972.)*

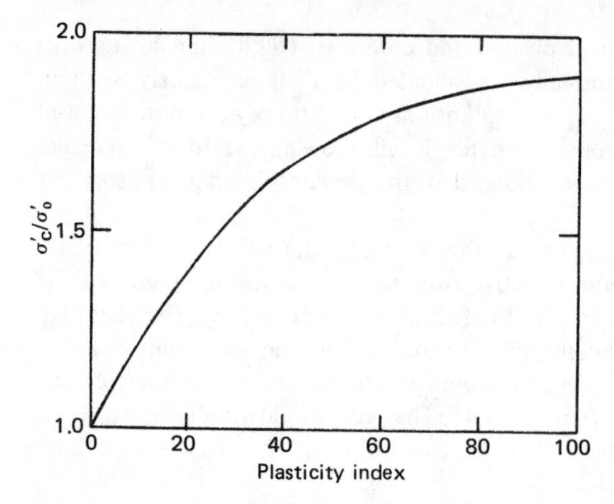

Fig. 5.47 Typical values of σ'_c/σ'_o observed in normally consolidated late-glacial and post-glacial clays. *(After L. Bjerrum, Embankments on Soft Ground,* Proc. Specialty Conference on Performance of Earth and Earth Supported Structures, ASCE, *vol. 2, 1972.)*

more pronounced in more plastic clays. Thus, it may be reasoned that under similar conditions the ratio of the quasi-preconsolidation pressure to the effective overburden pressure, σ'_c/σ'_o, will increase with the plasticity index of the soil. This effect is shown in Fig. 5.47, which is based on observations made on normally consolidated clays deposited during late-glacial and post-glacial periods.

5.1.10 Constant Rate-of-Strain Consolidation Tests

The standard one-dimensional consolidation test procedure discussed in Sec. 5.1.6 is time-consuming. Recently, at least two other one-dimensional consolidation test procedures have been developed which are much faster but yet give reasonably good results. The methods are (1) the constant rate-of-strain consolidation test and (2) the constant-gradient consolidation test. The fundamentals of these test procedures are described in this and the next sections.

The constant rate-of-strain method was developed by Smith and Wahls (1969). A soil specimen is taken in a fixed-ring consolidometer and saturated. For conducting the test, drainage is permitted at the top of the sample, but not at the bottom. A continuously increasing load is applied to the top of the specimen so as to produce a constant rate of compressive strain, and the excess pore water pressure u_b (generated by the continuously increasing stress σ at the top) is measured. Figure 5.48 shows a schematic diagram of the laboratory test setup.

Theory. The mathematical derivations developed by Smith and Wahls for obtaining the void ratio–effective pressure relationship and the corresponding coefficient of consolidation are given below.

The basic equation for continuity of flow through a soil element is given in Eq. (5.9) as

$$\frac{k}{\gamma_w}\frac{\partial^2 u}{\partial z^2} = \frac{1}{1+e}\frac{\partial e}{\partial t}$$

The coefficient of permeability at a given time is a function of the average void ratio $\bar{e}$ in the specimen. The average void ratio is, however, continuously changing due to the constant rate of strain. Thus,

$$k = k(\bar{e}) = f(t) \tag{5.88}$$

The average void ratio is given by

$$\bar{e} = \frac{1}{H}\int_0^H e\,dz$$

where $H\,(=H_t)$ is the sample thickness.

In the constant rate-of-strain type of test, the rate of change of volume is constant, or

$$\frac{dV}{dt} = -RA \tag{5.89}$$

where V = volume of specimen
A = area of cross section of specimen
R = constant rate of deformation of upper surface

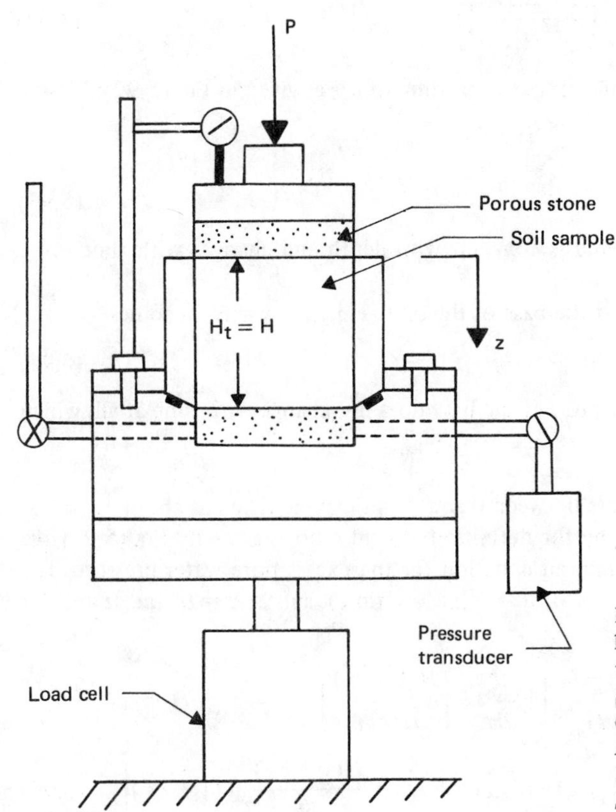

Fig. 5.48 Schematic diagram for laboratory test setup for controlled rate-of-strain type of test. *(After R. E. Smith and H. E. Wahls, Consolidation under Constant Rate of Strain,* J. Soil Mech. Found. Div., ASCE, *vol. 95, no. SM2, 1969.)*

The rate of change of average void ratio $\bar{e}$ can be given by

$$\frac{d\bar{e}}{dt} = \frac{1}{V_s}\frac{dV}{dt} = -\frac{1}{V_s}RA = -r \tag{5.90}$$

where r is a constant.

Based on the definition of $\bar{e}$ and Eq. (5.88), we can write

$$e_{(z,t)} = g(z)t + e_o \tag{5.91}$$

where $e_{(z,t)}$ = void ratio at depth z and time t
e_o = initial void ratio at beginning of test
$g(z)$ = a function of depth only

The function $g(z)$ is difficult to determine. We will assume it to be a linear function of the form

$$-r\left[1 - \frac{b}{r}\left(\frac{z - 0.5H}{H}\right)\right]$$

where b is a constant. Substitution of this into Eq. (5.91) gives

$$e_{(z,t)} = e_o - rt\left[1 - \frac{b}{r}\left(\frac{z - 0.5H}{H}\right)\right] \tag{5.92}$$

Let us consider the possible range of variation of b/r as given in Eq. (5.92):

1. If $b/r = 0$,

$$e_{(z,t)} = e_o - rt \tag{5.93}$$

This indicates that the void is constant with depth and changes with time only. In reality, this is not the case.

2. If $b/r = 2$, the void ratio at the base of the sample, i.e., at $z = H$, becomes

$$e_{(H,t)} = e_o \tag{5.94}$$

This means that the void ratio at the base does not change with time at all, which is not realistic.

So the value of b/r is somewhere between 0 and 2 and may be taken as about 1.

Assuming $b/r \neq 0$ and using the definition of void ratio as given by Eq. (5.92), we can integrate Eq. (5.9) to obtain an equation for the excess pore water pressure. The boundary conditions are: at $z = 0$, $u = 0$ (at any time); and at $z = H$, $\partial u/\partial z = 0$ (at any time). Thus,

$$u = \frac{\gamma_w r}{k}\left\{zH\left[\frac{1 + e_o - bt}{rt(bt)}\right] + \frac{z^2}{2rt} - \left[\frac{H(1 + e_o)}{rt(bt)}\right]\right.$$
$$\left.\times\left[\frac{H(1+e)}{bt}\ln(1+e) - z\ln(1+e_B) - \frac{H(1+e_T)}{bt}\ln(1+e_T)\right]\right\} \tag{5.95}$$

where $$e_B = e_o - rt\left(1 - \frac{1}{2}\frac{b}{r}\right) \tag{5.96}$$

$$e_T = e_o - rt\left(1 + \frac{1}{2}\frac{b}{r}\right) \tag{5.97}$$

Equation (5.95) is very complicated. Without loosing a great deal of accuracy, it is possible to obtain a simpler form of expression for u by assuming that the term $1 + e$ in Eq. (5.9) is approximately equal to $1 + \bar{e}$ (note that this is not a function of z). So, from Eqs. (5.9) and (5.92),

$$\frac{\partial^2 u}{\partial z^2} = \left[\frac{\gamma_w}{k(1+\bar{e})}\right]\frac{\partial}{\partial t}\left\{e_o - rt\left[1 - \frac{b}{r}\left(\frac{z - 0.5H}{H}\right)\right]\right\} \tag{5.98}$$

Using the boundary condition $u = 0$ at $z = 0$ and $\partial u/\partial t = 0$ at $z = H$, Eq. (5.98) can be integrated to yield

$$u = \left[\frac{\gamma_w r}{k(1+\bar{e})}\right]\left[\left(Hz - \frac{z^2}{2}\right) - \frac{b}{r}\left(\frac{z^2}{4} - \frac{z^3}{6H}\right)\right] \tag{5.99}$$

The pore pressure at the base of the specimen can be obtained by substituting $z = H$ in Eq. (5.99)

$$u_{z=H} = \frac{\gamma_w r H^2}{k(1+\bar{e})}\left(\frac{1}{2} - \frac{1}{12}\frac{b}{r}\right) \tag{5.100}$$

The average effective stress corresponding to a given value of $u_{z=H}$ can be obtained by writing

$$\sigma'_{av} = \sigma - \frac{u_{av}}{u_{z=H}}\, u_{z=H} \tag{5.101}$$

where σ'_{av} = average effective stress on specimen at any time
σ = total stress on sample
u_{av} = corresponding average pore water pressure

$$\frac{u_{av}}{u_{z=H}} = \frac{\dfrac{1}{H}\displaystyle\int_0^H u\, dz}{u_{z=H}} \tag{5.102}$$

Substitution of Eqs. (5.99) and (5.100) into Eq. (5.102) and further simplification gives

$$\frac{u_{av}}{u_{z=H}} = \frac{\frac{1}{3} - \frac{1}{24}(b/r)}{\frac{1}{2} - \frac{1}{12}(b/r)} \tag{5.103}$$

Note that for $b/r = 0$, $u_{av}/u_{z=H} = 0.667$; and for $b/r = 1$, $u_{av}/u_{z=H} = 0.700$. Hence, for $0 \leqslant b/r \leqslant 1$, the values of $u_{av}/u_{z=H}$ does not change significantly. So, from Eqs. (5.101) and (5.103),

$$\sigma'_{av} = \sigma - \left[\frac{\frac{1}{3} - \frac{1}{24}(b/r)}{\frac{1}{2} - \frac{1}{12}(b/r)}\right] u_{z=H} \tag{5.104}$$

Coefficient of consolidation. The coefficient of consolidation was defined previously as

$$C_v = \frac{k(1+e)}{a_v \gamma_w}$$

We can assume $1 + e \approx 1 + \bar{e}$, and from Eq. (5.100)

$$k = \frac{\gamma_w r H^2}{(1+\bar{e})u_{z=H}}\left(\frac{1}{2} - \frac{1}{12}\frac{b}{r}\right) \tag{5.105}$$

Substitution of these into the expression for C_v gives

$$C_v = \frac{rH^2}{a_v u_{z=H}}\left(\frac{1}{2} - \frac{1}{12}\frac{b}{r}\right) \tag{5.106}$$

Interpretation of experimental results. The following information can be obtained from a constant rate-of-strain consolidation test:

1. Initial height of sample, H_i.
2. A.
3. V_s.
4. Strain rate R.
5. A continuous record of $u_{z=H}$.
6. A corresponding record of σ (total stress applied at the top of the specimen).

The plot of e vs. σ'_{av} can be obtained in the following manner:

1. Calculate $r = RA/V_s$.
2. Assume $b/r \approx 1$.
3. For a given value of $u_{z=H}$, the value of σ is known (at time t from the start of the test), and so σ'_{av} can be calculated from Eq. (5.104).
4. Calculate $\Delta H = Rt$ and then the change in void ratio that has taken place during time t,

   $$\Delta e = \frac{\Delta H}{H_i}(1 + e_o)$$

   where H_i is the initial height of the sample.
5. The corresponding void ratio (at time t) is $e = e_o - \Delta e$.
6. After obtaining a number of points of σ'_{av} and the corresponding e, plot the graph of e vs. $\log \sigma'_{av}$.
7. For a given value of σ'_{av} and e, the coefficient of consolidation C_v can be calculated by using Eq. (5.106). (Note that H in Eq. (5.106) is equal to $H_i - \Delta H$.)

Figures 5.49 and 5.50 show the experimental results of constant rate-of-strain consolidation tests on two clays. Results obtained from standard consolidation tests are also shown for comparison. In the case of calcium montmorillonite, it can be seen that at higher strain rates the experimental values of Δe for a given σ'_{av} deviate considerably from the standard test results. This is probably due to the simplified assumptions introduced in Eq. (5.92). For that reason, it is recommended that the strain rate for a given test should be chosen such that the value of $u_{z=H}/\sigma$ at the end of the test does not exceed 0.5. However, the value should be high enough such that it can be measured with reasonable accuracy. (See pp. 536–537 of Smith and Wahl's paper for a tentative strain-rate selection procedure.)

5.1.11 Constant-Gradient Consolidation Test

The constant-gradient consolidation test was developed by Lowe et al. (1969). In this procedure, a saturated soil sample is taken in a consolidation ring. As in the case of the constant rate-of-strain type of test, drainage is allowed at the top of the sample and pore water pressure is measured at the bottom (Fig. 5.51). A load P is applied on the sample, which increases the excess pore water pressure in the specimen by an amount Δu (Fig. 5.52*a*). After a small lapse of time t_1, the excess pore water pressure at the top of the sample will be equal to zero (since drainage is permitted). However, at the bottom of the sample the excess pore water pressure will still be approximately Δu (Fig. 5.52*b*). From this point on, the load P is increased slowly in such a way that the difference between the pore water pressures at the top and bottom of the specimen remains constant, i.e., the difference is maintained at a constant Δu (Fig. 5.52*c* and *d*). When the desired value of P is reached, say at time t_3, the loading is stopped and the excess pore water pressure is allowed to dissipate. The elapsed time t_4 at which the pore water pressure at the bottom of the specimen reaches a value of 0.1 Δu is recorded. During the entire test, the compression ΔH that the specimen undergoes is recorded. For complete details of the laboratory test arrangement, the reader is referred to the original paper of Lowe et al. (1969).

Theory. From the basic equations (5.9) and (5.10), we have

$$\frac{k}{\gamma_w}\frac{\partial^2 u}{\partial z^2} = -\frac{a_v}{1+e}\frac{\partial \sigma'}{\partial t} \tag{5.107}$$

or

$$\frac{\partial \sigma'}{\partial t} = -\frac{k}{\gamma_w m_v}\frac{\partial^2 u}{\partial z^2} = -C_v\frac{\partial^2 u}{\partial z^2} \tag{5.108}$$

Since $\sigma' = \sigma - u$,

$$\frac{\partial \sigma'}{\partial t} = \frac{\partial \sigma}{\partial t} - \frac{\partial u}{\partial t} \tag{5.109}$$

For the controlled-gradient tests (i.e., during the time t_1 to t_3 in Fig. 5.52), $\partial u/\partial t = 0$. So,

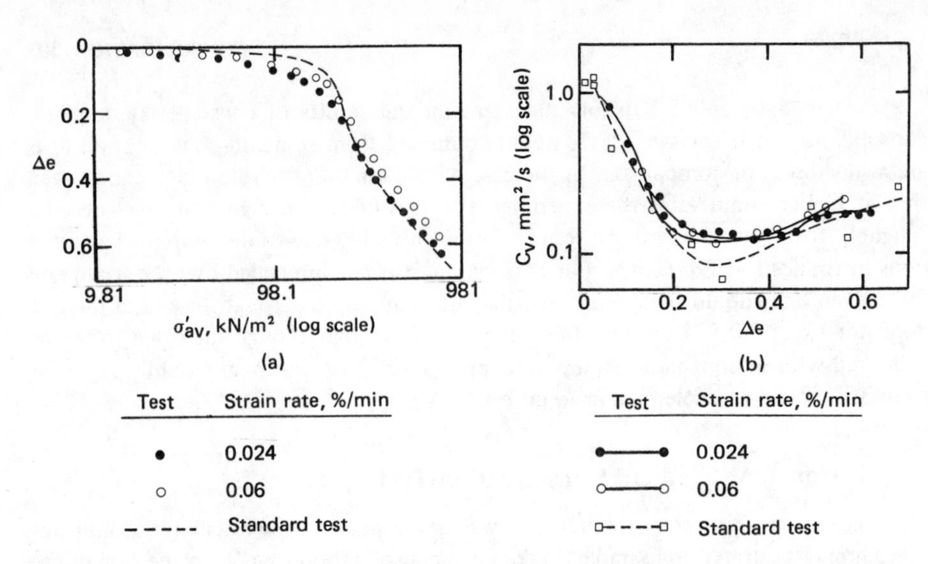

Fig. 5.49 Results of CRS tests on Massena clay and comparison with standard tests. *(Replotted from R. E. Smith and H. E. Wahls, Consolidation under Constant Rate of Strain,* J. Soil Mech. Found. Div., ASCE, *vol. 95, no. SM2, 1969.)*

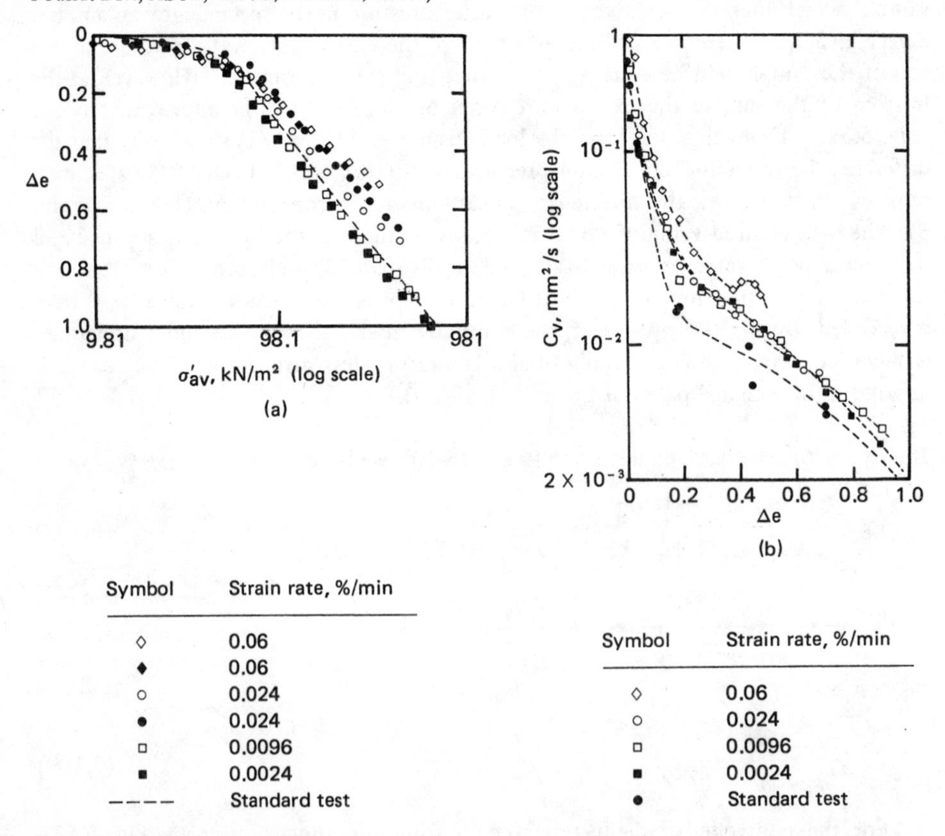

Fig. 5.50 Results of CRS tests on calcium montmorillonite and comparison with standard tests. *(After R. E. Smith and H. E. Wahls, Consolidation under Constant Rate of Strain,* J. Soil Mech. Found. Div., ASCE, *vol. 95, no. SM2, 1969.)*

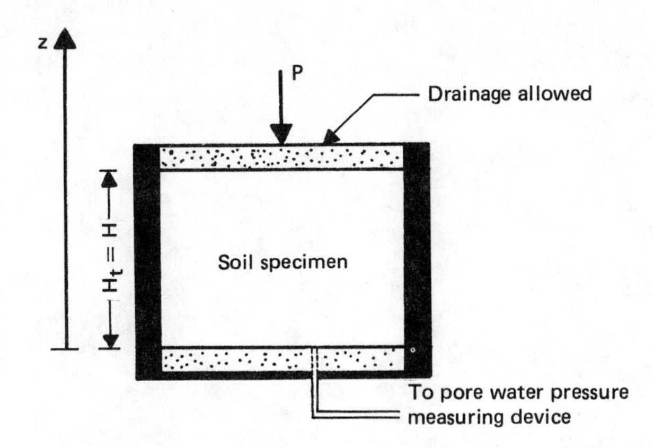

Fig. 5.51 Schematic diagram for constant-gradient consolidation test.

$$\frac{\partial \sigma'}{\partial t} = \frac{\partial \sigma}{\partial t} \tag{5.110}$$

Combining Eqs. (5.108) and (5.110),

$$\frac{\partial \sigma}{\partial t} = -C_v \frac{\partial^2 u}{\partial z^2} \tag{5.111}$$

Note that the left-hand side of Eq. (5.111) is independent of the variable z and the right-hand side is independent of the variable t. So both sides should be equal to a constant, say A_1. Thus,

$$\frac{\partial \sigma}{\partial t} = A_1 \tag{5.112}$$

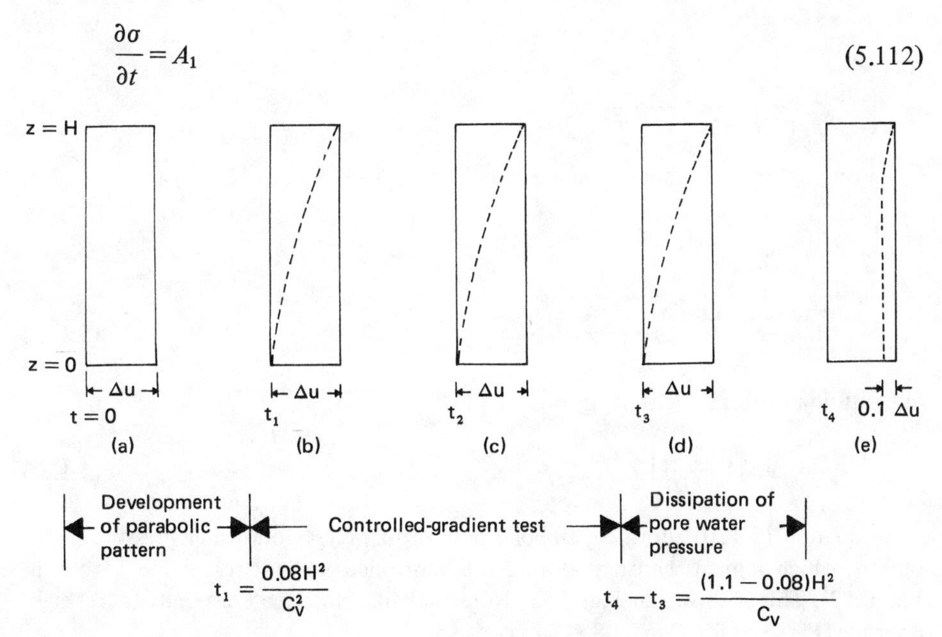

Fig. 5.52 Stages in controlled-gradient test. *(Redrawn after J. Lowe, E. Jonas, V. Obrician, Controlled Gradient Consolidation Test,* J. Soil Mech. Found. Div., ASCE, *vol. 95, no. SMI, 1965.)*

and $$\frac{\partial^2 u}{\partial z^2} = -\frac{A_1}{C_v} \tag{5.113}$$

Integration of Eq. (5.113) yields

$$\frac{\partial u}{\partial z} = -\frac{A_1}{C_v} z + A_2 \tag{5.114}$$

and $$u = -\frac{A_1}{C_v}\frac{z^2}{2} + A_2 z + A_3 \tag{5.115}$$

The boundary conditions are as follows:

1. At $z = 0$, $\partial u/\partial z = 0$.
2. At $z = H, u = 0$.
3. At $z = 0, u = \Delta u$.

From the first boundary condition and Eq. (5.114), we find that $A_2 = 0$. So,

$$u = -\frac{A_1}{C_v}\frac{z^2}{2} + A_3 \tag{5.116}$$

From the secondary boundary condition and Eq. (5.116),

$$A_3 = \frac{A_1 H^2}{2C_v} \tag{5.117}$$

or $$u = -\frac{A_1}{C_v}\frac{z^2}{2} + \frac{A_1}{C_v}\frac{H^2}{2} \tag{5.118}$$

From the third boundary condition and Eq. (5.118),

$$\Delta u = \frac{A_1}{C_v}\frac{H^2}{2}$$

or $$A_1 = \frac{2C_v\,\Delta u}{H^2} \tag{5.119}$$

Substitution of this value of A_1 into Eq. (5.118) yields

$$u = \Delta u\left(1 - \frac{z^2}{H^2}\right) \tag{5.120}$$

Equation (5.120) shows a parabolic pattern of excess pore water pressure distribution, which remains constant during the controlled-gradient test (time t_1 to t_3 in Fig. 5.52). This is shown in Fig. 5.53. Note that this closely corresponds to Terzaghi isocrone (Fig. 5.5) for $T_v = 0.08$.

Combining Eqs. (5.112) and (5.119), we obtain

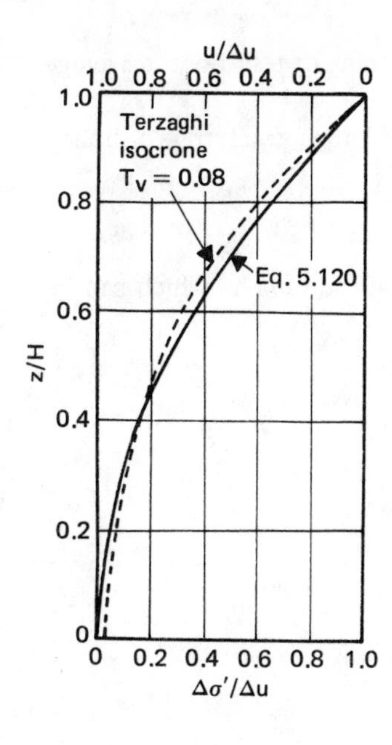

Fig. 5.53 Comparison of the parabolic pattern of excess pore water pressure distribution with Terzaghi isocrone. *(After J. Lowe, E. Jonas, and V. Obrician, Controlled Gradient Consolidation Test,* J. Soil Mech. Found. Div., ASCE, *vol. 95, no. SM1, 1965.)*

$$\frac{\partial \sigma}{\partial t} = A_1 = \frac{2C_v \, \Delta u}{H^2}$$

or
$$C_v = \frac{\partial \sigma}{\partial t} \frac{H^2}{2\Delta u} \tag{5.121}$$

Interpretation of experimental results. The following information will be available from the constant-gradient test:

1. Initial height of the sample, H_i, and height, H, at any time during the test.
2. The rate of application of the load P and thus the rate of application of stress $\partial\sigma/\partial t$ on the sample.
3. The differential pore pressure Δu.
4. Time t_1.
5. Time t_3.
6. Time t_4.

The plot of e vs. σ'_{av} can be obtained in the following manner:

1. Calculate the initial void ratio e_o.
2. Calculate the change in void ratio at any other time t during the test as

$$\Delta e = \frac{\Delta H}{H_i}(1 + e_o)$$

where ΔH is the total change in height from the beginning of test. So, the average void ratio at time t is $e = e_o - \Delta e$.

3. Calculate the average effective stress at time t using the known total stress σ applied on the sample at that time:

$$\sigma'_{av} = \sigma - u_{av}$$

where u_{av} is the average excess pore water pressure in the sample, which can be calculated from Fig. 5.53.

Calculation of the coefficient of consolidation is as follows:

1. At time t_1,

$$C_v = \frac{0.08H^2}{t_1}$$

2. At time $t_1 < t < t_3$,

$$C_v = \frac{\Delta\sigma}{\Delta t}\frac{H^2}{2\Delta u} \tag{5.121}$$

Note that $\Delta\sigma/\Delta t, H,$ and Δu are all known from the tests.

3. Between time t_3 and t_4,

$$C_v = \frac{(1.1 - 0.08)H^2}{t_3 - t_4} = \frac{1.02H^2}{t_3 - t_4}$$

Figure 5.54*a* shows the results of several controlled-gradient tests on an overconsolidated clay (marine clay from the vicinity of Portland, Maine). It appears that, in spite of the wide variation of Δu [1.21 to 3.05 lb/in^2 (8.35 to 21.05 kN/m^2)], the variation of $\Delta H/H_i$ with σ'_{av} falls within a narrow band. This compares well with the results of conventional consolidation tests.

Figure 5.54*b* compares the results obtained for the coefficient of consolidation with those obtained through the conventional square-root-of-time fitting and logarithm-of-time methods. In general, the C_v values obtained from the constant-gradient test are somewhere between those obtained from the two conventional methods.

The increase of effective stress σ' on a given sample obtained from the standard one-dimensional consolidation test is compared in Fig. 5.55 to that from the controlled-gradient test. The difference in the increase may also have some effect on the value of the coefficient of consolidation.

5.1.12 One-Dimensional Consolidation with Viscoelastic Models

The theory of consolidation we have studied thus far is based on the assumption that the effective stress and the volumetric strain can be described by linear elasticity. Since Terzaghi's founding work on the theory of consolidation, several investigators

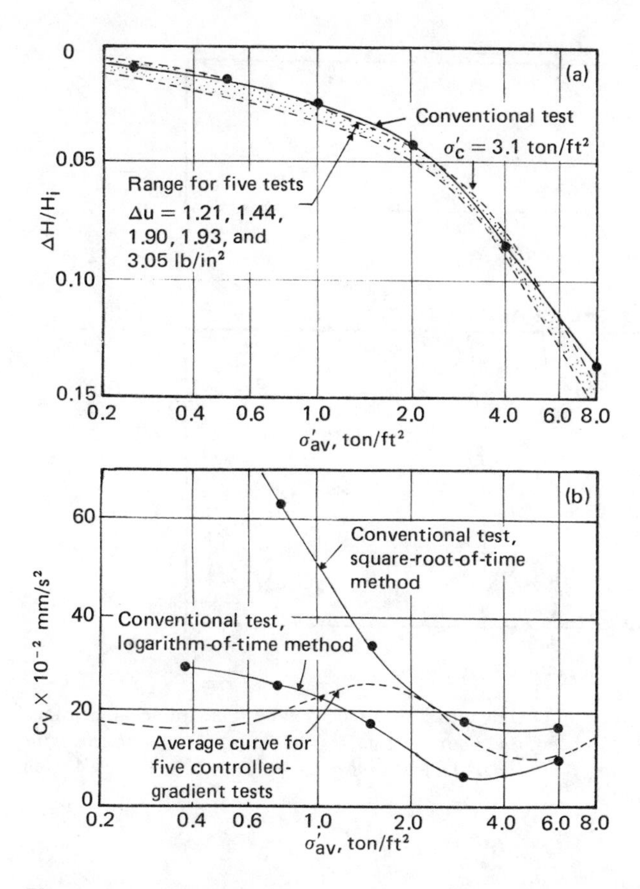

Fig. 5.54 Comparison of controlled-gradient tests with conventional consolidation tests on an overconsolidated marine clay from Portland, Maine, area. (Note: 1 ton/ft^2 = 95.8 kN/m^2). *(After J. Lowe, E. Jonas, and V. Obrician, Controlled Gradient Consolidation Test,* J. Soil Mech. Found. Div., ASCE, *vol. 95, no. SM1, 1965.)*

(Taylor and Merchant, 1940; Taylor, 1942; Tan, 1957; Gibson and Lo, 1961; Barden, 1965, 1968; Schiffman et al., 1964) have used viscoelastic models to study one-dimensional consolidation. This gives an insight into the secondary consolidation phenomenon which Terzaghi's theory does not explain. In this section, the work of Barden is briefly outlined.

The rheological model for soil chosen by Barden consists of a linear spring and nonlinear dashpot as shown in Fig. 5.56. The equation of continuity for one-dimensional consolidation is given in Eq. (5.9) as

$$\frac{k(1+e)}{\gamma_w}\frac{\partial^2 u}{\partial z^2} = \frac{\partial e}{\partial t}$$

Figure 5.57 shows the typical nature of the variation of void ratio with effective stress. From this figure we can write that

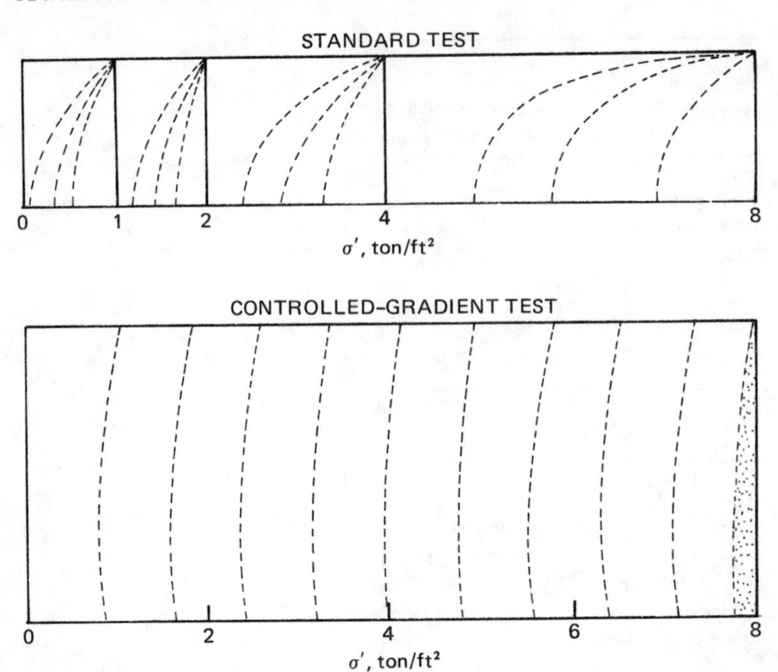

Fig. 5.55 Difference in the increase of effective stress on a specimen by standard one-dimensional test and controlled-gradient test procedures. (Note: $1 \text{ ton/ft}^2 = 95.8 \text{ kN/m}^2$). *(Replotted after J. Lowe, E. Jonas, and V. Obrician, Controlled Gradient Consolidation Test,* J. Soil Mech. Found. Div., ASCE, *vol. 95, no. SM1, 1965.)*

$$\frac{e_1 - e_2}{a_v} = \frac{e_1 - e}{a_v} + u + \tau \tag{5.122}$$

where $\dfrac{e_1 - e_2}{a_v} = \Delta\sigma' =$ total effective stress increase the soil will be subjected to at end of consolidation

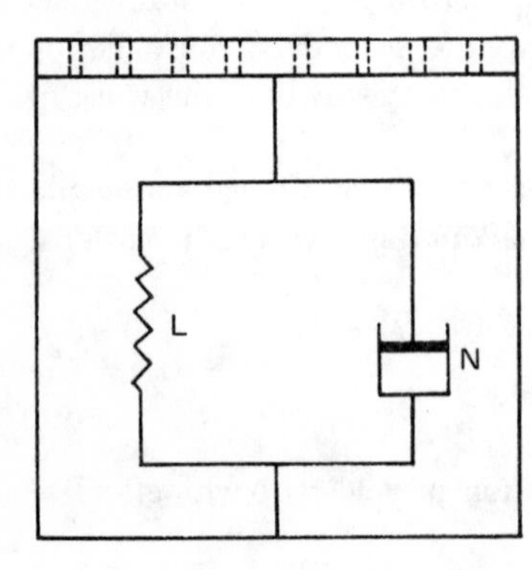

Fig. 5.56 Rheological model for soil. *L:* Linear spring; *N:* Nonlinear dashpot.

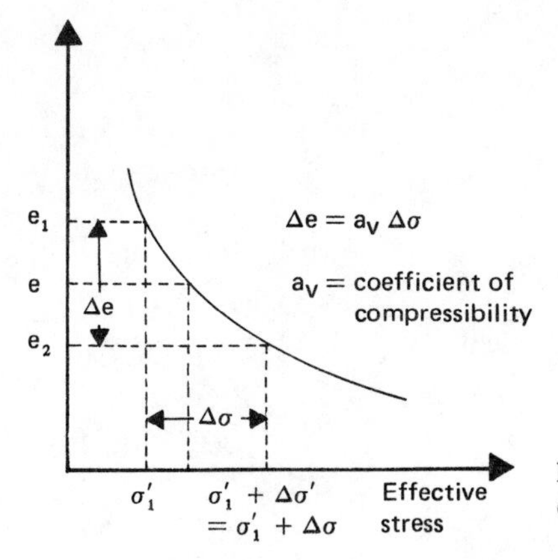

Fig. 5.57 Nature of variation of void ratio with effective stress.

$\dfrac{e_1 - e}{a_v}$ = effective stress increase in the soil at some stage of consolidation (i.e., the stress carried by the soil grain bond, represented by the spring in Fig. 5.56)

u = excess pore water pressure

τ = strain carried by film bond (represented by the dashpot in Fig. 5.56)

The strain τ can be given by a power-law relation:

$$\tau = b\left(\frac{\partial e}{\partial t}\right)^{1/n}$$

where $n > 1$, and b is assumed to be a constant over the pressure range $\Delta\sigma$. Substitution of the preceding power-law relation for τ in Eq. (5.122) and simplification gives

$$e - e_2 = a_v\left[u + b\left(\frac{\partial e}{\partial t}\right)^{1/n}\right] \tag{5.123}$$

Now let $e - e_2 = e'$. So,

$$\frac{\partial e'}{\partial t} = \frac{\partial e}{\partial t} \tag{5.124}$$

$$\bar{z} = \frac{z}{H} \tag{5.125}$$

where H is the length of maximum drainage path, and

$$\bar{u} = \frac{u}{\Delta\sigma'} \tag{5.126}$$

The degree of consolidation is

$$U_z = \frac{e_1 - e}{e_1 - e_2} \tag{5.127}$$

and
$$\lambda = 1 - U_z = \frac{e - e_2}{e_1 - e_2} = \frac{e'}{a_v \, \Delta\sigma'} \tag{5.128}$$

Elimination of u from Eqs. (5.9) and (5.123) yields

$$\frac{k(1+e)}{\gamma_w} \frac{\partial^2}{\partial z^2}\left[\frac{e'}{a_v} - b\left(\frac{\partial e'}{\partial t}\right)^{1/n}\right] = \frac{\partial e'}{\partial t} \tag{5.129}$$

Combining Eqs. (5.125), (5.128), and (5.129), we obtain

$$\frac{\partial^2}{\partial \bar{z}^2}\left\{\lambda - \left[a_v b^n (\Delta\sigma')^{1-n} \frac{\partial \lambda}{\partial t}\right]^{1/n}\right\} = \frac{a_v H^2 \gamma_w}{k(1+e)} \frac{\partial \lambda}{\partial t}$$

$$= \frac{m_v H^2 \gamma_w}{k} \frac{\partial \lambda}{\partial t} = \frac{H^2}{C_v} \frac{\partial \lambda}{\partial t} \tag{5.130}$$

where m_v is the volume coefficient of compressibility and C_v is the coefficient of consolidation.

The right-hand side of Eq. (5.130) can be written in the form

$$\frac{\partial \lambda}{\partial T_v} = \frac{H^2}{C_v} \frac{\partial \lambda}{\partial t} \tag{5.131}$$

where T_v is the nondimensional time factor and is equal to $C_v t/H^2$.

Similarly defining

$$T_s = \frac{t(\Delta\sigma')^{n-1}}{a_v b^n} \tag{5.132}$$

we can write

$$\left[a_v b^n (\Delta\sigma')^{1-n} \frac{\partial \lambda}{\partial t}\right]^{1/n} = \left(\frac{\partial \lambda}{\partial T_s}\right)^{1/n} \tag{5.133}$$

T_s in Eqs. (5.132) and (5.133) is defined as structural viscosity.

It is useful now to define a nondimensional ratio R as

$$R = \frac{T_v}{T_s} = \frac{C_v a_v}{H^2} \frac{b^n}{(\Delta\sigma')^{n-1}} \tag{5.134}$$

From Eqs. (5.130), (5.131), and (5.133),

$$\frac{\partial^2}{\partial \bar{z}^2}\left[\lambda - \left(\frac{\partial \lambda}{\partial T_s}\right)^{1/n}\right] = \frac{\partial \lambda}{\partial T_v} \tag{5.135}$$

Note that Eq. (5.136) is nonlinear. For that reason, Barden suggested solving the two simultaneous equations obtained from the basic equation (5.9).

$$\frac{\partial^2 \bar{u}}{\partial \bar{z}^2} = \frac{\partial \lambda}{\partial T_v} \tag{5.136}$$

and
$$-\frac{1}{R}(\lambda - \bar{u})^n = \frac{\partial \lambda}{\partial T_v} \tag{5.137}$$

Finite-difference approximation is employed for solving the above two equations. Figure 5.58 shows the variation of λ and $\bar{u}$ with depth for a clay layer of height $H_t = 2H$ and drained both at the top and bottom (for $n = 5, R = 10^{-4}$). Note that for a given value of T_v (i.e., time t) the nondimensional excess pore water pressure decreases more than λ (i.e., void ratio).

For a given value of T_v, R, and n, the average degree of consolidation can be determined as (Fig. 5.58)

$$U_{\text{av}} = 1 - \int_0^1 \lambda \, d\bar{z} \tag{5.138}$$

Figure 5.59 shows the variation of U_{av} with T_v (for $n = 5$). Similar results can be obtained for other values of n. Note that in this figure the beginning of secondary consolidation is assumed to start after the *midplane excess pore water pressure* falls below an arbitrary value of $u = 0.01\,\Delta\sigma$. Several other observations can be made concerning this plot:

1. Primary and secondary consolidation are continuous processes and depend on the structural viscosity (i.e., R or T_s).
2. The proportion of the total settlement associated with the secondary consolidation increases with the increase of R.
3. In the conventional consolidation theory of Terzaghi, $R = 0$. Thus, the average degree of consolidation becomes equal to 100% at the end of primary consolidation.
4. As defined in Eq. (5.134),

$$R = \frac{C_v a_v}{H^2} \frac{b^n}{(\Delta\sigma')^{n-1}}$$

The term b is a complex quantity and depends on the electrochemical environment and structure of clay. The value of b increases with the increase of effective pressure σ' on the soil. When the ratio $\Delta\sigma'/\sigma'$ is small it will result in an increase of R, and thus in the proportion of secondary to primary consolidation. Other factors remaining constant, R will also increase with decrease of H, which is the length of the maximum drainage path, and thus so will the ratio of secondary to primary consolidation.

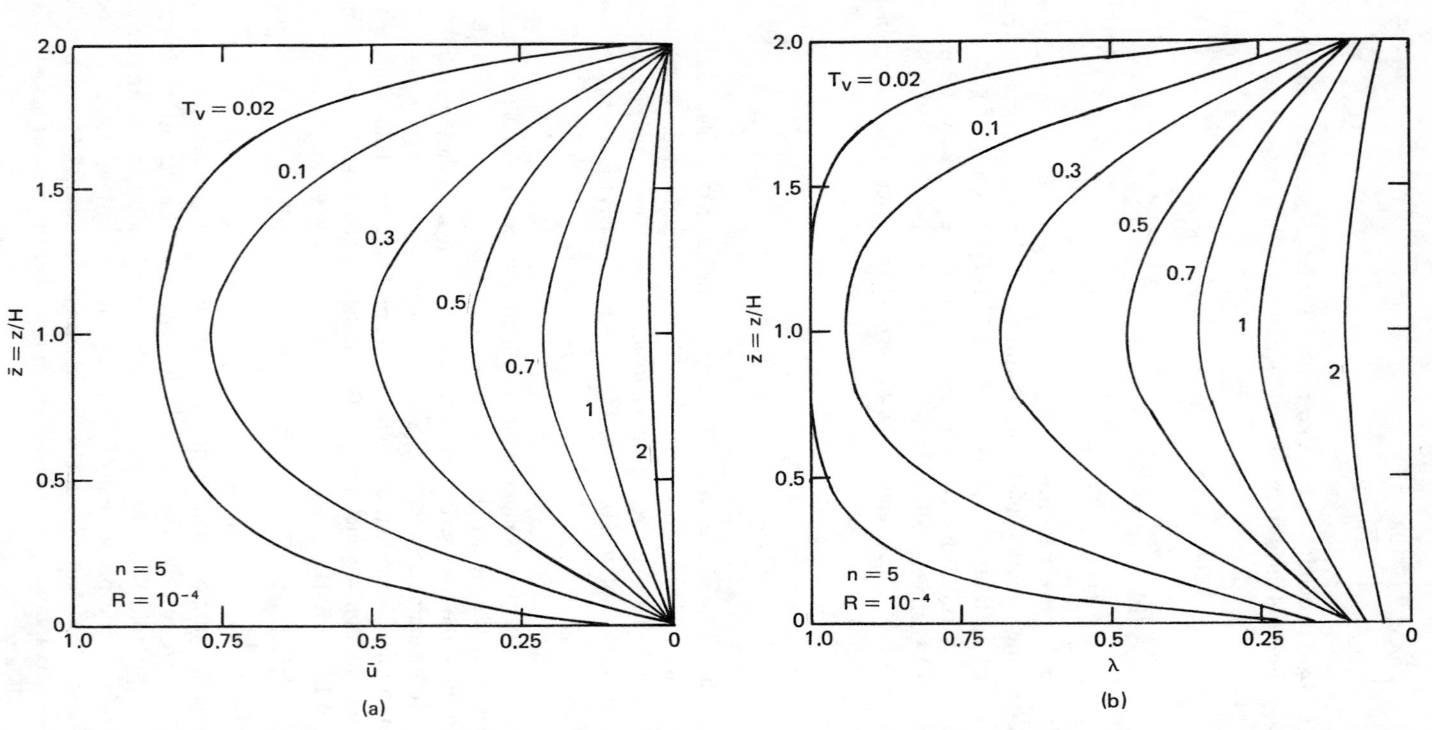

Fig. 5.58 Plot of $\bar{z}$ against $\bar{u}$ and λ for a two-way drained clay layer. *(Redrawn after L. Barden, Consolidation of Clay with Nonlinear Viscosity,* Geotechnique, *vol. 15, no. 4, 1965. By permission of the publisher, The Institution of Civil Engineers.)*

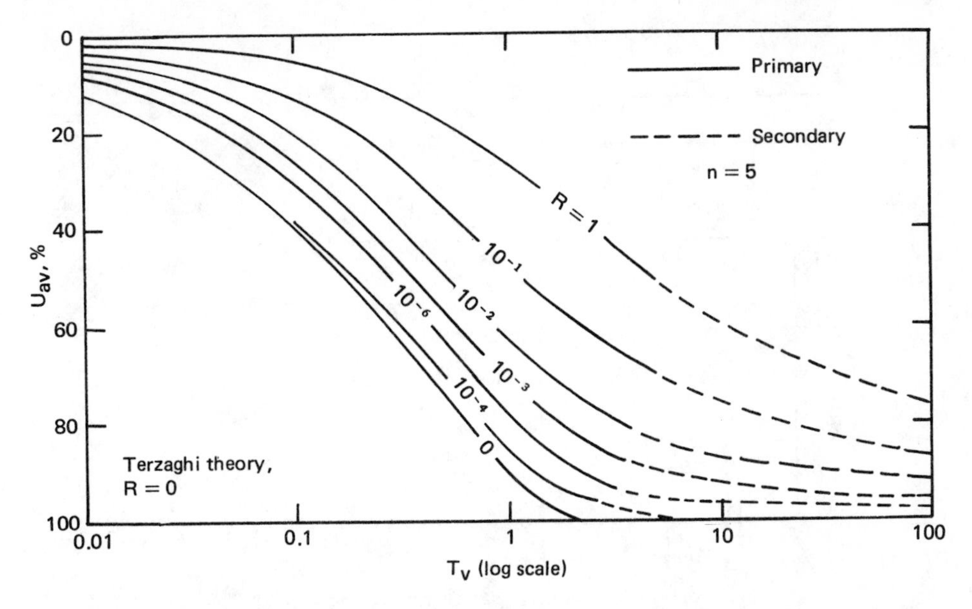

Fig. 5.59 Plot of degree of consolidation vs. T_v for various values of R ($n = 5$). (*After L. Barden, Consolidation of Clay with Nonlinear Viscosity,* Geotechnique, *vol. 15, no. 4, 1965. By permission of the publisher, The Institution of Civil Engineers.)*

5.2 CONSOLIDATION BY SAND DRAINS

5.2.1 Sand Drains

In order to accelerate the process of consolidation settlement for the construction of some structures, the useful technique of building sand drains can be used. Sand drains are constructed by driving down casings or hollow mandrels into the soil. The holes are then filled with sand, after which the casings are pulled out. When a surcharge is applied at ground surface, the pore water pressure in the clay will increase, and there will be drainage in the vertical and horizontal directions (Fig. 5.60). The horizontal drainage is induced by the sand drains. Hence, the process of dissipation of excess pore water pressure created by the loading (and hence the settlement) is accelerated.

The basic theory of sand drains was presented by Rendulic (1935) and Barron (1948) and later summarized by Richart (1959). In the study of sand drains, two fundamental cases arise;

1. Free-strain case: When the surcharge applied at the ground surface is of a flexible nature (Sec. 3.2.8), there will be equal distribution of surface load. This will result in an uneven settlement at the surface.
2. Equal-strain case: When the surcharge applied at the ground surface is rigid, the surface settlement will be the same all over. However, this will result in an unequal distribution of stress.

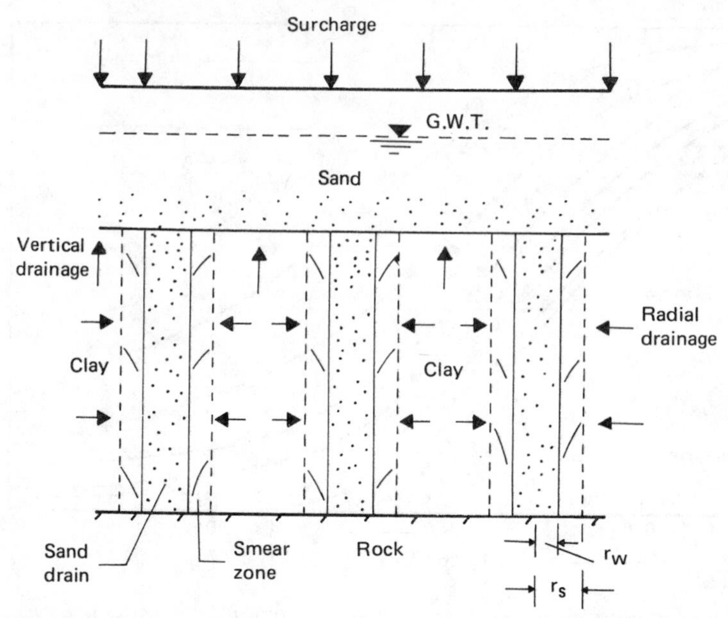

Fig. 5.60 Sand drains.

Another factor that must be taken into consideration is the effect of "smear." A "smear" zone in a sand drain is created by the remolding of clay during the drilling operation for building it (see Fig. 5.60). This remolding of the clay results in a decrease of the coefficient of permeability in the horizontal direction.

The theories for free-strain and equal-strain consolidation are treated separately in Secs. 5.2.2 and 5.2.3. In the development of these theories, it is assumed that drainage takes place *only in the radial direction*, i.e., *no dissipation of excess pore water pressure in the vertical direction.*

5.2.2 Free-Strain Consolidation with no Smear

Figure 5.61 shows the general pattern of the layout of sand drains. For triangular spacing of the sand drains, the zone of influence of each drain is hexagonal in plan. This hexagon can be approximated as an equivalent circle of diameter d_e. Other notations used in this section are as follows (Fig. 5.61*b*):

1. r_e = radius of the equivalent circle = $d_e/2$.
2. r_w = radius of the sand drain well.
3. r_s = radial distance from the center line of the drain well to the farthest point of the smear zone. Note that, in the no-smear case, $r_w = r_s$.

The basic differential equation of Terzaghi's consolidation theory for flow in the vertical direction is given in Eq. (5.14). For radial drainage, this equation can be

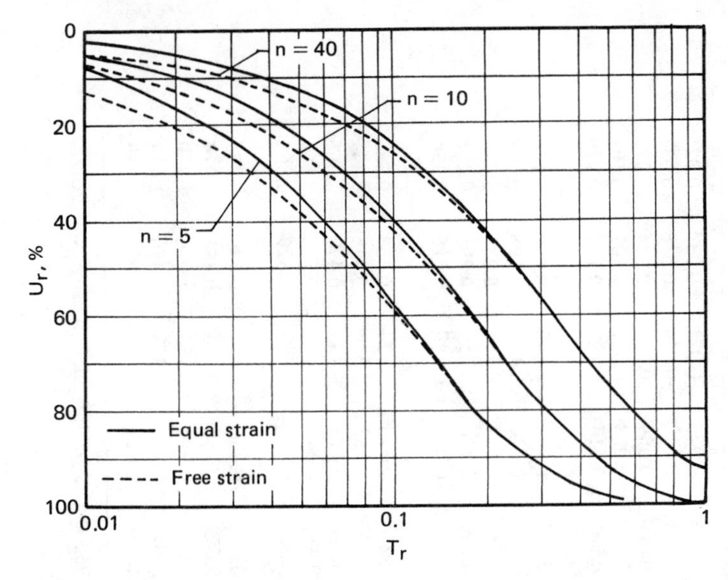

Fig. 5.62 Variation of degree of consolidation U_r with time factor T_r. *(After F. E. Richart, Review of the Theories for Sand Drains,* Trans. ASCE, *vol. 124, 1959.)*

$$\lambda = \frac{-8T_r}{F(n)} \tag{5.152}$$

The average degree of consolidation due to radial drainage is

$$U_r = 1 - \exp\left[\frac{-8T_r}{F(n)}\right] \tag{5.153}$$

Table 5.4 gives the values of the time factor T_r for various values of U_r (also see Fig. 5.62). For $r_e/r_w > 5$, the free-strain and equal-strain solutions give approximately the same results for the average degree of consolidation.

5.2.4 Effect of Smear Zone on Radial Consolidation

Barron (1948) also extended the analysis of equal-strain consolidation by sand drains to account for the smear zone explained in Sec. 5.2.1. The analysis is based on the assumption that the clay in the smear zone will have one boundary with zero excess pore water pressure and the other boundary with an excess pore water pressure which will be time dependent. Based on this assumption,

$$u = \frac{1}{m}\ u_{\text{av}}\left[\ln\left(\frac{r}{r_e}\right) - \frac{r^2 - r_s^2}{2r_e^2} + \frac{k_h}{k_s}\left(\frac{n^2 - S^2}{n^2}\right)\ln S\right] \tag{5.154}$$

where k_s = coefficient of permeability of smeared zone

$$S = \frac{r_s}{r_w} \tag{5.155}$$

Table 5.4 Solution of radial-flow equation: equal vertical-strain condition

Degree of consoli-dation U_r, %	Time factor T_r										
	$\frac{r_e}{r_w} = 5$	10	15	20	25	30	40	50	60	80	100
5	0.006	0.010	0.013	0.014	0.016	0.017	0.019	0.020	0.021	0.032	0.025
10	0.012	0.021	0.026	0.030	0.032	0.035	0.039	0.042	0.044	0.048	0.051
15	0.019	0.032	0.040	0.046	0.050	0.054	0.060	0.064	0.068	0.074	0.079
20	0.026	0.044	0.055	0.063	0.069	0.074	0.082	0.088	0.092	0.101	0.107
25	0.034	0.057	0.071	0.081	0.089	0.096	0.106	0.114	0.120	0.131	0.139
30	0.042	0.070	0.088	0.101	0.110	0.118	0.131	0.141	0.149	0.162	0.172
35	0.050	0.085	0.106	0.121	0.133	0.143	0.158	0.170	0.180	0.196	0.208
40	0.060	0.101	0.125	0.144	0.158	0.170	0.188	0.202	0.214	0.232	0.246
45	0.070	0.118	0.147	0.169	0.185	0.198	0.220	0.236	0.250	0.291	0.288
50	0.081	0.137	0.170	0.195	0.214	0.230	0.255	0.274	0.290	0.315	0.334
55	0.094	0.157	0.197	0.225	0.247	0.265	0.294	0.316	0.334	0.363	0.385
60	0.107	0.180	0.226	0.258	0.283	0.304	0.337	0.362	0.383	0.416	0.441
65	0.123	0.207	0.259	0.296	0.325	0.348	0.386	0.415	0.439	0.477	0.506
70	0.137	0.231	0.289	0.330	0.362	0.389	0.431	0.463	0.490	0.532	0.564
75	0.162	0.273	0.342	0.391	0.429	0.460	0.510	0.548	0.579	0.629	0.668
80	0.188	0.317	0.397	0.453	0.498	0.534	0.592	0.636	0.673	0.730	0.775
85	0.222	0.373	0.467	0.534	0.587	0.629	0.697	0.750	0.793	0.861	0.914
90	0.270	0.455	0.567	0.649	0.712	0.764	0.847	0.911	0.963	1.046	1.110
95	0.351	0.590	0.738	0.844	0.926	0.994	1.102	1.185	1.253	1.360	1.444
99	0.539	0.907	1.135	1.298	1.423	1.528	1.693	1.821	1.925	2.091	2.219

From *Foundation Engineering* by G. A. Leonards (Ed.). Copyright © 1962 McGraw-Hill Book Company, New York. Used with the permission of McGraw-Hill Book Company.

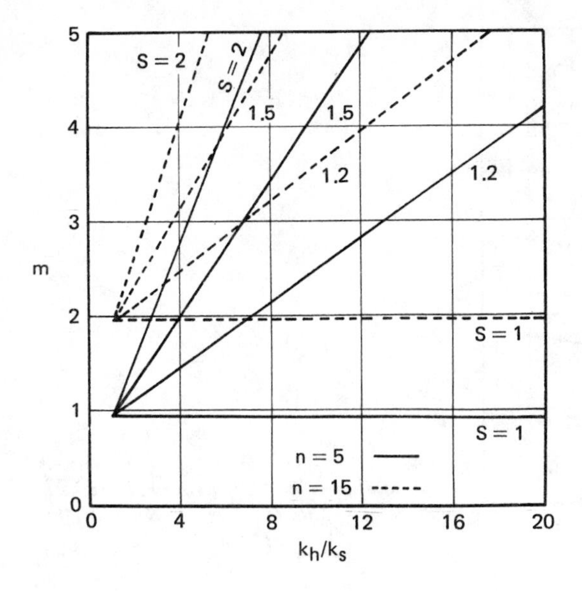

Fig. 5.63 Values of m for various values of k_h/k_s and S ($n = 5$ and 15). *(After F. E. Richart, Review of the Theories for Sand Drains,* Trans. ASCE, *vol. 124, 1959.)*

$$m = \frac{n^2}{n^2 - S^2} \ln\left(\frac{n}{S}\right) - \frac{3}{4} + \frac{S^2}{4n^2} + \frac{k_h}{k_s}\left(\frac{n^2 - S^2}{n^2}\right) \ln S \tag{5.156}$$

$$u_{av} = u_i \exp\left(\frac{-8T_r}{m}\right) \tag{5.157}$$

The average degree of consolidation is given by the relation

$$U_r = 1 - \frac{u_{av}}{u_i} = 1 - \exp\left(\frac{-8T_r}{m}\right) \tag{5.158}$$

Figure 5.63 gives the values of m [Eq. (5.156)] for various values of k_h/k_s and S (for $n = 5$ and 15). It must be pointed out that for the case where we assume no smear, i.e., $S = 1$, the expression for m [Eq. (5.156)] becomes equal to the expression for $F(n)$, i.e., Eq. (5.150). Figure 5.64 gives a plot of m against n for various values of S (for $k_h/k_s = 20$).

Olson (1977) gave a solution for the average degree of consolidation U_r for time-dependent loading (ramp load) similar to that for vertical drainage as described in Sec. 5.1.5.

5.2.5 Calculation of the Degree of Consolidation with Vertical and Radial Drainage

The relation for average degree of consolidation for *vertical drainage only* was presented in Sec. 5.1.2; also, the relations for the degree of consolidation due to *radial drainage*

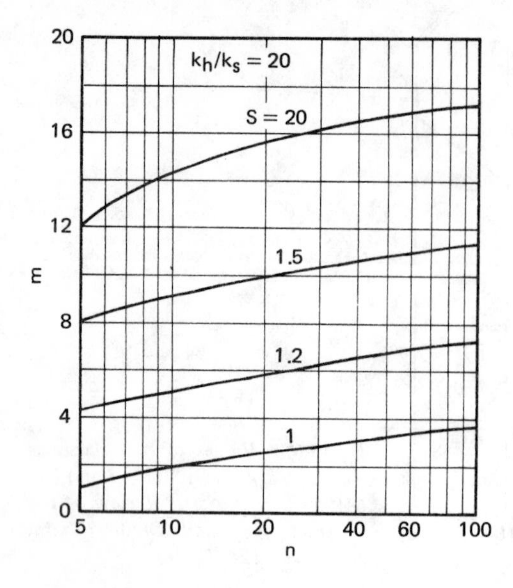

Fig. 5.64 Plot of m against n for various values of S ($k_h/k_s = 20$). *(After F. E. Richart, Review of the Theories for Sand Drains, Trans. ASCE, vol. 124, 1959.)*

only were given in Secs. 5.2.2 to 5.2.4. In reality, the drainage for the dissipation of excess pore water pressure takes place in both directions simultaneously. For such a case, Carrillo (1942) has shown that

$$U = 1 - (1 - U_v)(1 - U_r) \tag{5.159}$$

where U = average degree of consolidation for simultaneous vertical and radial drainage

U_v = average degree of consolidation calculated on the assumption that only vertical drainage exists (note we have used the notation U_{av} before in this chapter)

U_r = average degree of consolidation calculated on the assumption that only radial drainage exists

Example 5.9 A 20-ft-thick clay layer is drained at the top and bottom and has some sand drains. The given data are: C_v (for vertical drainage) = 0.055 ft²/day; $k_{\text{vertical}} = k_h$; $d_w = 18$ in; $d_e = 10$ ft; $r_w = r_s$ (i.e., no smear at the periphery of drain wells).

It has been estimated that a given uniform surcharge would cause a total consolidation settlement of 10 in without the sand drains. Calculate the consolidation settlement of the clay layer with the same surcharge and sand drains at time $t = 0, 0.2, 0.4, 0.6, 0.8$, and 1 yr.

SOLUTION *Vertical drainage:* $C_v = 0.055$ ft²/day = 20.08 ft²/yr.

$$T_v = \frac{C_v t}{H^2} = \frac{20.08 \times t}{(20/2)^2} = 0.2008t \cong 0.2t \tag{a}$$

Radial drainage:

$$\frac{r_e}{r_w} = \frac{5 \text{ ft}}{0.75 \text{ ft}} = 6.67 = n$$

$$F_n = \frac{n^2}{n^2 - 1} \ln(n) - \frac{3n^2 - 1}{4n^2} \qquad \text{(equal strain case)}$$

$$= \left[\frac{(6.67)^2}{(6.67)^2 - 1} \ln(6.67) - \frac{3(6.67)^2 - 1}{4(6.67)^2}\right]$$

$$= 1.94 - 0.744 = 1.196$$

Since $k_{\text{vertical}} = k_h$, $C_v = C_{vr}$. So,

$$T_r = \frac{C_{vr} t}{d_e^2} = \frac{20 \times t}{10^2} = 0.2t \tag{b}$$

The steps in the calculation of the consolidation settlement are shown in Table 5.5. From Table 5.5, the consolidation settlement at $t = 1$ yr is 8.70 in. Without the the sand drains, the consolidation settlement at the end of one year would have been only 5.05 in.

5.2.6 Numerical Solution for Radial Drainage

As shown before for vertical drainage (Sec. 5.1.4), we can adopt the finite-difference technique for solving consolidation problems in the case of radial drainage. From Eq. (5.139),

$$\frac{\partial u}{\partial t} = C_{vr} \left(\frac{\partial^2 u}{\partial r^2} + \frac{1}{r} \frac{\partial u}{\partial r}\right)$$

Let u_R, t_R, and r_R be any reference excess pore water pressure, time, and radial distance, respectively. So,

$$\text{Nondimensional excess pore water pressure} = \bar{u} = \frac{u}{u_R} \tag{5.160}$$

$$\text{Nondimensional time} = \bar{t} = \frac{t}{t_R} \tag{5.161}$$

$$\text{Nondimensional radial distance} = \bar{r} = \frac{r}{r_R} \tag{5.162}$$

Substituting Eqs. (5.160) to (5.162) into Eq. (5.139), we get

$$\frac{1}{t_R} \frac{\partial \bar{u}}{\partial \bar{t}} = \frac{C_{vr}}{r_R^2} \left(\frac{\partial^2 \bar{u}}{\partial \bar{r}^2} + \frac{1}{\bar{r}} \frac{\partial \bar{u}}{\partial \bar{r}}\right) \tag{5.163}$$

Table 5.5

t, yr	T_v [Eq. (*a*)]	U_v (Fig. 5.6)	$1 - U_v$	T_r [Eq. (*b*)]	$1 - \exp[-8T_r/F(n)] = U_r$	$1 - U_r$	$U = 1 - (1 - U_v)(1 - U_r)$	$S_c = 10 \times U$, in
0	0	0	1	0	0	1	0	0
0.2	0.04	0.22	0.78	0.04	0.235	0.765	0.404	4.04
0.4	0.08	0.32	0.68	0.08	0.414	0.586	0.601	6.01
0.6	0.12	0.39	0.61	0.12	0.552	0.448	0.727	7.27
0.8	0.16	0.45	0.55	0.16	0.657	0.343	0.812	8.12
1	0.2	0.505	0.495	0.2	0.738	0.262	0.870	8.70

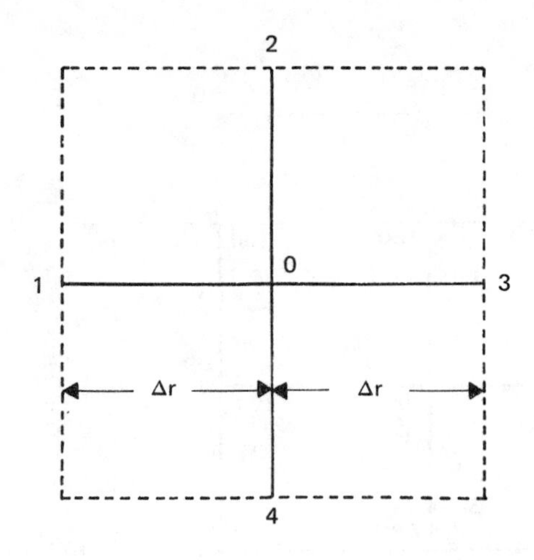

Fig. 5.65

Referring to Fig. 5.65,

$$\frac{\partial \bar{u}}{\partial \bar{t}} = \frac{1}{\Delta \bar{t}} (\bar{u}_{0,\bar{t}+\Delta \bar{t}} - \bar{u}_{0,\bar{t}}) \tag{5.164}$$

$$\frac{\partial^2 \bar{u}'}{\partial \bar{r}^2} = \frac{1}{(\Delta \bar{r})^2} (\bar{u}_{1,\bar{t}} + \bar{u}_{3,\bar{t}} - 2\bar{u}_{0,\bar{t}}) \tag{5.165}$$

and
$$\frac{1}{\bar{r}}\frac{\partial \bar{u}}{\partial \bar{r}} = \frac{1}{\bar{r}}\left(\frac{u_{3,\bar{t}} - u_{1,\bar{t}}}{2\Delta \bar{r}}\right) \tag{5.166}$$

If we adopt t_R in such a way that $1/t_R = C_{vr}/r_R^2$ and then substitute Eqs. (5.164) to (5.166) into Eq. (5.163), then

$$\bar{u}_{0,\bar{t}+\Delta \bar{t}} = \frac{\Delta \bar{t}}{(\Delta \bar{r})^2}\left[\bar{u}_{1,t} + \bar{u}_{3,t} + \frac{\bar{u}_{3,\bar{t}} - \bar{u}_{1,\bar{t}}}{2(\bar{r}/\Delta \bar{r})} - 2\bar{u}_{0,\bar{t}}\right] + \bar{u}_{0,\bar{t}} \tag{5.167}$$

Equation (5.167) is the basic finite-difference equation for solution of the excess pore water pressure (for radial drainage only).

Example 5.10 For a sand drain, the following data are given: $r_w = 1.25$ ft; $r_e = 5$ ft; $r_w = r_s$; and $C_{vr} = 0.05$ ft^2/day. A uniformly distributed load of 1000 lb/ft^2 is applied at the ground surface. Determine the distribution of excess pore water pressure after 10 days of load application assuming radial drainage only.

SOLUTION Let $r_R = 1.25$ ft, $\Delta r = 1.25$ ft, and $\Delta t = 5$ days. So, $\bar{r}_e = r_e/r_R = 5/1.25 = 4$; $\Delta \bar{r}/r_R = 1.25/1.25 = 1$

$$\Delta \bar{t} = \frac{C_v \, \Delta t}{r_R^2} = \frac{0.05 \times 5}{(1.25)^2} = 0.16$$

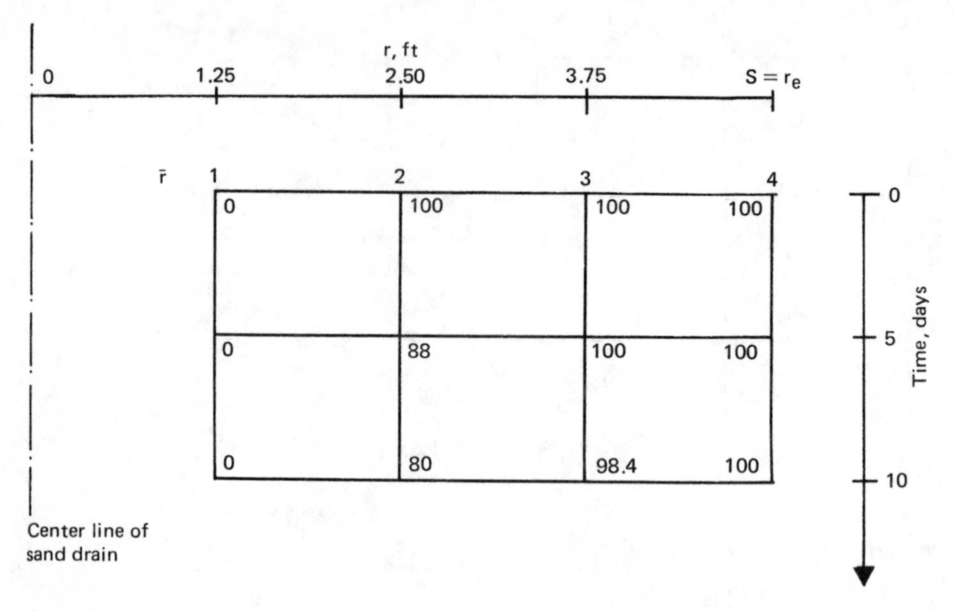

Fig. 5.66

$$\frac{\Delta \bar{t}}{(\Delta \bar{r})^2} = \frac{0.16}{(1)^2} = 0.16$$

Let $u_R = 10\,\text{lb/ft}^2$. So, immediately after load application, $\bar{u} = 1000/10 = 100$.

Figure 5.66 shows the initial nondimensional pore water pressure distribution at time $t = 0$. (Note that at $\bar{r} = 1, \bar{u} = 0$ due to the drainage face.)

$\bar{u}$ at 5 days: $\bar{u} = 0, \bar{r} = 1$. From Eq. (5.167),

$$\bar{u}_{0,\bar{t}+\Delta\bar{t}} = \frac{\Delta \bar{t}}{(\Delta \bar{r})^2}\left[\bar{u}_{1,\bar{t}} + \bar{u}_{3,\bar{t}} + \frac{\bar{u}_{3,\bar{t}} - \bar{u}_{1,\bar{t}}}{2(\bar{r}/\Delta\bar{r})} - 2\bar{u}_{0,\bar{t}}\right] + \bar{u}_{0,\bar{t}}$$

At $\bar{r} = 2$,

$$\bar{u}_{0,\bar{t}+\Delta\bar{t}} = 0.16\left[0 + 100 + \frac{100 - 0}{2(2/1)} - 2(100)\right] + 100 = 88$$

At $\bar{r} = 3$,

$$\bar{u}_{0,\bar{t}+\Delta\bar{t}} = 0.16\left[100 + 100 + \frac{100 - 100}{2(3/1)} - 2(100)\right] + 100 = 100$$

Similarly, at $\bar{r} = 4$,

$$\bar{u}_{0,\bar{t}+\Delta\bar{t}} = 100$$

(note that, here, $\bar{u}_{3,\bar{t}} = \bar{u}_{1,\bar{t}}$).

$\bar{u}$ *at 10 days:* At $\bar{r} = 1, \bar{u} = 0$.

At $\bar{r} = 2$,

$$\bar{u}_{0,\bar{t}+\Delta\bar{t}} = 0.16\left[0 + 100 + \frac{100 - 0}{2(2/1)} - 2(88)\right] + 88$$

$$= 79.84 \cong 80$$

At $\bar{r} = 3$,

$$\bar{u}_{0,\bar{t}+\Delta\bar{t}} = 0.16\left[88 + 100 + \frac{100 - 88}{2(3/1)} - 2(100)\right] + 100 = 98.4$$

At $\bar{r} = 4$,

$\bar{u} = 100$

$u = \bar{u} \times u_R = 10\bar{u}\,\text{lb/ft}^2$

The distribution of nondimensional excess pore water pressure is shown in Fig. 5.66.

PROBLEMS

5.1 Consider a clay layer drained at the top and bottom as shown in Fig. 5.4. For the case of constant initial excess pore water pressure ($u_i = u_o$) and $T_v = 0.4$, determine the degree of consolidation U_z at $z/H_t = 0, 0.2, 0.4, 0.5, 0.6, 0.8$, and 1. For the solution, start from Eq. (5.32).

5.2 Starting from Eq. (5.34), solve for average degree of consolidation for linearly varying initial excess pore water pressure distribution for a clay layer with two-way drainage (Fig. 5.8) for $T_v = 0$, 0.2, 0.6, 0.8, and 1. Plot a graph of U_{av} against T_v from your results.

5.3 Refer to Fig. 5.8. Starting from Eq. (5.39), derive the expression for the average degree of consolidation as,

$$U_{av} = 1 - \sum_{m=0}^{m=\infty} \frac{2}{M^2} \exp(-M^2 T_v)$$

5.4 Refer to Fig. 5.15*a*. For the 10-ft-thick clay layer, $C_v = 0.02\,\text{in}^2/\text{min}$ and $q = 3500\,\text{lb/ft}^2$. Plot the variation of excess pore water pressure in the clay layer with depth after 6 months of load application.

5.5 A 10-in total consolidation settlement of the two clay layers shown in Fig. P5.1 is expected due to the application of the uniform surcharge q. Find the duration after load application at which 5 in of total settlement would take place.

5.6 Repeat Prob. 5.5 assuming that a layer of rock is located at the bottom of the 5-ft-thick clay layer.

5.7 Due to a certain loading condition, the initial excess pore water distribution in a 4-m-thick clay layer is as shown in Fig. P5.2. Given that $C_v = 0.3\,\text{mm}^2/\text{s}$, determine the degree of consolidation after 100 days of load application.

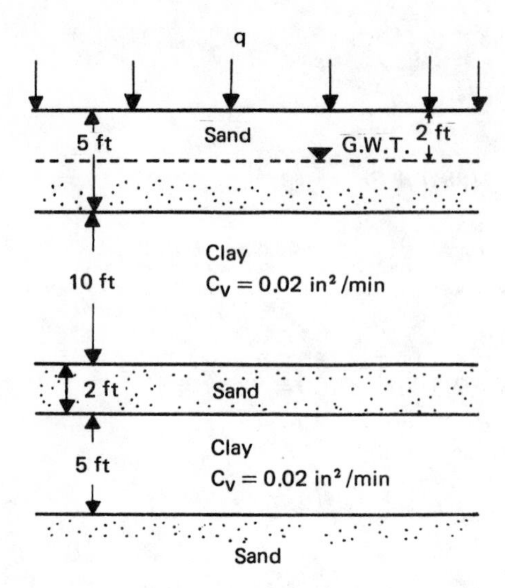

Fig. P5.1

5.8 A uniform surcharge of 2000 lb/ft^2 is applied at the ground surface of a soil profile as shown in Fig. P5.3. Determine the distribution of the excess pore water pressure in the 10-ft-thick clay layer after 1 yr of load application. Use the numerical method of calculation given in Sec. 5.1.4. Also calculate the average degree of consolidation at that time using the above results.

5.9 Refer to Fig. P5.3. Assume that there is an impervious rock layer at the bottom of the 10-ft clay layer. Using the numerical technique (Sec. 5.1.4), determine the excess pore water pressure distribution after 1 yr of load application.

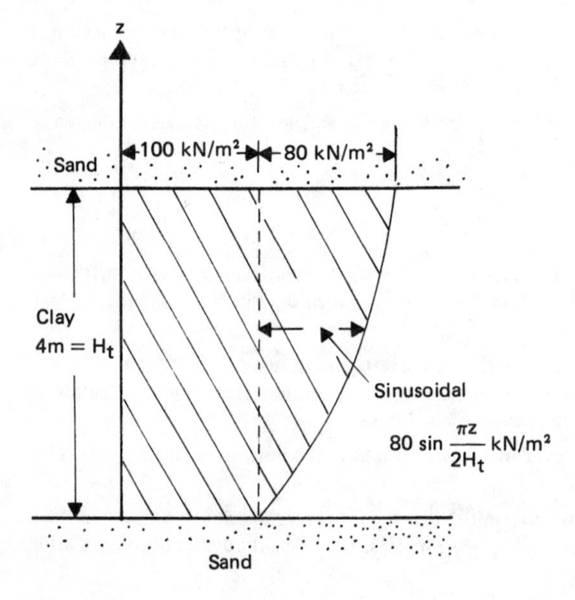

Fig. P5.2

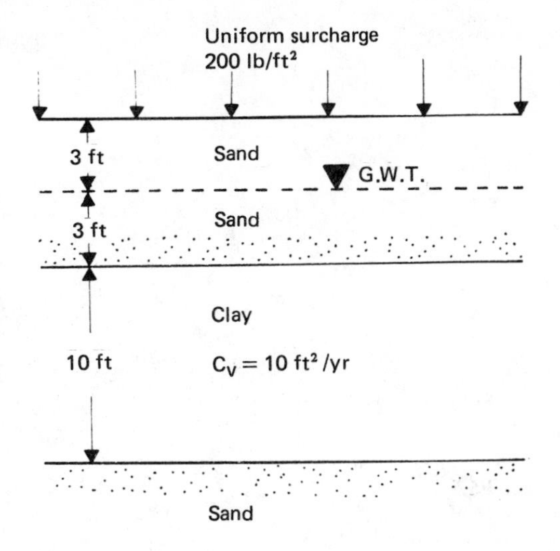

Fig. P5.3

5.10 A two-layered soil is shown in Fig. P5.4. At a given time $t = 0$, a uniform load was applied at the ground surface so as to increase the pore water pressure by 1200 lb/ft² at all depths. Divide the soil profile into six equal layers. Using the numerical analysis method, find the excess pore water pressure at depths of -10, -20, -30, -40, -50, and -60 ft at $t = 5$, 10, 15, 20, and 25 days.

5.11 Solve Prob. 5.10 assuming that there is a highly permeable sand layer below the clay layer II.

5.12 Solve Prob. 5.10 assuming that the initial excess pore water pressure u_i decreases linearly as shown in Fig. P5.5. Draw the excess pore water pressure diagram at $t = 25$ days and from that find the average degree of consolidation.

5.13 Refer to Fig. P5.6. A uniform surcharge q is applied at the ground surface. The variation of

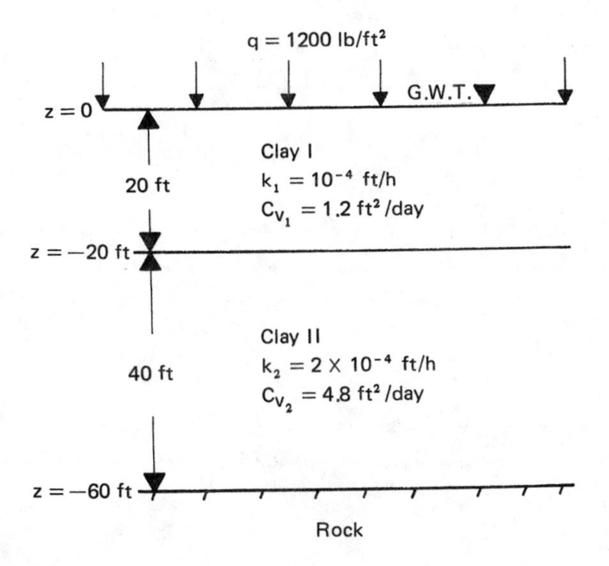

Fig. P5.4

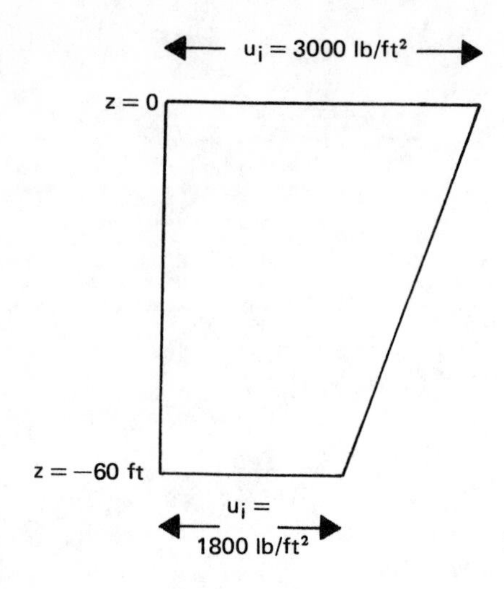

Fig. P5.5

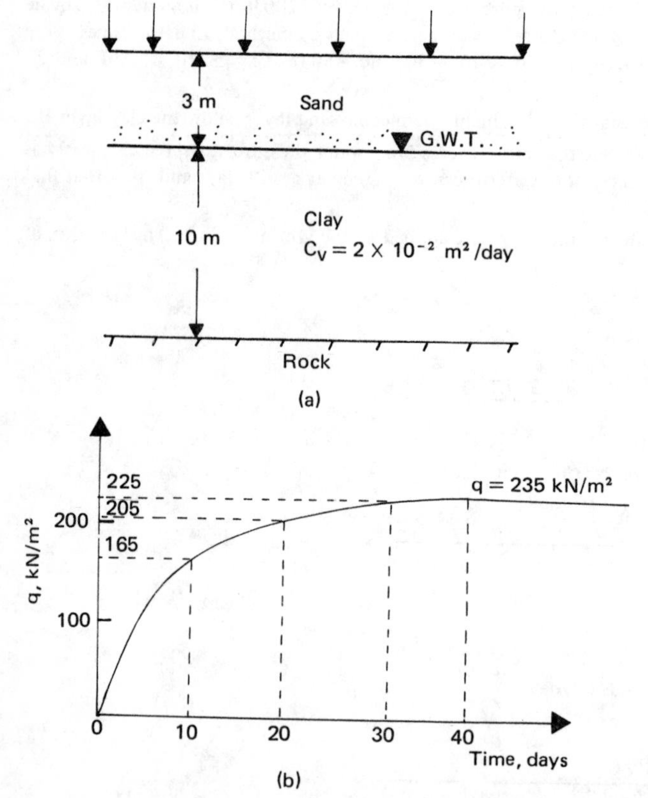

Fig. P5.6

q with time is shown in Fig. P5.6b. Divide the 10-m-thick clay layer into five layers, each 2 m thick. Determine the excess pore water pressure in the clay layer at $t = 100$ days by the numerical method.

5.14 Redo Prob. 5.13 assuming that the 10-m-thick clay layer is drained at the top and bottom.

5.15 Refer to Fig. P5.6a. The uniform surcharge is time dependent and is a single ramp load as shown in Fig. P5.7. Using Fig. 5.26, determine the average degree of consolidation for the clay layer at $t = 50$ days and $t = 1$ yr.

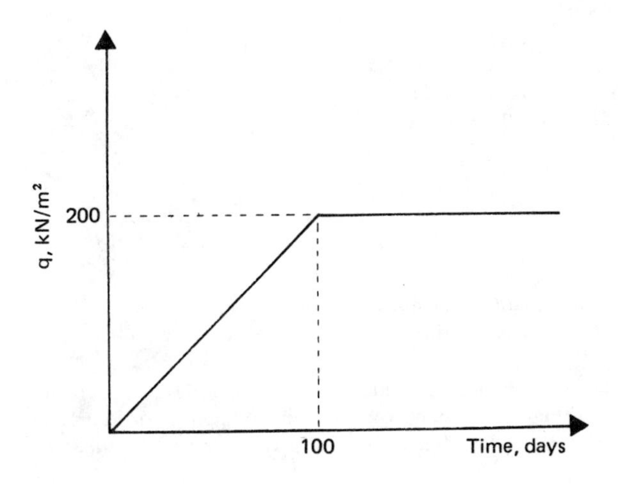

Fig. P5.7

5.16 The following are the results of a standard one-dimensional consolidation test on a remolded clay:

σ', kN/m²	Total height of sample at the end of consolidation, mm
24	17.82
48	17.40
96	17.03
192	16.56
384	16.15
768	15.88

The initial height of the specimen was 19 mm, $G_s = 2.68$, the mass of dry specimen was 95.2 g, and the area of the specimen was 31.70 cm². Make the necessary calculations to obtain the e vs. log σ' plot. Find the compression index C_c. Estimate the probable liquid limit of the clay.

5.17 Refer to Prob. 5.16. What would be the total height of the specimen at the end of consolidation under a pressure of 1536 kN/m²? What would the void ratio be at that time?

5.18 The following are the results of a standard consolidation test for a load increment from 1 to 2 ton/ft²:

Time	Dial reading, in	Time, min	Dial reading, in
0 s	0.1046	30.25	0.1177
4.5 s	0.1076	36	0.1179
15 s	0.1087	42.25	0.1180
34.5 s	0.1097	49	0.1181
1 min	0.1106	56.25	0.1182
2.25 min	0.1125	64	0.1183
4 min	0.1141	72.30	0.1184
6.25 min	0.1153	81	0.1185
9 min	0.1162	90.25	0.1186
12.25 min	0.1167	100	0.1186
16 min	0.1170	121	0.1187
20.25 min	0.1173	144	0.1188
25 min	0.1175		

(*a*) Calculate the time for 90% consolidation using the square-root-of-time method.

(*b*) If the average height of the specimen during the loading was 17 mm, calculate C_v in m^2/day.

5.19 The average effective overburden pressure on a 10-m-thick clay layer in the field is 136 kN/m², and the average void ratio is 0.98. If a uniform surcharge of 200 kN/m² is applied on the ground surface, what will be the consolidation settlement for the following cases, given $C_c = 0.35$ and $C_r = 0.08$?

(*a*) Preconsolidation pressure, $\sigma_c' = 350$ kN/m².

(*b*) $\sigma_c' = 100$ kN/m².

(*c*) $\sigma_c' = 200$ kN/m².

5.20 Refer to Prob. 5.19(*c*).

(*a*) What is the average void ratio at the end of 100% consolidation?

(*b*) If $C_v = 1.5$ mm²/min, how long will it take for the first 10 mm of settlement? Assume two-way drainage for the clay layer.

5.21 The results of an oedometer test on a clay layer are as follows:

σ', kN/m²	Void ratio e
385	0.95
770	0.87

The time for 90% consolidation was 10 min, and the average thickness of the clay was 23 mm (two-way drainage). Calculate the coefficient of permeability of clay in mm/s.

5.22 Refer to Fig. P5.8. For a uniform surcharge q, determine the time in days for half the consolidation settlement to take place.

5.23 A 5-m-thick clay layer, drained at the top only, has some sand drains. A uniform surcharge is applied at the top of the clay layer. Calculate the average degree of consolidation for combined vertical and radial drainage after 100 and 150 days of load application, given $C_{vr} = C_v = 4$ mm²/min, $d_e = 2$ m, and $r_w = 0.2$ m.

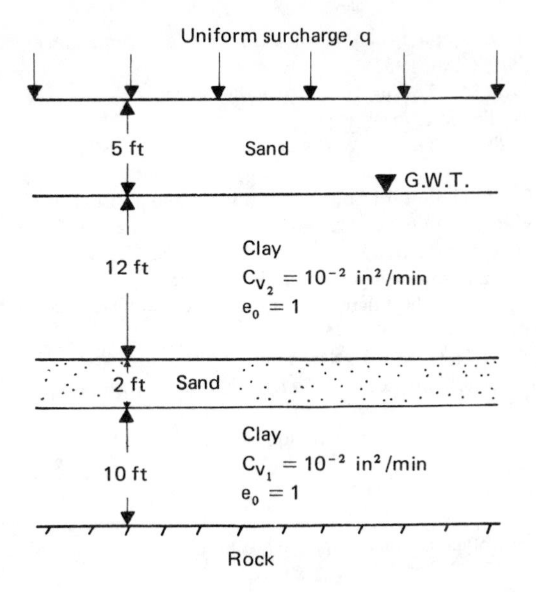

Fig. P5.8

5.24 Redo Prob. 5.23. Assume that there is some smear around the sand drains and that $r_s = 0.3$ m and $k_h/k_s = 4$. (This is an equal-strain case.)

5.25 For a sand drain problem, $r_w = 1$ ft, $r_s = 1$ ft, $r_e = 6$ ft, and $C_{vr} = 0.03$ ft²/day. If a uniform load of 2000 lb/ft² is applied at the ground surface, find the distribution of the excess pore water pressure after 50 days of load application. Use the numerical method. Consider that there is radial drainage only.

REFERENCES

American Society for Testing and Materials, "ASTM Standards," Part 19, 1979.

Azzouz, A. S., R. J. Krizek, and R. B. Corotis, Regression Analysis of Soil Compressibility, *Soils Found., Tokyo,* vol. 16, no. 2, pp. 19–29, 1976.

Barden, L., Consolidation of Clay with Non-Linear Viscosity, *Geotechnique,* vol. 15, no. 4, pp. 345–362, 1965.

Barden, L., Primary and Secondary Consolidation of Clay and Peat, *Geotechnique,* vol. 18, pp. 1–14, 1968.

Barron, R. A., Consolidation of Fine-Grained Soils by Drain Wells, *Trans. ASCE,* vol. 113, p. 1718, 1948.

Bjerrum, L., Engineering Geology of Norwegian Normally Consolidated Marine Clays and Related to Settlements of Buildings, *Geotechnique,* vol. 17, pp. 81–118, 1967.

Bjerrum, L., Embankments on Soft Ground, *Proc. Specialty Conf. Perform. Earth and Earth-Supported Struct., ASCE,* vol. 2, pp. 1–54, 1972.

Brummund, W. F., E. Jonas, and C. C. Ladd, "Estimating In-situ Maximum Past (Preconsolidation) Pressure of Saturated Clays from Results of Laboratory Consolidometer Tests," Transportation Research Board, Special Report 163, pp. 4–12, 1976.

Burmister, D. M., The Application of Controlled Test Methods in Consolidation Testing, *Symp. Consol. Test. Soils, ASTM Spec. Tech. Pub.,* vol. 126, p. 83, 1951.

Carrillo, N., Simple Two- and Three-Dimensional Cases in Theory of Consolidation of Soils, *J. Math. Phys.,* vol. 21, no. 1, 1962.

Casagrande, A., The Determination of the Preconsolidation Load and Its Practical Significance, *Proc. 1st Int. Conf. Soil Mech. Found. Eng.*, p. 60, 1936.

Casagrande, A., and R. E. Fadum, "Notes on Soil Testing for Engineering Purposes," Harvard Univ. Graduate School of Engineering, Publication No. 8, 1940.

Crawford, C. B., Interpretation of the Consolidation Tests, *J. Soil Mech. Found. Div., ASCE,* vol. 90, no. SM5, pp. 93–108, 1964.

Gibson, R. E., and K. Y. Lo, "A Theory of Consolidation for Soils Exhibiting Secondary Compression," Norwegian Geotechnical Institute, Publication No. 41, 1961.

Kenney, T. C., Sea Level Movements and the Geologic Histories of the Post-Glacial Marine Soils at Boston, Nicolet, Ottawa, and Oslo, *Geotechnique,* vol. 14, pp. 203–230, 1964.

Lambe, T. W., The Structure of Compacted Clay and the Engineering Behavior of Compact Clay, *J. Soil Mech. Found. Div., ASCE,* vol. 84, no. SM2, p. 68, 1958.

Leonards, G. A. (Ed.), "Foundation Engineering," McGraw-Hill, New York, 1962.

Leonards, G. A., and A. G. Altschaeffl, Compressibility of Clay, *J. Soil Mech. Found. Div., ASCE,* vol. 90, no. SM5, pp. 133–156, 1964.

Leonards, G. A., and B. K. Ramiah, "Time Effects in the Consolidation of Clays," ASTM Special Technical Publication 254, pp. 116–130, 1960.

Lowe, J., New Concepts in Consolidation and Settlement Analysis, *J. Geotech. Eng. Div., ASCE,* vol. 100, no. GT6, pp. 574–612, 1974.

Lowe, J. III, E. Jonas, and V. Obrician, Controlled Gradient Consolidation Test, *J. Soil Mech. Found. Div., ASCE,* vol. 95, no. SM1, pp. 77–98, 1969.

Luscher, U., Discussion, *J. Soil Mech. Found. Div., ASCE,* vol. 91, no. SM1, pp. 190–195, 1965.

Mesri, G., Coefficient of Secondary Compression, *J. Soil Mech. Found. Div., ASCE,* vol. 99, no. SM1, pp. 123–137, 1973.

Nacci, V. A., M. C. Wang, and K. R. Demars, Engineering Behavior of Calcareous Soils, *Proc. Civil Eng. Oceans III, ASCE,* vol. 1, pp. 380–400, 1975.

Olson, R. E., Consolidation under Time-Dependent Loading, *J. Geotech. Eng. Div., ASCE,* vol. 103, no. GT1, pp. 55–60, 1977.

Raju, A. A., "The Preconsolidation Pressure in Clay Soils," M. S. thesis, Purdue University, 1956.

Rendulic, L., 1935. Der Hydrodynamische Spannungsaugleich in Zentral Entwässerten Tonzylindern, *Wasserwirtsch. Tech.,* vol. 2, pp. 250–253, 269–273, 1935.

Richart, F. E., Review of the Theories for Sand Drains, *Trans. ASCE,* vol. 124, pp. 709–736, 1959.

Schiffman, R. L., C. C. Ladd, and A. T. Chen, The Secondary Consolidation of Clay, *I.U.T.A.M. Symposium on Rheological Soil Mechanics, Grenoble,* p. 273, 1964.

Schmertmann, J. H., Undisturbed Laboratory Behavior of Clay, *Trans. ASCE,* vol. 120, p. 1201, 1953.

Scott, R. F., "Principles of Soil Mechanics," Addison-Wesley, Reading, Mass., 1963.

Sivaram, B., and P. Swamee, A Computational Method for Consolidation Coefficient, *Soils Found., Tokyo, Jpn.,* vol. 17, no. 2, pp. 48–52, 1977.

Smith, R. E., and H. E. Wahls, Consolidation Under Constant Rate of Strain, *J. Soil Mech. Found. Div., ASCE,* vol. 95, no. SM2, pp. 519–538, 1969.

Su, H. L., Procedure for Rapid Consolidation Test, *J. Soil Mech. Found. Div., ASCE,* vol. 95, Proc. paper 1729, 1958.

Tan, T. K., Discussion, *Proc. 4th Int. Conf. Soil Mech. Found. Eng.,* vol. 3, p. 278, 1957.

Taylor, D. W., "Research on Consolidation of Clays," Massachusetts Institute of Technology, Publication No. 82, 1942.

Taylor, D. W., and W. Merchant, A Theory of Clay Consolidation Accounting for Secondary Compression, *J. Math. Phys.,* vol. 19, p. 167, 1940.

Terzaghi, K., "Erdbaumechanik auf Boden-physicalischen Grundlagen" Deuticke, Vienna, 1925.

Terzaghi, K., "Theoretical Soil Mechanics," Wiley, New York, 1943.

Terzaghi, K., and R. B. Peck, "Soil Mechanics in Engineering Practice," 2d ed., Wiley, New York, 1967.

CHAPTER

SIX

EVALUATION OF SOIL SETTLEMENT

6.1 INTRODUCTION

The increase of stress in soil layers due to the load imposed by various structures at the foundation level will always be accompanied by some strain, which will result in the settlement of the structures. The various aspects of settlement calculation are analyzed in this chapter.

In general, the total settlement S of a foundation can be given as

$$S = S_e + S_c + S_s \tag{6.1}$$

where S_e = immediate settlement
S_c = primary consolidation settlement
S_s = secondary consolidation settlement

The immediate settlement is sometimes referred to as the elastic settlement. In granular soils this is the predominant part of the settlement, whereas in saturated inorganic silts and clays the primary consolidation settlement probably predominates. The secondary consolidation settlement forms the major part of the total settlement in highly organic soils and peats. We will consider the analysis of each component of the total settlement separately in some detail.

6.2 IMMEDIATE SETTLEMENT

6.2.1 Immediate Settlement from Theory of Elasticity

Settlement due to a concentrated point load at the surface. For elastic settlement due to a concentrated point load (Fig. 6.1), the strain at a depth z can be given, in cylindrical coordinates, by

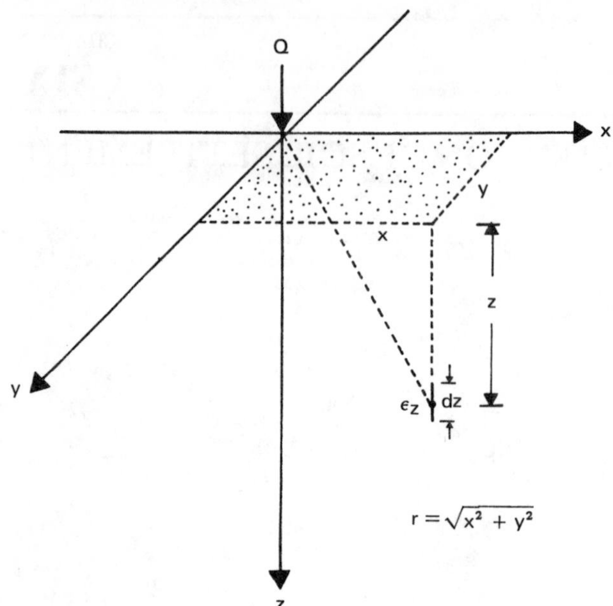

Fig. 6.1 Elastic settlement due to a concentrated point load.

$$\epsilon_z = \frac{1}{E}\left[\sigma_z - \nu(\sigma_r + \sigma_\theta)\right] \tag{6.2}$$

where E is the Young's modulus of the soil. The expressions for σ_z, σ_r, and σ_θ are given in Eqs. (3.53) to (3.55), respectively. Substitution of these in Eq. (6.2) and simplification yields

$$\epsilon_z = \frac{Q}{2\pi E}\left[\frac{3(1+\nu)r^2 z}{(r^2+z^2)^{5/2}} - \frac{3+\nu(1-2\nu)z}{(r^2+z^2)^{3/2}}\right] \tag{6.3}$$

The settlement at a depth z can be found by integrating Eq. (6.3):

$$S_e = \int \epsilon_z\, dz = \frac{Q}{2\pi E}\left[\frac{(1+\nu)z^2}{(r^2+z^2)^{3/2}} + \frac{2(1-\nu^2)}{(r^2+z^2)^{1/2}}\right]$$

The settlement at the surface can be evaluated by putting $z = 0$ in the above equation:

$$S_e(\text{surface}) = \frac{Q}{\pi E r}(1-\nu^2) \tag{6.4}$$

Settlement at the surface due to a uniformly loaded flexible circular area. The elastic settlement due to a uniformly loaded circular area (Fig. 6.2) can be determined by using the same procedure we used above for a point load, which involves determination of the strain ϵ_z from the equation

$$\epsilon_z = \frac{1}{E}\left[\sigma_z - \nu(\sigma_r + \sigma_\theta)\right]$$

and determination of the settlement by integration with respect to z.

The relationships for σ_z, σ_r, and σ_θ are given in Eqs. (3.66) to (3.68). Substitution of the relations for σ_z, σ_r, and σ_θ in the preceding equation for strain and simplification gives (Ahlvin and Ulery, 1962)

$$\epsilon_z = q\,\frac{1+\nu}{E}\left[(1-2\nu)A' + B'\right] \tag{6.5}$$

where q is the load per unit area. A' and B' are nondimensional and are functions of z/b and s/b; their values are given in Tables 3.7 and 3.8.

The vertical deflection at a depth z can be obtained by integration of Eq. (6.5) as

$$S_e = q\,\frac{1+\nu}{E}\,b\left[\frac{z}{b}I_1 + (1-\nu)I_2\right] \tag{6.6}$$

where $I_1 = A'$ (Table 3.7) and b is the radius of the circular loaded area. The numerical values of I_2 (which is a function of z/b and s/b) are given in Table 6.1.

From Eq. (6.6) it follows that the settlement at the surface (i.e., at $z = 0$) is

$$S_e(\text{surface}) = qb\,\frac{1-\nu^2}{E}\,I_2 \tag{6.7}$$

The term I_2 in Eq. (6.7) is usually referred to as the *influence number.* For saturated clays, we may assume $\nu = 0.5$. So, at the center of the loaded area (i.e., $s/b = 0$), $I_2 = 2$ and

$$S_e(\text{surface, center}) = \frac{1.5qb}{E} = \frac{0.75qB}{E} \tag{6.8}$$

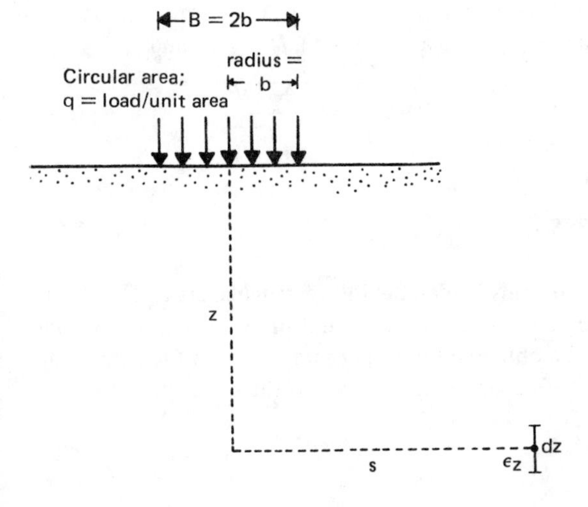

Fig. 6.2 Elastic settlement due to a uniformly loaded circular area.

Table 6.1 Values of I_2

	s/b								
z/b	0	0.2	0.4	0.6	0.8	1	1.2	1.5	2
0	2.0	1.97987	1.91751	1.80575	1.62553	1.27319	.93676	.71185	.51671
0.1	1.80998	1.79018	1.72886	1.61961	1.44711	1.18107	.92670	.70888	.51627
0.2	1.63961	1.62068	1.56242	1.46001	1.30614	1.09996	.90098	.70074	.51382
0.3	1.48806	1.47044	1.40979	1.32442	1.19210	1.02740	.86726	.68823	.50966
0.4	1.35407	1.33802	1.28963	1.20822	1.09555	.96202	.83042	.67238	.50412
0.5	1.23607	1.22176	1.17894	1.10830	1.01312	.90298	.79308	.65429	.49728
0.6	1.13238	1.11998	1.08350	1.02154	.94120	.84917	.75653	.63469	
0.7	1.04131	1.03037	.99794	.91049	.87742	.80030	.72143	.61442	.48061
0.8	.96125	.95175	.92386	.87928	.82136	.75571	.68809	.59398	
0.9	.89072	.88251	.85856	.82616	.77950	.71495	.65677	.57361	
1	.82843	.85005	.80465	.76809	.72587	.67769	.62701	.55364	.45122
1.2	.72410	.71882	.70370	.67937	.64814	.61187	.57329	.51552	.43013
1.5	.60555	.60233	.57246	.57633	.55559	.53138	.50496	.46379	.39872
2	.47214	.47022	.44512	.45656	.44502	.43202	.41702	.39242	.35054
2.5	.38518	.38403	.38098	.37608	.36940	.36155	.35243	.33698	.30913
3	.32457	.32403	.32184	.31887	.31464	.30969	.30381	.29364	.27453
4	.24620	.24588	.24820	.25128	.24168	.23932	.23668	.23164	.22188
5	.19805	.19785				.19455			.18450
6	.16554					.16326			.15750
7	.14217					.14077			.13699
8	.12448					.12352			.12112
9	.11079					.10989			.10854
10								.09900	.09820

After R. G. Ahlvin and H. R. Ulery, Tabulated Values for Determining the Complete Pattern of Stresses, Strains and Deflections beneath a Uniform Load on a Homogeneous Half Space, *Highway Research Board, Bulletin 342,* 1962.

where $B = 2b$ is the diameter of the loaded area.

At the edge of the loaded area (i.e., $z/b = 0$ and $s/b = 1$), $I_2 = 1.27$ and

$$S_e(\text{surface, edge}) = (1.27)(0.75)\frac{qb}{E} = 0.95\frac{qb}{E} = \frac{0.475qB}{E} \tag{6.9}$$

The average surface settlement is

$$S_e(\text{surface, average}) = 0.85S_e(\text{surface, center}) \tag{6.10}$$

Settlement at the surface due to a uniformly loaded flexible rectangular area. The elastic deformation in the vertical direction at the corner of a uniformly loaded rectangular area of size $L \times B$ (Fig. 3.24) can be obtained by proper integration of the expression for strain. The deformation at a depth z below the corner of the rectangular area can be expressed in the form (Harr, 1966)

	s/b								
z/b	3	4	5	6	7	8	10	12	14
0	.33815	.25200	.20045	.16626	.14315	.12576	.09918	.08346	.07023
0.1	.33794	.25184	.20081						
0.2	.33726	.25162	.20072	.16688	.14288	.12512			
0.3	.33638	.25124							
0.4									
0.5	.33293	.24996	.19982	.16668	.14273	.12493	.09996	.08295	.07123
0.6									
0.7									
0.8									
0.9									
1	.31877	.24386	.19673	.16516	.14182	.12394	.09952	.08292	.07104
1.2	.31162	.24070	.19520	.16369	.14099	.12350			
1.5	.29945	.23495	.19053	.16199	.14058	.12281	.09876	.08270	.07064
2	.27740	.22418	.18618	.15846	.13762	.12124	.09792	.08196	.07026
2.5	.25550	.21208	.17898	.15395	.13463	.11928	.09700	.08115	.06980
3	.23487	.19977	.17154	.14919	.13119	.11694	.09558	.08061	.06897
4	.19908	.17640	.15596	.13864	.12396	.11172	.09300	.07864	.06848
5	.17080	.15575	.14130	.12785	.11615	.10585	.08915	.07675	.06695
6	.14868	.13842	.12792	.11778	.10836	.09990	.08562	.07452	.06522
7	.13097	.12404	.11620	.10843	.10101	.09387	.08197	.07210	.06377
8	.11680	.11176	.10600	.09976	.09400	.08848	.07800	.06928	.06200
9	.10548	.10161	.09702	.09234	.08784	.08298	.07407	.06678	.05976
10	.09510	.09290	.08980	.08300	.08180	.07710			

$$S_e(\text{corner}) = \frac{qB}{2E}(1-\nu^2)\left[I_3 - \left(\frac{1-2\nu}{1-\nu}\right)I_4\right] \tag{6.11}$$

where $$I_3 = \frac{1}{\pi}\left[\ln\left(\frac{\sqrt{1+m_1^2+n_1^2}+m_1}{\sqrt{1+m_1^2+n_1^2}-m_1}\right)\right.$$

$$\left. + m\ln\left(\frac{\sqrt{1+m_1^2+n_1^2}+1}{\sqrt{1+m_1^2+n_1^2}-1}\right)\right] \tag{6.12}$$

$$I_4 = \frac{n_1}{\pi}\tan^{-1}\left(\frac{m_1}{n_1\sqrt{1+m_1^2+n_1^2}}\right) \tag{6.13}$$

$$m_1 = \frac{L}{B} \tag{6.14}$$

$$n_1 = \frac{z}{B} \tag{6.15}$$

Values of I_3 and I_4 are given in Table 6.2.

For elastic surface settlement at the corner of a rectangular area, we can substitute $z/B = n_1 = 0$ in Eq. (6.11) and make the necessary calculations; thus,

$$S_e(\text{corner}) = \frac{qB}{2E}(1 - \nu^2)I_3 \tag{6.16}$$

The settlement at the surface for the center of a rectangular area (Fig. 6.3) can be found by adding the settlement for the corner of four rectangular areas of dimension $L/2 \times B/2$. Thus, from Eq. (6.11),

$$S_e(\text{center}) = 4\left[\frac{q(B/2)}{2E}\right](1 - \nu^2)I_3 = \frac{qB}{E}(1 - \nu^2)I_3 \tag{6.17}$$

The average surface settlement can be obtained as

$$S_e(\text{average, surface}) = 0.848 S_e(\text{center, surface}) \tag{6.18}$$

Summary of elastic settlement at the ground surface ($z = 0$) due to uniformly distributed vertical loads on flexible areas. For circular areas:

$$S_e = \left[qB\left(\frac{1 - \nu^2}{2E}\right)\right]I_2$$

where B = diameter of circular loaded area
$I_2 = 2$ (at center)
$I_2 = 1.27$ (at edge)
$I_2 = 0.85 \times 2 = 1.7$ (average)

For rectangular areas, on the basis of Eqs. (6.16) to (6.18) we can write

$$S_e = \left[\frac{qB}{E}(1 - \nu^2)\right]I_5 \tag{6.19}$$

where $I_5 = I_3$ (at center)
$I_5 = \frac{1}{2}I_3$ (at edge)
$I_5 \approx 0.848 I_3$ (average)

Table 6.3 gives the values of I_5 for various L/B ratios.

Settlement of a flexible load area on an elastic layer of finite thickness. For the settlement equations presented in this section, it was assumed that the elastic soil layer extends to an infinite depth. However, if the elastic soil layer is underlain by

Table 6.2 Values of I_3 and I_4

n_1		m_1											
		1	1.5	2	3	5	7	10	15	20	30	50	100
0.0	I_3	1.122	1.358	1.532	1.783	2.105	2.318	2.544	2.802	2.985	3.243	3.568	4.010
	I_4	0.000	0.000	0.000	0.000	0.000	0.000	0.000	0.000	0.000	0.000	0.000	0.000
0.2	I_3	1.105	1.343	1.518	1.770	2.092	2.305	2.532	2.790	2.973	3.231	3.556	3.997
	I_4	0.082	0.085	0.086	0.087	0.087	0.087	0.087	0.087	0.087	0.087	0.087	0.087
0.4	I_3	1.057	1.301	1.479	1.733	2.056	2.270	2.497	2.755	2.938	3.196	3.521	3.962
	I_4	0.132	0.142	0.146	0.149	0.151	0.151	0.151	0.151	0.151	0.152	0.152	0.152
0.6	I_3	0.989	1.240	1.422	1.679	2.004	2.219	2.446	2.704	2.887	3.145	3.470	3.912
	I_4	0.158	0.176	0.184	0.191	0.195	0.196	0.196	0.197	0.197	0.197	0.197	0.197
0.8	I_3	0.914	1.169	1.354	1.615	1.943	2.158	2.386	2.644	2.827	3.086	3.411	3.852
	I_4	0.167	0.194	0.207	0.218	0.224	0.226	0.227	0.228	0.228	0.228	0.228	0.228
1.0	I_3	0.838	1.094	1.282	1.547	1.878	2.094	2.322	2.581	2.764	2.022	3.348	3.789
	I_4	0.167	0.200	0.218	0.234	0.244	0.247	0.248	0.249	0.250	0.250	0.250	0.250
1.2	I_3	0.768	1.020	1.209	1.478	1.812	2.029	2.258	2.517	2.701	2.959	3.284	3.726
	I_4	0.161	0.200	0.222	0.243	0.257	0.261	0.263	0.264	0.265	0.265	0.265	0.265
1.4	I_3	0.704	0.950	1.139	1.411	1.748	1.966	2.196	2.455	2.639	2.897	3.223	3.664
	I_4	0.154	0.196	0.221	0.247	0.265	0.270	0.273	0.275	0.276	0.276	0.276	0.276
1.6	I_3	0.647	0.886	1.073	1.346	1.686	1.906	2.136	2.396	2.580	2.839	3.164	3.605
	I_4	0.145	0.189	0.217	0.248	0.269	0.277	0.281	0.283	0.283	0.284	0.284	0.284
1.8	I_3	0.596	0.826	1.011	1.284	1.627	1.848	2.080	2.340	2.524	2.783	3.108	3.550
	I_4	0.136	0.181	0.212	0.246	0.272	0.281	0.286	0.288	0.289	0.290	0.290	0.290
2.0	I_3	0.552	0.773	0.954	1.226	1.571	1.794	2.026	2.287	2.471	2.730	3.056	3.497
	I_4	0.128	0.173	0.205	0.243	0.273	0.283	0.289	0.292	0.294	0.294	0.295	0.295
2.5	I_3	0.463	0.660	0.829	1.095	1.444	1.670	1.094	2.167	2.352	2.612	2.937	3.379
	I_4	0.110	0.153	0.186	0.230	0.269	0.284	0.293	0.298	0.300	0.302	0.302	0.303
3.0	I_3	0.396	0.572	0.728	0.984	1.332	1.561	1.798	2.063	2.249	2.509	2.835	3.277
	I_4	0.096	0.136	0.168	0.215	0.262	0.282	0.295	0.301	0.304	0.306	0.307	0.307

Table 6.2 *(Continued)*

n_1		m_1											
		1	1.5	2	3	5	7	10	15	20	30	50	100
3.5	I_3	0.346	0.503	0.647	0.890	1.234	1.465	1.705	1.971	2.158	2.419	2.745	3.187
	I_4	0.084	0.121	0.152	0.200	0.253	0.277	0.292	0.302	0.305	0.308	0.309	0.310
4.0	I_3	0.306	0.448	0.580	0.809	1.147	1.379	1.621	1.890	2.077	2.339	2.666	3.108
	I_4	0.075	0.109	0.138	0.186	0.243	0.270	0.289	0.301	0.306	0.309	0.311	0.312
4.5	I_3	0.274	0.404	0.525	0.741	1.070	1.301	1.545	1.816	2.005	2.267	2.594	3.036
	I_4	0.067	0.098	0.126	0.173	0.222	0.263	0.285	0.300	0.305	0.310	0.312	0.313
5.0	I_3	0.248	0.367	0.479	0.682	1.001	1.231	1.477	1.749	1.939	2.202	2.530	2.972
	I_4	0.061	0.090	0.116	0.161	0.221	0.255	0.281	0.298	0.305	0.310	0.313	0.314
6.0	I_3	0.208	0.309	0.406	0.586	0.884	1.109	1.355	1.631	1.823	2.088	2.417	2.860
	I_4	0.052	0.076	0.099	0.141	0.201	0.239	0.270	0.293	0.302	0.309	0.313	0.315
7.0	I_3	0.179	0.267	0.352	0.513	0.788	1.006	1.251	1.529	1.723	1.990	2.320	2.764
	I_4	0.045	0.066	0.087	0.124	0.183	0.223	0.259	0.286	0.298	0.308	0.313	0.315
8.0	I_3	0.158	0.235	0.310	0.455	0.710	0.918	1.160	1.440	1.635	1.904	2.236	2.680
	I_4	0.039	0.058	0.077	0.111	0.168	0.208	0.247	0.279	0.294	0.306	0.313	0.316
9.0	I_3	0.140	0.209	0.277	0.408	0.644	0.843	1.080	1.360	1.556	1.828	2.161	2.606
	I_4	0.035	0.052	0.069	0.100	0.154	0.194	0.235	0.272	0.289	0.304	0.312	0.316
10.0	I_3	0.126	0.189	0.251	0.370	0.589	0.778	1.009	1.287	1.485	1.759	2.093	2.539
	I_4	0.032	0.047	0.062	0.091	0.142	0.182	0.224	0.264	0.284	0.301	0.311	0.316
12.0	I_3	0.106	0.158	0.210	0.311	0.502	0.672	0.889	1.162	1.361	1.638	1.976	2.423
	I_4	0.026	0.039	0.052	0.077	0.122	0.160	0.203	0.248	0.272	0.295	0.309	0.315
14.0	I_3	0.091	0.136	0.180	0.268	0.436	0.590	0.792	1.057	1.255	1.534	1.875	2.325
	I_4	0.023	0.034	0.045	0.067	0.107	0.142	0.185	0.232	0.260	0.288	0.306	0.315
16.0	I_3	0.079	0.119	0.158	0.236	0.385	0.525	0.712	0.967	1.163	1.443	1.787	2.239
	I_4	0.020	0.030	0.039	0.059	0.095	0.127	0.168	0.217	0.248	0.280	0.303	0.314
18.0	I_3	0.071	0.106	0.141	0.210	0.345	0.472	0.646	0.890	1.082	1.362	1.709	2.164
	I_4	0.018	0.026	0.035	0.052	0.085	0.115	0.154	0.204	0.236	0.273	0.299	0.313
20.0	I_3	0.064	0.095	0.127	0.189	0.312	0.428	0.591	0.823	1.011	1.290	1.639	2.096
	I_4	0.016	0.024	0.032	0.047	0.077	0.105	0.142	0.191	0.225	0.265	0.295	0.312

30.0	I_3	0.042	0.064	0.085	0.127	0.210	0.292	0.410	0.591	0.751	1.011	1.363	1.831
	I_4	0.011	0.016	0.021	0.032	0.052	0.072	0.101	0.142	0.176	0.225	0.273	0.305
40.0	I_3	0.032	0.048	0.064	0.095	0.158	0.221	0.312	0.457	0.591	0.823	1.164	1.640
	I_4	0.008	0.012	0.016	0.024	0.039	0.055	0.077	0.112	0.142	0.191	0.249	0.295
50.0	I_3	0.025	0.038	0.051	0.076	0.127	0.177	0.251	0.371	0.485	0.690	1.011	1.488
	I_4	0.006	0.010	0.013	0.019	0.032	0.044	0.062	0.091	0.118	0.164	0.225	0.285
60.0	I_3	0.021	0.032	0.042	0.064	0.106	0.148	0.210	0.312	0.410	0.591	0.890	1.363
	I_4	0.005	0.008	0.011	0.016	0.026	0.037	0.052	0.077	0.101	0.142	0.204	0.273
70.0	I_3	0.018	0.027	0.036	0.055	0.091	0.127	0.181	0.269	0.354	0.516	0.793	1.256
	I_4	0.005	0.007	0.009	0.014	0.023	0.032	0.045	0.067	0.087	0.125	0.185	0.261
80.0	I_3	0.016	0.024	0.032	0.048	0.079	0.111	0.158	0.236	0.312	0.457	0.713	1.164
	I_4	0.004	0.006	0.008	0.012	0.020	0.028	0.039	0.059	0.077	0.112	0.169	0.249
90.0	I_3	0.014	0.021	0.028	0.042	0.071	0.099	0.141	0.210	0.278	0.410	0.647	1.083
	I_4	0.004	0.005	0.007	0.011	0.018	0.025	0.035	0.052	0.069	0.101	0.155	0.237
100.0	I_3	0.013	0.019	0.025	0.038	0.064	0.089	0.127	0.190	0.251	0.371	0.591	1.011
	I_4	0.003	0.005	0.006	0.010	0.016	0.022	0.032	0.047	0.062	0.091	0.142	0.225

From *Foundations of Theoretical Soil Mechanics,* by M. E. Harr. Copyright © 1966 McGraw-Hill Book Company, New York. Used with permission of McGraw-Hill Book Company.

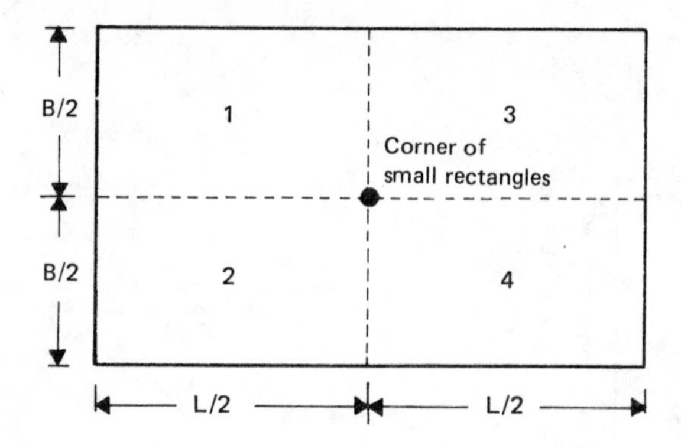

Fig. 6.3 Determination of settlement at the center of a rectangular area of dimensions $L \times B$.

a rigid incompressible base at a depth H (Fig. 6.4), the settlement can be approximately calculated as

$$S_e = S_{e(z=0)} - S_{e(z=H)} \tag{6.20}$$

where $S_{e(z=0)}$ and $S_{e(z=H)}$ are the settlements at the surface and at $z = H$, respectively.

Another condition for the calculation of immediate settlement also needs consideration. Foundations are almost never placed at the ground surface, but at some depth D_f (Fig. 6.5). Hence, a correction needs to be applied to the settlement values calculated on the assumption that the load is applied at the ground surface. Fox (1948) proposed a correction factor for this which is a function of D_f/B, L/B, and Poisson's ratio ν. Thus,

$$S'_e(\text{average}) = I_6 S_e(\text{average}) \tag{6.21}$$

where I_6 = correction factor for foundation depth, D_f
S'_e = corrected elastic settlement of foundation
S_e = elastic settlement of foundation calculated on assumption that load is applied at ground surface

By computer programming of the equation proposed by Fox, Bowles (1977)

Table 6.3 Values of I_5

	I_5		
L/B	Center	Corner	Average
1	1.122	0.561	0.951
2	1.532	0.766	1.299
3	1.783	0.892	1.512
5	2.105	1.053	1.785
10	2.544	1.272	2.157
20	2.985	1.493	2.531
50	3.568	1.784	3.026
100	4.010	2.005	3.400

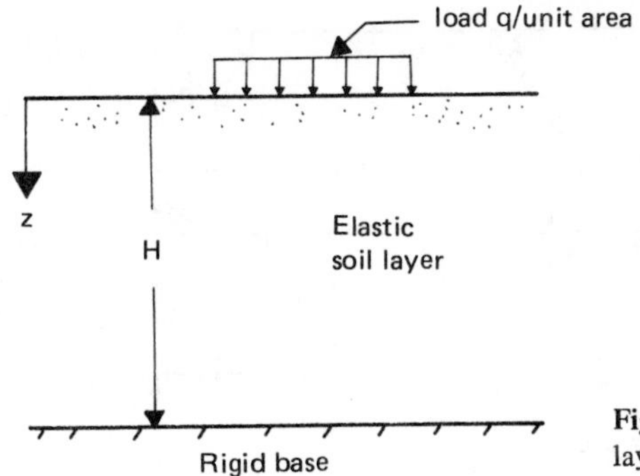

Fig. 6.4 Flexible loaded area over an elastic soil layer of finite thickness.

obtained the values of I_6 for various values of D_f/B, length-to-width ratio of the foundation, and Poisson's ratio of the soil layer. These values are shown in Fig. 6.6.

Janbu et al. (1956) proposed a generalized equation for average immediate settlement for uniformly loaded flexible footings in the form

$$S_e(\text{average}) = \mu_1 \mu_0 \frac{qB}{E} \qquad (\text{for } \nu = 0.5) \tag{6.22}$$

where μ_1 = correction factor for finite thickness of elastic soil layer, H, as shown in Fig. 6.5.
μ_0 = correction factor for depth of embedment of footing, D_f, as shown in Fig. 6.5
B = width of rectangular loaded area or diameter of circular loaded area

Christian and Carrier (1978) made a critical evaluation of Eq. (6.22), the details of which will not be presented here. However, they suggested that for $\nu = 0.5$, Eq. (6.22) could be retained for immediate settlement calculations with a modification of the values of μ_1 and μ_0. The modified values of μ_1 are based on the work of Giroud (1972) and those for μ_0 are based on the work of Burland (1970). These are shown in Fig. 6.7. It must be pointed out that the values of μ_0 and μ_1 given in Fig. 6.7 were actually obtained for flexible circular loaded areas. Christian and Carrier, after a careful

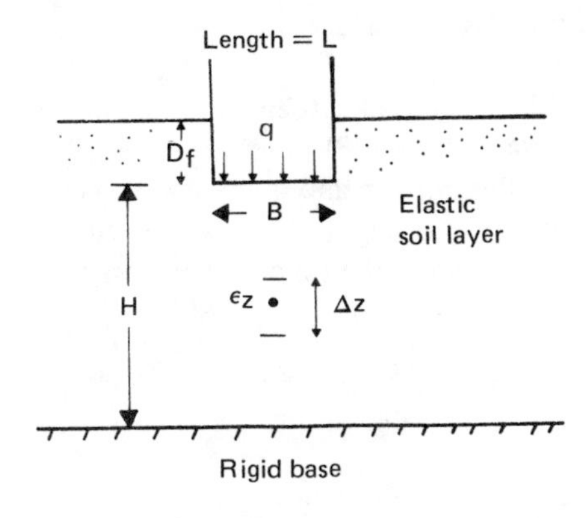

Fig. 6.5 Average immediate settlement for a flexible rectangular loaded area located at a depth D_f from the ground surface.

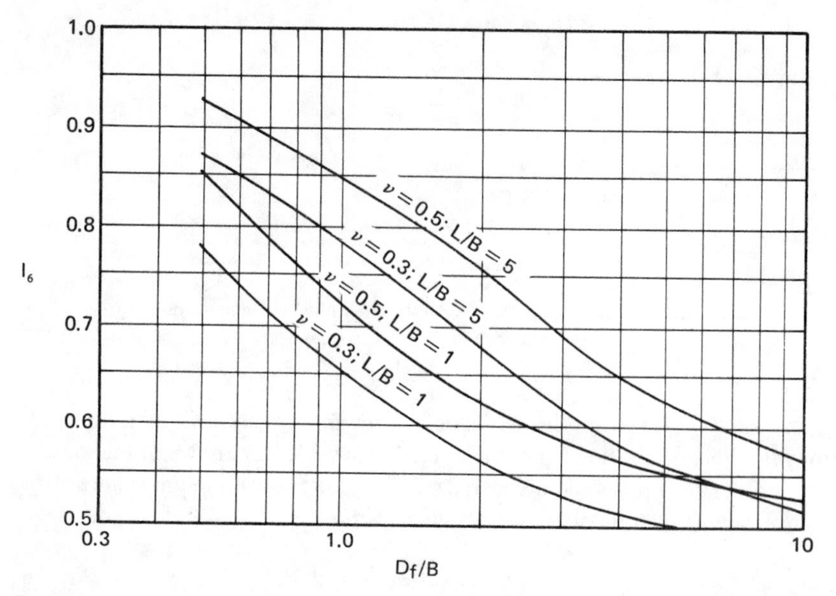

Fig. 6.6 Correction factor for the depth of embedment of the foundation. *(From* Foundation Analysis and Design *by J. E. Bowles. Copyright © 1977 McGraw-Hill Book Company, New York. Used with the permission of McGraw-Hill Book Company.)*

analysis, inferred that these values are generally adequate for circular and rectangular footings.

Another general method for estimation of immediate settlement is to divide the underlying soil into n layers of finite thicknesses (Fig. 6.5). If the strain at the middle of each layer can be calculated, we can obtain the total immediate settlement as

$$S_e = \sum_{i=1}^{i=n} \Delta z_{(i)} \, \epsilon_{z(i)} \tag{6.23}$$

where $\Delta z_{(i)}$ is the thickness of the ith layer and $\epsilon_{z(i)}$ is the vertical strain at the middle of the ith layer.

The method of using Eq. (6.23) is demonstrated in Example 6.2.

Settlement of rigid footings. In Sec. 3.2.8 the differences of settlement and contact pressure between flexible and rigid footings were explained. The immediate surface settlement of a uniformly loaded rigid footing (Fig. 6.8) is about 7% less than the average surface settlement of a flexible footing of similar dimensions (Schleicher, 1926). So, based on this simplified conclusion, we can write the following expressions:

For a uniformly loaded rigid circular footing of radius b (note $b = B/2$):

$$S_{e(z=0)} = \left[qb \left(\frac{1 - \nu^2}{E} \right) \right] I_7 \tag{6.24}$$

where $I_7 = 0.93 I_2$ (average settlement) $= (0.93)(1.7) = 1.58$.

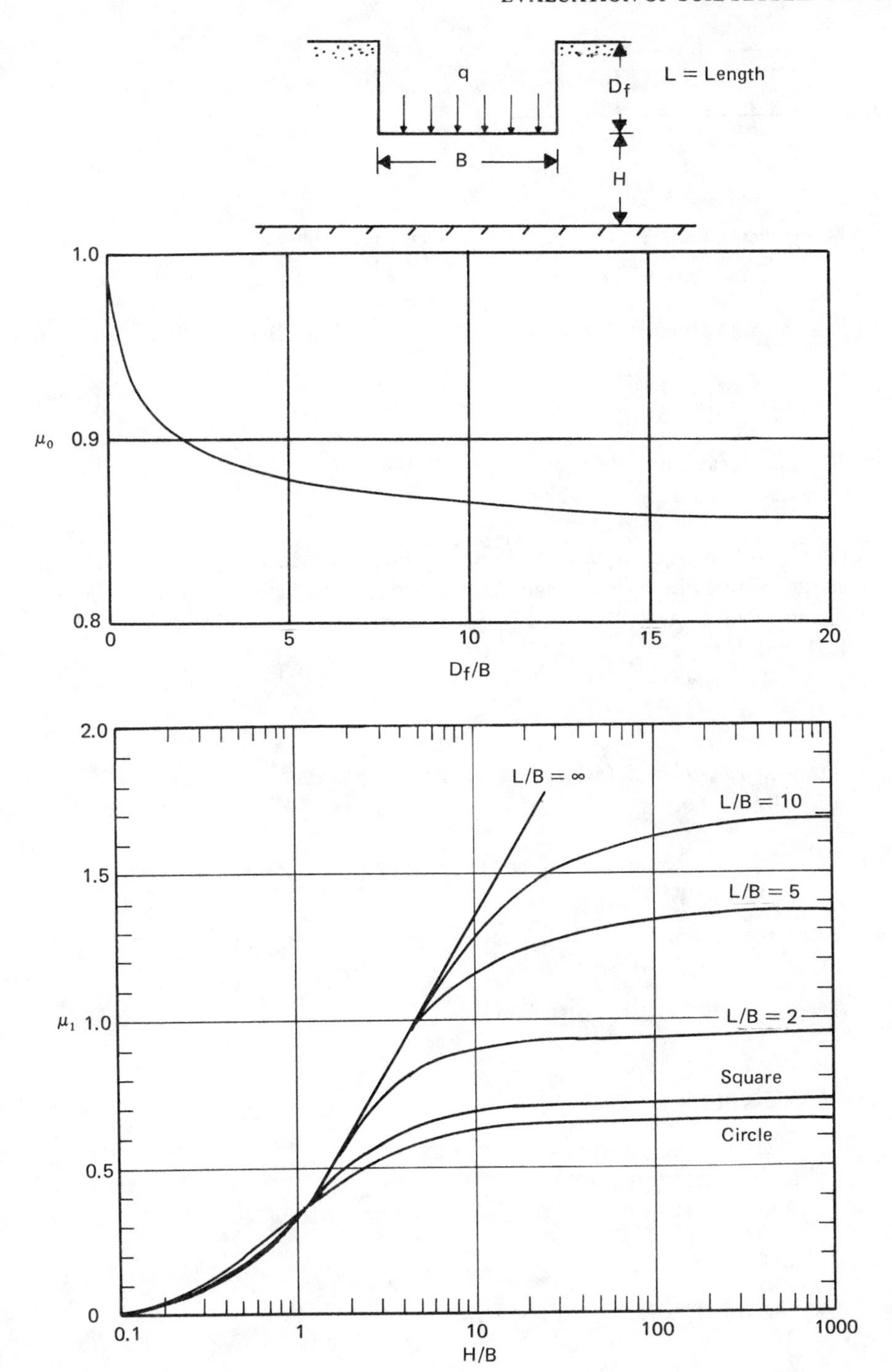

Fig. 6.7 Improved chart for use in Eq. (6.22). *(After J. T. Christian and W. D. Carrier, III, Janbu, Bjerrum, and Kjaernsli's Chart Reinterpreted,* Can. Geotech. J., *vol. 15, no. 1, 1978.)*

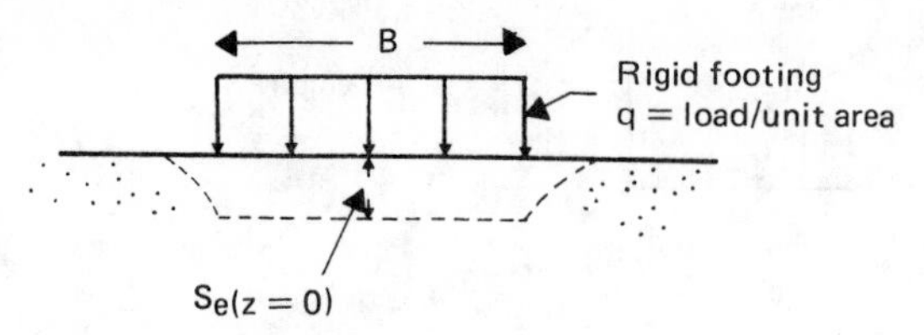

Fig. 6.8 Immediate settlement of rigid footing.

For a uniformly loaded rigid rectangular footing of dimension $L \times B$:

$$S_{e(z=0)} = \left[qB \left(\frac{1 - \nu^2}{E} \right) \right] I_8 \tag{6.25}$$

where $I_8 = 0.93 I_5$ (average settlement).

The values of I_8 are given in Table 6.4.

Example 6.1 A square tank is shown in Fig. 6.9. Assuming flexible loading conditions, find the average immediate (elastic) settlement of the tank for the following conditions:

(*a*) $D_f = 0, H = \infty$.
(*b*) $D_f = 1.5\,\text{m}, H = \infty$.
(*c*) $D_f = 1.5\,\text{m}, H = 10\,\text{m}$.

SOLUTION *Part (a):* $S_e(\text{average}) = (qB/E)(1 - \nu^2) I_5$; $B = 3\,\text{m}$; $L/B = 3/3 = 1$; and $I_5 = 0.951$. So,

$$S_e(\text{average}) = \frac{100(3)}{21{,}000}(1 - 0.3^2)\,0.951 = 0.0124\,\text{m} = 12.4\,mm$$

Part (b): From Eq. (6.18), $S'_e(\text{average}) = I_6 S_e(\text{average})$; $D_f/B = 1.5/3 = 0.5$; and $I_6 = 0.77$ (Fig. 6.6). So,

$$S'_e(\text{average}) = 0.77\,(12.4\,\text{mm}) = 9.55\,mm$$

Table 6.4 Values of I_8

L/B	I_8
1	0.884
2	1.208
3	1.406
5	1.660
10	2.006
20	2.353
50	2.814
100	3.162

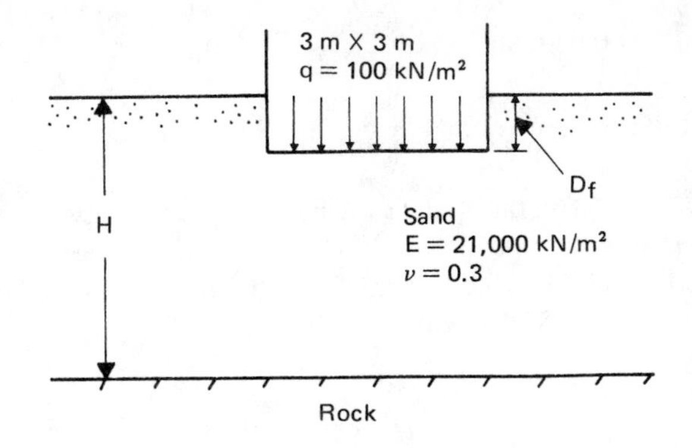

Fig. 6.9

Part (c): From Eq. (6.20),

$$S_e = S_{e(z=0)} - S_{e(z=H)}$$

From Eq. (6.11),

$$S_e(\text{corner}) = \frac{qB}{2E}(1-\nu^2)\left[I_3 - \left(\frac{1-2\nu}{1-\nu}\right) I_4\right]$$

We can determine the value of S_e below the corner at $z = 10$ m for one loaded area of dimension 1.5 X 1.5 m (Fig. 6.10) and then multiply that by 4 to obtain the displacement at the center of the tank at depth $z = 10$ m. So for a loaded area of 1.5 X 1.5 m,

$$m = \frac{1.5}{1.5} = 1 \qquad n = \frac{10}{1.5} = 6.67$$

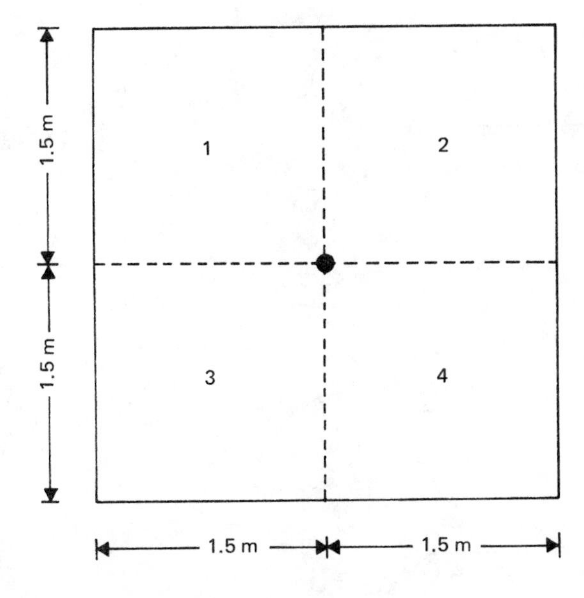

Fig. 6.10

From Table 6.2, $I_3 = 0.189$ and $I_4 = 0.047$. So,

$$S_e = \frac{100(1.5)}{2(21{,}000)}(1-0.3^2)\left[0.189 - \left(\frac{1-0.6}{1-0.3}\right)0.047\right] = 0.00053\text{ m}$$

For the whole loaded area of 3 X 3 m, the elastic settlement below the center at a depth of $z = 10$ m is equal to 4 X 0.00053 = 0.00212 m. Thus, S_e(average) without considering the depth effect is 12.4 − (0.848)(0.00212)(1000) = 10.6 mm. Now, S_e'(average) = $I_6 S_e$(average); also, $I_6 = 0.77$ from part (*b*). So,

$S_e' = (0.77)(10.6) = 8.16$ *mm*

Example 6.2 For the tank shown in Fig. 6.11,
(*a*) Determine the immediate settlement at the center of the tank by using Eq. (6.6).
(*b*) Determine the immediate settlement by using Eq. (6.23). Divide the underlying soil into three layers of equal thickness of 3 m.

SOLUTION *Part (a):* From Eq. (6.6),

$$S_e = \frac{q(1+\nu)}{E}b\left[\frac{z}{b}I_1 + (1-\nu)I_2\right]$$

From Eq. (6.20),

$$S_e = S_{e(z=0)} - S_{e(z=9\text{ m})}$$

For $z/b = 0$ and $s/b = 0$, $I_1 = 1$, and $I_2 = 2$ (Table 6.1); so,

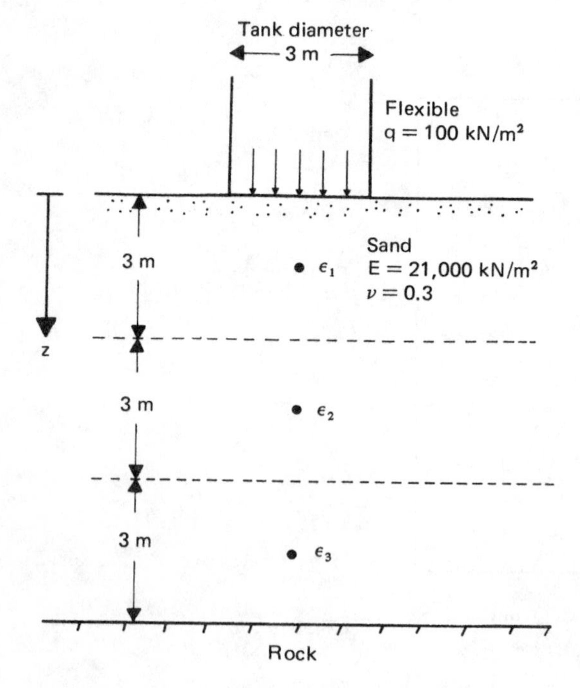

Fig. 6.11

$$S_{e(z=0)} = \frac{100(1+0.3)}{21{,}000}(1.5)\,[(1-0.3)2] = 0.013\text{ m} = 13\text{ mm}$$

For $z/b = 9/1.5 = 6$ and $s/b = 0$, $I_1 = 0.01361$ and $I_2 = 0.16554$; so,

$$S_{e(z=9\text{ m})} = \frac{100(1+0.3)(1.5)}{21{,}000}[6(0.01361) + (1-0.3)0.16554]$$

$$= 0.00183\text{ m} = 1.83\text{ mm}$$

Hence, $S_e = 13 - 1.83 = \mathit{11.17\,mm}$

Part (b): From Eq. (6.5),

$$\epsilon_z = \frac{q(1+\nu)}{E}[(1-2\nu)A' + B']$$

Layer 1: From Tables 3.7 and 3.8, for $z/b = 1.5/1.5 = 1$ and $s/b = 0$, $A' = 0.29289$ and $B' = 0.35355$:

$$\epsilon_{z(1)} = \frac{100(1+0.3)}{21{,}000}[(1-0.6)(0.29289) + 0.35355] = 0.00291$$

Layer 2: For $z/b = 4.5/1.5 = 3$ and $s/b = 0$, $A' = 0.05132$ and $B' = 0.09487$:

$$\epsilon_{z(2)} = \frac{100(1+0.3)}{21{,}000}[(1-0.6)(0.05132) + 0.09487] = 0.00071$$

Layer 3: For $z/b = 7.5/1.5 = 5$ and $s/b = 0$, $A' = 0.01942$ and $B' = 0.03772$:

$$\epsilon_{z(3)} = \frac{100(1+0.3)}{21{,}000}[(1-0.6)(0.01942) + 0.03772] = 0.00028$$

The final stages in the calculation are tabulated below.

Layer no. i	Layer thickness Δz_i m	Strain at the center of the layer, $\epsilon_{z(i)}$	$\epsilon_{z(i)}\Delta z_i$, m
1	3	0.00291	0.00873
2	3	0.00071	0.00213
3	3	0.00028	0.00084
			Σ *0.0117 m*
			= *11.7 mm*

6.2.2 Determination of Young's Modulus

The equations derived in Sec. 6.2.1 for calculation of immediate settlement require a value of the Young's modulus E for the soil layers involved. It is difficult to obtain

Table 6.5 Recommended values of E and ν

Type of Soil	Properties of soil*	Void ratio e 0.41 to 0.5	0.51 to 0.6	0.61 to 0.70
Sand (coarse)	Φ	43	40	38
$\nu = 0.15$	E (lb/in²)	6,550	5,700	4,700
	E (kN/m²)	45,200	39,300	32,400
Sand (medium	ϕ	40	38	35
coarse)	E (lb/in²)	6,550	5,700	4,700
$\nu = 0.2$	E (kN/m²)	45,200	39,300	32,400
Sand (fine	ϕ	38	36	32
grained)	E (lb/in²)	5,300	4,000	3,400
$\nu = 0.25$	E (kN/m²)	36,600	27,600	23,500
Sandy silt	ϕ	36	34	30
$\nu = 0.3$ to 0.35	E (lb/in²)	2,000	1,700	1,450
	E (kN/m²)	13,800	11,700	10,000

From *Foundations of Theoretical Soil Mechanics* by M. E. Harr. Copyright © 1966 McGraw-Hill Book Company, New York. Used with the permission of McGraw-Hill Book Company.

*Conversion factor: 1 lb/in² = 6.9 kN/m² (the values of kN/m² have been rounded off). ϕ is the drained friction angle.

the correct value of E since it increases with the depth of soil, i.e., the effective overburden pressure. Some approximate recommended values of E and Poisson's ratio ν for granular soils are given in Table 6.5.

More representative values of E and ν can be obtained from triaxial compression tests of undisturbed samples collected from a depth equal to the width of the foundation measured from the bottom of the proposed foundation elevation.

However, in cohesionless soils, it is difficult to obtain undisturbed samples. Whenever Young's modulus for a soil is quoted, it is usually the secant modulus from zero up to about half of the maximum deviator stress, i.e., $E = \Delta\sigma/\epsilon$, as shown in Fig. 6.12. Poisson's ratio ν can be calculated by measuring the axial compressive strain and the lateral strain during the triaxial testing. The deviator stress-strain curve can be approximately represented by a hyperbolic equation (Kondner, 1963):

$$\Delta\sigma = \frac{\epsilon}{a + b\epsilon} \tag{6.26}$$

where a and b are constants for a given soil.

For granular soils, Young's modulus determined from triaxial test is approximately proportional to σ_o^n (Fig. 6.12) or

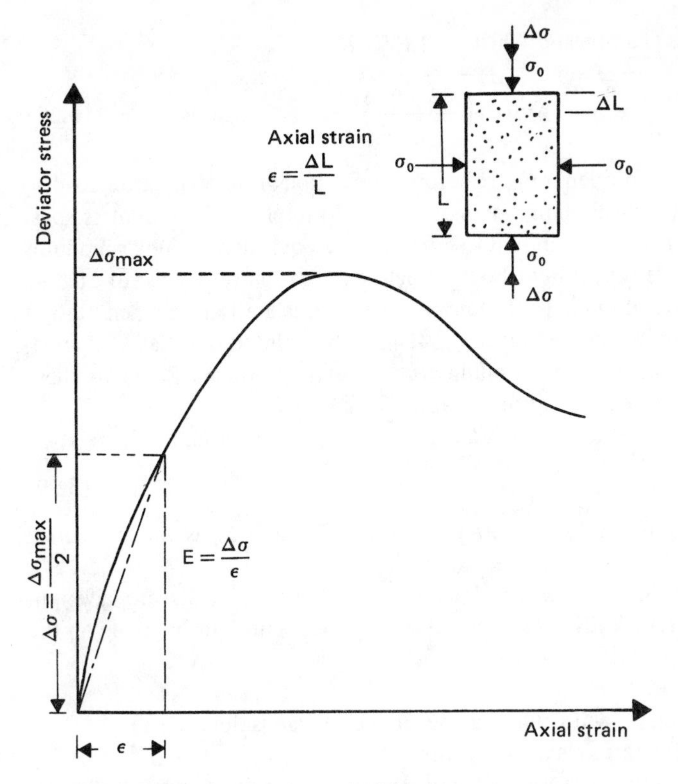

Fig. 6.12 Young's modulus from triaxial test.

$$E \propto \sigma_o^n \tag{6.27}$$

where σ_o is the hydrostatic confining pressure.

A reasonable average value of n is about 0.5 (Lambe and Whitman, 1969). However, in practical cases the stresses in soil before loading are not isotropic, as shown in Fig. 6.13. So Young's modulus is approximately proportional to the square root of

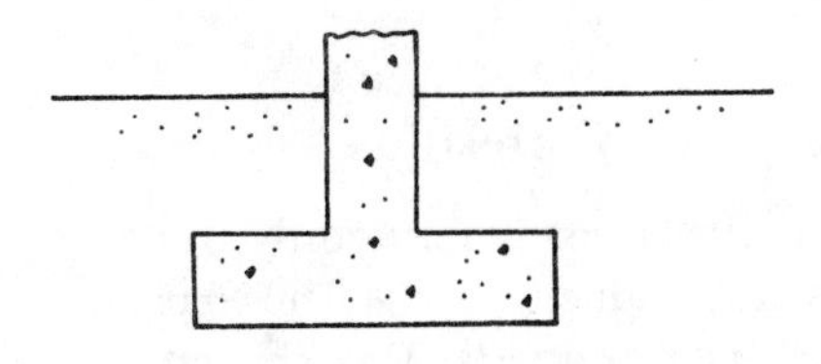

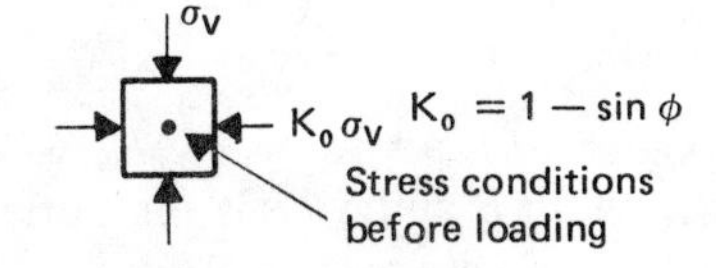

Fig. 6.13 Stress conditions in soil before loading.

the mean principal stress (Lambe and Whitman 1969), i.e.,

$$E \propto \sqrt{\frac{\sigma_v + K_o\sigma_v + K_o\sigma_v}{3}} = \sqrt{\sigma_v\left(\frac{1+2K_o}{3}\right)} \tag{6.28}$$

where σ_v is the effective overburden pressure before application of the foundation load.

Due to the difficulty in obtaining undisturbed soil samples in cohesionless soils, a number of investigators have attempted to correlate the equivalent Young's modulus E_s with the conventional results obtained during field exploration programs for calculation of static compression of sand. These conventional results are standard penetration number N and static dutch cone resistance q_c. It must be pointed out that E_s is somewhat equivalent to the constrained modulus (odeometer modulus). Some of these correlations of E_s with N and q_c are given in Tables 6.6 and 6.7.

In saturated clay soils the undrained Young's modulus can be given by the relation

$$E = \beta S_u \tag{6.29}$$

where β varies from about 500 to 1500 (Bjerrum 1972) and S_u is the undrained cohesion.

Some typical values of β determined from large-scale field tests are given in Table 6.8. Based on present knowledge, the following comments can be made on the values of β and E:

1. The value of β decreases with the increase of the overconsolidation ratio of the clay. This is shown for three clays in Fig. 6.14.
2. The value of β generally decreases with the increase of the plasticity index of the soil.
3. The value of β decreases with the organic content in the soil.
4. For highly plastic clays, consolidated-undrained tests yield E values generally indicative of field behavior.
5. The values of E determined from unconfined compression tests and unconsolidated-undrained triaxial tests are generally low (Liepens, 1957; Crawford and Burn, 1962; D'Appolonia et al., 1971).
6. For most cases, CIU or CK_0U types of tests on undisturbed samples yield values of E that are more representative of field behavior.

6.2.3 Settlement Prediction in Sand by Empirical Correlation

Conducting full-scale field load tests to estimate settlement is very costly; for that reason, on many occasions small plate load tests are conducted in the field to predict the behavior of actual footings. Based on several field load tests, Terzaghi and Peck (1967) suggested that for similar intensities of load q on a footing

$$S_e = \left(\frac{2B}{B+B_1}\right)^2 S_{e(1)} \tag{6.30}$$

where S_e is the settlement of a footing with width B and $S_{e(1)}$ is the settlement of a smaller footing with width B_1. The value of B_1 is usually taken as 1 ft.

Table 6.6 Young's modulus for vertical static compression of sand from standard penetration number

Reference	Relationship*	Soil types	Basis	Remarks
Schultze and Meizer (1965)	$E_s = \nu \sigma_o^{0.522}$ kg/cm² $\nu = 246.2 \log N - 263.4\, \sigma_o + 375.6 \pm 57.6$ $0 < \sigma_o < 1.2$ kg/cm² σ_o = effective overburden pressure	Dry sand	Penetration tests in field and in test shaft. Compressibility based on e, e_{max}, and e_{min} (Schultze and Moussa, (1961)	Correlation coefficient = 0.730 for 77 tests
Webb (1969)	$E_s = 5(N + 15)$ ton/ft² $E_s = 10/3(N + 5)$ ton/ft²	Sand Clayey sand	Screw plate tests	Below water table
Farrent (1963)	$E_s = 7.5(1 - \nu^2)N$ ton/ft² ν = Poisson's ratio	Sand	Terzaghi and Peck loading settlement curves	
Begemann (1974)	$E_s = 40 + C(N - 6)$ kg/cm² $N > 15$ $E_s = C(N + 6)$ kg/cm² $N < 15$ $C = 3$ (silt with sand) to 12 (gravel with sand)	Silt with sand to gravel with sand		Used in Greece
Trofimenkov (1974)	$E_s = (350$ to $500) \log N$ kg/cm²	Sand		U.S.S.R. practice

After J. K. Mitchell and W. S. Gardner, In Situ Measurement of Volume Characteristic, *Proc. Specialty Conference of the Geotechnical Engineering Division, ASCE*, vol. 2, 1975.

*N = standard penetration number. Note: 1 kgf/cm² = 98.1 kN/m²; 1 ton/ft² = 95.6 kN/m².

Table 6.7 Equivalent Young's modulus for vertical static compression of sand–static cone resistance

Reference	Relationship	Soil types	Remarks
Buisman (1940)	$E_s = 1.5q_c$	Sands	Overpredicts settlements by a factor of about 2
Trofimenkov (1964)	$E_s = 2.5q_c$	Sand	Lower limit
	$E_s = 100 + 5q_c$		Average
De Beer (1967)	$E_s = 1.5q_c$	Sand	Overpredicts settlements by a factor of 2
Schultze and Melzer (1965)	$E_s = \frac{1}{m_v} \nu \sigma_o{}^{0.522}$	Dry sand	Based on field and lab penetration tests—compressibility based on e, e_{max}, and e_{min}
	$\nu = 301.1 \log q_c - 382.3\sigma_o \pm 60.3 \pm 50.3$ σ_o = effective overburden pressure		Correlation coefficient = 0.778 for 90 tests valid for $\sigma_o = 0$ to $0.8\ kg/cm^2$
Bachelier and Parez (1965)	$E_s = \alpha q_c$		
	$\alpha = 0.8$ to 0.9	Pure sand	
	$\alpha = 1.3$ to 1.9	Silty sand	
	$\alpha = 3.8$ to 5.7	Clayey sand	
	$\alpha = 7.7$	Soft clay	
Thomas (1968)	$E_s = \alpha q_c$ $\alpha = 3$ to 12	3 sands	Based on penetration and compression tests in large chambers Lower values of α at higher values of q_c; attributed to grain crushing
Webb (1969)	$E_s = \frac{5}{2}(q_c + 30)\ ton/ft^2$	Sand below water table	Based on screw plate tests: correlated well with settlement of oil tanks
	$E_s = \frac{5}{3}(q_c + 15)\ lb/ft^2$	Clayey sand below water table	

Vesic (1970)	$E_s = 2(1 + D_R^2)q_c$ D_r = relative density	Sand	Based on pile load tests and assumptions concerning state of stress
Schmertmann (1970)	$E_s = 2q_c$	Sand	Based on screw plate tests
Bogdanović (1973)	$E_s = \alpha q_c$		Based on analysis of soil settlements over a period of 10 years
	$q_c > 40$ kg/cm² $\alpha = 1.5$	Sand, sandy gravels	
	$20 < q_c < 40$ $\alpha = 1.5$ to 1.8	Silty saturated sands	
	$10 < q_c < 20$ $\alpha = 1.8$ to 2.5 $5 < q_c < 10$ $\alpha = 2.5$ to 3.0	Clayey silts with silty sand and silty saturated sands with silt	
Schmertmann (1974)	$E_s = 2.5q_c$	NC sands	$L/B = 1$ to 2, axisymmetric
	$E_s = 3.5q_c$	NC sands	$L/B \geqslant 10$, plane strain
De Beer (1974)	$E_s = 1.6q_c - 8$	Sand	Bulgarian practice
	$E_s = 1.5q_c$, $q_c > 30$ kg/cm² $E_s = 3q_c$, $q_c < 30$ kg/cm²	Sand	Greek practice
	$E_s > 1.5q_c$ or $E_s = 2q_c$	Sand	Italian practice
	$E_s = 1.9q_c$	Sand	South African practice
	$E_s = \frac{5}{2}(q_c + 3200)$ kN/m²	Fine to medium sand	South African practice
	$E_s = \frac{5}{3}(q_c + 1600)$ kN/m²	Clayey sands, $PI < 15\%$	South African practice
	$E_s = \alpha q_c$, $1.5 < \alpha < 2$	Sands	U.K. practice
Trofimenkov (1974)	$E_s = 3q_c$	Sands	U.S.S.R. practice
	$E_s = 7q_c$	Clays	U.S.S.R. practice

After J. K. Mitchell and W. S. Gardner, In Situ Measurement of Volume Characteristics, *Proc. Specialty Conference of the Geotechnical Engineering Division, ASCE,* vol. 2, 1975.

Note: 1 kg/cm² = 98.1 kN/m²; 1 ton/ft² = 95.6 kN/m²; 1 lb/ft² = 47.8 N/m².

Table 6.8 Values of β from various case studies of immediate settlement

		Clay properties					
No.	Location of structure	Plasticity index	Sensitivity	Over-consolidation ratio	E_{field}, ton/m²	β	Source of S_u*
1	Oslo: Nine-story building	15	2	3.5	7,600	1,200	CIU
2	Asrum I: Circular load	16	100	2.5	990	1,000	Field vane
						1,200	CIU
3	Asrum II: Circular load test	14	100	1.7	880	1.000	Field vane
						1,100	CIU
4	Mastemyr: Circular load test	14	–	1.5	1,300	1,200	Field vane
						1,700	Bearing capacity
5	Portsmouth: Highway embankment	15	10	1.3	3,000	2,000	Field vane
						1,700	Bearing capacity
6	Boston: Highway embankment	24	5	1.5	10,000	1,600	Field vane
						1,200	CK_oU
				1.0	13,000	2,500	Field vane
						1,500	CK_oU
7	Drammen: Circular load test	28	10	1.4	3,200	1,400	Field vane
						1,100	CK_oU
8	Kawasaki: Circular load test	38	6 ± 3	1.0	2,200	400	Field vane
							CIU
9	Venezuela: Oil tanks	37	8 ± 2	1.0	5,00	800	CIU
10	Maine: Rectangular load test†	33 ± 2	4	1.5 to 4.5	100 to 200	80 to 160	UU and Bearing capacity

After D. J. D'Appolonia, H. G. Poulos, and C. C. Ladd, Initial Settlement of Structures on Clay, *J. Soil Mech. Found. Div., ASCE,* vol. 97, no. SM10, 1971.

*Average value at a depth equal to the width of foundation. CIU = isotropically consolidated undrained shear test; UU = consolidated undrained shear test; CK_oU = consolidated undrained shear test with sample consolidated in K_o condition.

†Slightly organic plastic clay.

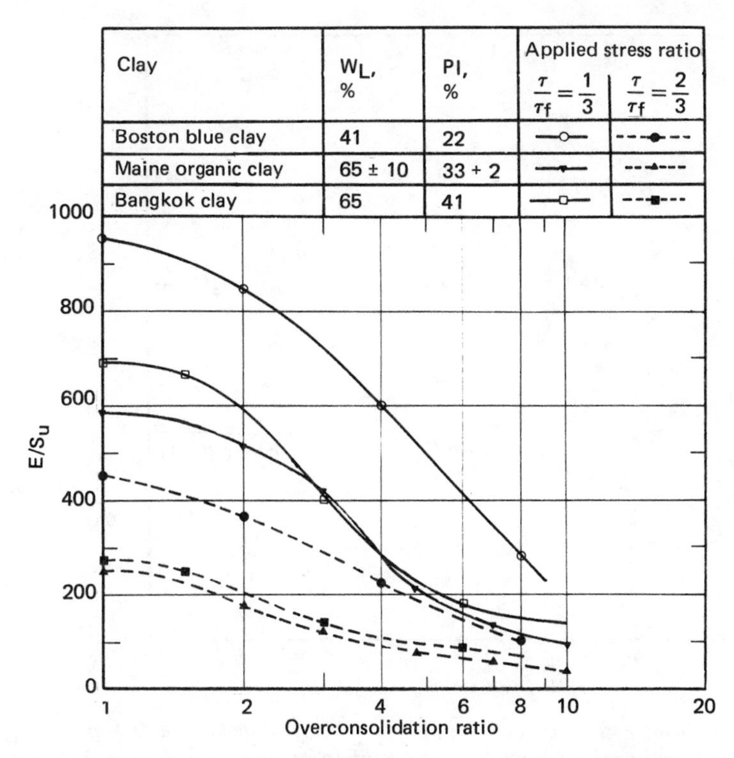

Fig. 6.14 Relationship between E/S_u and over consolidation ratio from CU tests on three clays determined from CK_0U-type direct shear tests. *(After D. J. D'Appolonia, H. G. Poulos, and C. C. Ladd, Initial Settlement of Structures on Clay,* J. Soil Mech. Found. Div., ASCE, *vol. 97, no. SM10, 1971.)*

Eq. (6.30) can be rewritten in the form

$$\frac{S_e}{S_{e(1)}} = \frac{4}{(1 + B_1/B)^2} \tag{6.31}$$

D'Appolonia et al. (1970) compared the above equation with several field experiments conducted by Bjerrum and Eggstad (1963) and Bazaraa (1967). The results of the comparison are shown in Fig. 6.15. It appears that the relationship gives the general trend; however, there appears to be a wide scattering of points.

Using the standard penetration resistance obtained from field explorations, Meyerhof (1965) proposed the following relationships for settlement calculations in sand:

$$S_e = \frac{4q}{N} \qquad \text{for } B \leqslant 4 \text{ ft} \tag{6.32a}$$

and

$$S_e = \frac{6q}{N}\left(\frac{B}{B+1}\right)^2 \qquad \text{for } B > 4 \text{ ft} \tag{6.32b}$$

where q = intensity of applied load, kip/ft^2

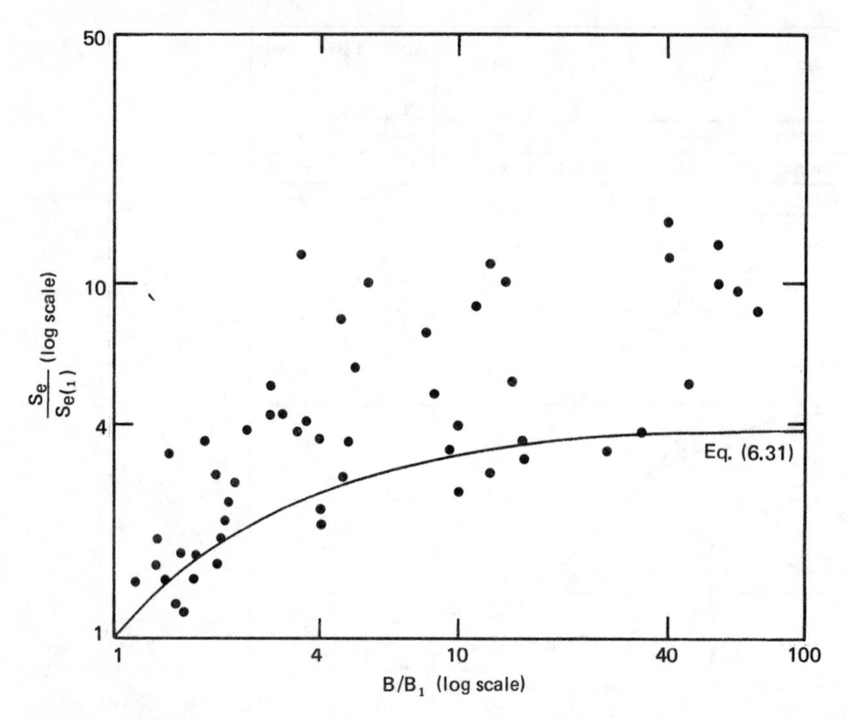

Fig. 6.15 Comparison of field test results with Eq. (6.31). *(After D. J. D'Appolonia, E. D'Appolonia, and R. F. Brisette, Discussion on Settlement of Spread Footings on Sand,* J. Soil Mech. Found. Div., ASCE, *vol. 96, 1970.)*

B = width of footing, ft
S_e = settlement, in
N = standard penetration number

Figure 6.16 shows a comparison of the observed settlements to those obtained through Eq. (6.32). It appears that the predicted settlements are rather conservative. Bowles (1977) suggested that for a more reasonable agreement Eq. (6.32) can be modified as

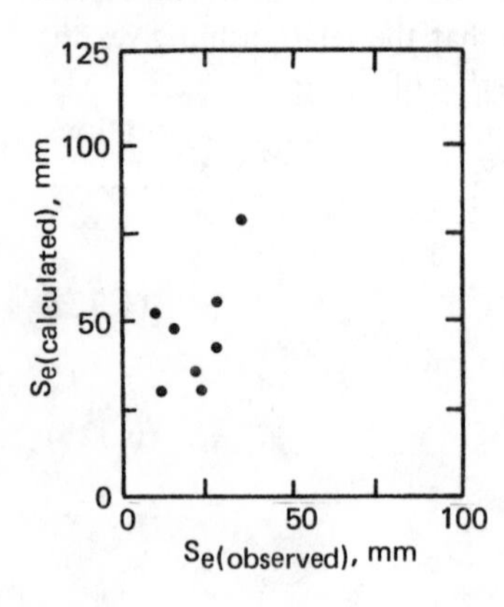

Fig. 6.16 Comparison of observed settlement to that calculated from Eq. (6.32). *(After G. G. Meyerhof, Shallow Foundations,* J. Soil Mech. Found. Div., ASCE, *vol. 91, no. SM2, 1965.)*

$$S_e = \frac{2.5q}{N} \qquad \text{for } B \leqslant 4 \text{ ft} \tag{6.33a}$$

and

$$S_e = \frac{4q}{N}\left(\frac{B}{B+1}\right)^2 \qquad \text{for } B > 4 \text{ ft} \tag{6.33b}$$

In a later work, based on the analysis of the field data of Schultze and Sherif (1973), Meyerhof (1974) gave the following empirical correlations for settlement of shallow foundations:

$$S_e = \frac{q\sqrt{B}}{2N} \qquad \text{(for sand and gravel)} \tag{6.34a}$$

$$S_e = \frac{q\sqrt{B}}{N} \qquad \text{(for silty sand)} \tag{6.34b}$$

where S_e = settlement, in
q = intensity of applied loading, ton/ft²
B = width of foundation, in

6.2.4 Calculation of Immediate Settlement in Granular Soil Using Simplified Strain Influence Factor

The equation for vertical strain ϵ_z under the center of a flexible circular load was given in Eq. (6.5) as

$$\epsilon_z = \frac{q(1+\nu)}{E}\,[(1-2\nu)A' + B']$$

or

$$I_z = \frac{\epsilon_z E}{q} = (1+\nu)\,[(1-2\nu)A' + B'] \tag{6.35}$$

where I_z is the strain influence factor.

Figure 6.17 shows the variation of I_z with depth based on Eq. (6.35) for ν equal to 0.4 and 0.5. Also shown in the same figure are the experimental results of Eggstad (1963). Based on these results, Schmertmann (1970) proposed a simplified distribution of I_z with depth. This is referred to as the $2B$-$0.6I_z$ distribution and is also shown in Fig. 6.17. According to this simplified strain-influence factor method, the immediate settlement of a foundation can be calculated as

$$S_e = C_1 C_2 q \sum_0^{2B} \frac{I_z}{E_s}\,\Delta z \tag{6.36}$$

where C_1 is the correction factor for the depth of embedment of foundation, and C_2 is a correction factor to account for the creep in soil. The factors C_1 and C_2 are given by the following equations:

$$C_1 = 1 - 0.5\left(\frac{q_o}{q}\right) \tag{6.37}$$

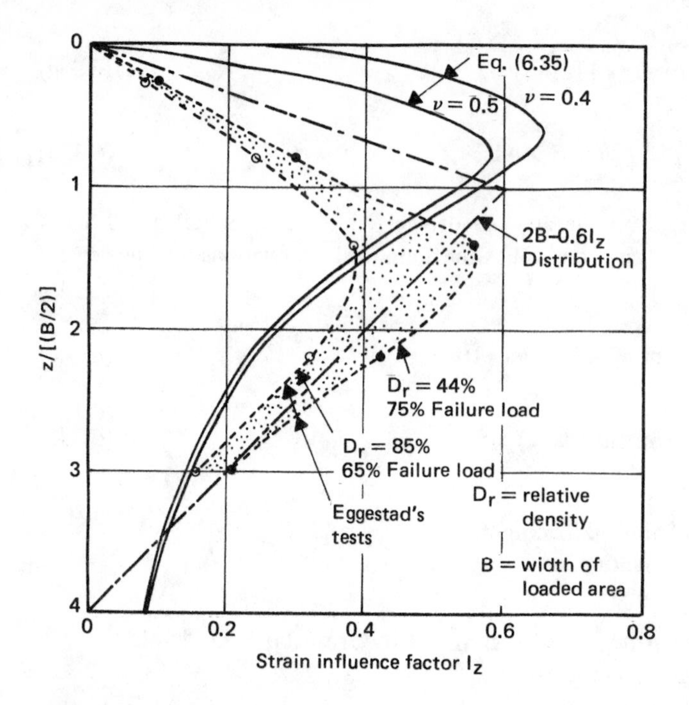

Fig. 6.17 Theoretical and experimental distribution of vertical strain influence factor below the center of a circular loaded area. *(Redrawn after J. H. Schmertmann, Static Cone to Compute Static Settlement over Sand,* J. Soil Mech. Found. Div., ASCE, *vol. 96, no. SM3, 1970.)*

where q_o = effective overburden pressure at foundation level

q = net foundation pressure increase $= q_1 - q_o$

q_1 = average pressure of foundation against soil

$$C_2 = 1 + 0.2 \log \left(\frac{t}{0.1}\right) \tag{6.38}$$

where t is time, in years.

Below is an example for using Eq. (6.36) which was given in Schmertmann's 1970 paper.

Referring to Fig. 6.18, a rectangular foundation ($L = 23$ m and $B = 2.6$ m) is placed at a depth of 2 m below the ground surface of a granular soil deposit. On the basis of the field exploration, a number of layers of soil having constant cone penetration has been assumed. Note that $q_1 = 178.54\ \text{kN/m}^2$ and q_o (overburden pressure at a depth of 2 m) $= 31.39\ \text{kN/m}^2$. The $2B$–$0.6I_z$ distribution is also shown in Fig. 6.18. Based on this I_z diagram, we can divide the soil below the foundation into six layers and obtain the value of $\Sigma(I_z/E_s)\Delta z$ as shown in Table 6.9. Now,

$$q = q_1 - q_o = 178.54 - 31.39 = 147.15\ \text{kN/m}^2$$

and $$C_1 = 1 - 0.5\frac{q_o}{q} = 1 - 0.5\frac{31.39}{147.15} = 0.893$$

$$C_2 = 1 + 0.2\log\frac{t\text{ yr}}{0.1} = 1 + 0.2\log\frac{5}{0.1} = 1.34 \qquad \text{(assumed time } t = 5\text{ yr)}$$

Hence the immediate settlement is

$$S = C_1C_2q\sum\frac{I_z}{E_s}\Delta z$$

$$= (0.893)(1.34)(147.5)(23.1 \times 10^{-5})$$

$$= 0.04077\text{ m} = 40.77\text{ mm}$$

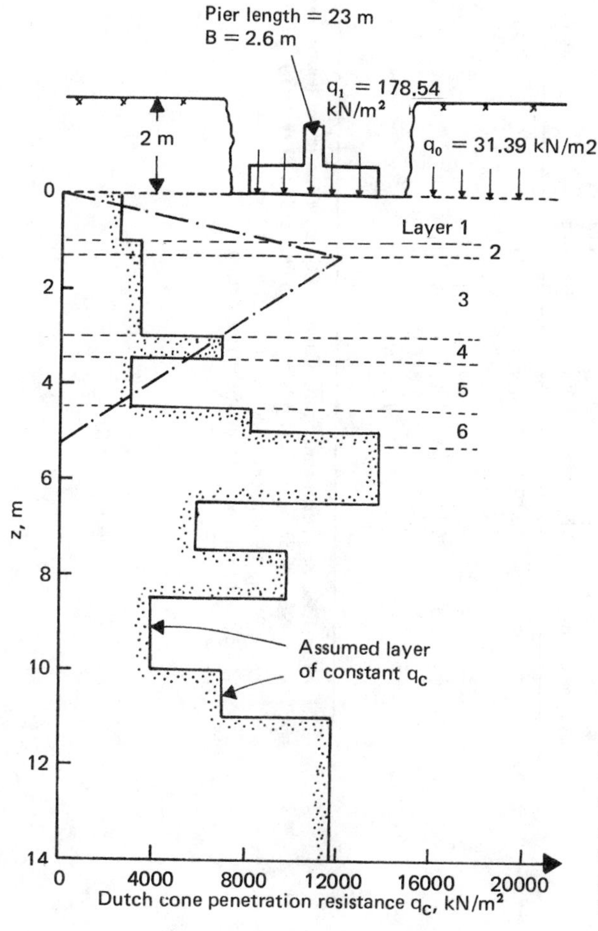

Fig. 6.18 Calculation of elastic settlement based on $2B - 0.6I_z$ distribution. *(Redrawn after J. H. Schmertmann, Static Cone to Compute Static Settlement over Sand,* J. Soil Mech. Found. Div., *vol. 96, no. SM3, 1970.)*

Table 6.9 Calculation of immediate settlement based on $2B$–$0.6I_z$ distribution

(1) Layer	(2) Δz, m	(3) q_c, kN/m²	(4) E_s,* kN/m²	(5) z† to the center of layer, m	(6) I_z at the center of layer	(7) $\frac{I_z}{E_s}\Delta z$, m³/kN
1	1.0	2,452.5	4,904.0	0.5	0.23	4.69×10^{-5}
2	0.3	3,433.5	6,867.0	1.15	0.53	2.32×10^{-5}
3	1.7	3,433.5	6,867.0	2.15	0.47	11.63×10^{-5}
4	0.5	6,867.0	13,734.0	3.25	0.30	1.09×10^{-5}
5	1.0	2,943.0	5,886.0	4.00	0.185	3.14×10^{-5}
6	0.7	8,338.5	16,676.0	4.85	0.055	0.23×10^{-5}
						$\Sigma = 23.1 \times 10^{-5}$

After J. H. Schmertmann, Static Cone to Compute Static Settlement over Sand, *J. Soil Mech. Found. Div., ASCE,* vol. 96, no. SM3, 1970. (Units have been changed to SI system.)

*$E_s = 2q_c$ (q_c = assumed cone penetration resistance).

†Measured from the bottom of the foundation.

6.3 PRIMARY CONSOLIDATION SETTLEMENT

6.3.1 One-Dimensional Consolidation Settlement Calculation

According to Eq. (5.76) in Sec. 5.1.6, the settlement for one-dimensional consolidation can be given by,

$$S_c = \frac{\Delta e}{1 + e_o} H_t \tag{5.76}$$

where $\Delta e = C_c \log \dfrac{\sigma_o' + \Delta\sigma}{\sigma_o'}$ (for normally consolidated clays) (5.77)

$$\Delta e = C_r \log \frac{\sigma_o' + \Delta\sigma}{\sigma_o'}$$

(for overconsolidated clays, $\sigma_o' + \Delta\sigma \leqslant \sigma_c'$) (5.78)

$$\Delta e = C_r \log \frac{\sigma_c'}{\sigma_o'} + C_c \log \frac{\sigma_o' + \Delta\sigma}{\sigma_c'} \quad \text{(for } \sigma_o' < \sigma_c' < \sigma_o' + \Delta\sigma\text{)} \tag{5.79}$$

where σ_c' is the preconsolidation pressure.

When a load is applied over a limited area, the increase of pressure due to the applied load will decrease with depth, as shown in Fig. 6.19. So, for a more realistic settlement prediction, we can use the following methods.

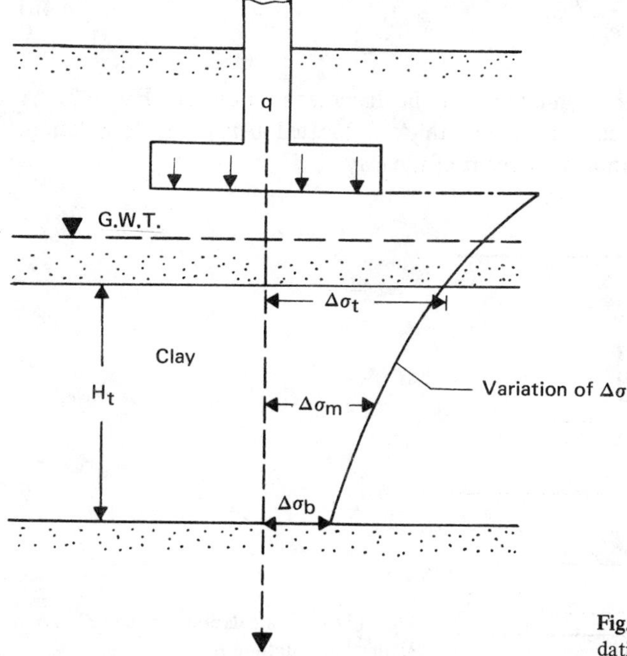

Fig. 6.19 Calculation of consolidation settlement – method A.

Method A

1. Calculate the average effective pressure σ_o' on the clay layer before the application of the load under consideration.
2. Calculate the increase of stress due to the applied load at the top, middle, and the bottom of the clay layer. This can be done by using theories developed in Chap. 3. The average increase of stress in the clay layer can be estimated by Simpson's rule,

$$\Delta\sigma_{av} = \tfrac{1}{6}(\Delta\sigma_t + 4\Delta\sigma_m + \Delta\sigma_b) \tag{6.39}$$

where $\Delta\sigma_t$, $\Delta\sigma_m$, and $\Delta\sigma_b$ are stress increases at the top, middle, and bottom of the clay layer, respectively.
3. Using the σ_o' and $\Delta\sigma_{av}$ calculated above, obtain Δe from Eqs. (5.77), (5.78), or (5.79), whichever is applicable.
4. Calculate the settlement by using Eq. (5.76).

Method B

1. Better results in settlement calculation may be obtained by dividing a given clay layer into n layers as shown in Fig. 6.20.
2. Calculate the effective stress $\sigma'_{o(i)}$ at the middle of each layer.
3. Calculate the increase of stress at the middle of each layer $\Delta\sigma_i$ due to the applied load.
4. Calculate Δe_i for each layer from Eqs. (5.77), (5.78), or (5.79), whichever is applicable.
5. Total settlement for the entire clay layer can be given by

$$S_c = \sum_{i=1}^{i=n} \Delta S_c = \sum_{i=1}^{n} \frac{\Delta e_i}{1 + e_o} \Delta H_i \tag{6.40}$$

Example 6.3 A circular foundation 2 m in diameter is shown in Fig. 6.21. A normally consolidated clay layer 5 m thick is located below the foundation. Determine the consolidation settlement of the clay.

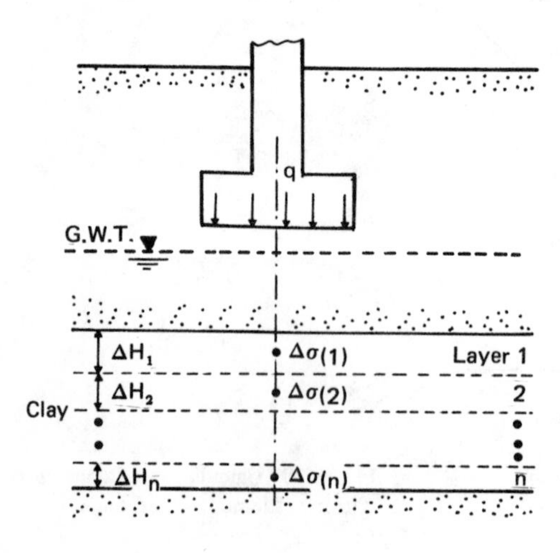

Fig. 6.20 Calculation of consolidation settlement–Method B.

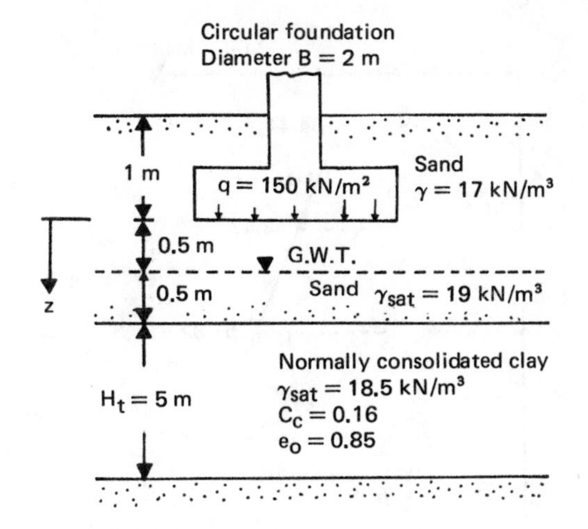

Fig. 6.21

SOLUTION We divide the clay layer into five layers each 1 m thick.

Calculation of $\sigma'_{o(i)}$: The effective stress at the middle of layer 1 is

$$\sigma'_{o(1)} = 17(1.5) + (19 - 9.81)(0.5) + (18.5 - 9.81)(0.5) = 34.44 \text{ kN/m}^2.$$

The effective stress at the middle of the second layer is

$$\sigma'_{o(2)} = 34.44 + (18.5 - 9.81)(1) = 34.44 + 8.69 = 43.13 \text{ kN/m}^2$$

Similarly

$$\sigma'_{o(3)} = 43.13 + 8.69 = 51.81 \text{ kN/m}^2$$

$$\sigma'_{o(4)} = 51.82 + 8.69 = 60.51 \text{ kN/m}^2$$

$$\sigma'_{o(5)} = 60.51 + 8.69 = 69.2 \text{ kN/m}^2$$

Calculation of $\Delta\sigma_i$: For a circular loaded area, the increase of stress below the center is given by Eq. (3.64), and so

$$\Delta\sigma_i = q\left\{1 - \frac{1}{[(b/z)^2 + 1]^{3/2}}\right\}$$

where b is the radius of the circular foundation, 1 m. Hence,

$$\Delta\sigma_1 = 150\left\{1 - \frac{1}{[(1/1.5)^2 + 1]^{3/2}}\right\} = 63.59 \text{ kN/m}^2$$

$$\Delta\sigma_2 = 150\left\{1 - \frac{1}{[(1/2.5)^2 + 1]^{3/2}}\right\} = 29.93 \text{ kN/m}^2$$

$$\Delta\sigma_3 = 150\left\{1 - \frac{1}{[(1/3.5)^2 + 1]^{3/2}}\right\} = 16.66 \text{ kN/m}^2$$

$$\Delta\sigma_4 = 150\left\{1 - \frac{1}{[(1/4.5)^2 + 1]^{3/2}}\right\} = 10.46 \text{ kN/m}^2$$

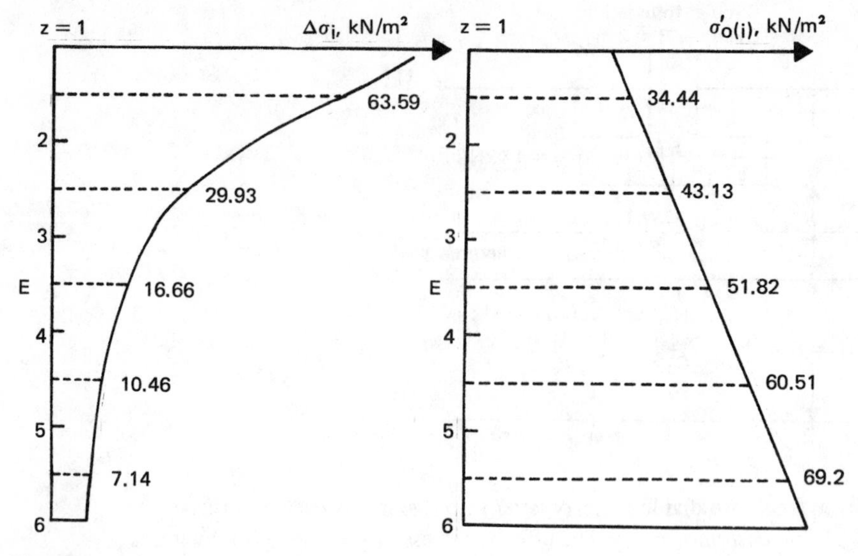

Fig. 6.22

$$\Delta\sigma_5 = 150\left\{1 - \frac{1}{[(1/5.5)^2 + 1]^{3/2}}\right\} = 7.14\,\text{kN/m}^2$$

Calculation of consolidation settlement S_c: The steps in the calculation are given in the following table (see also Fig. 6.22):

Layer no.	ΔH_i, m	$\sigma'_{o(i)}$, kN/m²	$\Delta\sigma_i$, kN/m²	Δe*	$\frac{\Delta e}{1+e_o}\Delta H_i$, m
1	1	34.44	63.59	0.0727	0.0393
2	1	43.13	29.93	0.0366	0.0198
3	1	51.82	16.66	0.0194	0.0105
4	1	60.51	10.46	0.0111	0.0060
5	1	69.2	7.14	0.00682	0.0037
					$\Sigma = 0.0793$

$*\ \Delta e = C_c \log \dfrac{\sigma'_{o(i)} + \Delta\sigma_i}{\sigma'_{o(i)}}$; $C_c = 0.16$

So, $S_c = 0.0793$ m = *79.3 mm.*

6.3.2 Skempton–Bjerrum Modification for Calculation of Consolidation Settlement

In one-dimensional consolidation tests, there is no lateral yield of the soil specimen and the ratio of the minor to major principal effective stresses, K_o, remains constant. In that case, the increase of pore water pressure due to an increase of vertical stress is equal in magnitude to the latter; or

$$\Delta u = \Delta\sigma \tag{6.41}$$

where Δu is the increase of pore water pressure and $\Delta\sigma$ is the increase of vertical stress.

However, in reality the final increase of major and minor principal stresses due to a given loading condition at a given point in a clay layer do not maintain a ratio equal to K_o. This causes a lateral yield of soil. The increase of pore water pressure at a point due to a given load is (Fig. 6.23)

$$\Delta u = \Delta\sigma_3 + A(\Delta\sigma_1 - \Delta\sigma_3) \tag{4.10}$$

Skempton and Bjerrum (1957) proposed that the vertical compression of a soil element of thickness dz due to an increase of pore water pressure Δu may be given by

$$dS_c = m_v\, \Delta u\, dz \tag{6.42}$$

where m_v is coefficient of volume compressibility (Sec. 5.1.2), or

$$dS_c = m_v\,[\Delta\sigma_3 + A(\Delta\sigma_1 - \Delta\sigma_3)]\,dz = m_v\,\Delta\sigma_1\left[A + \frac{\Delta\sigma_3}{\Delta\sigma_1}(1 - A)\right]dz$$

The preceding equation can be integrated to obtain the total consolidation settlement:

$$S_c = \int_0^{H_t} m_v\,\Delta\sigma_1\left[A + \frac{\Delta\sigma_3}{\Delta\sigma_1}(1 - A)\right]dz \tag{6.43}$$

For conventional one-dimensional consolidation (K_o condition)

$$S_{c(\text{oed})} = \int_0^{H_t} \frac{\Delta e}{1 + e_o}\,dz = \int_0^{H_t} \frac{\Delta e}{\Delta\sigma_1}\frac{1}{1 + e_o}\,\Delta\sigma_1\,dz = \int_0^{H_t} m_v\,\Delta\sigma_1\,dz \tag{6.44}$$

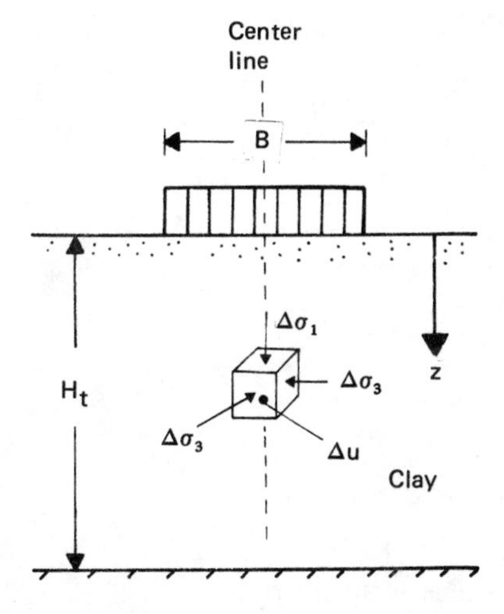

Fig. 6.23 Development of excess pore water pressure below the center line of a circular loaded area.

(Note that Eq. (6.44) is the same as the one we used for settlement calculation in Sec. 6.3.1). Thus

$$\text{Settlement ratio, } \rho_{\text{circle}} = \frac{S_c}{S_{c(\text{oed})}}$$

$$= \frac{\int_0^{H_t} m_v \, \Delta\sigma_1 \, [A + (\Delta\sigma_3/\Delta\sigma_1)(1-A)] \, dz}{\int_0^{H_t} m_v \, \Delta\sigma_1 \, dz}$$

$$= A + (1-A) \frac{\int_0^{H_t} \Delta\sigma_3 \, dz}{\int_0^{H_t} \Delta\sigma_1 \, dz}$$

$$= A + (1-A)M_1 \tag{6.45}$$

where $$M_1 = \frac{\int_0^{H_t} \Delta\sigma_3 \, dz}{\int_0^{H_t} \Delta\sigma_1 \, dz} \tag{6.46}$$

The values of M_1 for the stresses developed below the center of a uniformly loaded

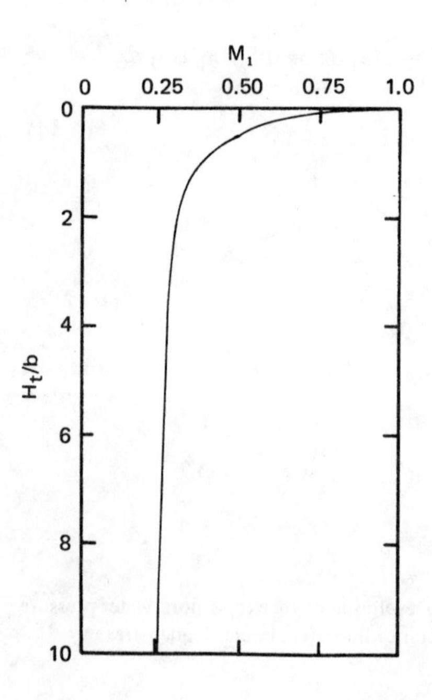

Fig. 6.24 Variation of M_1 with H_t/B.

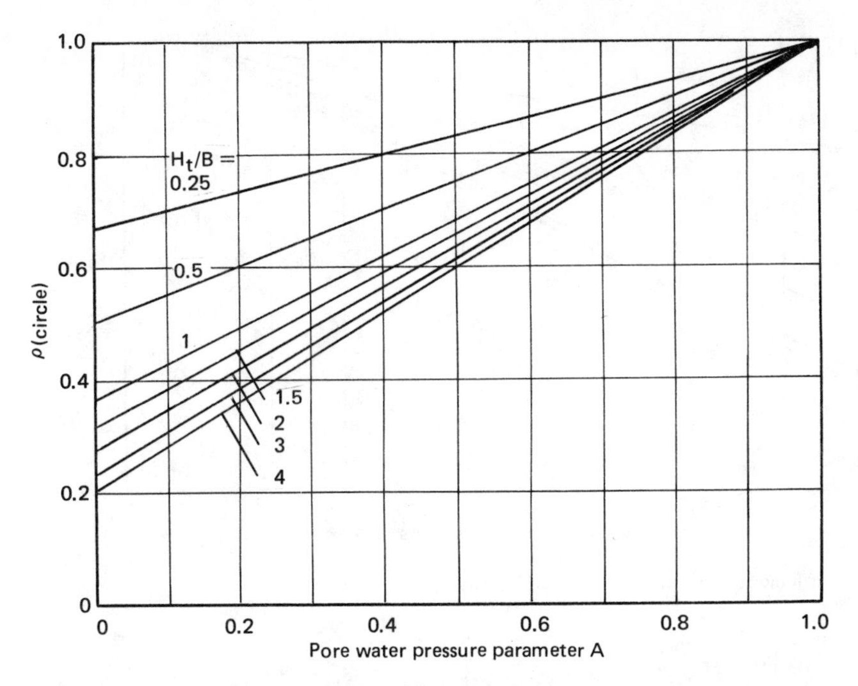

Fig. 6.25 Settlement ratio for circular loading [Eq. (6.45)].

circular area of diameter B are given in Fig. 6.24. The values of settlement ratio, ρ_{circle}, for various values of the pore water pressure parameter A are given in Fig. 6.25.

We can also develop an expression similar to Eq. (6.45) for consolidation under the center of a strip load (Scott, 1963) of width B (Fig. 6.26). From Eq. (4.16),

$$\Delta u = \Delta\sigma_3 + \left[\frac{\sqrt{3}}{2}\left(A - \frac{1}{3}\right) + \frac{1}{2}\right](\Delta\sigma_1 - \Delta\sigma_3) \qquad (\text{for } \nu = 0.5)$$

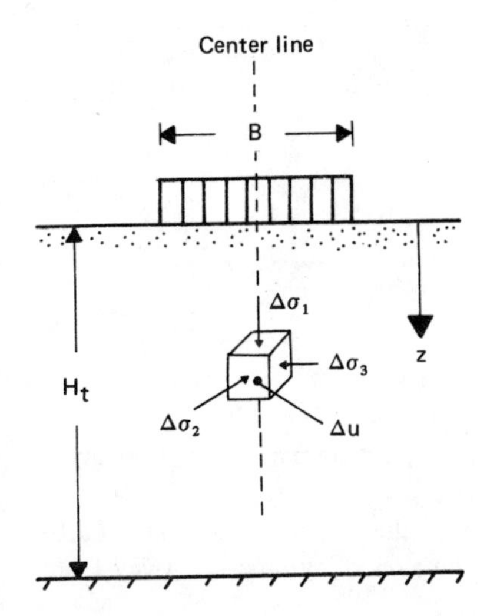

Fig. 6.26 Excess pore water pressure below the center line of a uniform strip load.

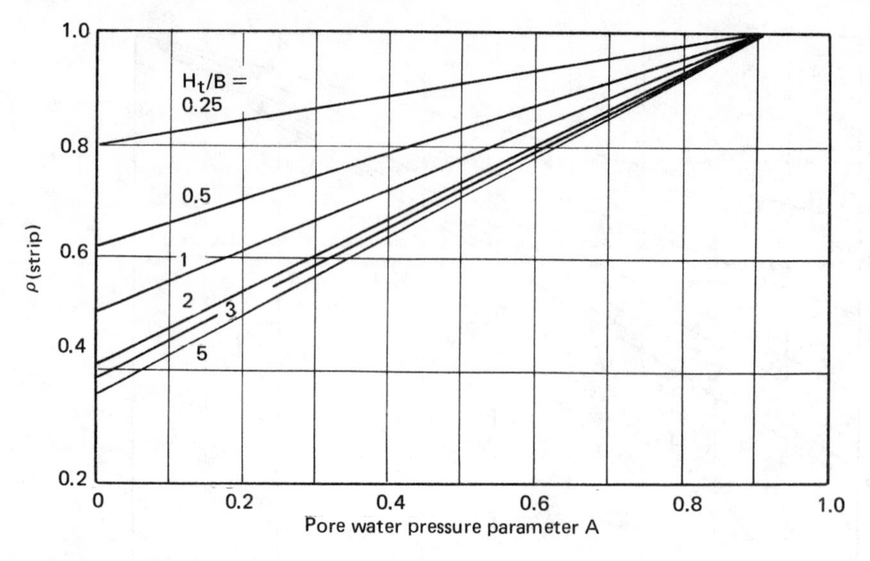

Fig. 6.27 Settlement ratio for strip loading [Eq. (6.48)].

So,
$$S_c = \int_0^{H_t} m_v \, \Delta u \, dz = \int_0^{H_t} m_v \, \Delta\sigma_1 \left[N + (1-N)\frac{\Delta\sigma_3}{\Delta\sigma_1} \right] dz \tag{6.47}$$

where
$$N = \frac{\sqrt{3}}{2}\left(A - \frac{1}{3}\right) + \frac{1}{2}$$

Hence,
$$\text{Settlement ratio, } \rho_{\text{strip}} = \frac{S_c}{S_{c(\text{oed})}}$$

$$= \frac{\int_0^{H_t} m_v \, \Delta\sigma_1 \, [N + (1-N)(\Delta\sigma_3/\Delta\sigma_1)] \, dz}{\int_0^{H_t} m_v \, \Delta\sigma_1 \, dz}$$

$$= N + (1-N)M_2 \tag{6.48}$$

where
$$M_2 = \frac{\int_0^{H_t} \Delta\sigma_3 \, dz}{\int_0^{H_t} \Delta\sigma_1 \, dz} \tag{6.49}$$

The values of ρ_{strip} for different values of the pore pressure parameter A are given in Fig. 6.27.

It must be pointed out that the settlement ratio obtained in Eqs. (6.45) and (6.48) can only be used for settlement calculation along the axes of symmetry. Away from the axes of symmetry, the principal stresses are no longer in vertical and horizontal directions.

Example 6.4 The average representative value of the pore water pressure parameter A (as determined from triaxial tests on undisturbed samples) for the clay layer shown in Fig. 6.28 is about 0.6. Estimate the consolidation settlement of the circular tank.

SOLUTION The average effective overburden pressure for the 20-ft-thick clay layer is $\sigma'_o = (20/2)(122.4 - 62.4) = 600\ \text{lb/ft}^2$. We will use Eq. (6.39) to obtain the average pressure increase:

$$\Delta\sigma_{av} = \tfrac{1}{6}(\Delta\sigma_t + 4\Delta\sigma_m + \Delta\sigma_b)$$

$$\Delta\sigma_t = 2000\ \text{lb/ft}^2$$

From Eq. (3.64),

$$\Delta\sigma_m = 2000\left\{1 - \frac{1}{[(5/10)^2 + 1]^{3/2}}\right\} = 568.9\ \text{lb/ft}^2$$

$$\Delta\sigma_b = 2000\left\{1 - \frac{1}{[(5/20)^2 + 1]^{3/2}}\right\} = 173.85\ \text{lb/ft}^2$$

$$\Delta\sigma_{av} = \tfrac{1}{6}[2000 + 4(568.9) + 173.85] = 741.6\ \text{lb/ft}^2$$

$$\Delta e = C_c \log \frac{\sigma'_o + \Delta\sigma_{av}}{\Delta\sigma_{av}} = 0.2 \log\left(\frac{600 + 741.6}{600}\right) = 0.07$$

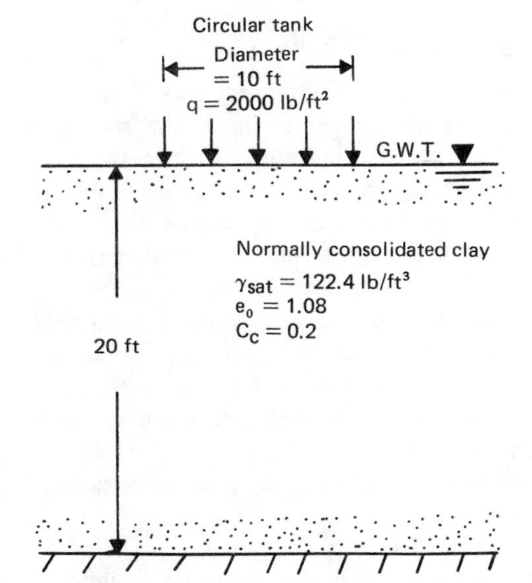

Fig. 6.28

$e_o = 1.08$

$$S_{c(\text{oed})} = \frac{\Delta e\, H_t}{1 + e_o} = \frac{0.07 \times 20}{1 + 1.08} = 0.673 \text{ ft} = 8.08 \text{ in}$$

From Fig. 6.25, the settlement ratio ρ_{circular} is approximately 0.73 (note that $H_t/B = 2$), so

$$S_c = \rho_{\text{circular}} S_{c(\text{oed})} = 0.73(8.08) = 5.9 \textit{ in}$$

6.3.3 Settlement of Overconsolidated Clays

Settlement of structures founded on overconsolidated clay can be calculated by dividing the clay layer into a finite number of layers of smaller thicknesses as outlined in method B in Sec. 6.3.1. Thus,

$$S_{c(\text{oed})} = \sum \frac{C_r\, \Delta H_i}{1 + e_o} \log \frac{\sigma'_{o(i)} + \Delta\sigma_i}{\sigma'_{o(i)}} \tag{6.50}$$

To account for the small departure from one-dimensional consolidation as discussed in Sec. 6.3.2, Leonards (1976) proposed a correction factor, α:

$$S_c = \alpha S_{c(\text{oed})} \tag{6.51}$$

The values of the correction factor α are given in Fig. 6.29*b* and are a function of the average value of σ'_c/σ'_o and B/H_t (B is the width of the foundation and H_t is the thickness of the clay layer, as shown in Fig. 6.29*a*). According to Leonards, if $B > 4H_t$, $\alpha = 1$ may be used. Also, if the depth to the top of the clay stratum exceeds twice the width of the loaded area, $\alpha = 1$ should be used in Eq. (6.51).

6.3.4 Precompression for Improving Foundation Soils

In instances when it appears that too much consolidation settlement is likely to occur due to the construction of foundations, it may be desirable to apply some surcharge loading before foundation construction in order to eliminate or reduce the post-construction settlement. This technique has more recently been used with success in many large construction projects (Johnson, 1970). In this section the fundamental concept of surcharge application for elimination of primary consolidation of compressible clay layers is presented.

Let us consider the case where a given construction will require a permanent uniform loading of intensity σ_f as shown in Fig. 6.30. The total primary consolidation settlement due to loading is estimated to be equal to $S_{c(f)}$. If we want to eliminate the expected settlement due to primary consolidation, we will have to apply a total uniform load of intensity $\sigma = \sigma_f + \sigma_s$. This load will cause a faster rate of settlement of the underlying compressible layer; when a total settlement of $S_{c(f)}$ has been reached, the surcharge can be removed for actual construction.

For a quantitative evaluation of the magnitude of σ_s and the time it should be kept on, we need to recognize the nature of the variation of the degree of consolidation

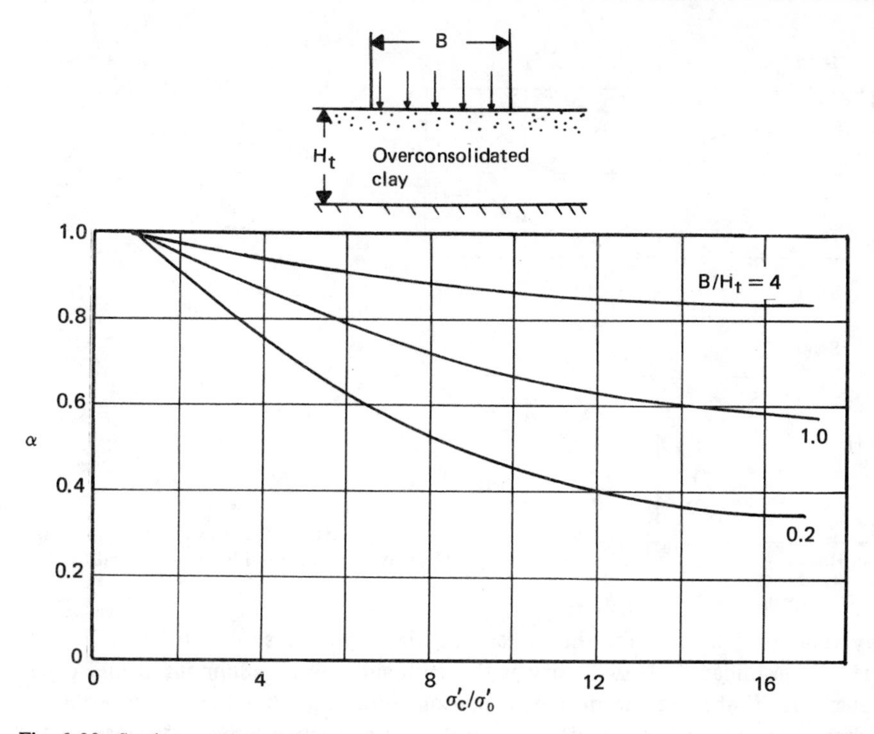

Fig. 6.29 Settlement ratio in overconsolidated clay. *(After G. A. Leonards, Estimating Consolidation Settlement of Shallow Foundations on Overconsolidated Clay,* Transportation Research Board, Special Report 163, *1976.)*

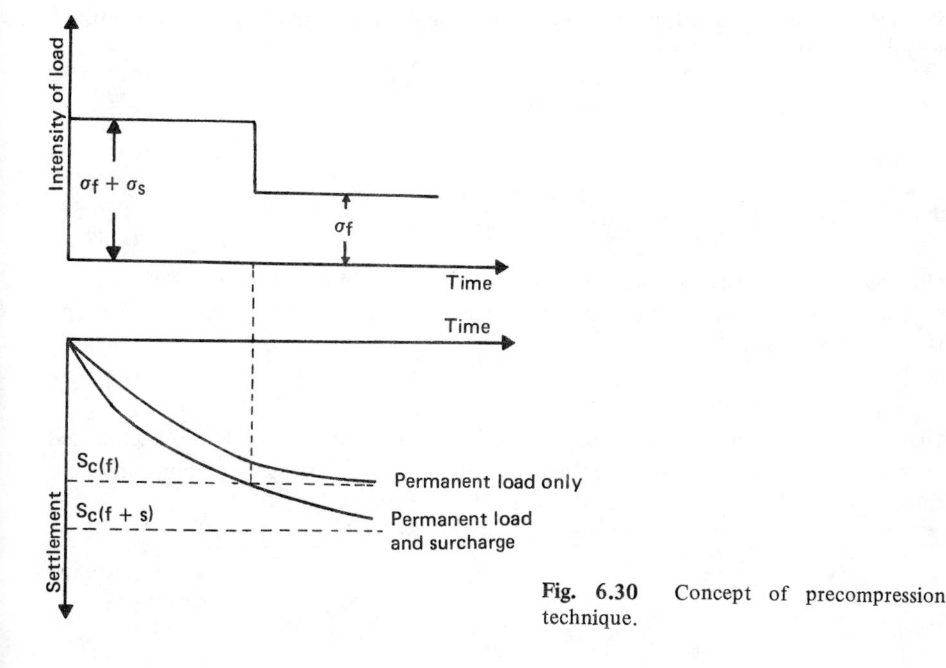

Fig. 6.30 Concept of precompression technique.

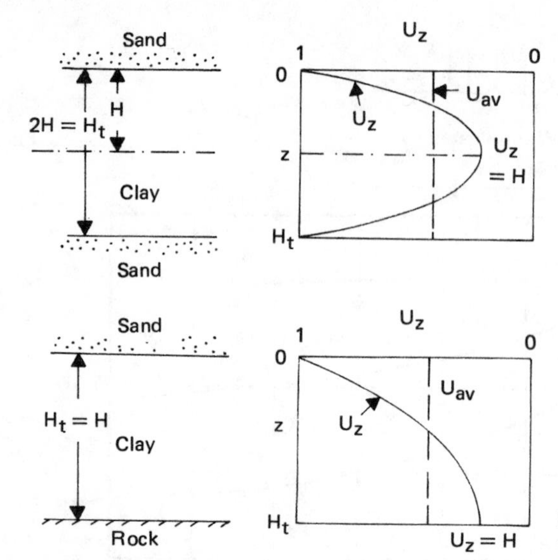

Fig. 6.31 Choice of degree of consolidation for calculation of precompression.

at any time after loading for the underlying clay layer, as shown in Fig. 6.31. The degree of consolidation U_z will vary with depth and will be minimum at mid plane, i.e., at $z = H$. If the average degree of consolidation U_{av} is used as the criterion for surcharge load removal, then after removal of the surcharge the clay close to the mid-plane will continue to settle and the clay close to the pervious layer(s) will tend to swell. This will probably result in a net consolidation settlement. To avoid this problem, we need to take a more conservative approach and use the mid-plane degree of consolidation $U_{z=H}$ as the criterion for our calculation. Using the procedure outlined by Johnson (1970),

$$S_{c(f)} = \left(\frac{H_t}{1+e_o}\right) C_c \log\left(\frac{\sigma'_o + \sigma_f}{\sigma'_o}\right) \tag{6.52}$$

and

$$S_{c(f+s)} = \left(\frac{H_t}{1+e_o}\right) C_c \log\left(\frac{\sigma'_o + \sigma_f + \sigma_s}{\sigma'_s}\right) \tag{6.53}$$

where σ'_o is the initial average in situ effective overburden pressure and $S_{c(f)}$ and $S_{c(f+s)}$ are the primary consolidation settlements due to load intensities of σ_f and $\sigma_f + \sigma_s$, respectively. But,

$$S_{c(f)} = U_{(f+s)} S_{c(f+s)} \tag{6.54}$$

where $U_{(f+s)}$ is the degree of consolidation due to the loading of $\sigma_f + \sigma_s$. As explained before, this is conservatively taken as the mid-plane ($z = H$) degree of consolidation. Thus,

$$U_{(f+s)} = \frac{S_{c(f)}}{S_{c(f+s)}} \tag{6.55}$$

Combining Eqs. (6.52), (6.53), and (6.55),

$$U_{(f+s)} = \frac{\log\,[1 + (\sigma_f/\sigma'_o)]}{\log\,\{1 + (\sigma_f/\sigma'_o)\,[1 + (\sigma_s/\sigma_f]\}} \tag{6.56}$$

The values of $U_{(f+s)}$ for several combinations of σ_f/σ_o and σ_s/σ_f are given in Fig. 6.32. Once $U_{(f+s)}$ is known, we can evaluate the nondimensional time factor T_v from Fig. 5.5. (Note that $U_{(f+s)} = U_z$ at $z = H$ of Fig. 5.5 based on our assumption.) For convenience, a plot of $U_{(f+s)}$ against T_v is given in Fig. 6.33. So the time for surcharge load removal, t, is

$$t = \frac{T_v H^2}{C_v} \tag{6.57}$$

where C_v is the coefficient of consolidation and H is the length of the maximum drainage path.

A similar approach may be adopted to estimate the intensity of the surcharge fill and the time for its removal to eliminate or reduce post-construction settlement due to secondary consolidation.

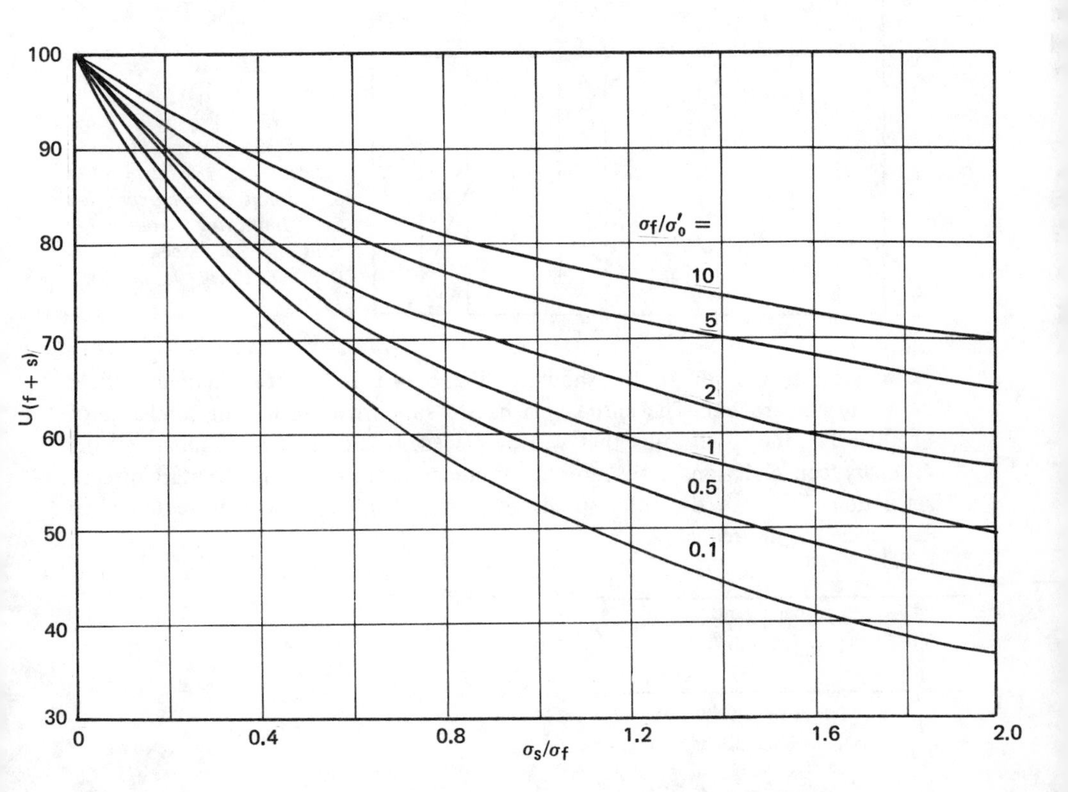

Fig. 6.32 Variation of $U_{(f+s)}$ with σ_s/σ_f and σ_f/σ'_0. *(Redrawn after S. J. Johnson, Precompression for Improving Foundation Soils,* J. Soil Mech. Found. Div., ASCE, *vol. 96, no. SM1, 1970.)*

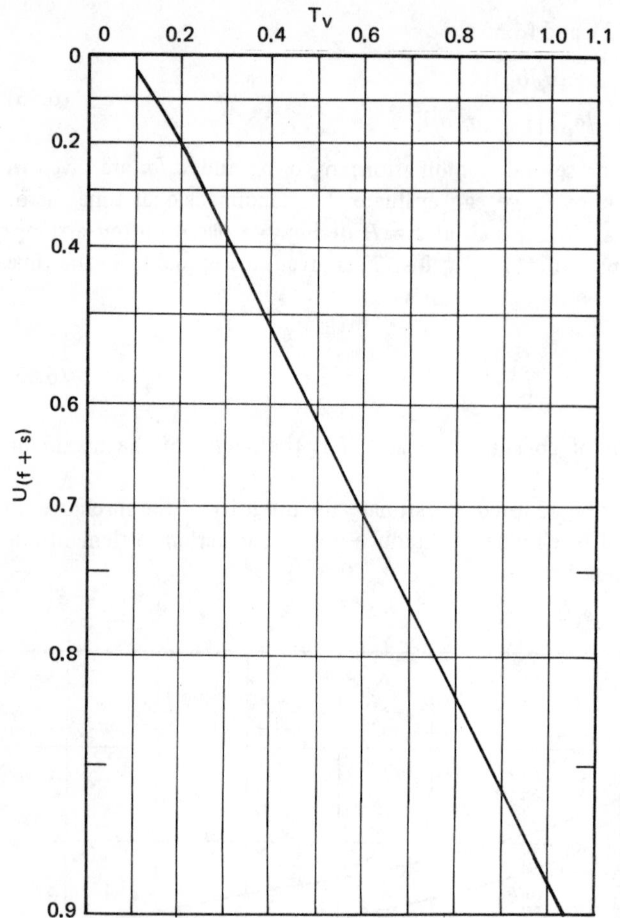

Fig. 6.33 Plot of $U_{(f+s)}$ against T_v. *(Redrawn after S. J. Johnson, Precompression for Improving Foundation Soils,* J. Soil Mech. Found. Div., ASCE, *vol. 96, no. SM1, 1970.)*

Example 6.5 The soil profile shown in Fig. 6.34 is in an area where an airfield is to be constructed. The entire area has to support a permanent surcharge of 1200 lb/ft^2 due to the fills that will be placed. It is desired to eliminate all the primary consolidation in six months by precompression before the start of construction. Estimate the total surcharge ($q = q_s + q_f$) that will be required for achieving the desired goal.

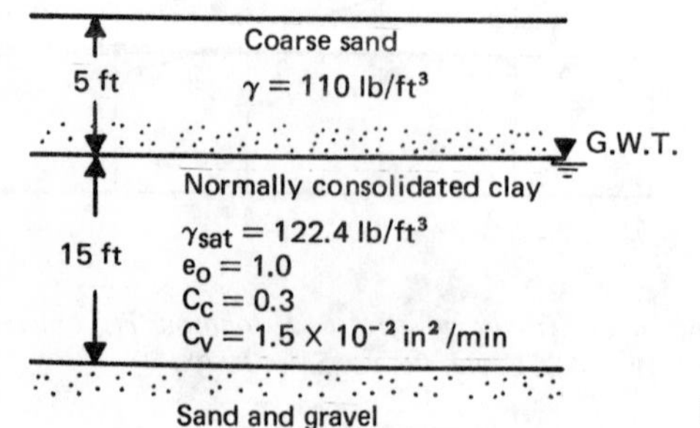

Fig. 6.34

SOLUTION

$$t = \frac{T_v H^2}{C_v} \qquad \text{or} \qquad T_v = \frac{tC_v}{H^2}$$

For two-way drainage,

$H = H_t/2 = 7.5 \text{ ft} = 90 \text{ in}$

We are given that

$t = 6 \times 30 \times 24 \times 60 \text{ min}$

So,

$$T_v = \frac{(6 \times 30 \times 24 \times 60)(1.5 \times 10^{-2})}{90^2} = 0.48$$

From Fig. 6.33, for $T_v = 0.48$ and $U_{(f+s)} \approx 0.62$,

$\sigma'_o = 110(5) + 7.5(122.4 - 62.4) = 1000 \text{ lb/ft}^2$

$\sigma_f = 1200 \text{ lb/ft}^2$ (given)

So

$$\frac{\sigma_f}{\sigma'_o} = \frac{1200}{1000} = 1.2$$

From Fig. 6.32, for $U_{(f+s)} = 0.62$ and $\sigma_f/\sigma'_o = 1.2$,

$\sigma_s/\sigma_f = 1.17$

So,

$\sigma_s = 1.17\sigma_f = 1.17\,(1200) = 1404 \text{ lb/ft}^2$

Thus,

$\sigma = \sigma_f + \sigma_s = 1200 + 1404 = \mathit{2604\ lb/ft^2}$

6.4 SECONDARY CONSOLIDATION SETTLEMENT

The coefficient of secondary consolidation C_α was defined in Sec. 5.1.7 as

$$C_\alpha = \frac{\Delta H_t/H_t}{\Delta \log t}$$

where t is time and H_t is the thickness of the clay layer.

It has been reasonably established that C_α decreases with time in a logarithmic manner and is directly proportional to the total thickness of the clay layer at the beginning of secondary consolidation. Thus, secondary consolidation settlement can be given by

$$S_s = C_\alpha H_{ts} \log \frac{t}{t_p} \tag{6.58}$$

where H_{ts} = thickness of clay layer at beginning of secondary consolidation $= H_t - S_c$
t = time at which secondary compression is required
t_p = time at end of primary consolidation

Actual field measurements of secondary settlements are relatively scarce. However, good agreement of measured and estimated settlements have been reported by some observers, e.g., Horn and Lambe (1964), Crawford and Sutherland (1971), and Su and Prysock (1972).

6.5 STRESS-PATH METHOD OF SETTLEMENT CALCULATION

Lambe (1964) proposed a technique for calculation of settlement in clay which takes into account both the immediate and the primary consolidation settlements. This is called the *stress-path method.*

6.5.1 Definition of Stress Path

In order to understand what a stress path is, consider a normally consolidated clay specimen subjected to a consolidated drained triaxial test (Fig. 6.35*a*). At any time

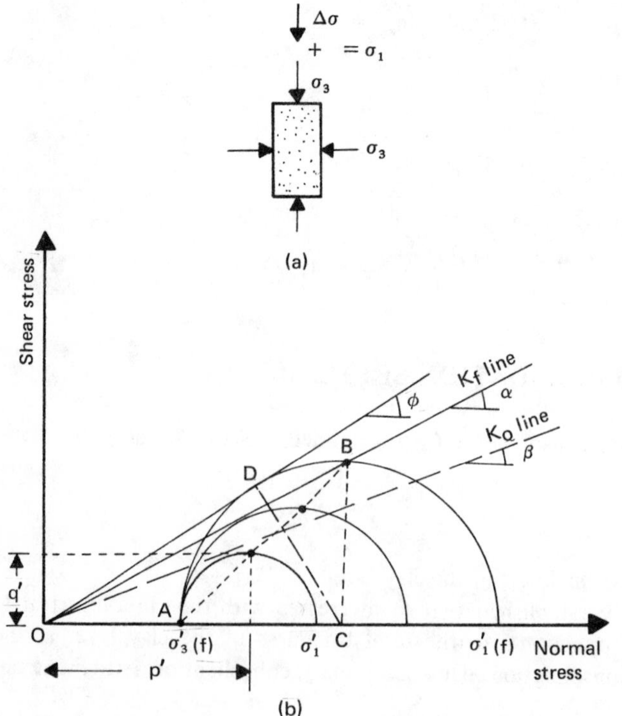

Fig. 6.35 Definition of stress path.

during the test, the stress condition in the specimen can be represented by a Mohr's circle (Fig. 6.35*b*). Note here that, in a drained test, total stress is equal to effective stress. So,

$$\sigma_3 = \sigma_3' \qquad \text{(minor principal stress)}$$

$$\sigma_1 = \sigma_3 + \Delta\sigma = \sigma_1' \qquad \text{(major principal stress)}$$

At failure, the Mohr's circle will touch a line that is the Mohr-Coulomb failure envelope; this makes an angle ϕ with the normal stress axis (ϕ is the soil friction angle).

We now consider another concept; without drawing the Mohr's circles, we may represent each one by a point defined by the coordinates

$$p' = \frac{\sigma_1' + \sigma_3'}{2} \tag{6.59}$$

and

$$q' = \frac{\sigma_1' - \sigma_3'}{2} \tag{6.60}$$

This is shown in Fig. 6.35*b* for the smaller of the Mohr's circles. If the points with p' and q' coordinates of all the Mohr's circles are joined, this will result in the line *AB*. This line is called a *stress path*. The straight line joining the origin and the point *B* will be defined here as the K_f line. The K_f line makes an angle α with the normal stress axis. Now,

$$\tan\alpha = \frac{BC}{OC} = \frac{(\sigma_{1(f)}' - \sigma_{3(f)}')/2}{(\sigma_{1(f)}' + \sigma_{3(f)}')/2} \tag{6.61}$$

where $\sigma_{1(f)}'$ and $\sigma_{3(f)}'$ are the effective major and minor principal stresses at failure. Similarly,

$$\sin\phi = \frac{DC}{OC} = \frac{(\sigma_{1(f)}' - \sigma_{3(f)}')/2}{(\sigma_{1(f)}' + \sigma_{3(f)}')/2} \tag{6.62}$$

From Eqs. (6.61) and (6.62), we obtain

$$\tan\alpha = \sin\phi \tag{6.63}$$

Again let us consider a case where a soil specimen is subjected to an oedometer (one-dimensional consolidation) type of loading (Fig. 6.36). For this case, we can write

$$\sigma_3' = K_o\sigma_1' \tag{6.64}$$

where K_o is the at-rest earth pressure coefficient and can be given by the expression (Jaky, 1944)

$$K_o = 1 - \sin\phi \tag{6.65}$$

For the Mohr's circle shown in Fig. 6.36, the coordinates of point *E* can be given by

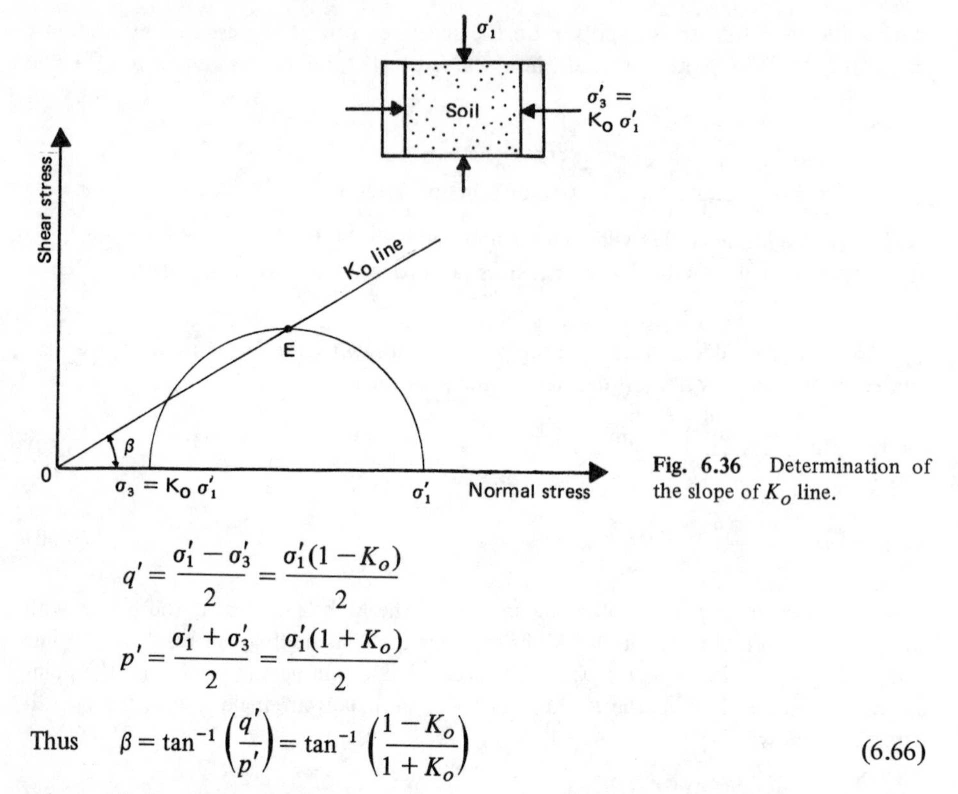

Fig. 6.36 Determination of the slope of K_o line.

$$q' = \frac{\sigma_1' - \sigma_3'}{2} = \frac{\sigma_1'(1 - K_o)}{2}$$

$$p' = \frac{\sigma_1' + \sigma_3'}{2} = \frac{\sigma_1'(1 + K_o)}{2}$$

Thus $$\beta = \tan^{-1}\left(\frac{q'}{p'}\right) = \tan^{-1}\left(\frac{1 - K_o}{1 + K_o}\right) \tag{6.66}$$

where β is the angle that the line OE (K_o line) makes with the normal stress axis. For purposes of comparison, the K_o line is also shown in Fig. 6.35*b*.

In any particular problem, if a stress path is given in a p' vs. q' plot, we should be able to determine the values of the major and minor principal stresses for any given point on the stress path. This is demonstrated in Fig. 6.37, in which ABC is an effective stress path.

6.5.2 Stress and Strain Path for Consolidated Undrained Triaxial Tests

Consider a clay specimen consolidated under an isotropic stress $\sigma_3 = \sigma_3'$ in a triaxial test. When a deviator stress $\Delta\sigma$ is applied on the specimen and drainage is not permitted,

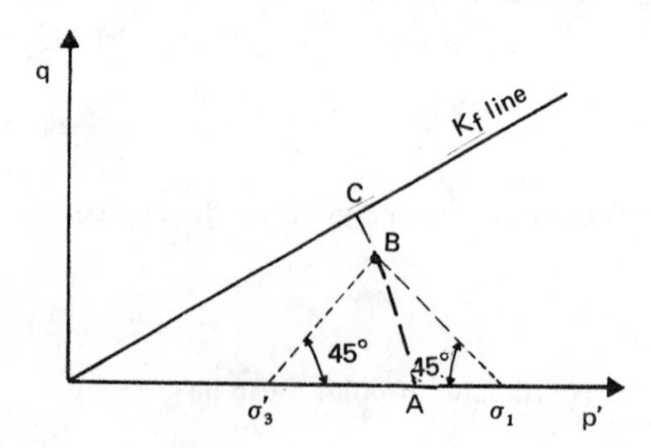

Fig. 6.37 Determination of major and minor principal stresses for a point on a stress path.

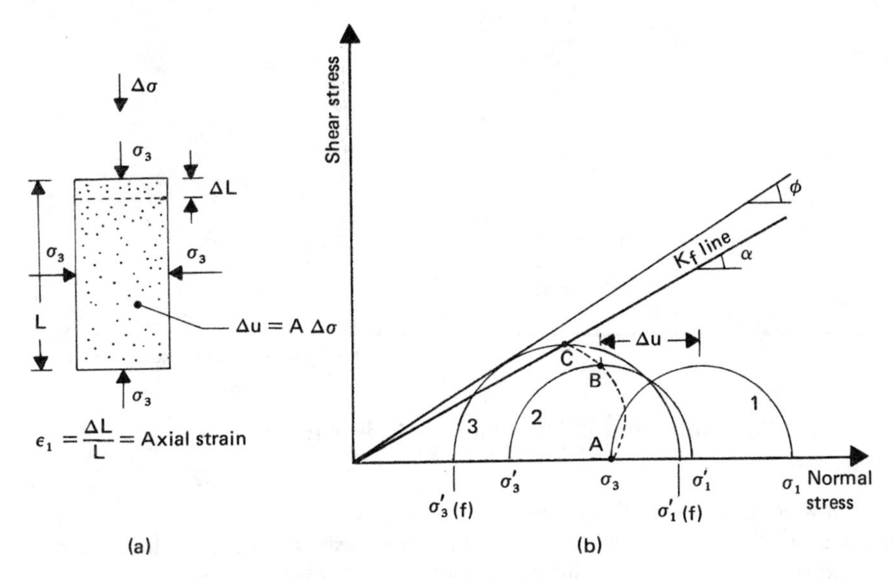

Fig. 6.38 Stress path for consolidated undrained triaxial test.

there will be an increase in the pore water pressure, Δu (Fig. 6.38*a*):

$$\Delta u = A\ \Delta\sigma \tag{6.67}$$

where A is the pore water pressure parameter (Chap. 4).

At this time, the effective major and minor principal stresses can be given by:

$$\text{Minor effective principal stress} = \sigma_3' = \sigma_3 - \Delta u$$

and

$$\text{Major effective principal stress} = \sigma_1' = \sigma_1 - \Delta u = (\sigma_3 + \Delta\sigma) - \Delta u$$

Mohr's circles for the total and effective stress at any time of deviator stress application are shown in Fig. 6.38*b*. (Mohr's circle no. 1 is for total stress and no. 2 is for effective stress.) Point B on the effective-stress Mohr's circle has the coordinates p' and q'. If the deviator stress is increased until failure occurs, the effective-stress Mohr's circle at failure will be represented by circle No. 3 as shown in Fig. 6.38*b*, and the effective stress path will be represented by the line *ABC*.

The general nature of the effective-stress path will depend on the value of the pore pressure parameter A. This is shown in Fig. 6.39.

Two more important aspects of effective-stress paths are as follows:

1. The stress paths for a given normally consolidated soil are geometrically similar. These facts are shown for three clays in p' vs. q' plots in Fig. 6.40.
2. The axial strain in a CU test may be defined as $\epsilon_1 = \Delta L/L$ as shown in Fig. 6.38*a*. For a given soil, if the points representing equal strain in a number of stress paths are joined, they will be approximately straight lines passing through the origin. This is also shown in Fig. 6.40 for all three clays.

These two basic observations for effective-stress paths will be helpful in the calculation of settlement in clays.

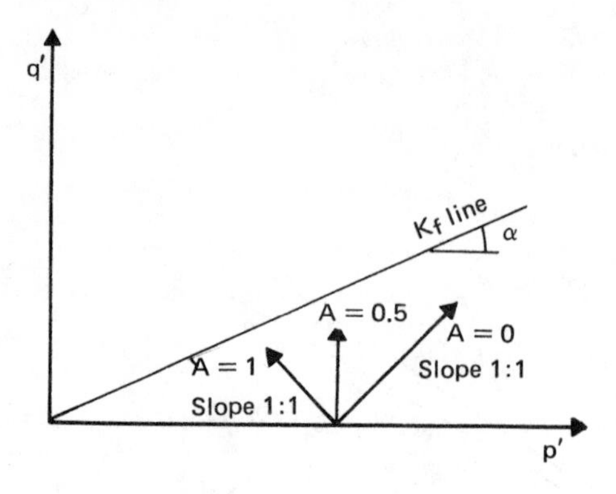

Fig. 6.39 Dependence of effective stress path on pore water pressure parameter A.

6.5.3 Stress Path and Sample Distortion for Similar Increase of Axial Stress

To obtain a general idea of the nature of distortion in soil samples derived from the application of an axial stress, we consider a specimen as shown in Fig. 6.41*a*. If $\sigma_1' = \sigma_3'$ (i.e., hydrostatic compression) and the sample is subjected to a hydrostatic stress increase of $\Delta\sigma$ under drained conditions (i.e., $\Delta\sigma = \Delta\sigma'$), then the drained stress path would be EF as shown in Fig. 6.41*b*. There would be uniform strain in all directions. If $\sigma_3' = K_o\sigma_1'$ (at-rest pressure) and the sample is subjected to an axial stress increase of $\Delta\sigma$ under drained condition (i.e., $\Delta\sigma = \Delta\sigma'$), the sample deformation would depend on the stress path it follows. For stress path AC (Fig. 6.41*b*) which is along the K_o line, there will be axial deformation only and no lateral deformation. For stress path AB, there will be lateral expansion and so the axial strain at B will be greater than that at C. For stress path AD, there will be some lateral compression and the axial strain at D will be more than that at F but less than that at C.

Note that the axial strain is gradually increasing as we go from F to B in Fig. 6.41*b*. In all cases the effective major principal stress is $\sigma_1 + \Delta\sigma'$. However, the lateral strain is compressive at F, zero at C, and we get lateral expansion at B. This is due to the nature of the lateral effective stress to which the sample is subjected during the loading.

6.5.4 Calculation of Settlement from Stress Paths

In the calculation of settlement from stress paths it is assumed that, for normally consolidated clays, the volume change between any two points on a p' vs. q' plot is independent of the path followed. This is explained in Fig. 6.42. For a soil sample, the volume changes between stress paths AB, GH, CD, and CI, for example, are all the same. However, the axial strains will be different. With this basic assumption, we can now proceed to determine the settlement.

For ease in understanding, the procedure for settlement calculation will be explained with the aid of an example. For settlement calculation in a normally consolidated clay, undisturbed samples from representative depths are obtained. Consolidated undrained triaxial tests on these samples at several confining pressures, σ_3, are conducted, along with a standard one-dimensional consolidated test. The stress-strain

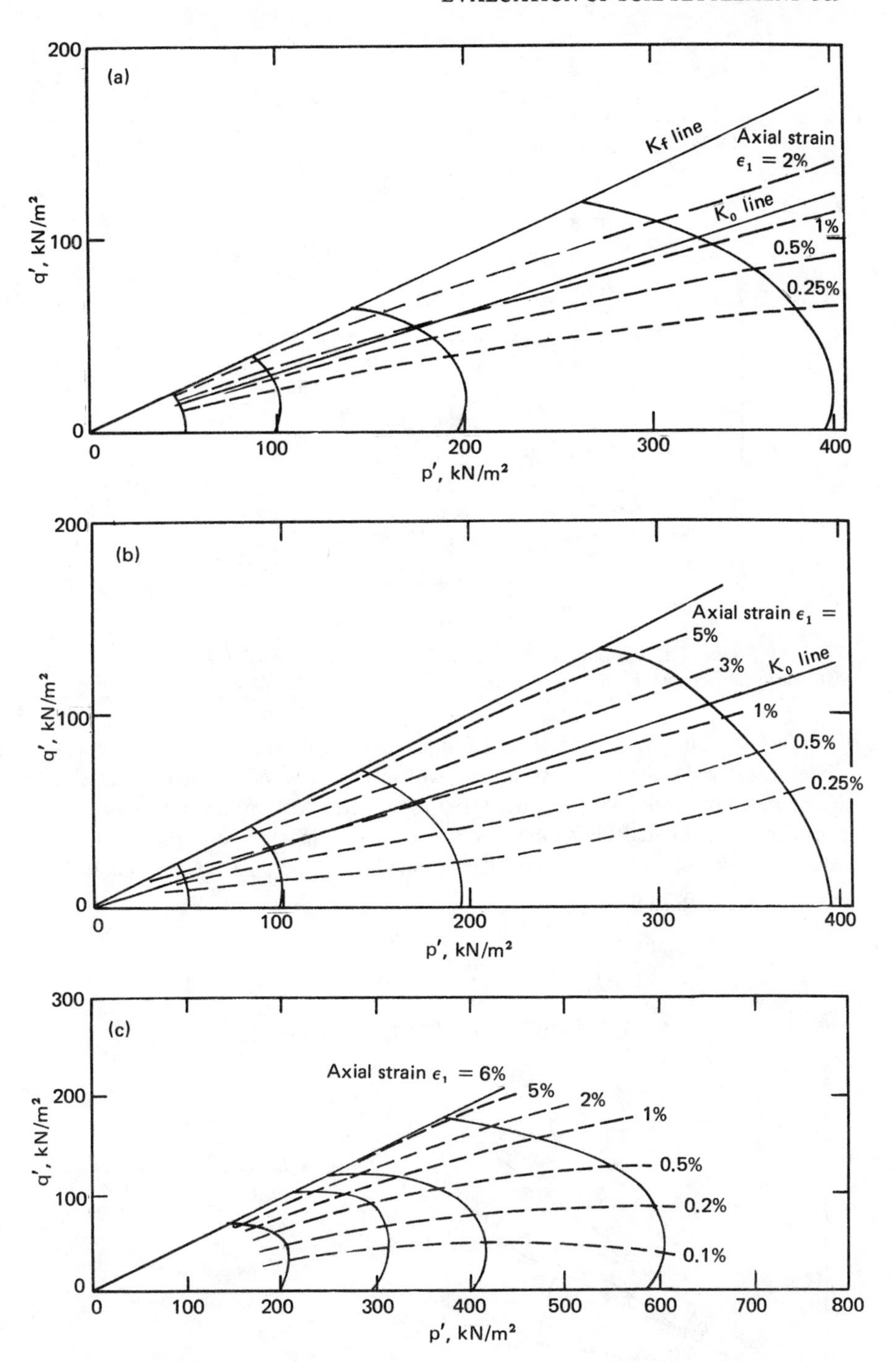

Fig. 6.40 Stress path of three clays. (*a*) Lagunillas clay; (*b*) Amuay clay; (*c*) Boston blue clay. *(Redrawn after T. W. Lambe, Methods of Estimating Settlement,* J. Soil Mech. Found. Div., ASCE, *vol. 90, no. SM5, 1964.)*

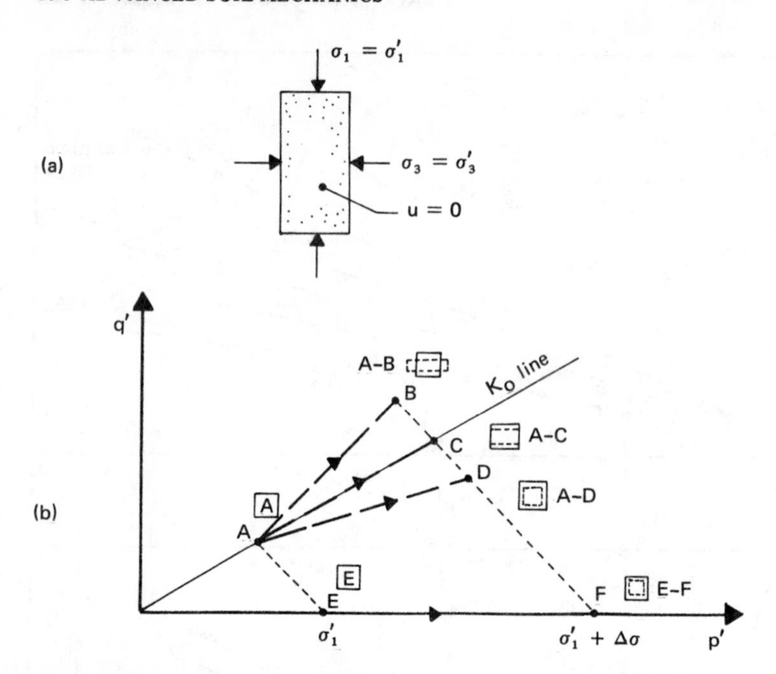

Fig. 6.41 Stress path and sample distortion. *(Figure b redrawn after T. W. Lambe, Methods of Estimating Settlement,* J. Soil Mech. Found. Div., ASCE, *vol. 90, no. SM5, 1964.)*

contours are plotted on the basis of the CU triaxial test results. The standard one-dimensional consolidation test results will give us the values of compression index C_c. For an example, let Fig. 6.43 represent the stress-strain contours for a given normally consolidated clay sample obtained from an *average depth* of a clay layer. Also let $C_c = 0.25$ and $e_o = 0.9$. The drained friction angle ϕ (determined from CU tests) is 30°. From Eq. (6.66),

$$\beta = \tan^{-1}\left(\frac{1-K_o}{1+K_o}\right)$$

and $K_o = 1 - \sin\phi = 1 - \sin 30° = 0.5$. So

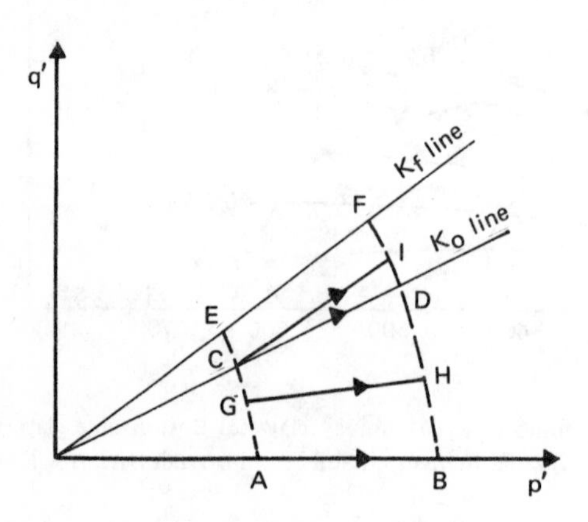

Fig. 6.42 Volume change between two points of a p' vs. q' plot.

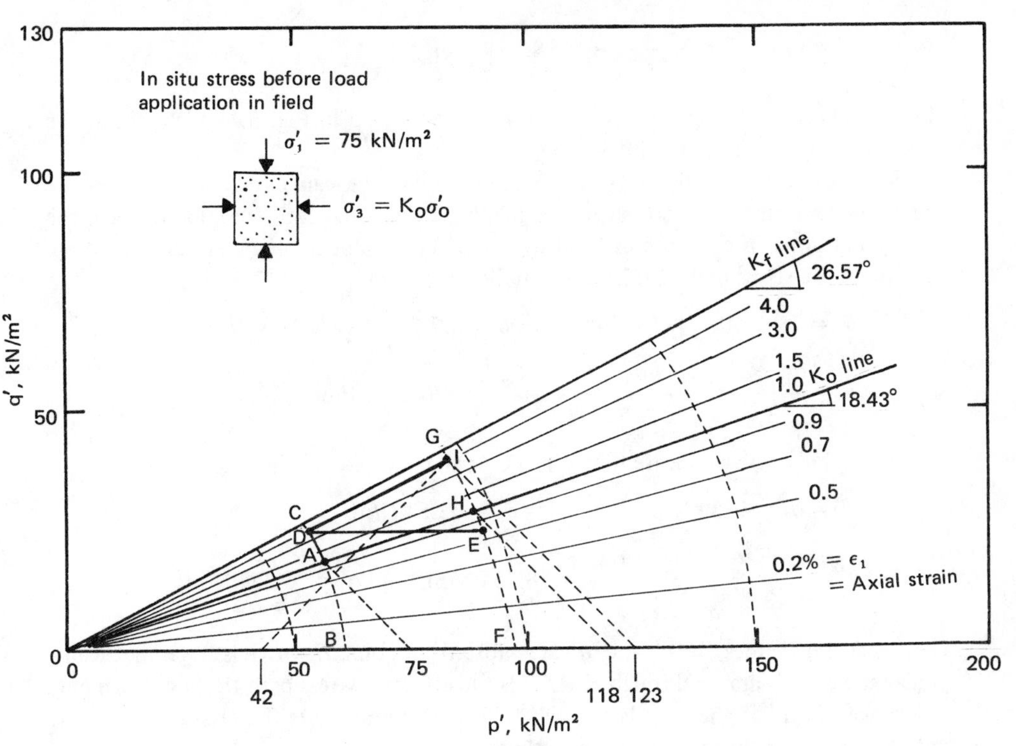

Fig. 6.43

$$\beta = \tan^{-1}\left(\frac{1-0.5}{1+0.5}\right) = 18.43°$$

Knowing the value of β, we can now plot the K_o line in Fig. 6.43. Also note that $\tan\alpha = \sin\phi$. Since $\phi = 30°$, $\tan\alpha = 0.5$. So $\alpha = 26.57°$. Let us calculate the settlement in the clay layer for the following conditions (Fig. 6.43):

1. In situ average effective overburden pressure $= \sigma_1' = 75\ \text{kN/m}^2$.
2. Total thickness of clay layer $= H_t = 3$ m.

Due to the construction of a structure, the increase of the total major and minor principal stresses at an average depth are:

$$\Delta\sigma_1 = 40\ \text{kN/m}^2$$

$$\Delta\sigma_3 = 25\ \text{kN/m}^2$$

(assuming that the load is applied instantaneously.) The in situ minor principal stress (at-rest pressure) is $\sigma_3 = \sigma_3' = K_o\sigma_1' = 0.5(75) = 37.5\ \text{kN/m}^2$.

So, before loading,

$$p' = \frac{\sigma_1' + \sigma_3'}{2} = \frac{75 + 37.5}{2} = 56.25\ \text{kN/m}^2$$

$$q' = \frac{\sigma_1' - \sigma_3'}{2} = \frac{75 - 37.5}{2} = 18.75 \text{ kN/m}^2$$

The stress conditions before loading can now be plotted in Fig. 6.43 from the above values of p' and q'. This is point A.

Since the stress paths are geometrically similar, we can plot BAC, which is the stress path through A. Also, since the loading is instantaneous (i.e., undrained), the stress conditions in clay, represented by the p' vs. q' plot immediately after loading, will fall on the stress path BAC. Immediately after loading,

$$\sigma_1 = 75 + 40 = 115 \text{ kN/m}^2 \qquad \sigma_3 = 37.5 + 25 = 62.5 \text{ kN/m}^2$$

So,
$$q' = \frac{\sigma_1' - \sigma_3'}{2} = \frac{\sigma_1 - \sigma_3}{2} = \frac{115 - 62.5}{2} = 26.25 \text{ kN/m}^2$$

With this value of q', we locate the point D. At the end of consolidation,

$$\sigma_1' = \sigma_1 = 115 \text{ kN/m}^2 \qquad \sigma_3' = \sigma_3 = 62.5 \text{ kN/m}^2$$

So,
$$p' = \frac{\sigma_1' + \sigma_3'}{2} = \frac{115 + 62.5}{2} = 88.75 \text{ kN/m}^2 \text{ and } q' = 26.25 \text{ kN/m}^2$$

The preceding values of p' and q' are plotted as point E. FEG is a geometrically similar stress path drawn through E. ADE is the effective stress path that a soil element, at average depth of the clay layer, will follow. AD represents the elastic settlement, and DE represents the consolidation settlement.

For elastic settlement (stress path A to D),

$$S_e = [(\epsilon_1 \text{ at } D) - (\epsilon_1 \text{ at } A)]H_t = (0.04 - 0.01)3 = 0.09\text{m}$$

For consolidation settlement (stress path D to E), based on our previous assumption the volumetric strain between D and E is the same as the volumetric strain between A and H. Note that H is on the K_o line. For point A, $\sigma_1' = 75 \text{ kN/m}^2$; and for point H, $\sigma_1' = 118 \text{ kN/m}^2$. So the volumetric strain, ϵ_v, is

$$\epsilon_v = \frac{\Delta e}{1 + e_o} = \frac{C_c \log(118/75)}{1 + 0.9} = \frac{0.25 \log(118/75)}{1.9} = 0.026$$

The axial strain ϵ_1 along a horizontal stress path is about one-third the volumetric strain along the K_o line, or

$$\epsilon_1 = \tfrac{1}{3}\epsilon_v = \tfrac{1}{3}(0.026) = 0.0087$$

So, the consolidation settlement is

$$S_c = 0.0087 H_t = 0.0087(3) = 0.0261 \text{ m}$$

and hence the total settlement is

$$S_e + S_c = 0.09 + 0.0261 = \mathit{0.116\,m}$$

Another type of loading condition is also of some interest. Suppose that the stress increase at the average depth of the clay layer was carried out in two steps: (1) instantaneous load application, resulting in stress increases of $\Delta\sigma_1 = 40\ \text{kN/m}^2$ and $\Delta\sigma_3 = 25\ \text{kN/m}^2$ (stress path *AD*), followed by (2) a gradual load increase, which results in a stress path *DI* (Fig. 6.43). As before, the undrained shear along stress path *AD* will produce an axial strain of 0.03. The volumetric strains for stress paths *DI* and *AH* will be the same; so $\epsilon_v = 0.026$. The axial strain ϵ_1 for the stress path *DI* can be given by the relation (based on the theory of elasticity)

$$\frac{\epsilon_1}{\epsilon_v} = \frac{1 + K_o - 2KK_o}{(1 - K_o)(1 + 2K)} \tag{6.68}$$

where $K = \sigma_3'/\sigma_1'$ for the point *I*. In this case, $\sigma_3' = 42\ \text{kN/m}^2$ and $\sigma_1' = 123\ \text{kN/m}^2$. So,

$$K = \tfrac{42}{123} = 0.341$$

$$\frac{\epsilon_1}{\epsilon_v} = \frac{\epsilon_1}{0.026} = \frac{1 + 0.5 - 2(0.341)(0.5)}{(1 - 0.5)\,[1 + 2(0.341)]} = 1.38$$

or

$$\epsilon_1 = (0.026)(1.38) = 0.036$$

Hence, the total settlement due to the loading is equal to

$$S = [(\epsilon_1 \text{ along } AD) + (\epsilon_1 \text{ along } DI)]H_t$$

$$= (0.03 + 0.036)H_t = 0.066H_t$$

6.5.5 Comparison of Primary Consolidation Settlement Calculation Procedures

It is of interest at this point to compare the primary settlement calculation procedures outlined in Secs. 6.3.1 and 6.3.2 with the stress path technique described in Sec. 6.5.4 (Fig. 6.44).

Based on the one-dimensional consolidation procedure outlined in Sec. 6.3.1, essentially we calculate the settlement along the stress path *AE*, i.e., along the K_o line. *A* is the initial at-rest condition of the soil, and *E* is the final stress condition (at rest) of soil at the end of consolidation. According to the Skempton-Bjerrum modification, the consolidation settlement is calculated for stress path *DE*. *AB* is the immediate settlement. However, Lambe's stress path method gives the consolidation settlement for stress path *BC*. *AB* is the immediate or elastic settlement. Although the stress path technique provides us with a better insight into the fundamentals of settlement calculation, it is more time-consuming because of the elaborate laboratory tests involved.

A number of works have been published that compare the observed and predicted settlements of various structures. Terzaghi and Peck (1967) pointed out that the field consolidation settlement is approximately one-dimensional when a comparatively thin layer of clay is located between two stiff layers of soil. Peck and Uyanik (1955) analyzed the settlement of eight structures in Chicago located over thick deposits of soft clay. The settlements of these structures were predicted by the method outlined

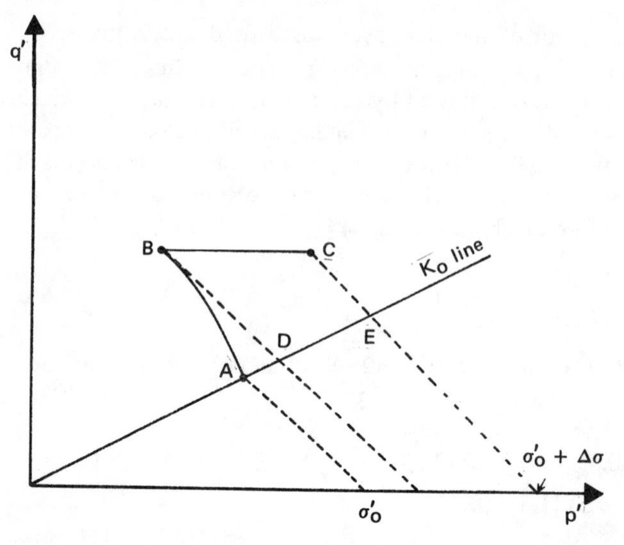

Fig. 6.44 Comparison of consolidation settlement calculation procedures.

in Sec. 6.3.1. Immediate settlements were not calculated. For this investigation, the ratio of the settlements observed to that calculated had an average value of 0.85. Skempton and Bjerrum (1957) also analyzed the settlements of four structures in the Chicago area (Auditorium, Masonic temple, Monadnock block, Isle of Grain oil tank) located on overconsolidated clays. The predicted settlements included the immediate settlements and the consolidation settlements (by the method given in Sec. 6.3.2). The ratio of the observed to the predicted settlements varied from 0.92 to 1.17. Settlement analysis of Amuya Dam, Venezuela (Lambe, 1963) by the stress path method showed very good agreement with the observed settlement.

However, there are several instances where the predicted settlements vary widely from the observed settlements. The discrepancies can be attributed to deviation of the actual field conditions from those assumed in the theory, difficulty in obtaining undisturbed samples for laboratory tests, and so forth.

PROBLEMS

6.1 An oil-storage tank that is circular in plan is to be constructed over a layer of sand, as shown in Fig. P6.1. Calculate the following settlements due to the uniformly distributed load q of the storage tank:

(*a*) The elastic settlement below the center of the tank at $z = 0, 5, 10, 15$, and 20 ft.

(*b*) The elastic settlement at (1) $z = 5$ ft, $s = 0$; (2) $z = 5$ ft, $s = 5$ ft; (3) $z = 5$ ft, $s = 10$ ft; (4) $z = 5$ ft, $s = 15$ ft.

(*c*) The average surface elastic settlement.

Assume that $H = \infty$, $\nu = 0.3$, $E = 5200$ lb/in², $B = 20$ ft, $q = 3000$ lb/ft², and $D_f = 0$.

6.2 Refer to the flexible circular tank foundation shown in Fig. P6.1. Given that $H = \infty$, $\nu = 0.3$, $\gamma = 16.8$ kN/m³, $E = 28{,}000$ kN/m², $B = 4.5$ m, $q = 230$ kN/m², and $D_f = 0$, calculate the elastic surface settlement at the center and the edge of the tank. Also calculate the average surface settlement.

6.3 If the circular foundation in Prob. 6.2 were rigid, what would be the surface elastic settlement?

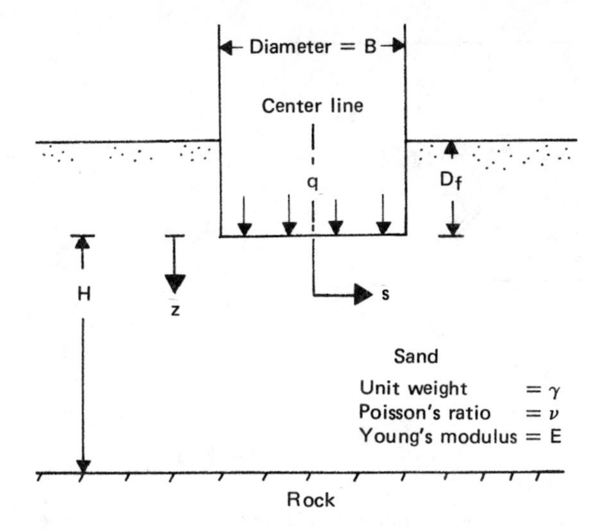

Fig. P6.1

6.4 Refer to Prob. 6.2. If $H = 8$ m, what would be the surface elastic settlement at the center of the tank?

6.5 Refer to Prob. 6.2. Assume that $H = 9.0$ m and $D_f = 1.5$ m but all other factors remain the same.

(*a*) Calculate the elastic settlement at the center of the tank by dividing the sand layer into nine 1-m-thick layers.

(*b*) Repeat part (*a*) for the edge of the tank.

6.6 A circular footing that is 10 ft in diameter rests on a 20-ft-thick saturated clay layer as shown in Fig. P6.2*a*. An approximation of the variation of the unconfined compression strength of clay with depth is also shown in Fig. P6.2*b*. Calculate the elastic settlement (assuming a flexible area) at

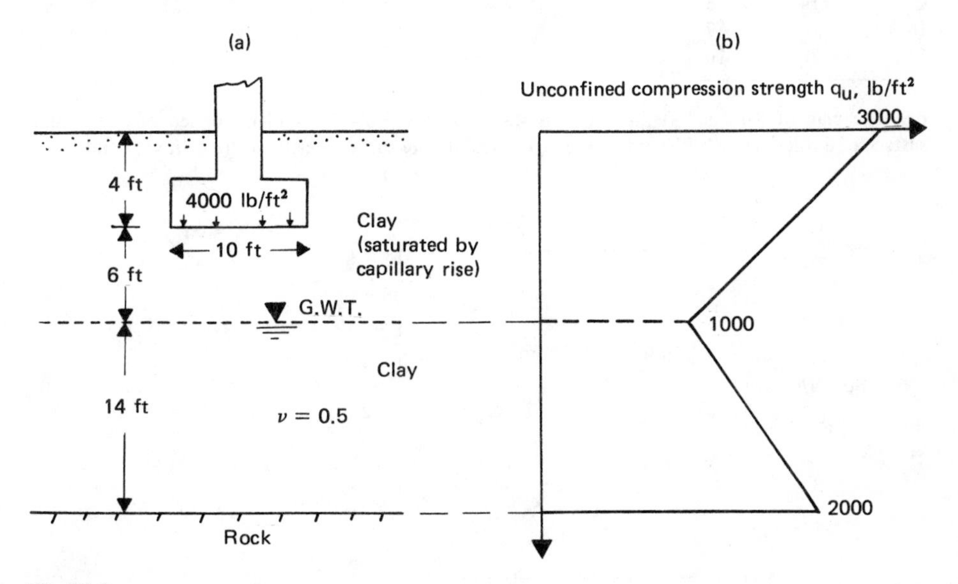

Fig. P6.2

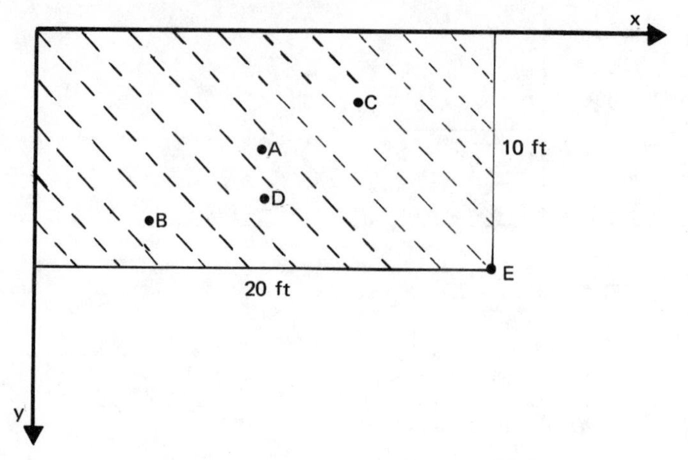

Fig. P6.3

the center and edge of the footing by dividing the 20-ft-thick clay layer below the footing into ten 2-ft-thick layers. Assume that $E = 200\,q_u$.

6.7 Refer to the circular footing given in Fig. P6.2. Estimate an average value of E. Determine the average elastic settlement of the footing by using Fig. 6.7.

6.8 The plan of a uniformly loaded flexible rectangular area is shown in Fig. P6.3. The loaded area is located at the ground surface over a thick deposit of silty clay. If $\nu = 0.5$, $E = 900\,\text{lb/in}^2$, and $q = 1500\,\text{lb/ft}^2$, determine the elastic settlement at points A, B, C, D, and E. The coordinates of the above points are as follows

Point	x, ft	y, ft
A	10	5
B	5	8
C	14	3
D	10	7
E	20	10

6.9 The plan of a loaded flexible area is shown in Fig. P6.4. If load is applied on the ground surface of a thick deposit of sand ($\nu = 0.25$), calculate the surface elastic settlement at A and B.

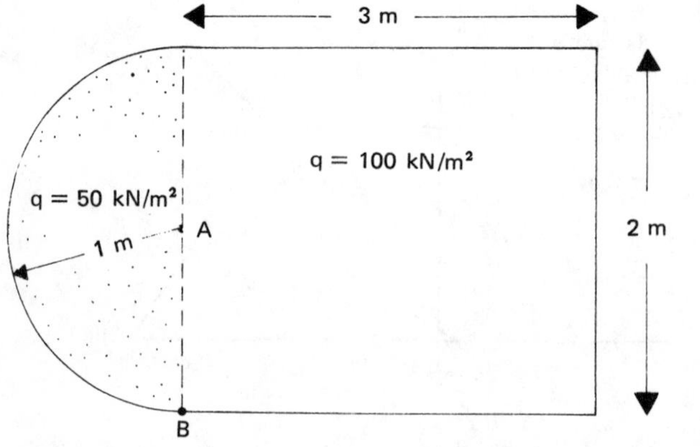

Fig. P6.4

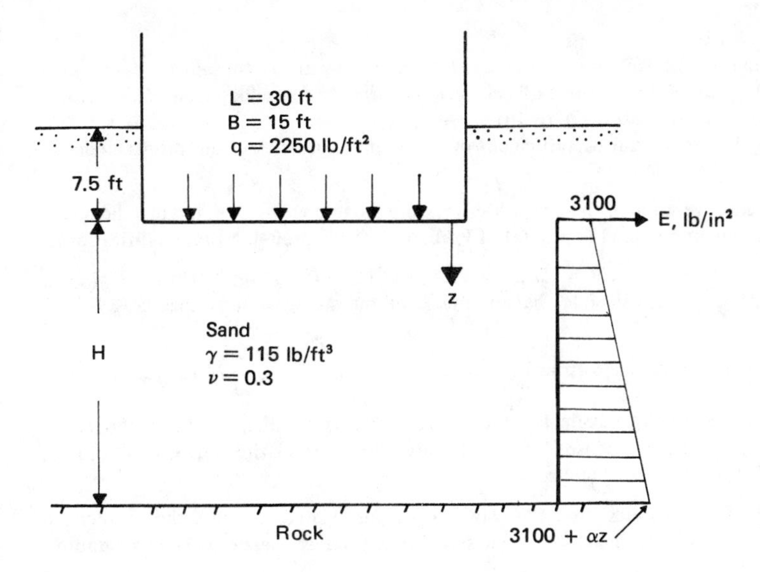

Fig. P6.5

6.10 A uniformly loaded flexible area (mat foundation) is shown in Fig. P6.5. Estimate the average elastic settlement of the loaded area. Assume that $\alpha = 0$ for Young's modulus and $H = 17.5$ ft.

6.11 Refer to Fig. P6.5. If $\alpha = 30$ and $H = 50$ ft, estimate the elastic settlement of the loaded area after five years of load application. Use the $2B$–$0.6I_z$ type of diagram (Sec. 6.2.4).

6.12 Assume that the circular foundation shown in Fig. P6.6 is flexible.

(*a*) Plot a diagram showing the variation of stress increase (due to the foundation load, $q = 130$ kN/m²) below the center of the foundation with depth from $z = 0$ to $z = 6$ m.

(*b*) Assume that $E_{\text{sand}} = 25{,}200$ kN/m² and $E_{\text{clay}} = 3500$ kN/m². Calculate the elastic settlement at the center of the foundation using Eq. (6.23).

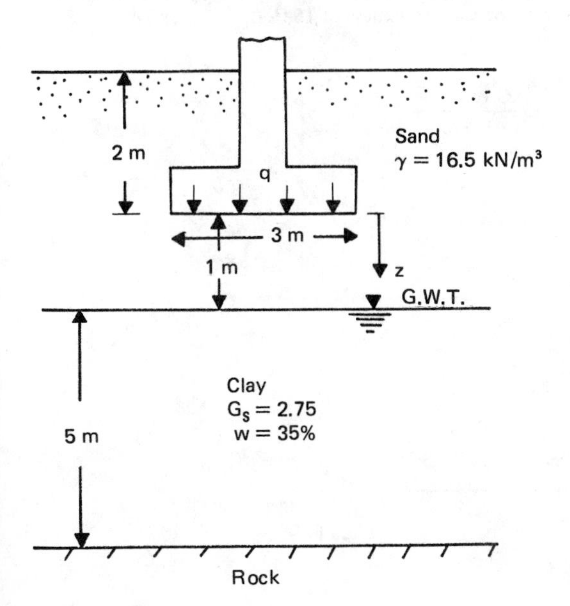

Fig. P6.6

6.13 Refer to Prob. 6.12.

(*a*) Based on the stress calculation of part (*a*), determine the primary consolidation settlement of the foundation. Use Eq. (6.39). Assume that the clay is normally consolidated and $C_c = 0.18$.

(*b*) Repeat the calculation using Eq. (6.40).

(*c*) Determine the probable total settlement of the foundation (elastic and primary consolidation).

6.14 Using a representative value of the pore water pressure parameter $A = 0.6$, modify the magnitude of the total settlement obtained in part (*c*) of Prob. 6.13 (Skempton–Bjerrum modification).

6.15 Refer to Fig. P6.1. Given that $H = 75$ ft, $\nu = 0.25$, $\gamma = 115\ \text{lb/ft}^3$, $E = 5200\ \text{lb/in}^2$, $B = 20$ ft, $q = 4000\ \text{lb/ft}^2$, and $D_f = 5$ ft, calculate the settlement of the storage tank using the $2B$–$0.6I_z$ type of diagram.

6.16 A rectangular foundation is shown in Fig. P6.7; $B = 2$ m, $L = 4$ m, $q = 240\ \text{kN/m}^2$, $H = 6$ m, and $D_f = 2$ m.

(*a*) Assuming $E = 3800\ \text{kN/m}^2$, calculate the average immediate settlement. Use Fig. 6.7.

(*b*) If the clay is normally consolidated, calculate the consolidation settlement. Use Eq. (6.40). $\gamma_{sat} = 17.5\ \text{kN/m}^2$, $C_c = 0.12$ and $e_o = 1.1$.

6.17 Refer to Prob. 6.16. Assume that the clay is overconsolidated and that the overconsolidation pressure is $50\ \text{kN/m}^2$. Estimate the total settlement (average elastic plus consolidation), given $C_r = \frac{1}{5}C_c$.

6.18 Refer to Prob. 6.16. Assume that the clay is overconsolidated and that the overconsolidation pressure is $140\ \text{kN/m}^2$. Calculate the consolidation settlement, given $C_r = 0.05$. Use the correction factor α given in Fig. 6.29.

6.19 Refer to Prob. 6.16(*a*). If the variation of Young's modulus is given by $E = 3500 + 100z\ \text{kN/m}^2$ and all other factors remain constant, calculate the elastic settlement at the center of the foundation ($\nu = 0.5$).

6.20 A permanent surcharge of $2000\ \text{lb/ft}^2$ is to be applied on the ground surface of the soil profile shown in Fig. P6.8. It is required to eliminate all of the primary consolidation in three months. Estimate the total surcharge $q = q_s + q_f$ needed to achieve the goal.

6.21 The p' vs. q' diagram for a normally consolidated clay is shown in Fig. P6.9. The sample was obtained from an average depth of a clay layer of total thickness of 15 ft. $C_c = 0.3$ and $e_o = 0.8$.

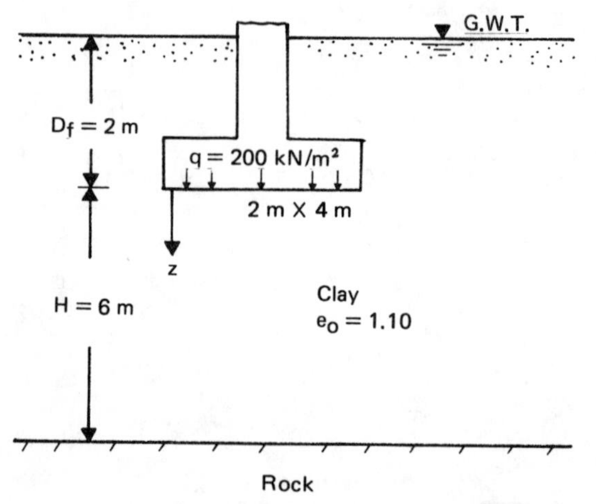

Fig. P6.7

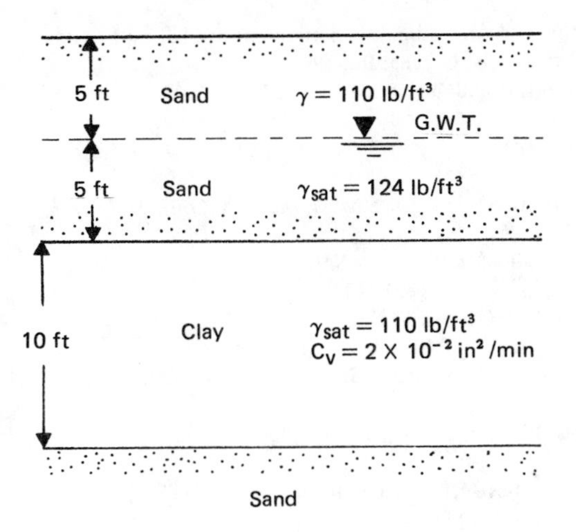

Fig. P6.8

(*a*) Calculate the total settlement (elastic and consolidation) for a loading following stress path *ABC*.

(*b*) Calculate the total settlement for a loading following stress path *ABD*.

6.22 Refer to Prob. 6.21. What would be the consolidation settlement according to the Skempton–Bjerrum method for the stress path *ABC*?

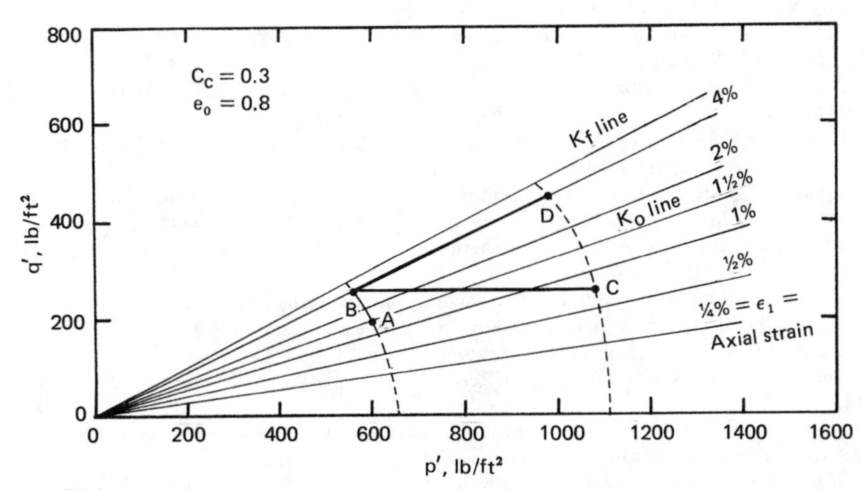

Fig. P6.9

REFERENCES

Ahlvin, R. G., and H. H. Ulery, "Tabulated Values for Determining the Complete Pattern of Stresses, Strains, and Deflections beneath a Uniform Circular Load on a Homogeneous Half Space," Highway Research Board, Bulletin 342, pp. 1–13, 1962.

Bachelier, M., and L. Parez, Contribution to the Study of Soil Compressibility by Means of a Cone Penetrometer, *Proc. 6th Int. Conf. Soil Mech. Found. Eng., Montreal*, vol. 2, pp. 3–7, 1965.

Bazaraa, A. R. S., "Use of Standard Penetration Test for Estimating Settlement of Shallow Foundations on Sand," Ph.D. thesis, University of Illinois, Urbana, Ill., 1967.

Begemann, H. K. S., General Report for Central and Western Europe, *Proc. Eur. Symp. Pen. Test., Stockholm,* 1974.

Bjerrum, L., Embankments on Soft Ground, *Proc. Specialty Conf. Perform. Earth and Earth-Supported Struct., ASCE,* vol. II, pp. 1–54, 1972.

Bjerrum, L., and A. Eggstad, Interpretation of Load Test on Sand, *Proc. Eur. Conf. Soil Mech. Found. Eng., Weisbaden,* vol. 1, p. 199, 1963.

Bogdanovi'c, L., Settlement of Stiff Structures (Silos) Founded on Soft Soil with Low Penetration Resistance, *Trans. SR Inst. Test. Mater., Belgrade,* no. 34, 1973.

Bowles, J. E., "Foundation Analysis and Design," McGraw-Hill, New York, 1977.

Buisman, A. S. K., "Groundmechania," Waltman, Delft, 1940.

Burland, J. B., Discussion, Session A, *Proc. Conf. In-Situ Invest. Soil Rocks, Brit. Geotech. Soc., London, England,* pp. 61–62, 1970.

Christian, J. T., and W. D. Carrier, III, Janbu, Bjerrum and Kjaernsli's Chart Reinterpreted, *Can. Geotech. J.,* vol. 15, no. 1, pp. 124–128, 1978.

Crawford, C. B., and K. N. Burn, Settlement Studies of Mt. Siani Hospital, *Eng. J. Can.,* vol. 45, no. 12, 1962.

Crawford, C. B., and J. G. Sutherland, The Empress Hotel, Victoria, British Columbia, Sixty-five Years of Foundation Settlements, *Can. Geotech. J.,* vol. 8, no. 1, pp. 77–93, 1971.

D'Appolonia, D. J., E. D'Appolonia, and R. F. Brissette, Settlement of Spread Footings on Sand: Closure, *J. Soil Mech. Found. Div., ASCE,* vol. 96, no. SM2, pp. 754–762, 1970.

D'Appolonia, D. J., H. G. Poulos, and C. C. Ladd, "Initial Settlement of Structures on Clay," *J. Soil Mech. Found. Div.,* ASCE, vol. 97, no. SM10, pp. 1359–1378, 1971.

DeBeer, E. E., Bearing Capacity and Settlement of Shallow Foundations on Sand, *Symp. Bearing Capacity Settle. Found., Duke University, Durham, N.C.,* pp. 15–33, 1967.

DeBeer, E. E., Interpretation of the Results of Static Penetration Tests, *Group IV Report, Eur. Symp. Pen. Test., Stockholm,* 1974.

Eggstad, A., Deformation Measurements below a Model Footing on the Surface of Dry Sand, *Proc. Eur. Conf. Soil Mech. Found. Eng., Weisbaden, W. Germany,* vol. 1, pp. 233, 1963.

Ferrent, T. A., The Prediction of Field Vertification of Settlements on Cohesionless Soils, *Proc. 4th Austr.-New Zeal. Conf. Soil Mech. Found. Eng.,* pp. 11–17, 1963.

Fox, E. N., The Mean Elastic Settlement of a Uniformly Loaded Area at a Depth below the Ground Surface, *Proc. 2d Int. Conf. Soil Mech. Found. Eng.,* vol. 1, pp. 129–132, 1948.

Giroud, J. P., Settlement of Rectangular Foundations on Soil Layer, *J. Soil Mech. Found. Div., ASCE,* vol. 98, no. SM1, pp. 149–154, 1972.

Harr, M. E., "Foundations of Theoretical Soil Mechanics," McGraw-Hill, New York, 1966.

Horn, H. M., and T. W. Lambe, Settlement of Buildings on the MIT Campus, *J. Soil Mech. Found. Div., ASCE,* vol. 90, no. SM5, pp. 181–195, 1964.

Jaky, J., The Coefficient of Earth Pressure at Rest, *J. Soc. Hungarian Arch. Fng.,* pp. 355–358, 1944.

Janbu, N., L. Bjerrum, and B. Kjaernsli, "Veiledning ved Losning av Fundamenteringsoppgaver," Norwegian Geotechnical Institute, Publication 16, Oslo, pp. 30–32, 1956.

Johnson, S. J., Precompression for Improving Foundation Soils, *J. Soil Mech. Found. Div., ASCE,* vol. 96, no. SM1, pp. 111–144, 1970.

Kondner, R. L., Hyperbolic Stress-Strain Response: Cohesive Soils, *J. Soil Mech. Found. Div., ASCE,* vol. 89, no. SM1, pp. 115–143, 1963.

Lambe, T. W., An Earth Dam for Storage of Fuel Oil, *Proc. 2d Pan Amer. Conf. Soil Mech. Found. Eng.,* vol. 1, p. 257, 1963.

Lambe, T. W., Methods of Estimating Settlement, *J. Soil Mech. Found. Div., ASCE,* vol. 90, no. SM5, p. 43, 1964.

Lambe, T. W., and R. V. Whitman, "Soil Mechanics," Wiley, New York, 1969.

Leonards, G. A., "Estimating Consolidation Settlement of Shallow Foundations on Overconsolidated Clay," Transportation Research Board, Special Report 163, Washington, D.C., pp. 13–16, 1976.

Liepens, A., "Settlement Analysis of a Seventeen-Story Reinforced Concrete Building," S. B. Thesis, Massachusetts Institute of Technology, Cambridge, Mass, 1957.

Meyerhof, G. G., Shallow Foundations, *J. Soil Mech. Found. Div., ASCE,* vol. 91, no. SM2, pp. 21–31, 1965.

Meyerhof, G. G., General Report: State-of-the-Art of Penetration Testing in Countries Outside Europe, *Proc. Eur. Symp. Pen. Test.,* 1974.

Mitchell, J. K., and W. S. Gardner, In Situ Measurement of Volume Characteristics, *Proc. Specialty Conf. Geotech. Eng. Div., ASCE,* vol. 2, pp. 279–345, 1975.

Peck, R. B., and M. E. Uyanik, "Observed and Computed Settlements of Structures in Chicago," University of Illinois Engineering Experiment Station, Bulletin No. 429, 1955.

Schleicher, F., Zur Theorie des Baugrundes, *Bauingenieur,* vol. 7, pp. 931–935, 949–952, 1926.

Schmertmann, J. H., Static Cone to Compute Static Settlement over Sand, *J. Soil Mech. Found. Div., ASCE,* vol. 96, no. SM3, pp. 1011–1043, 1970.

Schmertmann, J. H., "Guidelines for Design Using CPT Data," Prepared for Furgo-Cesco, Leidschendam, The Netherlands, 1974.

Schultze, E., and K. T. Melzer, The Determination of the Modulus of Compressibility of Noncohesive Soils by Soundings, *Proc. 6th Int. Conf. Soil Mech. Found. Eng., Montreal,* vol. 1, pp. 354–358, 1965.

Schultze, E., and G. Sherif, Prediction of Settlement from Evaluation of Settlement Observations for Sand., *Proc. 8th Int. Conf. Soil Mech. Found. Eng., Moscow, U.S.S.R.,* vol. 1, part 3, p. 225, 1973.

Scott, R. F., "Principles of Soil Mechanics," Addison-Wesley, Reading, Mass., 1963.

Skempton, A. W., and L. Bjerrum, A Contribution to Settlement Analysis of Foundations in Clay, *Geotechnique,* vol. 7, p. 168, 1957.

Su, H. H., and R. H. Prysock, Settlement Analysis of Two Highway Embankments, *Proc. Specialty Conf. Perform. Earth and Earth Supported Struct., ASCE,* vol. 1, pp. 465–488, 1972.

Terzaghi, K., and R. B. Peck, "Soil Mechanics in Engineering Practice," 2d ed., Wiley, New York, 1967.

Thomas, D., Deep Sounding Test Results and the Settlement on Normally Consolidated Sand, *Geotechnique,* vol. 18, pp. 472–488, 1968.

Trofimenkov, J. G., Penetration Testing in Western Europe, *Proc. Eur. Symp. Pen. Res., Stockholm,* 1974.

Trofimenkov, Yu, G., "Field Methods for Testing the Structural Properties of Soils," Building Literature Publishing House, Moscow, 1964.

Vesic, A. S., Tests on Instrumented Piles, Ogeechee River Site, *J. Soil Mech. Found. Div., ASCE,* vol. 96, no. SM2, pp. 561–584, 1970.

Webb, D. L., Settlement of Structures on Deep Alluvial Sandy Sediments in Durban, South Africa, *Proc. Conf. In-Situ Behav. Soil Rock, Inst. Civil Eng., London,* pp. 181–188, 1969.

CHAPTER

SEVEN

SHEAR STRENGTH OF SOILS

The shear strength of soils is an important aspect in many foundation engineering problems such as the bearing capacity of shallow foundations and piles, the stability of the slopes of dams and embankments, and lateral earth pressure on retaining walls. In this chapter, we will discuss the shear strength characteristics of granular and cohesive soils and the factors that control them.

7.1 MOHR–COULOMB FAILURE CRITERIA

In 1910, Mohr presented a theory for rupture in materials. According to this theory, failure along a plane in a material occurs by a critical combination of normal and shear stresses, and not by normal or shear stress alone. The functional relation between normal and shear stress on the failure plane can be given by

$$s = f(\sigma) \tag{7.1}$$

where s is the shear stress at failure and σ is the normal stress on the failure plane. The failure envelope defined by Eq. (7.1) is a curved line, as shown in Fig. 7.1.

In 1776, Coulomb defined the function $f(\sigma)$ as

$$s = c + \sigma \tan \phi \tag{7.2}$$

where c is cohesion and ϕ is the angle of friction of the soil.

Equation (7.2) is generally referred to as the Mohr-Coulomb failure criteria. The significance of the failure envelope can be explained using Fig. 7.1. If the normal and shear stresses on a plane in a soil mass are such that they plot as point A, shear failure

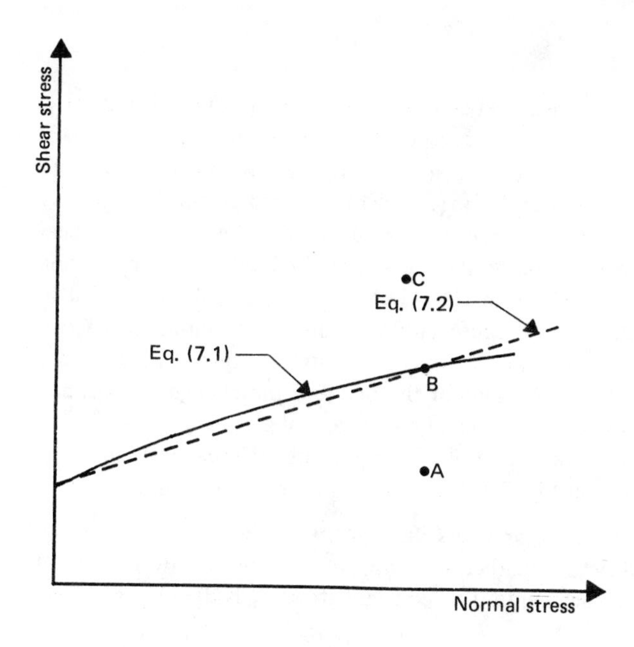

Fig. 7.1 Mohr–Coulomb failure criteria.

will not occur along that plane. Shear failure along a plane will occur if the stresses plot as point B, which falls on the failure envelope. A state of stress plotting as point C cannot exist, since this falls above the failure envelope; shear failure would have occurred before this condition was reached.

In saturated soils, the stress carried by the soil solids is the effective stress and so Eq. (7.2) must be modified:

$$s = c + (\sigma - u) \tan \phi = c + \sigma' \tan \phi \tag{7.3}$$

where u is the pore water pressure and σ' is the effective stress on the plane.

The term ϕ is also referred to as the drained friction angle. For sand, inorganic silts, and normally consolidated clays, $c \approx 0$. The value of c is greater than zero for overconsolidated clays.

The shear strength parameters of granular and cohesive soils will be treated separately in this chapter.

7.2 SHEARING STRENGTH OF GRANULAR SOILS

According to Eq. (7.3), the shear strength of a soil can be defined as $s = c + \sigma' \tan \phi$. For granular soils with $c = 0$,

$$s = \sigma' \tan \phi \tag{7.4}$$

The determination of the friction angle ϕ is commonly accomplished by one of two methods; the direct shear test or the triaxial test. The test procedures are given below.

7.2.1 Direct Shear Test

A schematic diagram of the direct shear test equipment is shown in Fig. 7.2. Basically, the test equipment consists of a metal shear box into which the soil specimen is placed. The specimen can be square or circular in plan, about 3 to 4 in^2 (19.35 to 25.80 cm^2) in area, and about 1 in (25.4 mm) in height. The box is split horizontally into two halves. Normal force on the specimen is applied from the top of the shear box by dead weights. The normal stress on the specimens obtained by the application of dead wèights can be as high as 150 lb/in^2 (1035 kN/m^2). Shear force is applied to the side of the top half of the box to cause failure in the soil specimen. (The two porous stones shown in Fig. 7.2 are not required for tests on dry soil.) During the test, the shear displacement of the top half of the box and the change in specimen thickness are recorded by the use of horizontal and vertical dial gauges.

Figure 7.3 shows the nature of the results of typical direct shear tests in loose, medium, and dense sands. Based on Fig. 7.3, the following observations can be made:

1. In dense and medium sands, shear stress increases with shear displacement to a maximum or peak value τ_m and then decreases to an approximately constant value τ_{cv} at large shear displacements. This constant stress τ_{cv} is the ultimate shear stress.
2. For loose sands, the shear stress increases with shear displacement to a maximum value and then remains constant.
3. For dense and medium sands, the volume of the specimen initially decreases and then increases with shear displacement. At large values of shear displacement, the volume of the specimen remains approximately constant.
4. For loose sands, the volume of the specimen gradually decreases to a certain value and remains approximately constant thereafter.

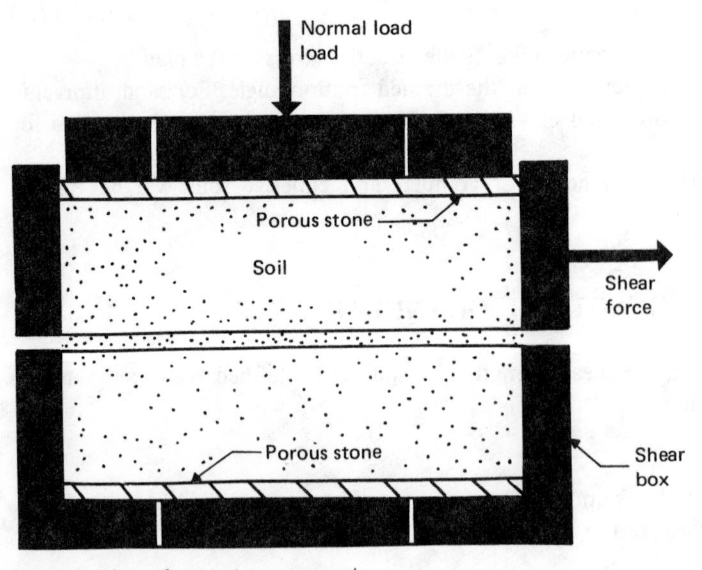

Fig. 7.2 Direct shear test arrangement.

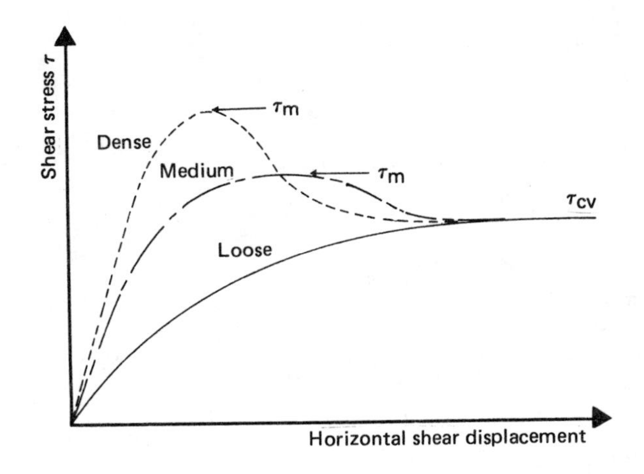

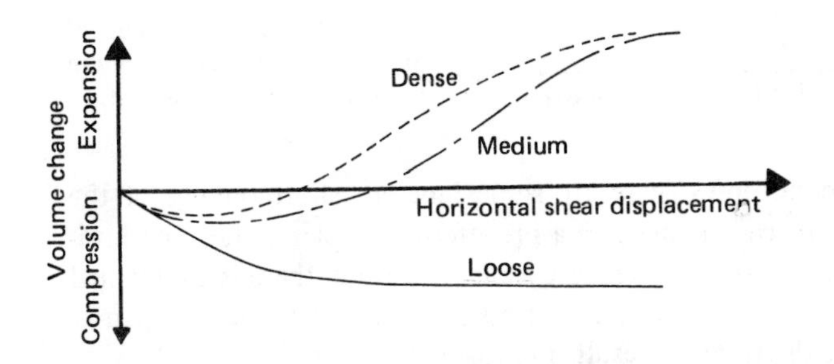

Fig. 7.3 Direct shear test results in loose, medium, and dense sands.

If dry sand is used for the test, the pore water pressure u is equal to zero, and so the total normal stress σ is equal to the effective stress σ'. The test may be repeated for several normal stresses. The angle of friction ϕ for the sand can be determined by plotting a graph of the maximum or peak shear stresses vs. the corresponding normal stresses, as shown in Fig. 7.4. The Mohr–Coulomb failure envelope can be determined by drawing a straight line through the origin and the points representing the experimental results. The slope of this line will give the peak friction angle ϕ of the soil. Similarly, the ultimate friction angle ϕ_{cv} can be determined by plotting the ultimate shear stresses τ_{cv} vs. the corresponding normal stresses, as shown in Fig. 7.4. The ultimate friction angle ϕ_{cv} represents a condition of shearing at constant volume of the specimen. For loose sands, the peak friction angle is approximately equal to the ultimate friction angle.

If the direct shear test is being conducted on a saturated granular soil, time between the application of the normal load and the shearing force should be allowed for drainage from the soil through the porous stones. Also, the shearing force should be applied at a slow rate to allow complete drainage. Since granular soils are highly permeable, this will not pose a problem. If complete drainage is allowed, the excess pore water pressure is zero, and so $\sigma = \sigma'$.

Some typical values of ϕ and ϕ_{cv} for granular soils are given in Table 7.1.

The strains in the direct shear test take place in two directions, i.e., in the vertical direction, and in the direction parallel to the applied horizontal shear force. This is

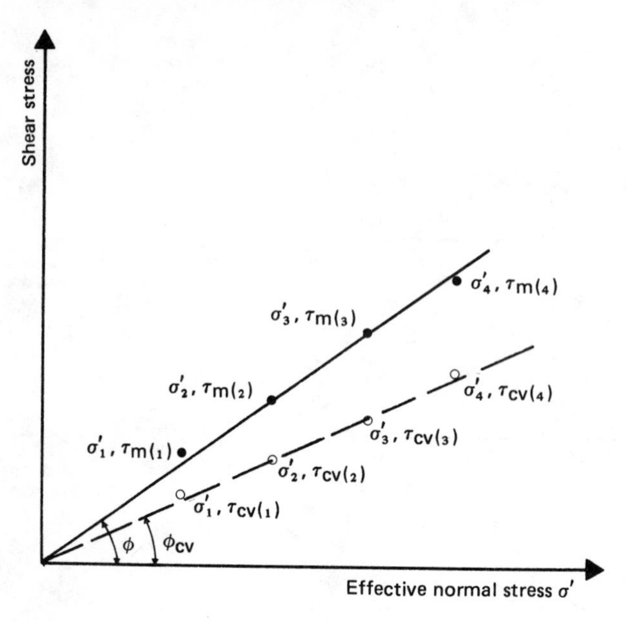

Fig. 7.4 Determination of peak and ultimate friction angle from direct shear test.

similar to the plane strain condition. There are some inherent shortcomings of the direct shear test. The soil is forced to shear in a predetermined plane—i.e., the horizontal plane—which is not necessarily the weakest plane. Secondly, there is an unequal distribution of stress over the shear surface. The stress is greater at the edges than at the center. This type of stress distribution results in progressive failure (Fig. 7.5).

Several attempts have been made to improve the direct shear test so that the shearing stresses over the failure surface will be uniform. A modified form of the direct shear test is the *simple shear test* (Fig. 7.6*a*) in which a rectangular specimen enclosed in a rubber membrane is placed inside a hinged box. Although the simple shear test is definitely an improvement over the direct shear test, the shear stresses in the specimen are still not uniformly distributed. This can be seen from Figs. 7.6*b* and *c*. Figure

Table 7.1 Typical values of ϕ and ϕ_{cv} for granular soils

Type of soil	ϕ, deg	ϕ_{cv}, deg
Sand: round grains		
Loose	28 to 30	
Medium	30 to 35	26 to 30
Dense	35 to 38	
Sand: angular grains		
Loose	30 to 35	
Medium	35 to 40	30 to 35
Dense	40 to 45	
Sandy gravel	34 to 48	33 to 36

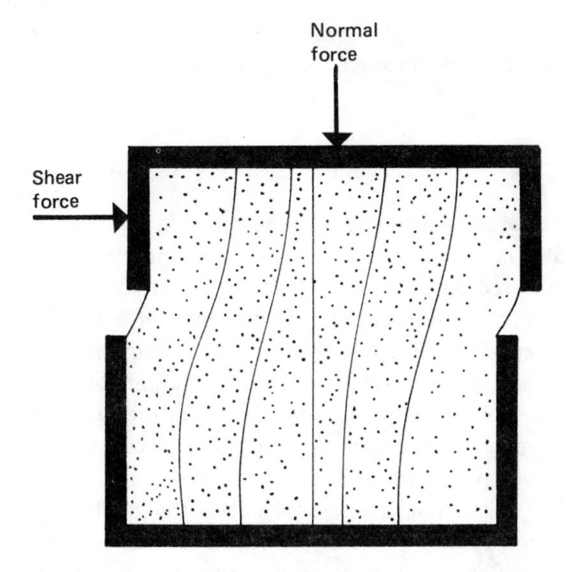

Fig. 7.5 Unequal stress distribution in direct shear equipment.

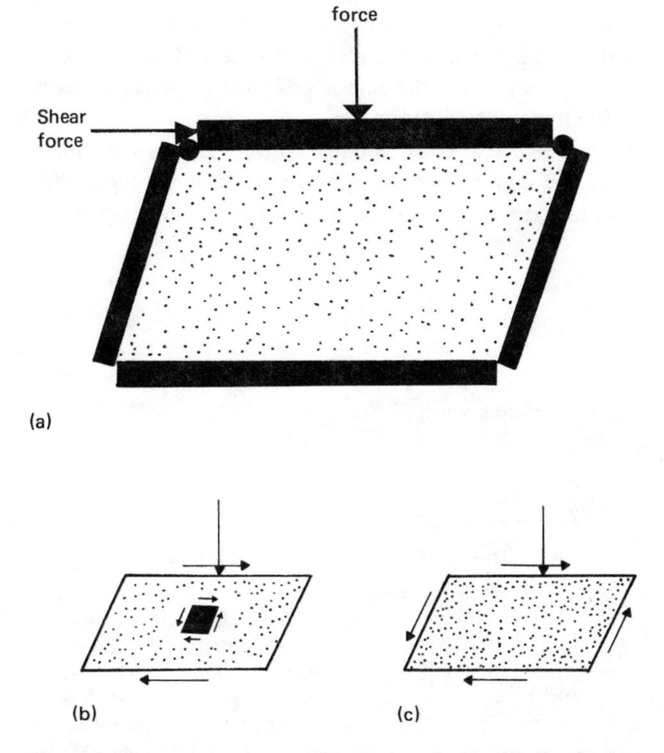

Fig. 7.6 Simple shear test. (*a*) Simple shear device. (*b*) Simple shear. (*c*) Pure shear.

7.6*c* shows the condition of pure shear. In simple shear (Fig. 7.6*b*), a pure shear condition exists at the center of the specimen; however, this is not true at the edges of the specimen.

7.2.2 Triaxial Test

A schematic diagram of a triaxial test equipment is shown in Fig. 7.7 In this type of test, a soil specimen about 1.5 in (38.1 mm) in diameter and 3 in (76.2 mm) in length is generally used. The specimen is enclosed inside a thin rubber membrane and placed inside a cylindrical plastic chamber. For conducting the test, the chamber is usually filled with water or glycerine. The specimen is subjected to a confining pressure σ_3 by application of pressure to the fluid in the chamber. (Air can sometimes be used as a medium for applying the confining pressure.) Connections to measure drainage into or out of the specimen or pressure in the pore water are provided. To cause shear failure in the soil, an axial stress $\Delta\sigma$ is applied through a vertical loading ram. This is also referred to as deviator stress. The axial strain is measured during the application of the deviator stress. For determination of ϕ, dry or fully saturated soil can be used. If saturated soil is used, the drainage connection is kept open during the application of the confining pressure and the deviator stress. Thus, during the test the excess pore water pressure in the specimen is equal to zero. The volume of the water drained from the specimen during the test provides a measure of the volume change of the specimen.

For *drained tests,* the total stress is equal to the effective stress. Thus, the major effective principal stress is $\sigma_1' = \sigma_1 = \sigma_3 + \Delta\sigma$; the minor effective principal stress is $\sigma_3' = \sigma_3$; and the intermediate effective principal stress is $\sigma_2' = \sigma_3'$.

At failure, the major effective principal stress is equal to $\sigma_3 + \Delta\sigma_f$, where $\Delta\sigma_f$ is the deviator stress at failure, and the minor effective principal stress is σ_3. Figure 7.8 shows the nature of the variation of $\Delta\sigma$ with axial strain for loose and dense granular

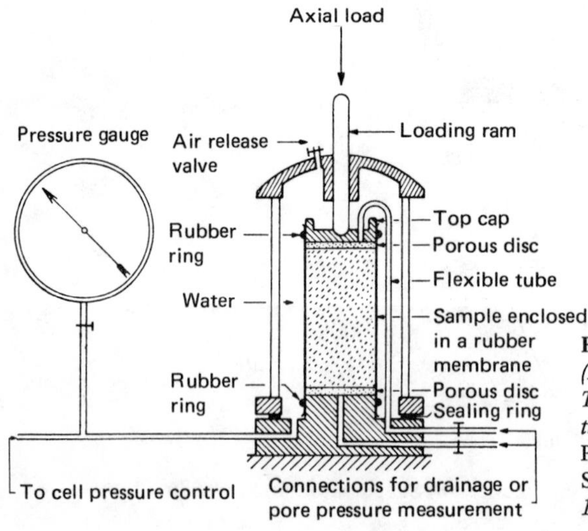

Fig. 7.7 Triaxial test equipment. *(After A. W. Bishop and L. Bjerrum, The Relevance of the Triaxial Test to the Solution of Stability Problems,* Proc. Research Conference on Shear Strength of Cohesive Soils, ASCE, *1960.)*

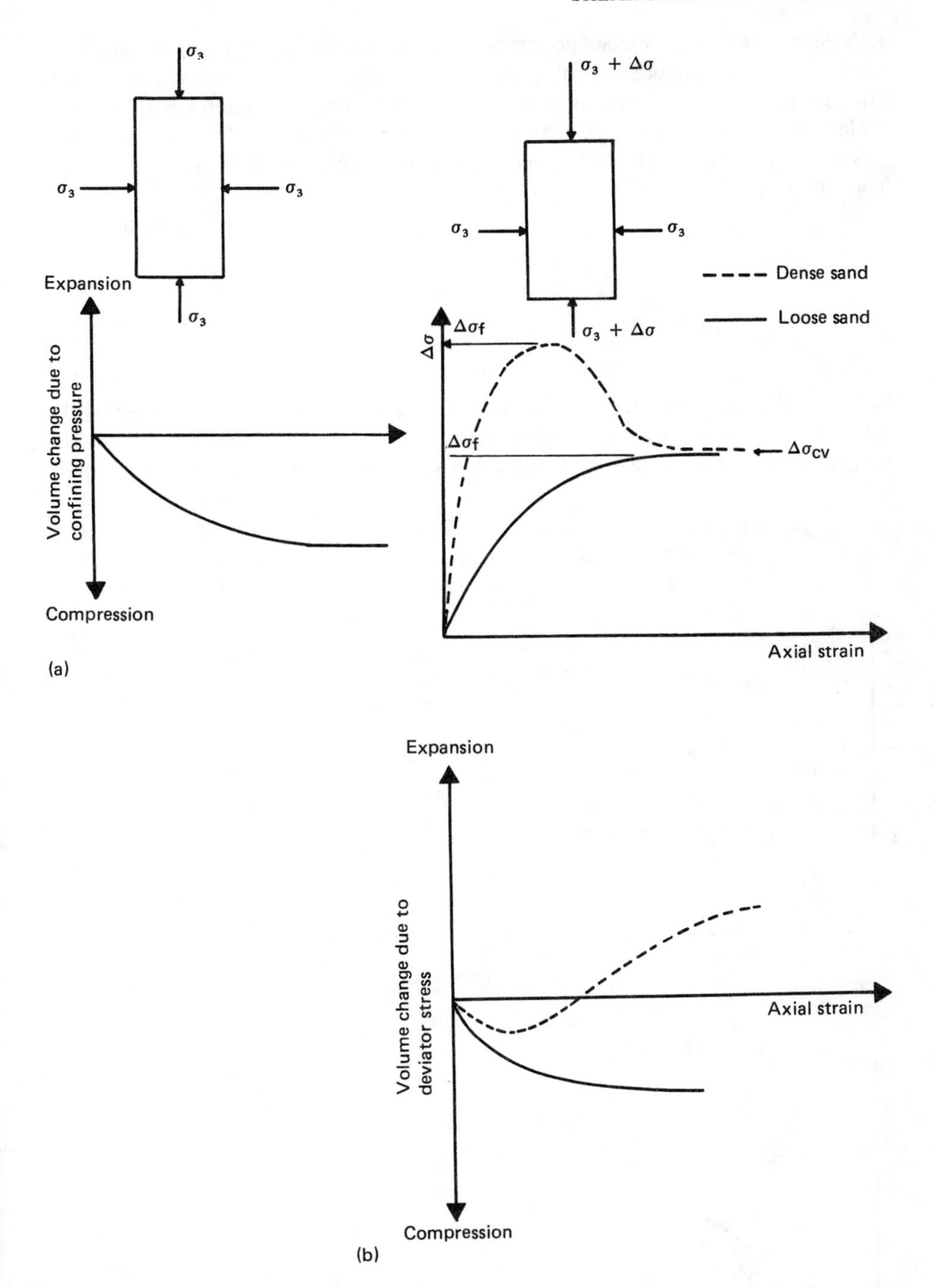

Fig. 7.8 Drained triaxial test in granular soils. (*a*) Application of confining pressure. (*b*) Application of deviator stress.

soils. Several tests with similar specimens can be conducted by using different confining pressures σ_3. The value of the soil peak friction angle ϕ can be determined by plotting effective-stress Mohr's circles for various tests and drawing a common tangent to these Mohr's circles passing through the origin. This is shown in Fig. 7.9*a*. The angle that this envelope makes with the normal stress axis is equal to ϕ. It can be seen from Fig. 7.9*b* that

$$\sin\phi \frac{\overline{ab}}{\overline{oa}} = \frac{(\sigma_1' - \sigma_3')/2}{(\sigma_1' + \sigma_3')/2}$$

or

$$\phi = \sin^{-1}\left(\frac{\sigma_1' - \sigma_3'}{\sigma_1' + \sigma_3'}\right)_{\text{failure}} \tag{7.5}$$

However, it must be pointed out that in Fig. 7.9*a* the failure envelope defined by the equation $s = \sigma' \tan\phi$ is an approximation to the actual curved failure envelope. The ultimate friction angle ϕ_{cv} for a given test can also be determined from the equation

$$\phi_{cv} = \sin^{-1}\left[\frac{\sigma_{1(cv)}' - \sigma_3'}{\sigma_{1(cv)}' + \sigma_3'}\right] \tag{7.6}$$

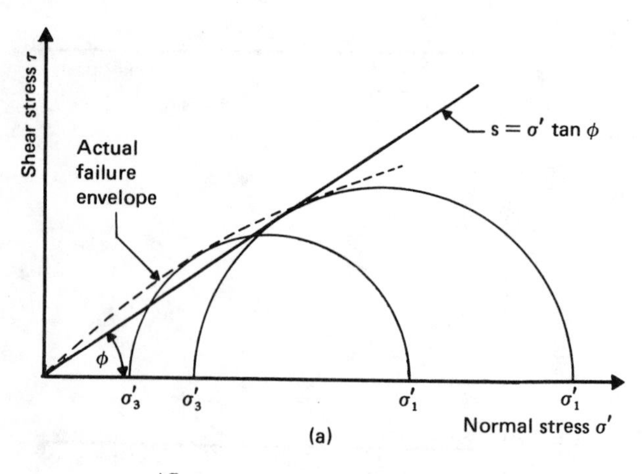

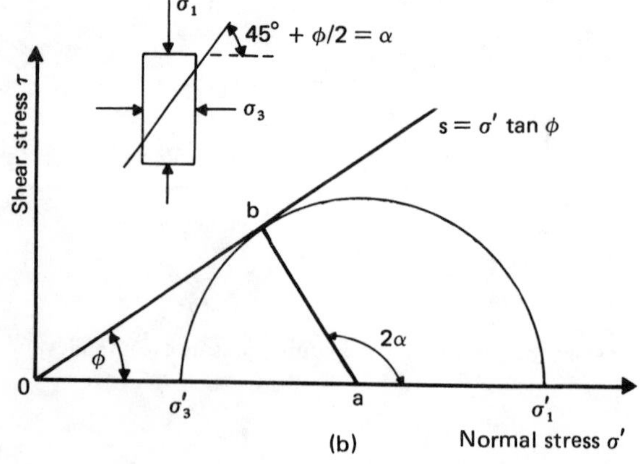

Fig. 7.9 Drained triaxial test results.

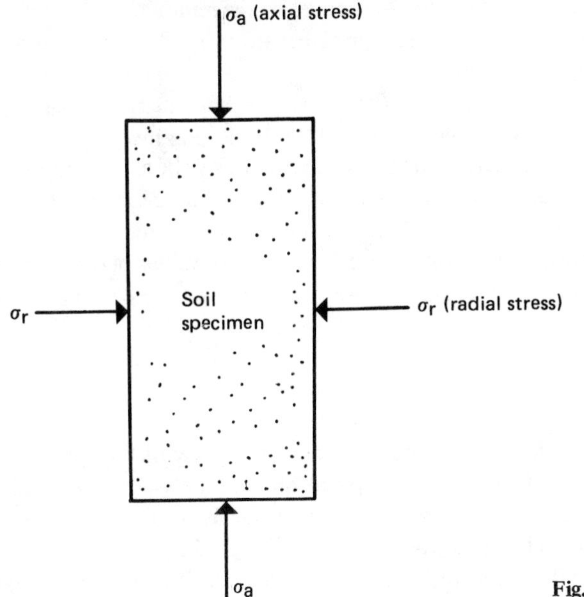

Fig. 7.10

where $\sigma'_{1(cv)} = \sigma'_3 + \Delta\sigma_{(cv)}$. For similar soils, the friction angle ϕ determined by triaxial tests is slightly lower (0 to 3°) than that obtained from direct shear tests.

The axial compression triaxial test described above is the conventional type. However, the loading process on the specimen in a triaxial chamber can be varied in several ways. In general, the tests can be divided into two major groups: axial compression tests and axial extension tests. The following is a brief outline of each type of test (refer to Fig. 7.10).

Axial compression tests

1. Radial confining stress σ_r constant and axial stress σ_a increased. This is the test procedure described above.
2. Axial stress σ_a constant and radial confining stress σ_r decreased.
3. Mean principal stress constant and radial stress decreased.

For drained compression tests, σ_a is equal to the major effective principal stress σ'_1, and σ_r is equal to the minor effective principal stress σ'_3 which is equal to the intermediate effective principal stress σ'_2. For the test listed under item 3, the mean principal stress, i.e., $(\sigma'_1 + \sigma'_2 + \sigma'_3)/3$, is kept constant. Or, in other words, $\sigma'_1 + \sigma'_2 + \sigma'_3 = J = \sigma_a + 2\sigma_r$ is kept constant by increasing σ_a and decreasing σ_r.

Axial extension tests

1. Radial stress σ_r kept constant and axial stress σ_a decreased.
2. Axial stress σ_a constant and radial stress σ_r increased.
3. Mean principal stress constant and radial stress increased.

For all *drained* extension tests at failure, σ_a is equal to the minor effective principal stress σ_3', and σ_r is equal to the major effective principal stress σ_1' which is equal to the intermediate effective principal stress σ_2'.

The detailed procedures for conducting these tests are beyond the scope of this text, and readers are referred to Bishop and Henkel (1969). Several investigations have been carried out to compare the peak friction angles determined by the axial compression tests to those obtained by the axial extension tests. A summary of these investigations is given by Roscoe et al. (1963). Some investigators found no difference in the value of ϕ from compression and extension tests; however, others reported values of ϕ determined from the extension tests which were several degress greater than those obtained by the compression tests.

7.2.3 Critical Void Ratio

We have seen that for shear tests in dense sands there is a tendency of the specimen to dilate as the test progresses. Similarly, in loose sand the volume gradually decreases (Figs. 7.3 and 7.8). An increase or decrease of volume means a change in the void ratio of soil. The nature of the change of the void ratio with strain for loose and dense sands is shown in Fig. 7.11. The void ratio for which the change of volume remains constant during shearing is called the *critical void ratio*. Figure 7.12 shows the results of some drained triaxial tests on washed Fort Peck sand. The void ratio after the application of σ_3 is plotted in the ordinate, and the change of volume, ΔV, at the peak point of the stress-strain plot, is plotted along the abscissa. For a given σ_3, the void ratio corresponding to $\Delta V = 0$ is the critical void ratio. Note that the critical void ratio is a function of the confining pressure σ_3. It is, however, necessary to recognize that, whether the volume of the soil specimen is increasing or decreasing, the critical void ratio is reached only in the shearing zone, even if it is generally calculated on the basis of the total volume change of the specimen.

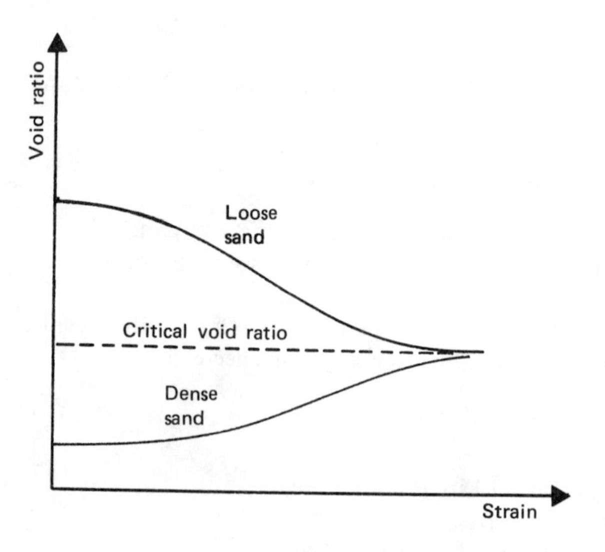

Fig. 7.11 Definition of critical void ratio.

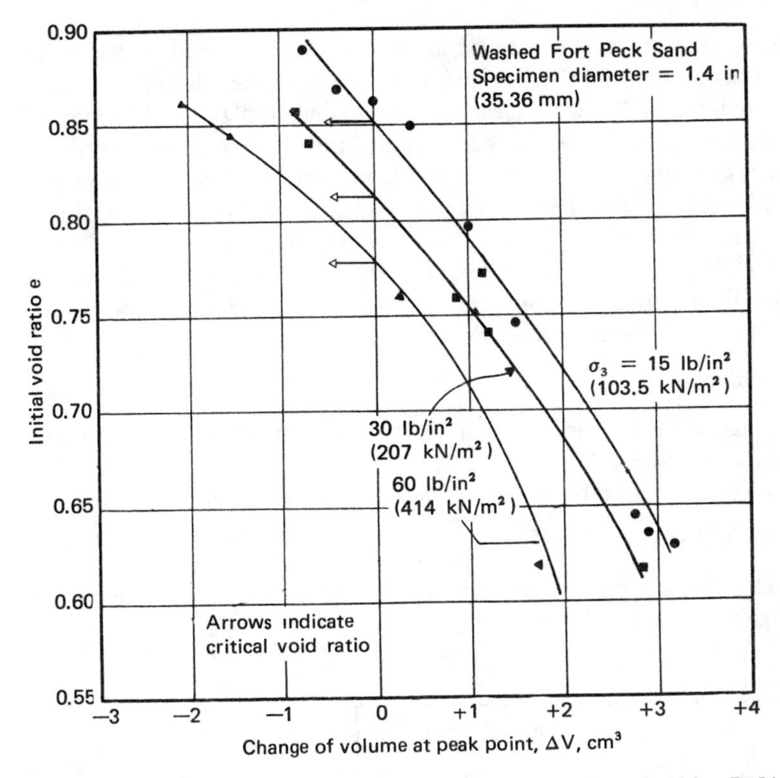

Fig. 7.12 Critical void ratio from triaxial tests on Fort Peck sand. *(After D. W. Taylor, "Fundamentals of Soil Mechanics," Wiley, New York, 1948.)*

The concept of critical void ratio was first introduced in 1938 by A. Casagrande to study liquefaction of granular soils. When a natural deposit of saturated sand that has a void ratio greater than the critical void ratio is subjected to a sudden shearing stress (due to an earthquake or to blasting, for example), the sand will undergo a decrease in volume. This will result in an increase of pore water pressure u. At a given depth, the effective stress is given by the relation $\sigma' = \sigma - u$. If σ (i.e., the total stress) remains constant and u increases, the result will be a decrease in σ'. This, in turn, will reduce the shear strength of the soil. If the shear strength is reduced to a value which is less than the applied shear stress, the soil will fail. This is called soil liquefaction. An advanced study of soil liquefaction can be obtained from the work of Seed and Lee (1966).

7.2.4 Curvature of the Failure Envelope

It was shown in Fig. 7.1 that Mohr's failure envelope [Eq. (7.1)] is actually curved, and the shear strength equation ($s = c + \sigma \tan \phi$) is only a straight-line approximation for the sake of simplicity. For a drained direct shear test on sand, $\phi = \tan^{-1} (\tau_{\max}/\sigma')$. Since Mohr's envelope is actually curved, a higher effective normal stress will yield lower values of ϕ. This fact is demonstrated in Fig. 7.13, which is a plot of the results of direct shear tests on standard Ottawa Sand. For loose sand, the value of ϕ decreases from about 30° to less than 27° when the normal stress is increased from 0.5 to 8 ton/ft^2 (47.9 to 766.4 kN/m^2). Similarly, for dense sand (initial void ratio approximately 0.56), ϕ decreases from about 36° to about 30.5° due to a sixteen-fold increase of σ'.

For high values of confining pressure (greater than about 4000 ton/ft^2), Mohr's failure envelope sharply deviates from the assumption given by Eq. (7.3). This is shown in Fig. 7.14. Skempton (1960, 1961) introduced the concept of *angle of intrinsic friction* for a formal relation between shear strength and effective normal stress. Based on Fig. 7.14, the shear strength can be defined as

$$s = k + \sigma' \tan \psi \tag{7.7}$$

where ψ is the angle of intrinsic friction. For quartz, Skempton (1961) gave the values of $k \approx 9500$ ton/ft^2 and $\psi \approx 13°$.

We have seen from Eq. (1.60) that the intergranular stress σ_{ig} can be given by

$$\sigma_{ig} = \sigma - u(1-a) + A' - R' \tag{1.60}$$

For granular soils, silts, and clays of low plasticity, A' and R' are small, and so we can approximate Eq. (1.60):

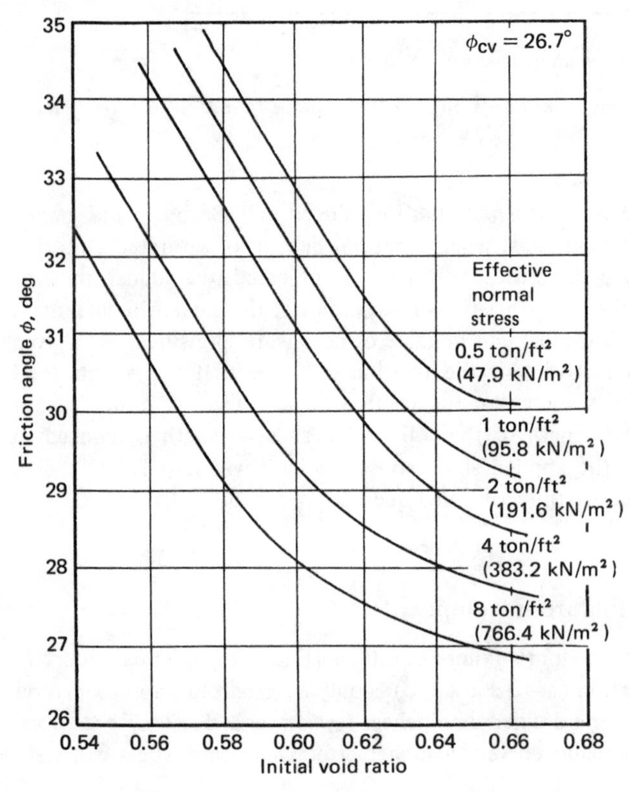

Fig. 7.13 Variation of peak friction angle, ϕ, with effective normal stress for direct shear tests on standard Ottawa sand. *(Redrawn after D. W. Taylor, "Fundamentals of Soil Mechanics," Wiley, New York, 1948.)*

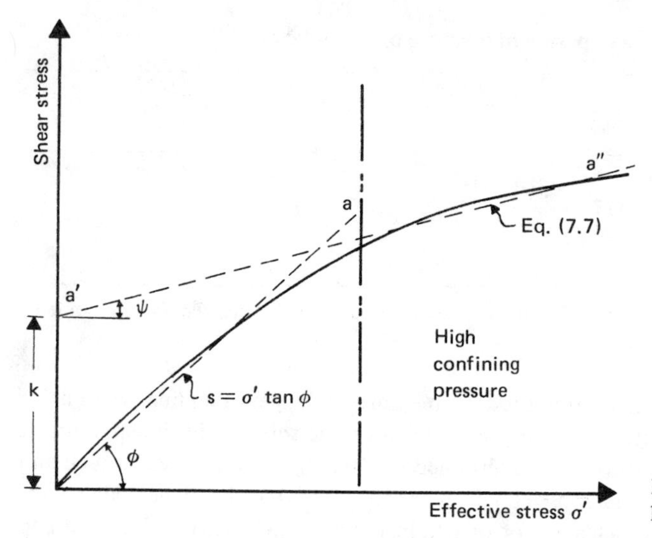

Fig. 7.14 Failure envelope at high confining pressure.

$$\sigma' = \sigma_{ig} = \sigma - u(1 - a) \tag{7.8}$$

Based on the analysis of intrinsic friction angle, Skempton suggested a modification of Eq. (7.8):

$$\sigma' = \sigma - u\left(1 - a\,\frac{\tan\psi}{\tan\phi}\right) \tag{7.9}$$

Using values of $\psi = 13°$, $\phi = 30°$, and $a = 0.1$,

$$\sigma' = \sigma - u\left(1 - 0.1\,\frac{\tan 13°}{\tan 30°}\right) = \sigma - 0.94u \tag{7.10}$$

Equation (7.10) is not significantly different from Eq. (1.58) for all practical considerations.

7.2.5 Some Comments on the Friction Angle of Granular Soils

The soil friction angle determined by the laboratory tests is influenced by two major factors. The energy applied to a soil by the external load is used both to overcome the frictional resistance between the soil particles and also to expand the soil against the confining pressure. The soil grains are highly irregular in shape and have to be lifted over one another for sliding to occur. This behavior is called *dilatency*. [A detailed study of the *stress dilatency* theory was presented by Rowe (1962)]. Hence, the angle of friction ϕ can be expressed as

$$\phi = \phi_\mu + \beta \tag{7.11}$$

where ϕ_μ is the angle of sliding friction between the mineral surfaces and β is the effect of interlocking.

Table 7.2 Comparison of the experimental values of ϕ_μ and ϕ_{cv}

Material	ϕ_μ, deg	ϕ_{cv}, deg
Steel ball, $\frac{3}{32}$-in diameter	7	14
Glass ballotini	17	24
Medium-to-fine quartz sand	26	32
Feldspar, 25 to 200 sieves	37	42

After I. K. Lee, Stress-Dilatency Performance of Feldspar, *J. Soil Mech. Found. Div., ASCE,* vol. 92, no. SM2, 1966.

Several investigators have attempted to measure the angle of sliding friction for quartz, feldspar, and calcite, which make up most of the soils of silt-size and larger. Bromwell (1966) and Dickey (1966) determined that tan ϕ_μ for quartz with rough and very rough surfaces lies in the approximate range of 0.3 to 0.5 (the higher value is for very rough surfaces). This yields ϕ_μ for quartz in the range of 17 to 27°. Horne and Deere (1962) gave the following values of tan ϕ_μ for saturated surface conditions:

Mineral	tan ϕ_μ	ϕ_μ, deg
Quartz	0.45	24
Feldspar	0.77	38
Calcite	0.68	34

Rowe (1962) obtained a value of ϕ_μ for medium-to-fine quartz sand in the order of 26°.

We saw in Table 7.1 that the friction angle of granular soils varies with the nature of the packing of the soil: the denser the packing, the higher the value of ϕ. If ϕ_μ for a given soil remains constant, from Eq. (7.11) the value of β has to increase with the increase of the denseness of soil packing. This is obvious, of course, because in a denser soil more work has to be done to overcome the effect of interlocking.

Table 7.2 provides a comparison of experimental values of ϕ_μ and ϕ_{cv}. From this, it can be seen that even at constant volume the value of ϕ_μ is less than ϕ_{cv}. This means that there must be some degree of interlocking even when the overall volume change is zero at very high strains.

Effect of angularity of soil particles. Other factors remaining constant, a soil possessing angular soil particles will show a higher friction angle ϕ than one with rounded grains because the angular soil particles will have a greater degree of interlocking and, thus, cause a higher value of β [Eq. (7.11)].

Effect of rate of loading during the test. The value of tan ϕ in triaxial compression tests is not greatly affected by the rate of loading. For sand, Whitman and Healy (1963) compared tests conducted in 5 min and in 5 ms and found that tan ϕ decreases at the most by about 10%.

7.2.6 Shear Strength of Granular Soils under Plane Strain Condition

The results obtained from triaxial tests are widely used for the design of structures. However, under structures such as continuous wall footings the soils are actually subjected to a plane strain type of loading, i.e., the strain in the direction of the intermediate principal stress is equal to zero. Several investigators have attempted to evaluate the effect of plane strain type of loading (Fig. 7.15) on the angle of friction of granular soils. A summary of the results obtained was compiled by Lee (1970) and is presented in Table 7.3. To discriminate the plane strain drained friction angle from the triaxial drained friction angle, the following notations have been used in Table 7.3:

ϕ_p = drained friction angle obtained from plane strain tests
ϕ_t = drained friction angle obtained from triaxial tests

Lee (1970) also conducted some drained shear tests on a uniform sand collected from the Sacramento River near Antioch, California. Drained triaxial tests were conducted with specimens of diameter 1.4 in (35.56 mm) and height 3.4 in (86.96 mm). Plane strain tests were carried out with rectangular specimens 2.4 in (60.96 mm) high and 1.1 × 2.8 in (27.94 × 71.12 mm) in cross section. The plane strain condition was obtained by the use of two lubricated rigid side plates. Loading of the plane strain specimens was achieved by placing them inside a triaxial chamber. All specimens, triaxial and plane strain, were anisotropically consolidated with a ratio of major to minor principal stress of 2:

$$k_c = \frac{\sigma'_{1(\text{consolidation})}}{\sigma'_{3(\text{consolidation})}} = 2$$

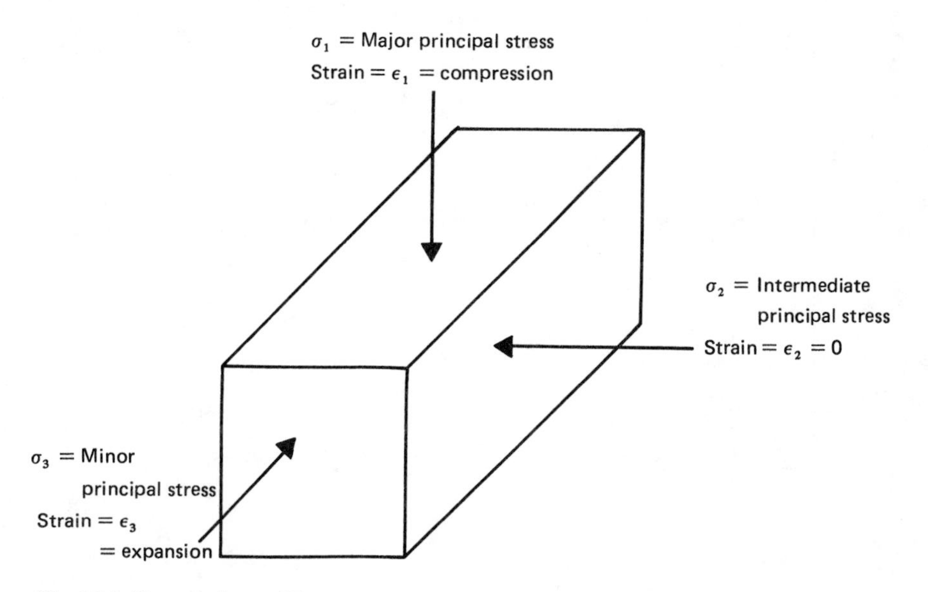

Fig. 7.15 Plane strain condition.

Table 7.3 Comparison of results from plane strain and triaxial compression tests

Soil	$\Phi_p - \Phi_t$, deg	Plane strain apparatus and comments	Reference
Sand	+8	Cube–vary all three stresses as desired	Kjellman (1936)
Dense sand Loose sand	+4 −1	Direct shear; critical void ratio is higher in plane strain	Taylor (1939)
Sand	+5	Direct shear	Hennes (1952)
Dense sand Loose sand	+4 −2	Direct shear Direct shear	Nash (1953)
Sand, gravel, and lead shot	+2 to +7	Direct shear	Bishop (1954)
Sand	+8	Direct shear	Peltier (1957)
Sand	+2	Hollow cylinder; failed by increasing internal radial pressure	Kirkpatrick (1957)
Compacted clay	+2 to +4 effective-stress basis	2 × 4 × 16 in plain strain apparatus, effective-stress basis; plane strain gives lower strain at failure and higher modulus; plane strain gives higher pore pressure at failure	Bishop (1957, 1961)
Dense sand	+4	Plane strain apparatus	Bishop (1961)
Loose sand	+0	Drained tests	Cornforth (1964)
Sand	+4 to +5	Active earth pressure on model retaining wall	Christensen (1961)

Ottawa sand	+2 to +5	Bearing capacity of model strip footings	Selig and McKee (1961)
Sand	+3 to +4	Vacuum compression on long rectangular specimens; modulus greater and strain to failure less in plane strain	Bjerrum and Kummeneje (1961)
Ottawa sand	+6	Hollow cylinders failed by increasing outside radial pressure	Whitman and Luscher (1962)
Ottawa sand	+5	Hollow cylinders failed by increasing outside or axial stress while measuring the other two stresses	Wu et al. (1963)
Ottawa sand	–4 to –6	Torsion tests on very thin annular rings of soil at various rates of strain	Healey (1963)
Compacted clay	+2 to +4 effective-stress basis	Rectangular plane strain apparatus; effective-stress basis modulus greater and strain to failure less in plane strain	Finn and Mittal (1964)
Glass spheres	$\Phi_p > \Phi_t$	Rectangular plane strain apparatus	Leussink and Wittke (1964)
Dense sand Loose sand	+5 +3	Bishop plane strain apparatus Plane strain gives lower strain to failure and less dilatant volume change at failure	Wade (1963)

Table 7.3 *(Continued)*

Soil	$\Phi_p - \Phi_t$, deg	Plane strain apparatus and comments	Reference
Saturated NC silty clay	$+3\frac{1}{2}$ effective stresses	Plane strain apparatus; S_u/p' greater for plane strain	Duncan and Seed (1965, 1966)
Saturated clay	+1 effective-stress basis	Bishop plane strain apparatus, ACU tests; pore pressure at failure approximately the same	Henkel and Wade (1966)
Ottawa		Vacuum plane strain and vacuum triaxial	Sultan and Seed (1967)
dense	+3		
loose	+1		
Monterey sand			
dense	+3		
loose	$+\frac{1}{2}$		
Dense fine sand			
low pressure	+2	Direct shear	Lee (1970
elevated pressure	0		

After K. L. Lee, Comparison of Plane Strain and Triaxial Tests on Sand, *J. Soil Mech. Fcund. Div., ASCE,* vol. 96, no. SM3, 1970.

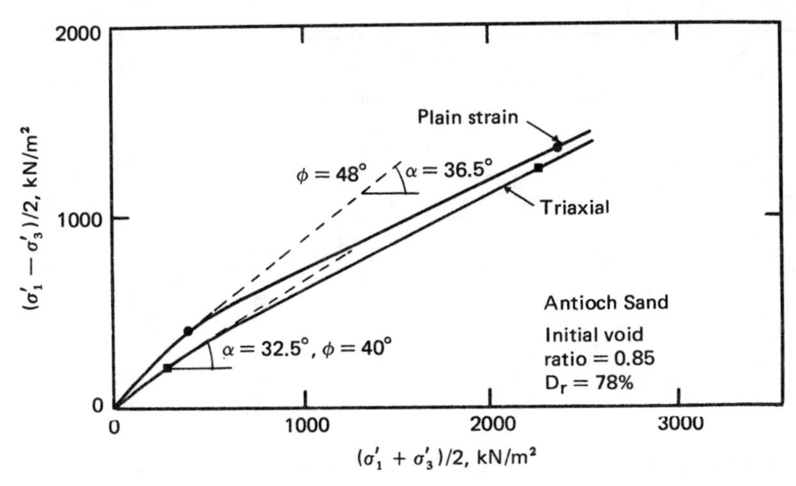

Fig. 7.16 Strength of Antioch sand under drained condition. *(Redrawn after K. L. Lee, Comparison of Plane Strain and Triaxial Tests on Sand,* J. Soil Mech. Found. Div., ASCE, *vol. 96, no. SM3, 1970.)*

The results of this study are very instructive, and highlights are illustrated in Figs. 7.16 to 7.18. Figure 7.16 shows the modified Mohr's diagrams (this is the p' vs. q' plot at failure or the K_f line explained in Fig. 6.35*b*) for triaxial and plane strain tests. Also note that the angle α in Fig. 7.16 is defined by Eq. (6.63) as $\alpha = \tan^{-1} (\sin \phi)$. The failure envelopes are somewhat curved, as explained in Sec. 7.2.4. At low confining pressures, σ_3', the value of ϕ_p is higher by as much as 8° and as the confining pressure

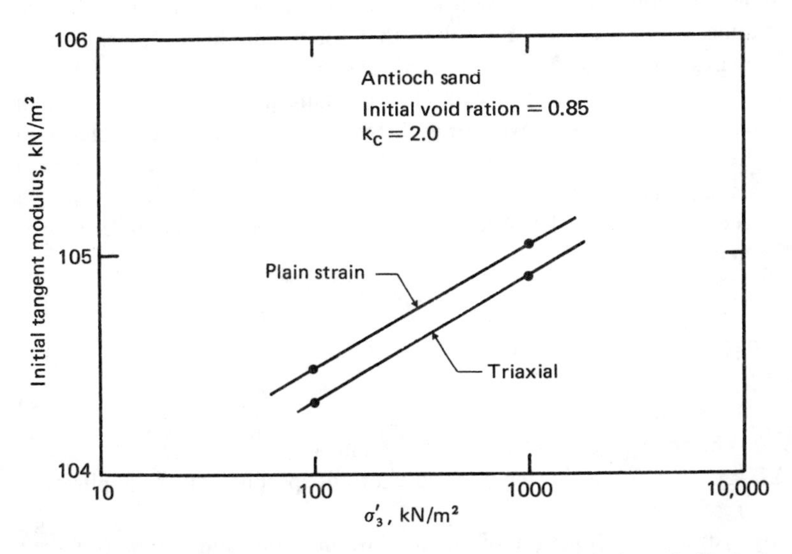

Fig. 7.17 Initial tangent modulus from drained tests on Antioch sand. *(Redrawn after K. L. Lee, Comparison of Plane Strain and Triaxial Tests on Sand,* J. Soil Mech. Found. Div., ASCE, *vol. 96, no. SM3, 1970.)*

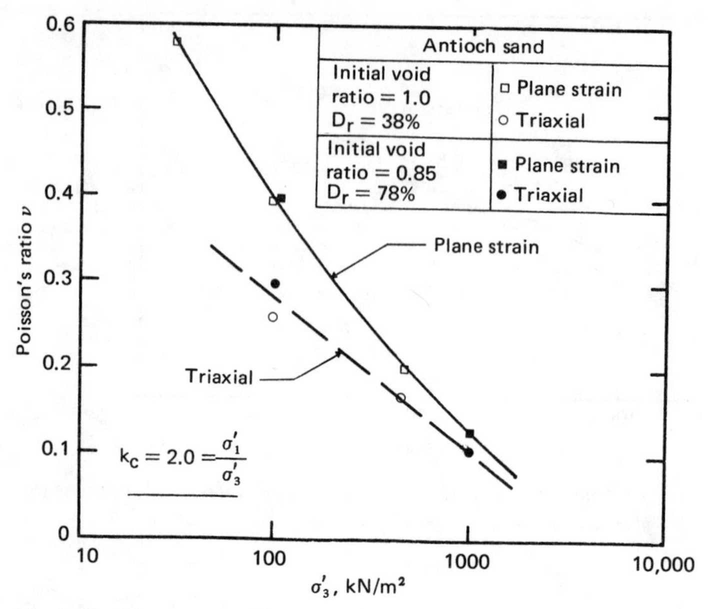

Fig. 7.18 Poisson's ratio from drained tests on Antioch sand. *(Redrawn after K. L. Lee, Comparison of Plane Strain and Triaxial Tests on Sand,* J. Soil Mech. Found. Div., ASCE, *vol. 96, no. SM3, 1970.)*

increases the envelopes flatten. At high values of σ'_3, the slopes of the two envelopes are approximately the same.

Figure 7.17 shows the results of the initial tangent modulus ($E = \Delta\sigma'/\Delta\epsilon_1$) for various confining pressures. For given values of σ'_3, the initial tangent modulus for plane strain loading shows a higher value than that for triaxial loading, although in both cases E increases exponentially with the confining pressure.

The variation of Poisson's ratio ν with the confining pressure for plane strain and triaxial loading conditions is shown in Fig. 7.18. The values of ν were calculated by measuring the change of the volume of specimens and the corresponding axial strains during loading. The derivation of the equations used for finding ν can be explained with the aid of Fig. 7.19. Assuming compressive strain to be positive, for the stresses shown in Fig. 7.19

$$\Delta H = H\epsilon_1 \tag{7.12}$$

$$\Delta B = B\epsilon_2 \tag{7.13}$$

$$\Delta L = L\epsilon_3 \tag{7.14}$$

where H, L, B = height, length, and width of specimen
$\Delta H, \Delta B, \Delta L$ = change in height, length, and width of specimen due to application of stresses
$\epsilon_1, \epsilon_2, \epsilon_3$ = strains in direction of major, intermediate, and minor principal stresses

The volume of the specimen before load application is equal to $V = LBH$, and the volume of the specimen after the load application is equal to $V - \Delta V$. Thus,

$$\Delta V = V - (V - \Delta V) = LBH - (L - \Delta L)(B - \Delta B)(H - \Delta H)$$
$$= LBH - LBH(1 - \epsilon_1)(1 - \epsilon_2)(1 - \epsilon_3) \tag{7.15}$$

where ΔV is change in volume. Neglecting the higher order terms such as $\epsilon_1\epsilon_2$, $\epsilon_2\epsilon_3$, $\epsilon_3\epsilon_1$, and $\epsilon_1\epsilon_2\epsilon_3$, Eq. (7.15) gives

$$v = \frac{\Delta V}{V} = \epsilon_1 + \epsilon_2 + \epsilon_3 \tag{7.16}$$

where v is the change in volume per unit volume of the specimen.

For triaxial tests, $\epsilon_2 = \epsilon_3$, and they are expansions (negative sign). So, $\epsilon_2 = \epsilon_3 = -\nu\epsilon_1$. Substituting this into Eq. (7.16), we get $v = \epsilon_1(1 - 2\nu)$, or

$$\nu = \frac{1}{2}\left(1 - \frac{v}{\epsilon_1}\right) \qquad \text{(for triaxial test conditions)} \tag{7.17}$$

With plane strain loading conditions, $\epsilon_2 = 0$ and $\epsilon_3 = -\nu\epsilon_1$. Hence, from Eq. (7.16), $v = \epsilon_1(1 - \nu)$, or

$$\nu = 1 - \frac{v}{\epsilon_1} \qquad \text{(for plane strain conditions)} \tag{7.18}$$

Figure 7.18 shows that for a given value of σ_3' the Poisson's ratio obtained from plane strain loading is higher than that obtained from triaxial loading.

Hence, based on the available information at this time, it can be concluded that ϕ_p exceeds the value of ϕ_t by 0 to 8°. The greatest difference is associated with dense sands at low confining pressures. The smaller differences are associated with loose

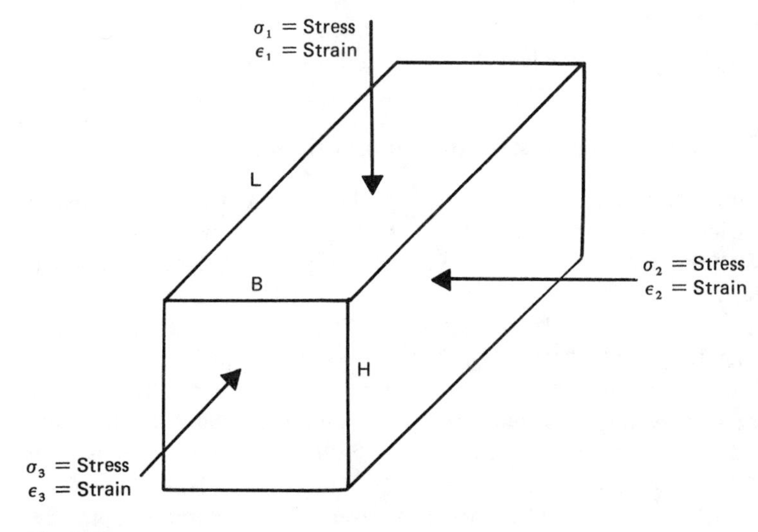

Fig. 7.19

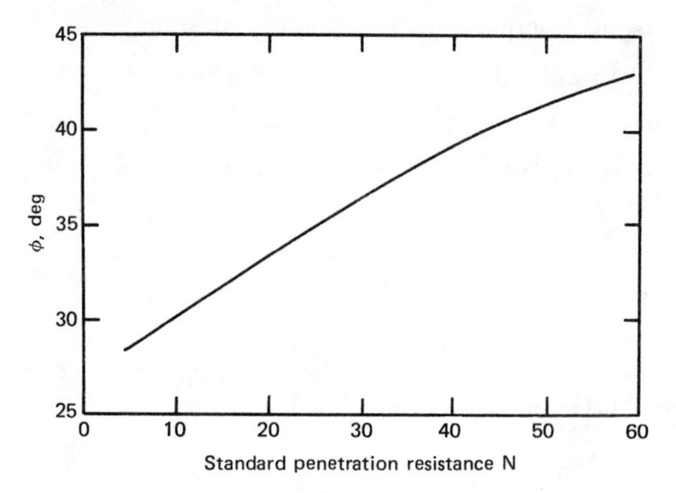

Fig. 7.20 Plot of standard penetration resistance vs. angle of friction for granular soil. *(Replotted after R. B. Peck, W. E. Hanson, and T. H. Thornburn, "Foundation Engineering," 2d ed., Wiley, New York, 1974.)*

sands at all confining pressures, or dense sand at high confining pressures. Although still disputed, based on the studies described in this section several suggestions have been made to use a value of $\phi \approx \phi_p = 1.1\phi_t$ for calculation of the bearing capacity of strip foundations. For rectangular foundations, the stress conditions on the soil cannot be approximated by either triaxial or plane strain loadings. Meyerhof (1963) suggested for this case that the friction angle to be used for calculation of the ultimate bearing capacity should be approximated as

$$\phi = \left(1.1 - 0.1\frac{B_f}{L_f}\right)\phi_t \tag{7.19}$$

where L_f is the length of foundation and B_f the width of foundation.

7.2.7 Other Correlations for Determination of Friction Angle

It is difficult in practice to obtain undisturbed samples of sand and gravelly soils for determination of shear strength parameters. For that reason, several approximate correlations have been developed over the years for approximate determination of the friction angle, ϕ.

Peck et al. (1974) gave a correlation of ϕ with the standard penetration resistance N obtained during field exploration (Fig. 7.20). The friction angle increases from about 27° to about 42° for N varying from about 5 to 60. The actual friction angle for a soil depends on several factors, and hence it may deviate by about ±3° from that shown in Fig. 7.20. The relationship shown in Fig. 7.20 is applicable for soils up to a depth of about 40 to 50 ft (about 12 to 15 m).

Meyerhof (1956), based on the observations of several field explorations, provided

Table 7.4 Relationship between relative density, penetration resistance, and angle of friction of cohesionless soils

State of packing	Relative density	Standard penetration resistance N, blows/ft	Static cone resistance q_c ton/ft²	Angle of friction ϕ, deg
Very Loose	< 0.2	< 4	< 20	< 30
Loose	0.2 to 0.4	4 to 10	20 to 40	30 to 35
Compact	0.4 to 0.6	10 to 30	40 to 120	35 to 40
Dense	0.6 to 0.8	30 to 50	120 to 200	40 to 45
Very Dense	> 0.8	> 50	> 200	> 45

After G. G. Meyerhoff, Penetration Test and the Bearing Capacity of Cohesionless Soils, *J. Soil Mech. Found. Div., ASCE,* vol. 82, no. 1, proc. paper 866, 1956.

the relationship between the soil friction angle, standard penetration resistance, and static cone penetration resistance. His results are given in Table 7.4.

Figure 7.21, which is based on the results of D'Appolonia et al. (1968), shows a relationship between the soil friction angle with relative density for a dune sand along the south shore of Lake Michigan. The minimum and maximum dry unit weights of this dune sand are 88.5 lb/ft³ (13.91 kN/m³) and 110 lb/ft³ (17.29 kN/m³), respectively.

The empirical relationships provided in this section should be used with caution. However, they are useful for initial estimation of the shear strength of granular soils.

7.3 SHEAR STRENGTH OF COHESIVE SOILS

The shear strength of cohesive soils can, generally, be determined in the laboratory by either direct shear test equipment or triaxial shear test equipment; however, the

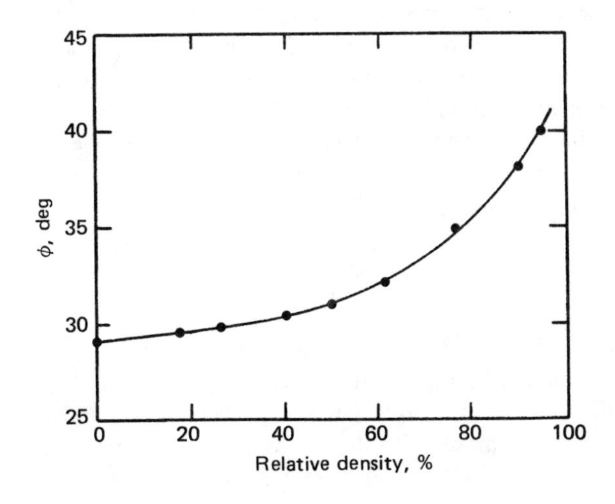

Fig. 7.21 Relation between angle of friction and relative density for Dune sand on the south shore of Lake Michigan. *(Plotted from results of D. J. D'Appolonia, E. D'Appolonia, and R. F. Brissette, Settlement of Spread Footings on Sand,* J. Soil Mech. Found. Div., ASCE, *vol. 94, no. SM3, 1968.)*

triaxial test is more commonly used. Only the shear strength of saturated cohesive soils will be treated here. The shear strength based on the effective stress can be given by [Eq. (7.3)] $s = c + \sigma' \tan \phi$. For normally consolidated clays, $c \approx 0$; and, for overconsolidated clays, $c > 0$.

7.3.1 Triaxial Testing in Clays

The basic features of the triaxial test equipment were shown in Fig. 7.7. Three conventional types of tests are conducted with clay soils in the laboratory:

1. Consolidated drained test or drained test (CD test or D test).
2. Consolidated undrained test (CU test).
3. Unconsolidated undrained test (UU test).

Each of these tests will be separately considered in the following sections.

Consolidated drained test. For the consolidated drained test the saturated soil specimen is first subjected to a confining pressure σ_3 through the chamber fluid; as a result, the pore water pressure of the sample will increase by u_c. The connection to the drainage is kept open for complete drainage so that u_c becomes equal to zero. Then the deviator stress (piston stress) $\Delta\sigma$ is increased at a very slow rate, keeping the drainage valve open to allow complete dissipation of the resulting pore water pressure u_d. Figure 7.22 shows the nature of the variation of the deviator stress with axial strain. From Fig. 7.22, it must also be pointed out that, during the application of the deviator stress, the volume of the specimen gradually reduces for normally consolidated clays. However, overconsolidated clays go through some reduction of volume initially but then expand. In a consolidated drained test, the total stress is equal to the effective stress since the excess pore water pressure is zero. At failure, the maximum *effective* principal stress is $\sigma_1' = \sigma_1 = \sigma_3 + \Delta\sigma_f$, where $\Delta\sigma_f$ is the deviator stress at failure. *The minimum* effective principal stress is $\sigma_3' = \sigma_3$.

From the results of a number of tests conducted using several specimens, the Mohr's circles at failure can be plotted as shown in Fig. 7.23. The values of c and ϕ are obtained by drawing a common tangent to these Mohr's circles, which is the Mohr-Coulomb envelope. For normally consolidated clays (Fig. 7.23*a*), we can see that $c = 0$. Thus the equation of the Mohr-Coulomb envelope can be given by $s = \sigma' \tan \phi$. The slope of the failure envelope will give us the angle of friction of the soil. As shown by Eq. (7.5), for these soils

$$\sin \phi = \left(\frac{\sigma_1' - \sigma_3'}{\sigma_1' + \sigma_3'}\right)_{\text{failure}} \qquad \text{or} \qquad \sigma_1' = \sigma_3' \tan^2\left(45° + \frac{\phi}{2}\right)$$

The plane of failure makes an angle of $45° + \phi/2$ with the major principal plane.

The procedure for drawing a modified Mohr's failure envelope [i.e., the K_f line from the plot of $q' = (\sigma_1' - \sigma_3')/2$ vs. $p' = (\sigma_1' + \sigma_3')/2$] was explained in Sec. 6.5.1 using Fig. 6.35. The angle that the K_f line makes with the p' axis is equal to α and can be given by the relation [Eq. (6.63)] $\alpha = \tan^{-1}(\sin \phi)$. Figure 7.24 shows the range

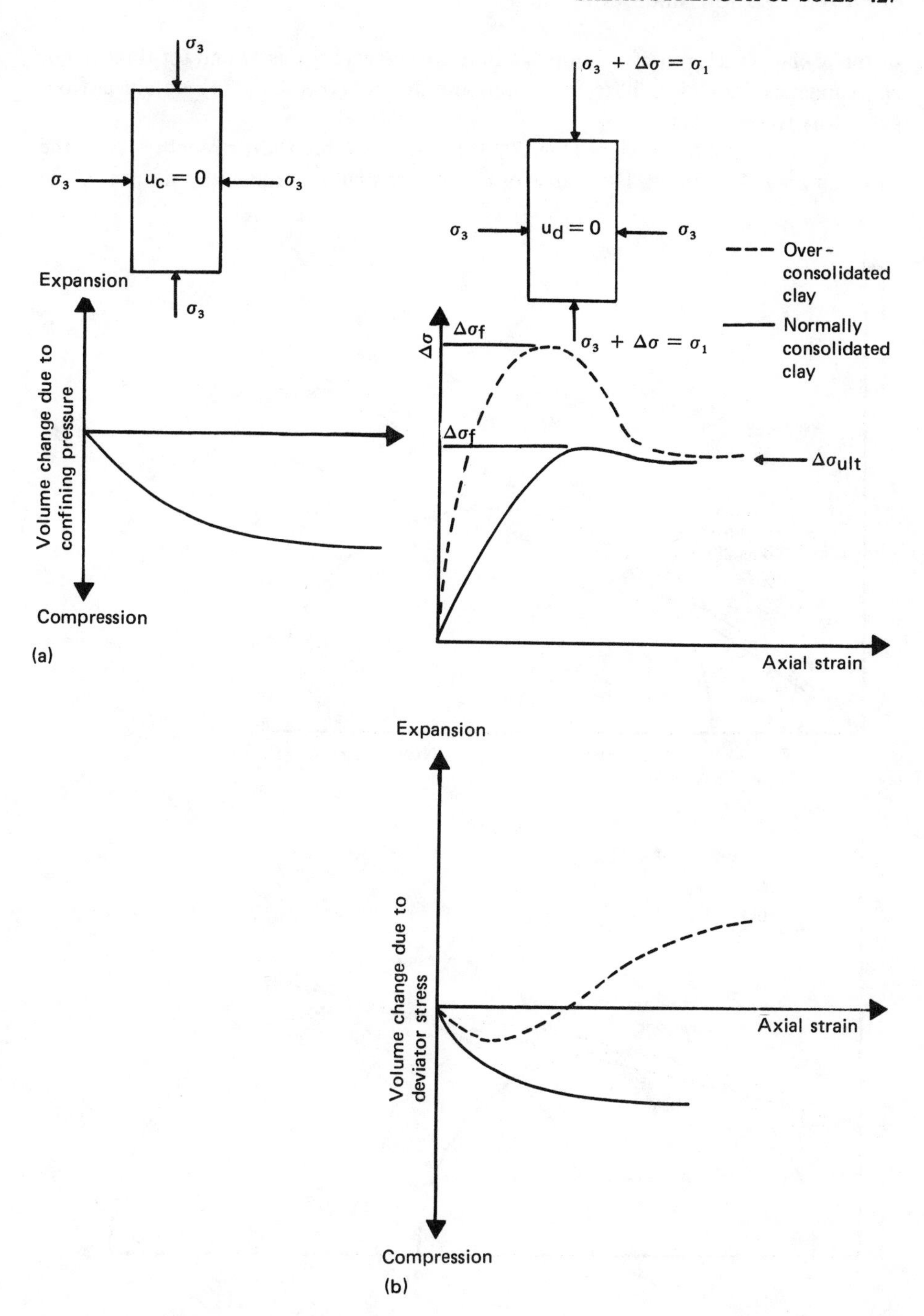

Fig. 7.22 Consolidated drained triaxial tests in clay. (*a*) Application of confining pressure. (*b*) Application of deviator stress.

of the K_f lines (i.e., modified Mohr's failure envelopes) for quartz and for three major clay minerals–kaolinite, illite, and montmorillonite–(Olson, 1974) obtained from laboratory triaxial tests.

For overconsolidated clays (Fig. 7.23*b*), $c \neq 0$. So, the shear strength follows the equation $s = c + \sigma' \tan \phi$. The values of c and ϕ can be determined by measuring the

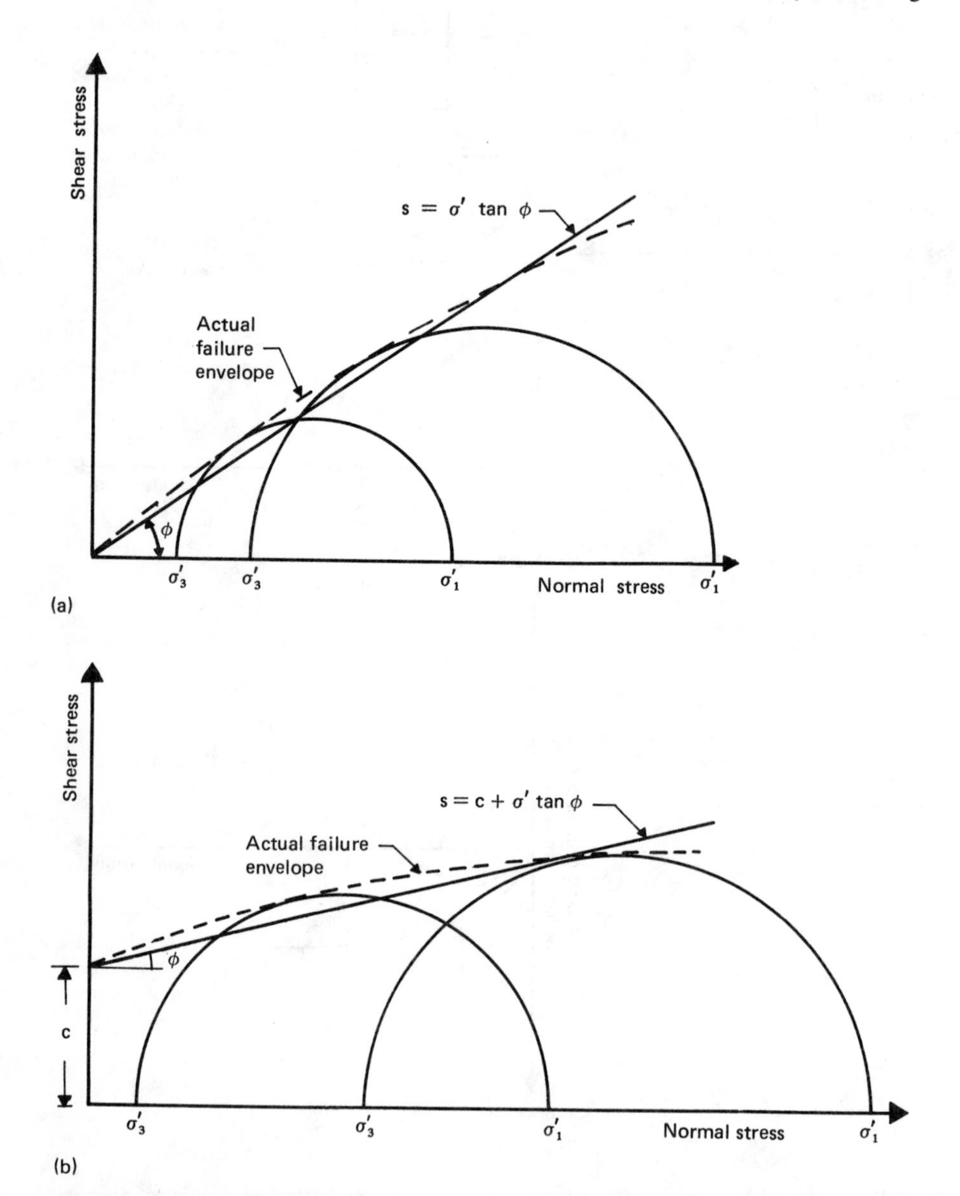

Fig. 7.23 Failure envelope for (*a*) normally consolidated and (*b*) overconsolidated clays from consolidated drained triaxial tests.

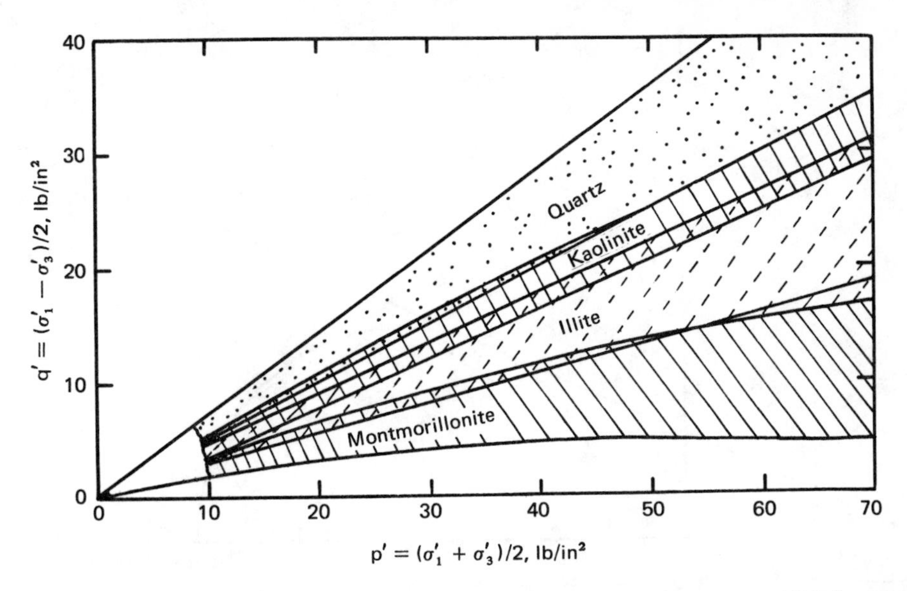

Fig. 7.24 Modified Mohr's failure envelope for quartz and clay minerals. (Note: 1 lb/in² = 6.9 kN/m².) *(Replotted after R. E. Olson, Shearing Strength of Kaolinite, Illite and Montmorillonite,* J. Geotech. Eng. Div., ASCE, *vol. 100, no. GT11, 1974.)*

intercept of the failure envelope on the shear stress axis and the slope of the failure envelope, respectively. To obtain a general relation between σ_1', σ_3', c, and ϕ, we refer to Fig. 7.25, from which

$$\sin\phi = \frac{\overline{ac}}{\overline{bO} + \overline{Oa}} = \frac{(\sigma_1' - \sigma_3')/2}{c\cot\phi + (\sigma_1' + \sigma_3')/2} \tag{7.20}$$

or

$$\sigma_1'(1 - \sin\phi) = 2c\cos\phi + \sigma_3'(1 + \sin\phi)$$

$$\sigma_1' = \sigma_3 \frac{1 + \sin\phi}{1 - \sin\phi} + \frac{2c\cos\phi}{1 - \sin\phi}$$

$$\sigma_1' = \sigma_3 \tan^2\left(45° + \frac{\phi}{2}\right) + 2c\tan\left(45° + \frac{\phi}{2}\right) \tag{7.21}$$

Note that the plane of failure makes an angle of $45° + \phi/2$ with the major principal plane.

If a clay is initially consolidated by an encompassing chamber pressure of $\sigma_c = \sigma_c'$ and allowed to swell under a reduced chamber pressure of $\sigma_3 = \sigma_3'$, the specimen will be overconsolidated. The failure envelope obtained from consolidated drained triaxial tests of these types of specimens has two distinct branches, as shown in Fig. 7.26. Portion ab of the failure envelope has a flatter slope with a cohesion intercept, and the portion bc represents a normally consolidated stage following the equation $s = \sigma' \tan\phi_{bc}$.

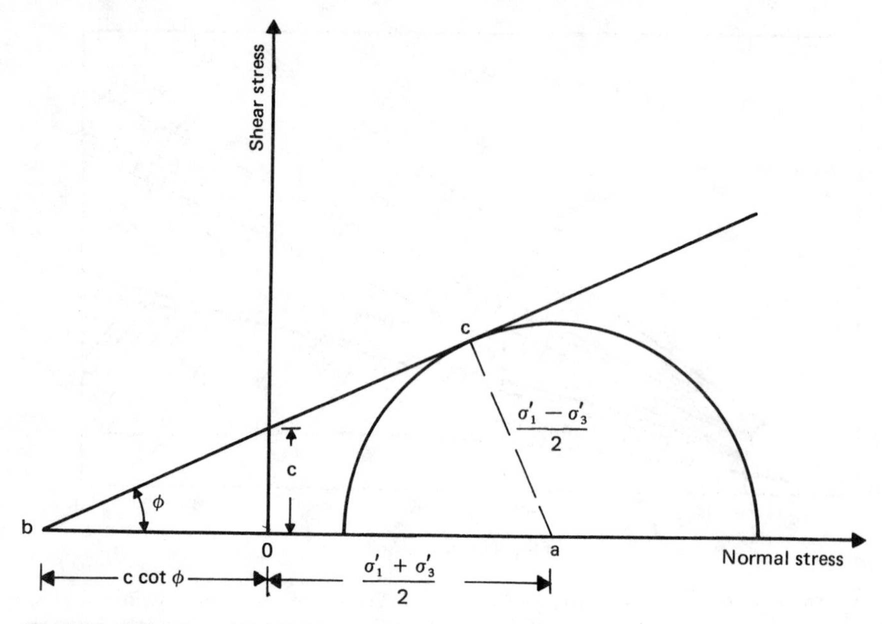

Fig. 7.25 Derivation of Eq. (7.21).

It may also be seen from Fig. 7.22 that at very large strains the deviator stress reaches a constant value. The shear strength of clays at very large strains is referred to as *residual shear strength* (i.e., the ultimate shear strength). It has been proved that the residual strength of a given soil is independent of past stress history, and it can be given by the equation (see Fig. 7.27)

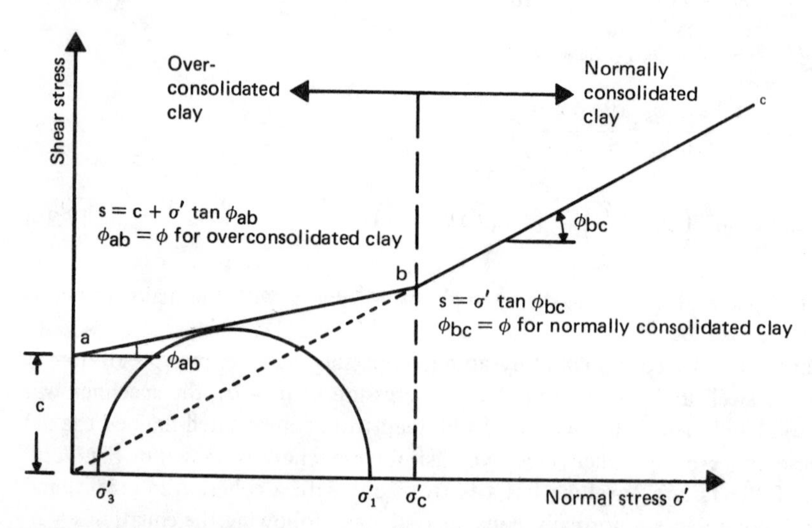

Fig. 7.26 Failure envelope of a clay with preconsolidation pressure $= \sigma_c'$.

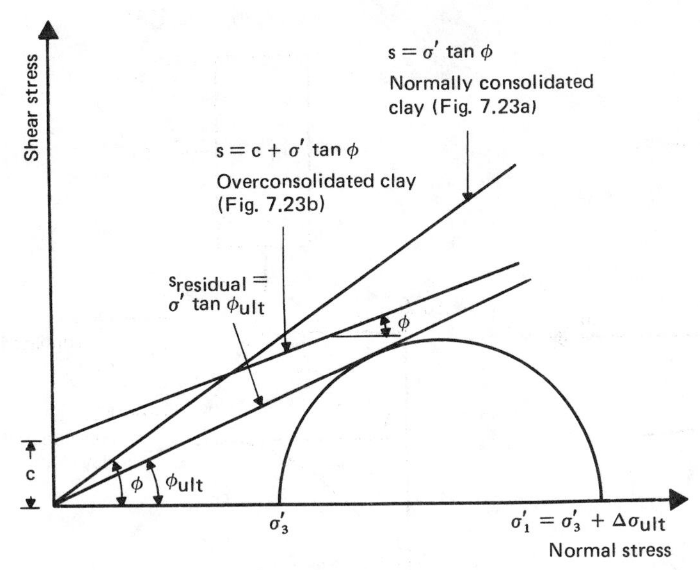

Fig. 7.27 Residual shear strength of clay.

$$s_{residual} = \sigma' \tan \phi_{ult} \tag{7.22}$$

(i.e., the c component is 0). For triaxial tests,

$$\phi_{ult} = \sin^{-1}\left(\frac{\sigma_1' - \sigma_3'}{\sigma_1' + \sigma_3'}\right)_{residual} \tag{7.23}$$

where $\sigma_1' = \sigma_3' + \Delta\sigma_{ult}$.

The residual friction angle in clays is of importance in subjects such as the long-term stability of slopes.

The consolidated drained triaxial test procedure described above is the conventional type. However, failure in the soil specimens can be produced by any one of the methods of axial compression or axial extension as described in Sec. 7.2.2 (with reference to Fig. 7.10) allowing full drainage condition.

Consolidated undrained test. In the consolidated undrained test, the soil specimen is first consolidated by a chamber confining pressure σ_3; full drainage from the specimen is allowed. After complete dissipation of excess pore water pressure, u_c, generated by the confining pressure, the deviator stress $\Delta\sigma$ is increased to cause failure of the specimen. During this phase of loading, the drainage line from the specimen is closed. Since drainage is not permitted, the pore water pressure (pore water pressure due to deviator stress, u_d) in the specimen increases. Simultaneous measurements of $\Delta\sigma$ and u_d are made during the test. Figure 7.28 shows the nature of the variation of $\Delta\sigma$ and u_d with axial strain; also shown is the nature of the variation of the pore water pressure parameter A [$A = u_d/\Delta\sigma$; see Eq. (4.5)] with axial strain. The value of A at failure, A_f, is positive for normally consolidated clays and becomes negative for overconsoli-

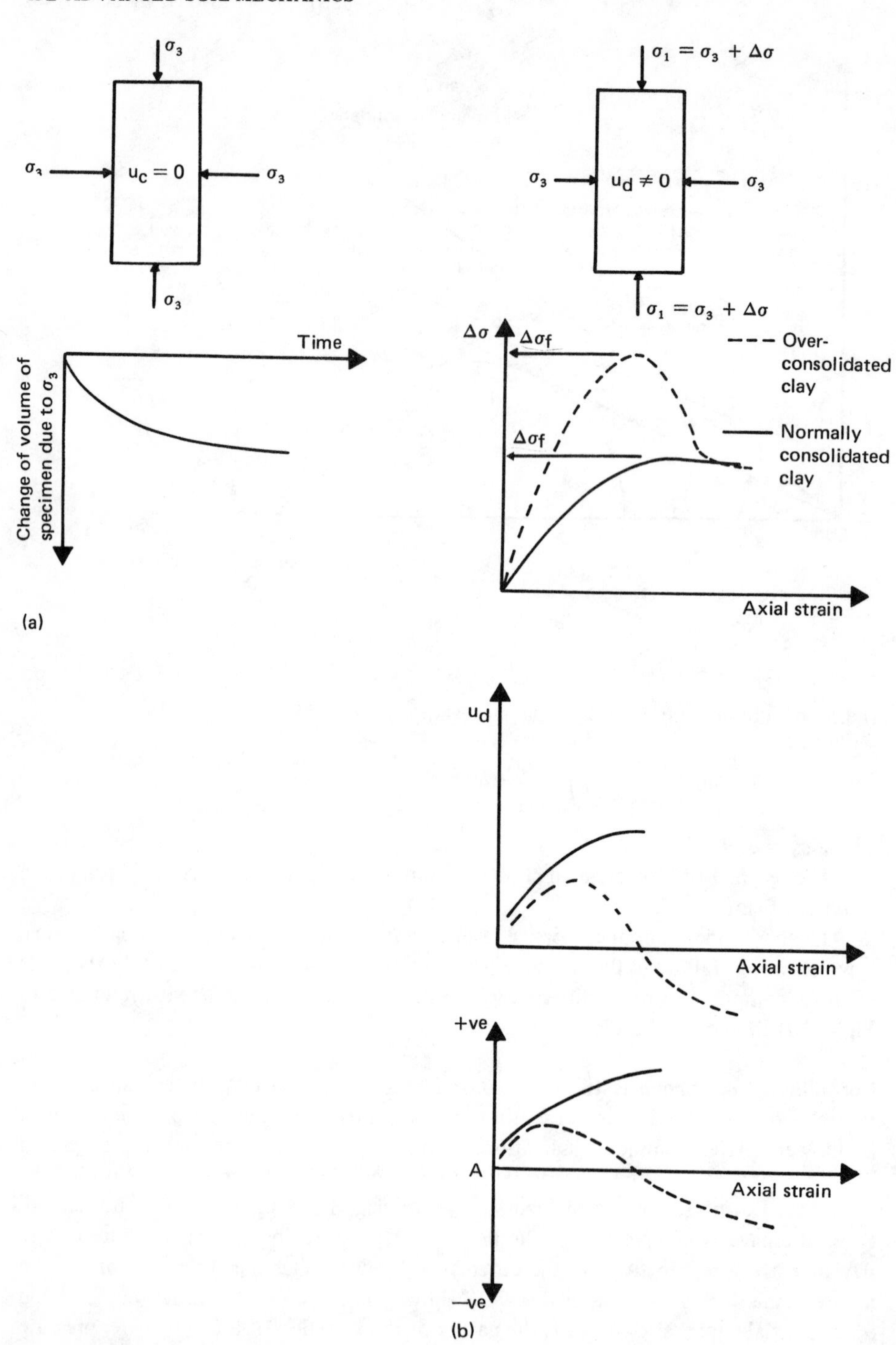

Fig. 7.28 Consolidated undrained triaxial test. (*a*) Application of confining pressure. (*b*) Application of deviator stress.

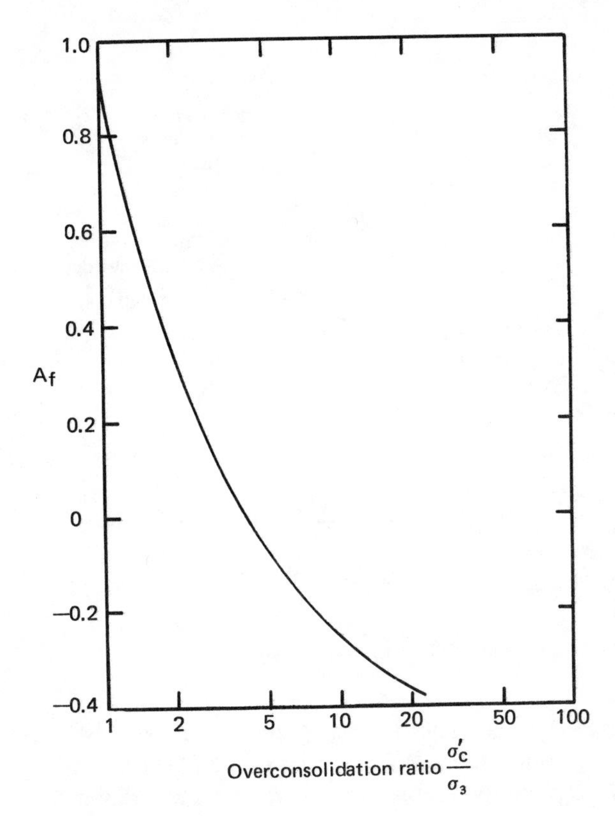

Fig. 7.29 Variation of A_f with overconsolidation ratio for Weald clay. *(Redrawn after N. E. Simons, The Effect of Overconsolidation on the Shear Strength of an Undisturbed Oslo Clay,* Proc. Research Conference on Shear Strength of Cohesive Soils, ASCE, *1960.)*

dated clays (also see Table 7.6). Thus, A_f is dependent on the overconsolidation ratio. The overconsolidation ratio, OCR, for triaxial test conditions may be defined as

$$\text{OCR} = \frac{\sigma_c'}{\sigma_3} \tag{7.24}$$

where $\sigma_c' = \sigma_c$ is the maximum chamber pressure at which the specimen is consolidated and then allowed to rebound under a chamber pressure of σ_3.

The typical nature of the variation of A_f with the overconsolidation ratio for Weald clay is shown in Fig. 7.29.

At failure,

total major principal stress $= \sigma_1 = \sigma_3 + \Delta\sigma_f$
total minor principal stress $= \sigma_3$
pore water pressure at failure $= u_{d(\text{failure})} = A_f\, \Delta\sigma_f$
effective major principal stress $= \sigma_1 - A_f\, \Delta\sigma_f = \sigma_1'$
effective minor principal stress $= \sigma_3 - A_f\, \Delta\sigma_f = \sigma_3'$

Consolidated undrained tests on a number of specimens can be conducted to determine the shear strength parameters of a soil, as shown for the case of a *normally consolidated*

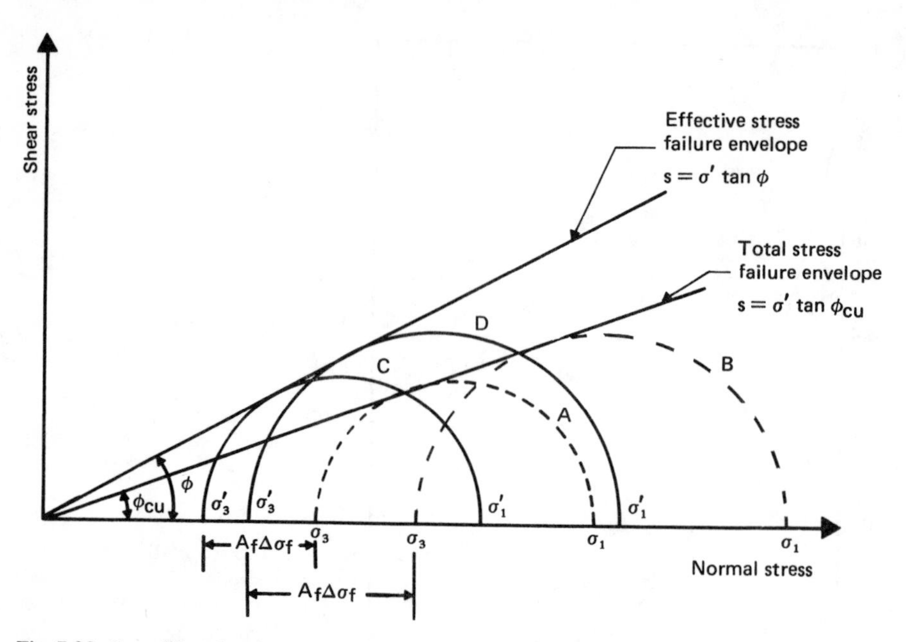

Fig. 7.30 Consolidated undrained test results—normally consolidated clay.

clay in Fig. 7.30. The total-stress Mohr's circles (circles A and B) for two tests are shown by the broken lines. The *effective*-stress Mohr's circles C and D correspond to the *total*-stress circles A and B, respectively. Since C and D are *effective*-stress circles at failure, a common tangent drawn to these circles will give the Mohr–Coulomb failure envelope given by the equation $s = \sigma' \tan \phi$. If we draw a common tangent to the *total*-stress circles, it will be a straight line passing through the origin. This is the *total*-stress failure envelope, and it may be given by

$$s = \sigma \tan \phi_{cu} \tag{7.25}$$

where ϕ_{cu} is the consolidated undrained angle of friction.

The *total*-stress failure envelope for an overconsolidated clay will be of the nature shown in Fig. 7.31 and can be given by the relation

$$s = c_{cu} + \sigma \tan \phi_{cu} \tag{7.26}$$

where c_{cu} is the intercept of the *total*-stress failure envelope along the shear stress axis.

The shear strength parameters for overconsolidated clay based on effective stress, i.e., c and ϕ, can be obtained by plotting the effective-stress Mohr's circle and then drawing a common tangent to c and ϕ.

As in consolidated drained tests, shear failure in the specimen can be produced by axial compression or extension by changing the loading conditions (Sec. 7.2.2).

Unconsolidated undrained test. In unconsolidated undrained triaxial tests, drainage from the specimen is not allowed at any stage. First, the chamber confining pressure

σ_3 is applied, after which the deviator stress $\Delta\sigma$ is increased until failure occurs. For these tests,

total major principal stress $= \sigma_3 + \Delta\sigma_f = \sigma_1$
total minor principal stress $= \sigma_3$

Tests of this type can be performed quickly since drainage is not allowed. For a saturated soil, the deviator stress at failure, $\Delta\sigma_f$, is practically the same irrespective of the confining pressure σ_3 (Fig. 7.32). So, the total-stress failure envelope can be assumed to be a horizontal line, and $\phi = 0$. The undrained shear strength can be expressed as

$$s = S_u = \frac{\Delta\sigma_f}{2} \tag{7.27}$$

This generally referred to as the shear strength based on $\phi = 0$ concept.

The fact that the strength of saturated clays in unconsolidated undrained loading conditions is the same irrespective of the confining pressure σ_3 can be explained with the help of Fig. 7.33. If a saturated clay specimen A is consolidated under a chamber confining pressure of σ_3 and then sheared to failure under undrained conditions, the Mohr's circle at failure will be represented by circle no. 1. The effective-stress Mohr's circle corresponding to circle no. 1 is circle no. 2, which touches the effective-stress failure envelope. If a similar soil specimen B, consolidated under a chamber confining pressure of σ_3, is subjected to an additional confining pressure of $\Delta\sigma_3$ without allowing drainage, the pore water pressure will increase by Δu_c. We saw in Chap. 4 that $\Delta u_c = B\,\Delta\sigma_3$ and, for saturated soils, $B = 1$. So, $\Delta u_c = \Delta\sigma_3$.

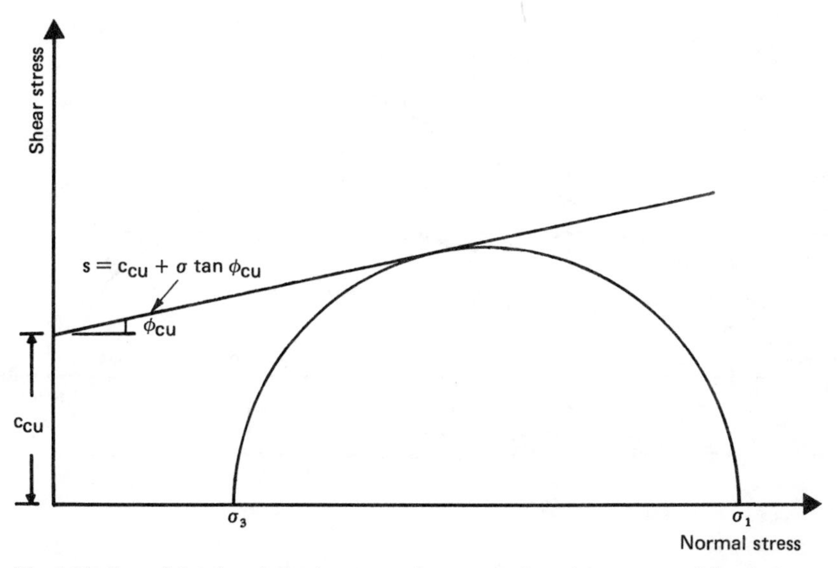

Fig. 7.31 Consolidated undrained test – total stress envelope for overconsolidated clay.

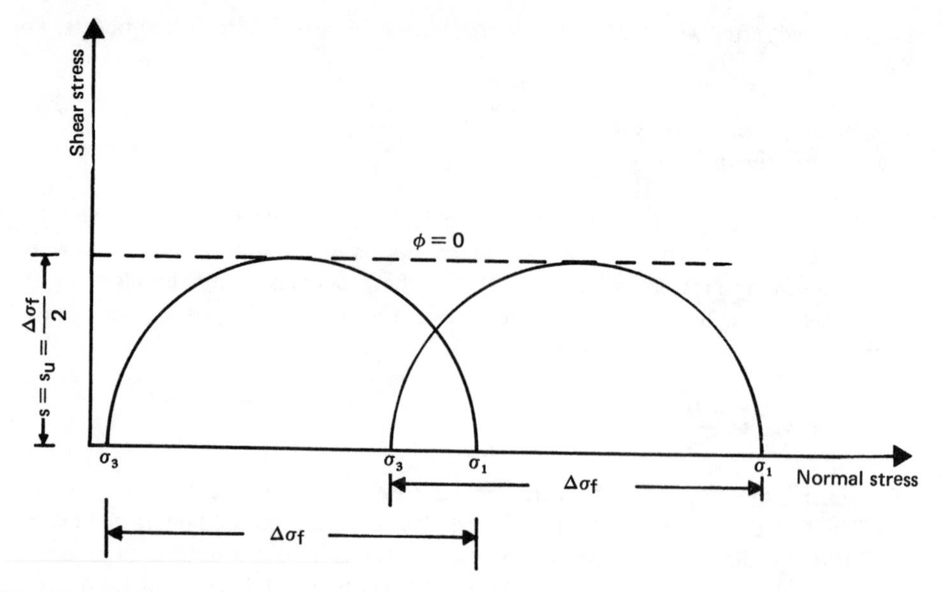

Fig. 7.32 Unconsolidated undrained triaxial test.

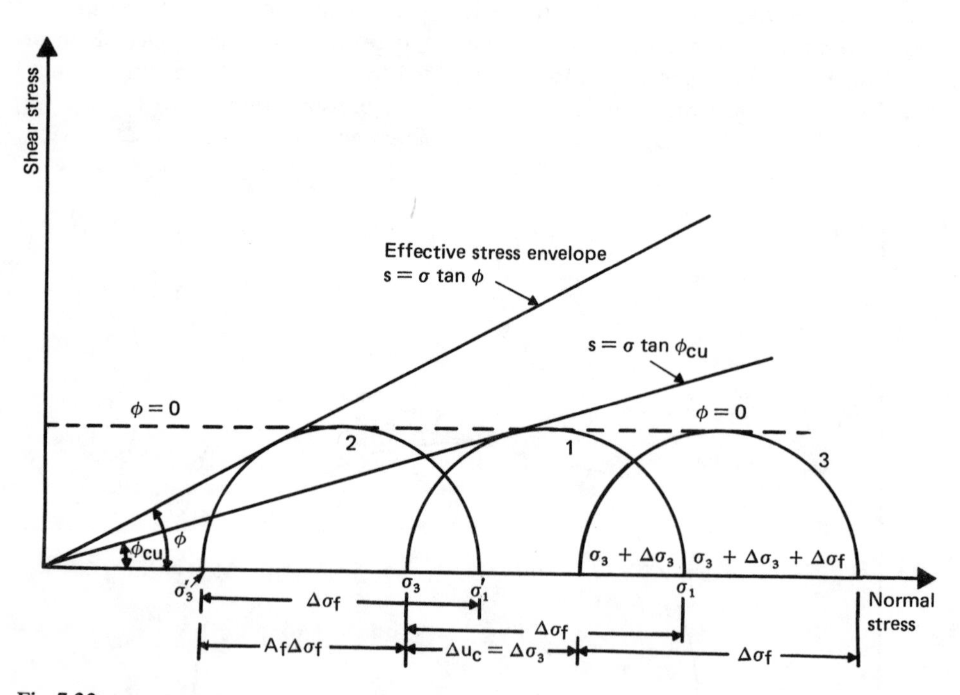

Fig. 7.33

Since the effective confining pressure of specimen B is the same as specimen A, it will fail with the same deviator stress, $\Delta\sigma_f$. The total-stress Mohr's circle for this specimen (i.e., B) at failure can be given by circle no. 3. So, at failure, for specimen B

$$\text{Total minor principal stress} = \sigma_3 + \Delta\sigma_3$$

$$\text{Total major principal stress} = \sigma_3 + \Delta\sigma_3 + \Delta\sigma_f$$

The effective stresses for the specimen are as follows:

$$\begin{aligned}\text{Effective major principal stress} &= (\sigma_3 + \Delta\sigma_3 + \Delta\sigma_f) - (\Delta u_c + A_f\,\Delta\sigma_f)\\ &= (\sigma_3 + \Delta\sigma_f) - A_f\,\Delta\sigma_f\\ &= \sigma_1 - A_f\,\Delta\sigma_f = \sigma_1'\end{aligned}$$

$$\begin{aligned}\text{Effective minor principal stress} &= (\sigma_3 + \Delta\sigma_3) - (\Delta u_c + A_f\,\Delta\sigma_f)\\ &= \sigma_3 - A_f\,\Delta\sigma_f = \sigma_3'\end{aligned}$$

The above principal stresses are the same as those we had for specimen A. Thus, the effective-stress Mohr's circle at failure for specimen B will be the same as that for specimen A, i.e., circle no. 1.

The value of $\Delta\sigma_3$ could be of any magnitude in specimen B; in all cases, $\Delta\sigma_f$ would be the same.

Example 7.1 Consolidated drained triaxial tests on two specimens of a soil gave the following results:

Test no.	Confining pressure σ_3, kN/m²	Deviator stress at failure $\Delta\sigma_f$, kN/m²
1	70	440.4
2	92	474.7

Determine the values of c and ϕ for the soil.

SOLUTION From Eq. (7.21), $\sigma_1 = \sigma_3 \tan^2(45° + \phi/2) + 2c\tan(45° + \phi/2)$. For test 1, $\sigma_3 = 70\ \text{kN/m}^2$; $\sigma_1 = \sigma_3 + \Delta\sigma_f = 70 + 440.4 = 510.4\ \text{kN/m}^2$. So,

$$510.4 = 70\tan^2\left(45° + \frac{\phi}{2}\right) + 2c\tan\left(45° + \frac{\phi}{2}\right) \qquad (a)$$

Similarly, for test 2, $\sigma_3 = 92\ \text{kN/m}^2$; $\sigma_1 = 92 + 474.7 = 566.7\ \text{kN/m}^2$. Thus,

$$566.7 = 92\tan^2\left(45° + \frac{\phi}{2}\right) + 2c\tan\left(45° + \frac{\phi}{2}\right) \qquad (b)$$

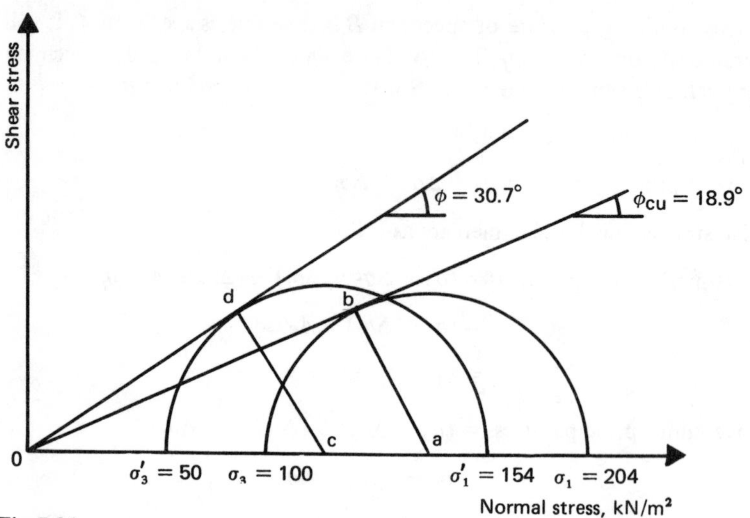

Fig. 7.34

Subtracting Eq. (*a*) from Eq. (*b*),

$$56.3 = 22 \tan^2\left(45° + \frac{\phi}{2}\right)$$

$$\phi = 2\left[\tan^{-1}\left(\frac{56.3}{22}\right)^{1/2} - 45°\right] = 26°$$

Substituting $\phi = 26°$ in Eq. (*a*),

$$c = \frac{510.4 - 70 \tan^2(45° + 26/2)}{2 \tan(45° + 26/2)} = \frac{510.4 - 70(2.56)}{2(1.6)} = 103.5 \text{ kN/m}^2$$

Example 7.2 A normally consolidated clay specimen was subjected to a consolidated undrained test. At failure, $\sigma_3 = 100\,\text{kN/m}^2$; $\sigma_1 = 204\,\text{kN/m}^2$; and $u_d = 50\,\text{kN/m}^2$. Determine ϕ_{cu} and ϕ.

SOLUTION Referring to Fig. 7.34,

$$\sin\phi_{cu} = \frac{\overline{ab}}{\overline{Oa}} = \frac{(\sigma_1 - \sigma_3)/2}{(\sigma_1 + \sigma_3)/2} = \frac{\sigma_1 - \sigma_3}{\sigma_1 + \sigma_3} = \frac{204 - 100}{204 + 100} = \frac{96}{304}$$

hence

$$\phi_{cu} = 18.9°$$

Again,

$$\sin\phi = \frac{\overline{cd}}{\overline{Oc}} = \frac{\sigma_1' - \sigma_3'}{\sigma_1' + \sigma_3'}$$

$\sigma_3' = 100 - 50 = 50\,\text{kN/m}^2$

$\sigma_1' = 204 - 50 = 154\,\text{kN/m}^2$

So,

$$\sin\phi = \frac{154 - 50}{154 + 50} = \frac{104}{204}$$

hence

$\phi = 30.7°$

Example 7.3 For a saturated clay soil, the following are the results of some consolidated drained triaxial tests at failure:

Test no.	$p' = \frac{\sigma_1' + \sigma_3'}{2}$, lb/in²	$q' = \frac{\sigma_1' - \sigma_3'}{2}$, lb/in²
1	60	25.6
2	90	36.5
3	110	44.0
4	180	68.0

Draw a p' vs. q' diagram, and from that determine c and ϕ for the soil.

SOLUTION The diagram of q' vs. p' is shown in Fig. 7.35; this is a straight line, and the equation of it may be written in the form

$$q' = m + p' \tan\alpha \tag{a}$$

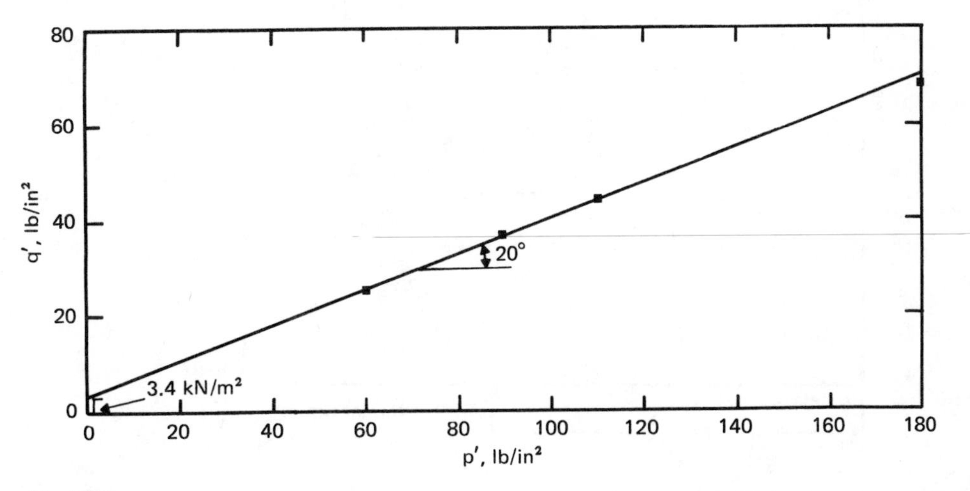

Fig. 7.35

Now, Eq. (7.20) can be written in the form

$$\frac{\sigma_1' - \sigma_3'}{2} = c \cos \phi + \frac{\sigma_1' + \sigma_3'}{2} \sin \phi \tag{b}$$

Comparing Eqs. (*a*) and (*b*), we find $m = c \cos \phi$ or $c = m/\cos \phi$ and $\tan \alpha = \sin \phi$. From Fig. 7.35, $m = 3.4 \text{ lb/in}^2$ and $\alpha = 20°$. So

$$\phi = \sin^{-1} (\tan 20°) = \mathit{21.34°}$$

and

$$c = \frac{m}{\cos \alpha} = \frac{3.4}{\cos 21.34°} = \mathit{3.65\ lb/in^2}$$

7.3.2 Unconfined Compression Test

The unconfined compression test is a special case of the unconsolidated undrained triaxial test. In this case, no confining pressure to the specimen is applied (i.e., $\sigma_3 = 0$). For such conditions, for saturated clays, the pore water pressure in the specimen at the beginning of the test is negative (capillary pressure). Axial stress on the specimen is gradually increased until the specimen fails (Fig. 7.36). At failure, $\sigma_3 = 0$ and so

$$\sigma_1 = \sigma_3 + \Delta\sigma_f = \Delta\sigma_f = q_u \tag{7.28}$$

where q_u is the unconfined compression strength.

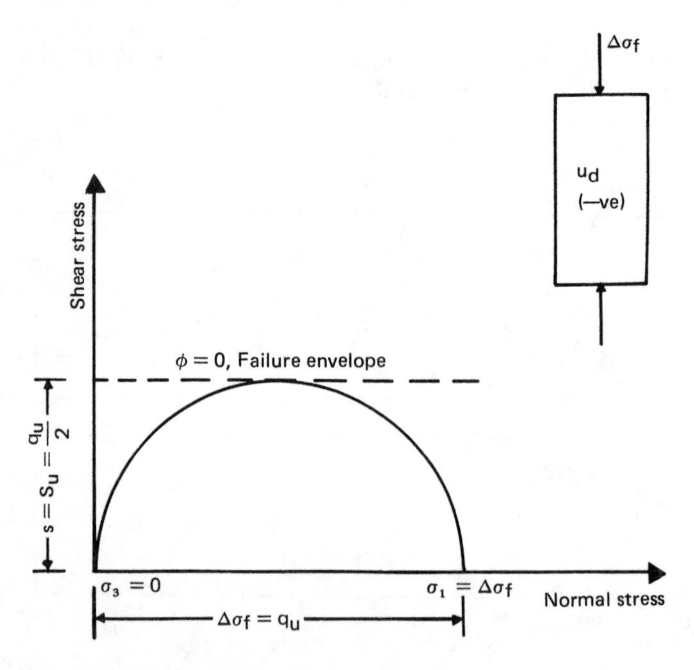

Fig. 7.36 Unconfined compression strength.

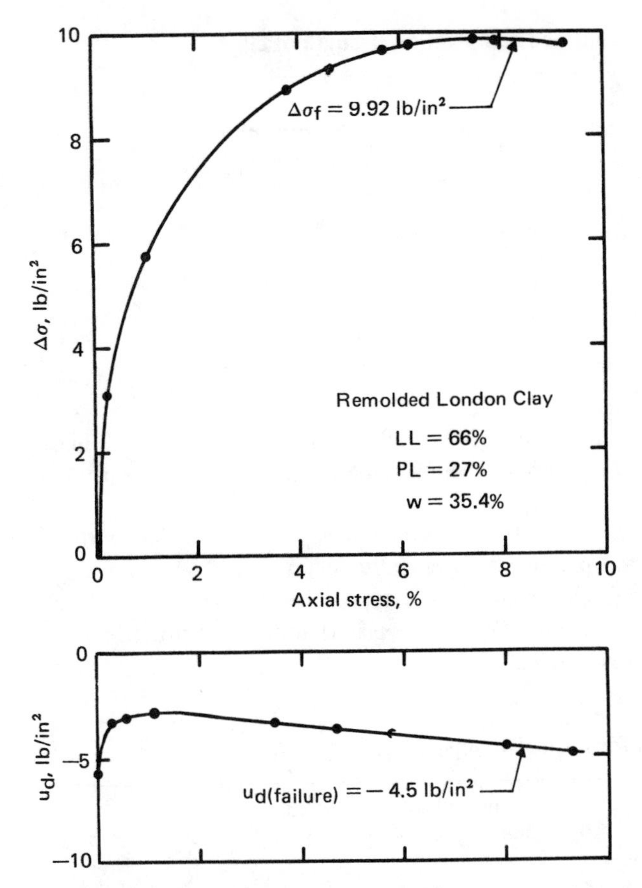

Fig. 7.37 Pore pressure change during shear in an unconfined compression test. (Note: 1 lb/in² = 6.9 kN/m².) *(Redrawn after A. W. Bishop and L. Bjerrum, The Relevance of the Triaxial Test to the Solution of Stability Problems,* Proc. Research Conference on Shear Strength of Cohesive Soils, ASCE, *1960.)*

Theoretically, the value of $\Delta\sigma_f$ of a saturated clay should be the same as that obtained from unconsolidated undrained tests using similar specimens. Thus $s = S_u = q_u/2$. However, this seldom provides high-quality results.

Figure 7.37 shows the results of an unconfined compression test on a specimen of remolded London clay. Note that the pore water pressure u_d in the specimen during the application of the axial stress $\Delta\sigma$ always remains negative.

The general relationship between consistency and unconfined compression strength of clays is given in Table 7.5.

7.3.3 Some Observations for the Values of ϕ and ϕ_{ult}

Typical values of the drained friction angle ϕ for some normally consolidated natural and remolded soils are given in Table 7.6. It can be seen that, in general, the value

Table 7.5 Consistency and unconfined compression strength of clays

	q_u	
Consistency	ton/ft²	kN/m²
Very soft	0 to 0.25	0 to 24
Soft	0.25 to 0.5	24 to 48
Medium	0.5 to 1	48 to 96
Stiff	1 to 2	96 to 192
Very stiff	2 to 4	192 to 383
Hard	> 4	> 383

of ϕ decreases with the plasticity index of the soil. Figure 7.38 also shows a linear correlation of $\sin \phi$ with $\log (PI)$. Although there is some scatter, the plot demonstrates a general validity of the fact.

Figure 7.39 shows the variation of the magnitude of ϕ_{ult} for several clays with the percentage of clay-size fraction present. ϕ_{ult} gradually decreases with the increase of clay-size fraction. At very high clay content, ϕ_{ult} approached the value of ϕ_μ (angle of sliding friction) for sheet minerals. For highly plastic sodium montmorillonites, the value of ϕ_{ult} can be as low as 3 to 4°.

Table 7.6 Values of ϕ and A_f for normally consolidated clays

Clay	Type	Liquid limit	Plasticity index	Sensi-tivity	A_f	ϕ
Natural soils						
Toyen	Marine	47	25	8	1.50	28.5
		47	25	8	1.48	
Drammen	Marine	36	16	4	1.2	35.0
		36	16	4	2.4	
Saco River	Marine	46	17	10	0.95	32.5
Boston	Marine	–	–	–	0.85	34.6
Bersimis	Estuarine	39	18	6	0.63	38.7
Chew Stoke	Alluvial	28	10	–	0.59	33
Kapuskasing	Lacustrine	39	23	4	0.46	30.0
Decomposed Talus	Residual	50	18	1	0.29	34.6
St. Catharines	Till (?)	49	28	3	0.26	25.6
Remolded soils						
London	Marine	78	52	1	0.97	19
Weald	Marine	43	25	1	0.95	23
Beauharnois	Till (?)	44	24	1	0.73	30.5
Boston	Marine	48	24	1	0.69	30.7
Beauharnois	Estuarine	70	42	1	0.65	32.8
Bersimis	Estuarine	33	13	1	0.38	39.0

After T. C. Kenney, Discussion, *J. Soil Mech. Found. Div., ASCE*, vol. 85, no. SM3, 1959.

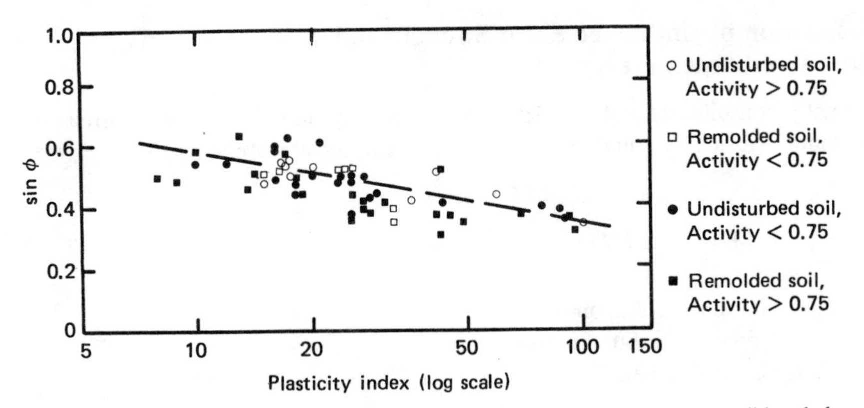

Fig 7.38 Relationship between sin ϕ and plasticity index for normally consolidated clays. *(After T. C. Kenney, Discussion,* J. Soil Mech. Found. Div., ASCE, *vol. 85, no. SM3, 1959.)*

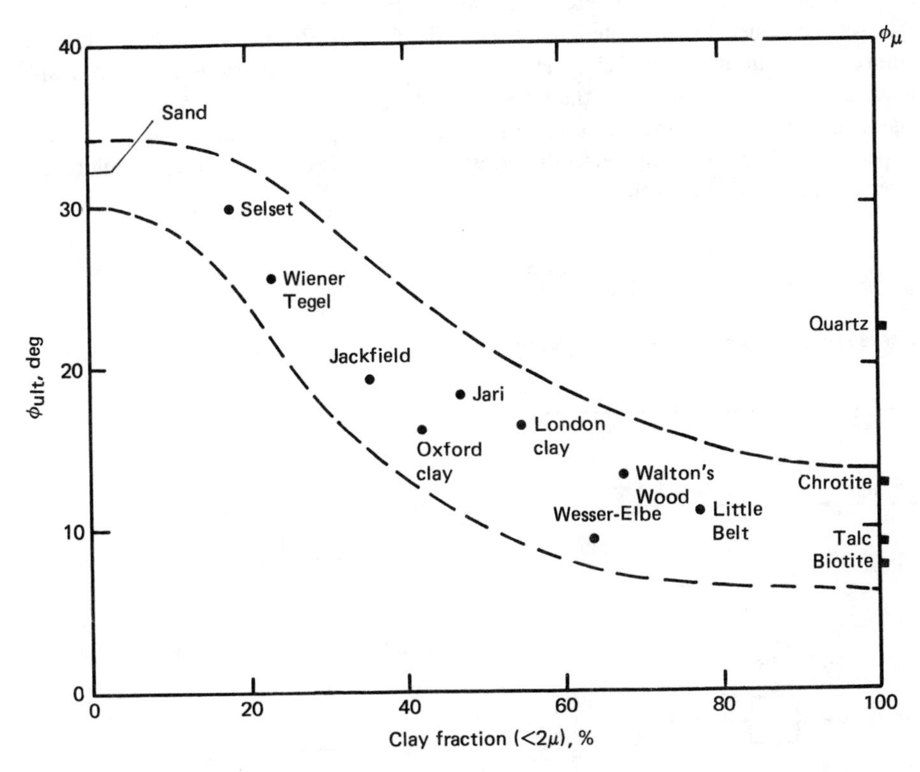

Fig. 7.39 Variation of ϕ_{ult} with percentage of clay content. *(Redrawn after A. W. Skempton, Long Term Stability of Clay Slopes,* Geotechnique, *vol. 14, no. 1, 1964. By permission of the publisher, The Institution of Civil Engineers.)*

7.3.4 Relation of Undrained Shear Strength and Effective Overburden Pressure

For normally consolidated natural clay deposits, Skempton (1957) gave a correlation for the undrained shear strength with the overburden pressure which may be expressed as

$$\frac{S_u}{p'} = 0.11 + 0.0037(PI) \tag{7.29}$$

where S_u = undrained cohesion of soil
p' = effective overburden pressure
PI = plasticity index

A relation between S_u, p', and the drained friction angle can also be derived as follows (e.g., see Leonards, 1962, p. 210). Referring to Fig. 7.40*a*, consider a soil specimen at A. The major and minor effective principal stresses at A can be given by p' and $K_o p'$, respectively (where K_o is the coefficient of at-rest earth pressure). Let this soil specimen be subjected to an UU triaxial test. As shown in Fig. 7.40*b*, at failure the *total* major principal stress is $\sigma_1 = p' + \Delta\sigma_1$; the *total* minor principal stress is $\sigma_3 = K_o p' + \Delta\sigma_3$; and the *excess* pore water pressure is Δu. So, the *effective* major and minor principal stresses can be given by $\sigma_1' = \sigma_1 - \Delta u$ and $\sigma_3' = \sigma_3 - \Delta u$, respectively. The total- and effective-stress Mohr's circles for this test, at failure, are shown in Fig. 7.41. From this, we can write

$$\frac{S_u}{c \cot \phi + (\sigma_1' + \sigma_3')/2} = \sin \phi$$

where ϕ is the drained friction angle, or

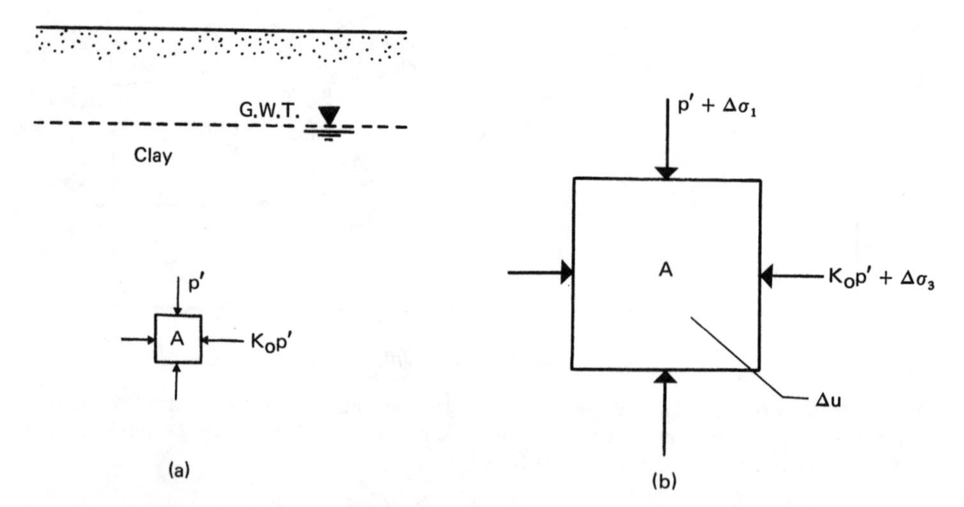

Fig. 7.40 Relation between the undrained strength of clay and the effective overburden pressure [Eq. (7.35)].

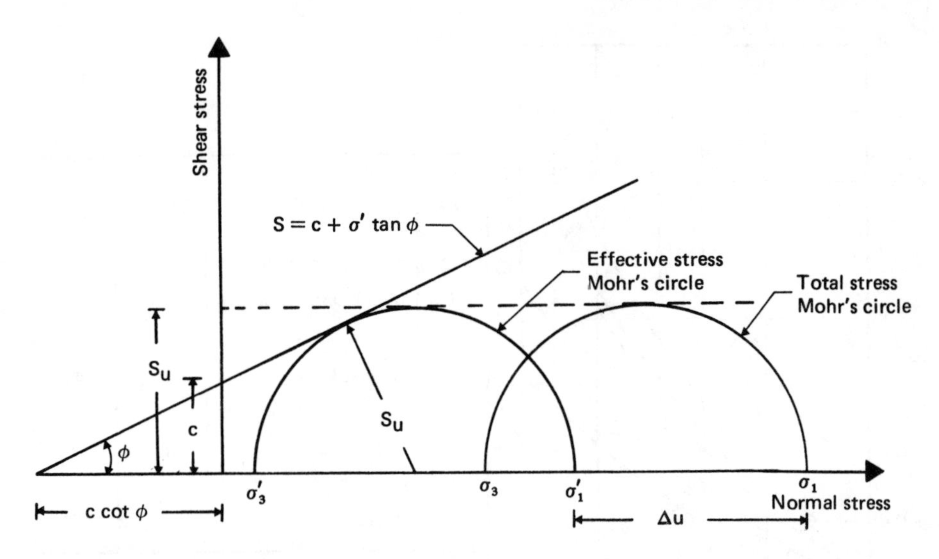

Fig. 7.41 Total and effective stress Mohr's circles for the soil element shown in Fig. 7.40*b*.

$$S_u = c \cos \phi + \frac{\sigma_1' + \sigma_3'}{2} \sin \phi$$

$$= c \cos \phi + \left(\frac{\sigma_1' + \sigma_3'}{2} - \sigma_3'\right) \sin \phi + \sigma_3' \sin \phi$$

But $$\frac{\sigma_1' + \sigma_3'}{2} - \sigma_3' = \frac{\sigma_1' - \sigma_3'}{2} = S_u$$

So, $$S_u = c \cos \phi + S_u \sin \phi + \sigma_3' \sin \phi$$

$$S_u(1 - \sin \phi) = c \cos \phi + \sigma_3' \sin \phi \tag{7.30}$$

$$\sigma_3' = \sigma_3 - \Delta u = K_o p' + \Delta\sigma_3 - \Delta u \tag{7.31}$$

We now make use of Eq. (4.9):

$$\Delta u = B\, \Delta\sigma_3 + A_f(\Delta\sigma_1 - \Delta\sigma_3) \tag{4.9}$$

For saturated clays, $B = 1$. Substituting Eq. (4.9) into Eq. (7.31),

$$\sigma_3' = K_o p' + \Delta\sigma_3 - [\Delta\sigma_3 + A_f(\Delta\sigma_1 - \Delta\sigma_3)]$$

$$= K_o p' - A_f(\Delta\sigma_1 - \Delta\sigma_3) \tag{7.32}$$

Again, from Fig. 7.40,

$$S_u = \frac{\sigma_1 - \sigma_3}{2} = \frac{(p' + \Delta\sigma_1) - (K_o p' + \Delta\sigma_3)}{2}$$

or $$2S_u = (\Delta\sigma_1 - \Delta\sigma_3) + (p' - K_o p')$$

or $$(\Delta\sigma_1 - \Delta\sigma_3) = 2S_u - (p' - K_o p') \tag{7.33}$$

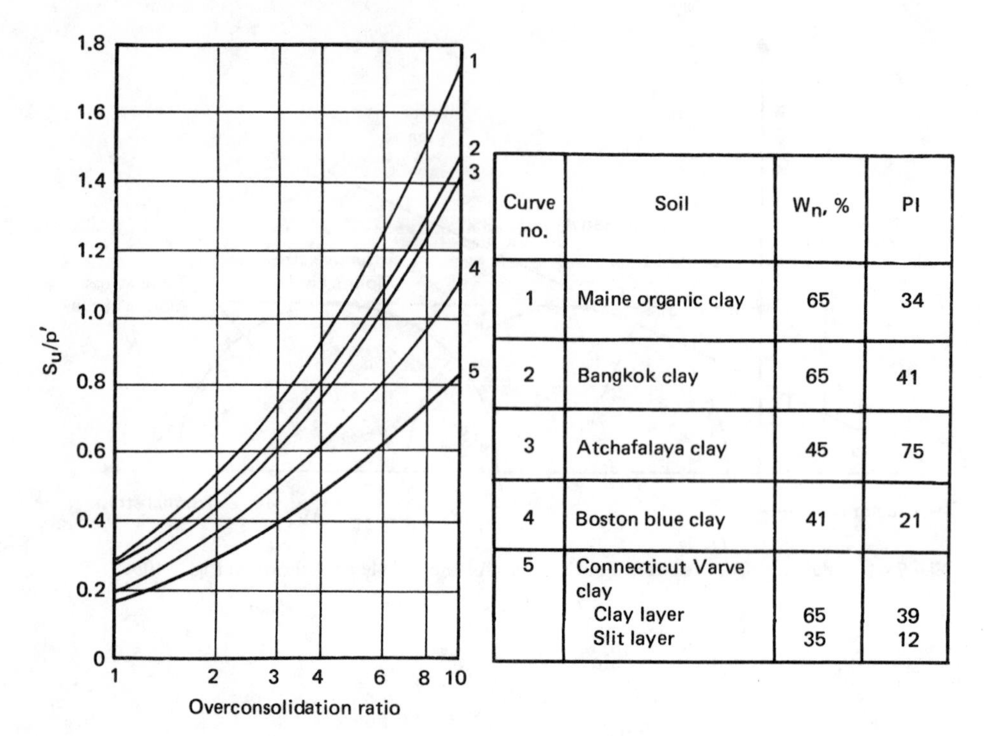

Curve no.	Soil	W_n, %	PI
1	Maine organic clay	65	34
2	Bangkok clay	65	41
3	Atchafalaya clay	45	75
4	Boston blue clay	41	21
5	Connecticut Varve clay		
	Clay layer	65	39
	Slit layer	35	12

Fig. 7.42 S_u/p' for several clays. *(Redrawn after C. C. Ladd and R. Foot, New Design Procedures for Stability of Soft Clays,* J. Geotech. Eng. Div., ASCE, *vol. 100, no. GT7, 1974.)*

Substituting Eq. (7.33) into Eq. (7.32), we obtain

$$\sigma_3' = K_o p' - 2S_u A_f + A_f p'(1 - K_o) \tag{7.34}$$

Substitution of Eq. (7.34) into the right-hand side of Eq. (7.30) and simplification yields

$$S_u = \frac{c \cos\phi + p' \sin\phi \, [K_o + A_f(1 - K_o)]}{1 + (2A_f - 1)\sin\phi} \tag{7.35}$$

For normally consolidated clays, $c = 0$; hence, Eq. (7.35) becomes

$$\frac{S_u}{p'} = \frac{\sin\phi \, [K_o + A_f(1 - K_o)]}{1 + (2A_f - 1)\sin\phi} \tag{7.36}$$

Equations such as (7.29) and (7.36) are very useful as approximate checks to evaluate if a given clay is normally consolidated or not.

Karlsson and Viberg (1967) also gave a correlation between S_u/p' and the liquid limit of soil. Ladd and Foott (1974) presented experimental results for the variation of S_u/p' with overconsolidation ratio for five clays (Fig. 7.42). Using the results of Fig. 7.42, a nondimensional plot of

$$\beta = \frac{(S_u/p')_{\text{overconsolidated}}}{(S_u/p')_{\text{normally consolidated}}}$$

is shown in Fig. 7.43. The upper and lower limits are not too far apart. The average curve may be used for estimation of undrained shear strength of overconsolidated clays.

Example 7.4 A soil profile is shown in Fig. 7.44. From a laboratory consolidation test, the preconsolidation pressure of a soil specimen obtained from a depth of 23 ft below the ground surface was found to be 2520 lb/ft^2. Estimate the undrained shear strength of the clay at that depth.

SOLUTION

$$\gamma_{\text{sat (clay)}} = \frac{G_s\gamma_w + wG_s\gamma_w}{1 + wG_s} = \frac{(2.7)(62.4)(1 + 0.3)}{1 + 0.3(2.7)}$$

$$= 121.0 \text{ lb/ft}^3$$

The effective overburden pressure at A is

$$p' = 8(110) + 15(121 - 62.4) = 880 + 879 = 1759 \text{ lb/ft}^2$$

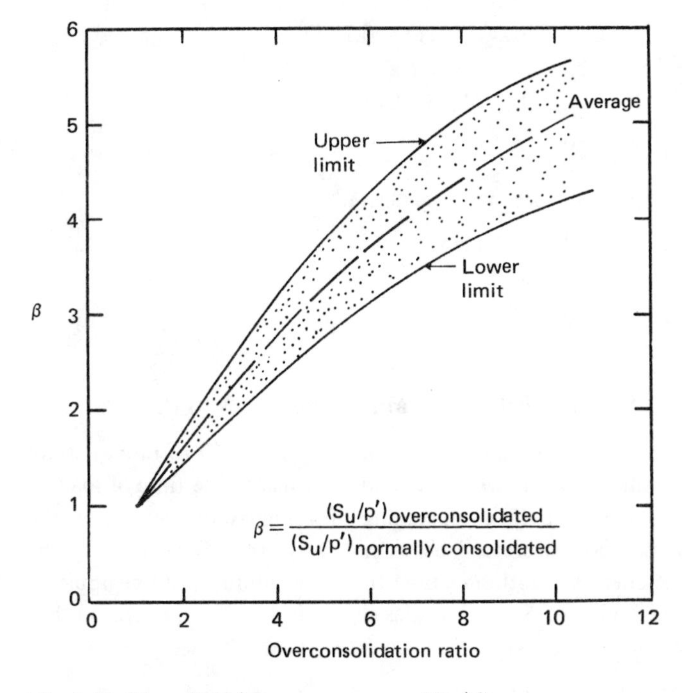

Fig. 7.43 Plot of $(S_u/p')_{\text{overconsolidated}}/(S_u/p')_{\text{normally consolidated}}$ against overconsolidation ratio based on the results of Fig. 7.42.

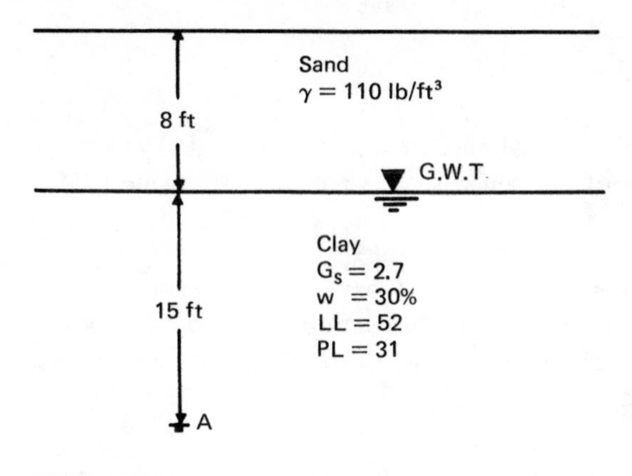

Fig. 7.44

The overconsolidation ratio is

$$\frac{p'_{\max}}{p'} = \frac{2520}{1759} = 1.43$$

If we assume that the clay was normally consolidated, then Eq. (7.29) can be applied: $S_u/p' = 0.11 + 0.0037(PI)$. But $PI = LL - PL = 52 - 31 = 21$, and so

$$\left(\frac{S_u}{p'}\right)_{\text{normally consolidated}} = 0.11 + 0.0037(21) = 0.1877$$

From Fig. 7.43, for OCR = 1.43, $\beta_{\text{av}} \approx 1.35$. Hence,

$$\left(\frac{S_u}{p'}\right)_{\text{overconsolidated}} = 1.35\left(\frac{S_u}{p'}\right)_{\text{normally consolidated}}$$

$$= (1.35)\,(1.43) = 1.93$$

or

$$S_u = (1.93)\,(1795) = \mathit{3464\,lb/ft^2}$$

7.3.5 Effect of Rate of Strain on the Undrained Shear Strength

Casagrande and Wilson (1949, 1951) studied the problem of the effect of rate of strain on the undrained shear strength of saturated clays and clay shales. The time of loading ranged from 1 to 10^4 min. Using a time of loading of 1 min as the reference, the undrained strength of some clays decreased by as much as 20%. The nature of the variation of the undrained shear strength and time to cause failure, t, can be approximated by a straight line in a plot of S_u vs. log t, as shown in Fig. 7.45. Based on this, Hvorslev (1960) gave the following relation:

$$S_{u(t)} = S_{u(a)}\left[1 - \rho_a \log\left(\frac{t}{t_a}\right)\right] \tag{7.37}$$

where $S_{u(t)}$ = undrained shear strength with time, t, to cause failure
$S_{u(a)}$ = undrained shear strength with time, t_a, to cause failure
ρ_a = coefficient for decrease of strength with time

In view of the time duration, Hvorslev suggested that the reference time be taken as 1000 min. In that case,

$$S_{u(t)} = S_{u(m)} \left[1 - \rho_m \log\left(\frac{t \text{ min}}{1000 \text{ min}}\right)\right] \tag{7.38}$$

where $S_{u(m)}$ = undrained shear strength at time 1000 min
ρ_m = coefficient for decrease of strength with reference time of 1000 min

The relation between ρ_a in Eq. (7.37) and ρ_m in Eq. (7.38) can be given by

$$\rho_m = \frac{\rho_a}{1 - \rho_a \log[(1000 \text{ min})/(t_a \text{ min})]} \tag{7.39}$$

For $t_a = 1$ min, Eq. (7.39) gives

$$\rho_m = \frac{\rho_1}{1 - 3\rho_1} \tag{7.40}$$

Hvorslev's analysis of the results of Casagrande and Wilson (1951) yielded the following results: general range: $\rho_1 = 0.04$ to 0.09, $\rho_m = 0.05$ to 0.13; Cucaracha clay-shale: $\rho_1 = 0.07$ to 0.19, $\rho_m = 0.09$ to 0.46. The study of the strength-time relationship of Bjerrum et al. (1958) for a normally consolidated marine clay (consolidated undrained test) yielded a value of ρ_m in the range of 0.06 to 0.07.

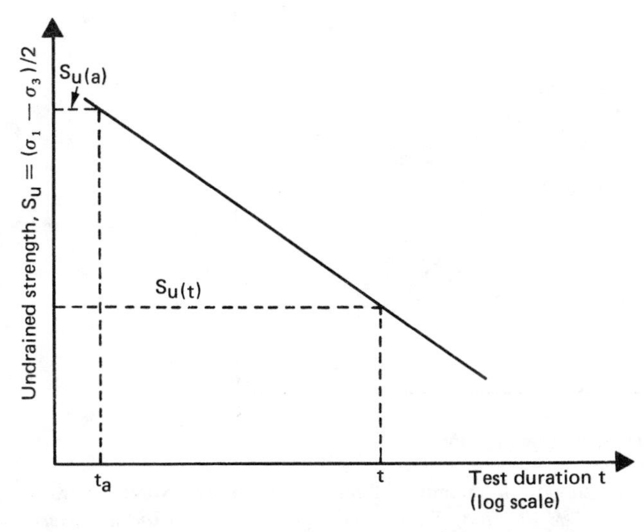

Fig. 7.45 Effect of rate of strain on undrained shear strength.

Whitman (1957) reported that the undrained shear strength of saturated soils shows approximately a two-fold increase when the time for failure is 5 ms over that with a time for failure of an hour. Richardson and Whitman (1964) concluded that the change in undrained shear strength is due to the difference in induced pore water pressure. With smaller time to cause failure (i.e., higher rate of strain), the induced pore water pressure is smaller.

7.3.6 Effect of Temperature on Shear Strength of Clay

A number of investigations have been conducted to determine the effect of temperature on the shear strength of saturated clay. Most studies indicate that a decrease of temperature will cause an increase of shear strength. Figure 7.46 shows the variation of the unconfined compression strength ($q_u = 2S_u$) of kaolinite with temperature. Note that for a given moisture content the value of q_u decreases with increase of temperature. A similar trend has been observed for San Francisco Bay mud (Mitchell, 1964), as shown in Fig. 7.47. The undrained shear strength $[S_u = (\sigma_1 - \sigma_3)/2]$ varies linearly with temperature. The results are for specimens with equal mean effective stress and similar structure. From these tests,

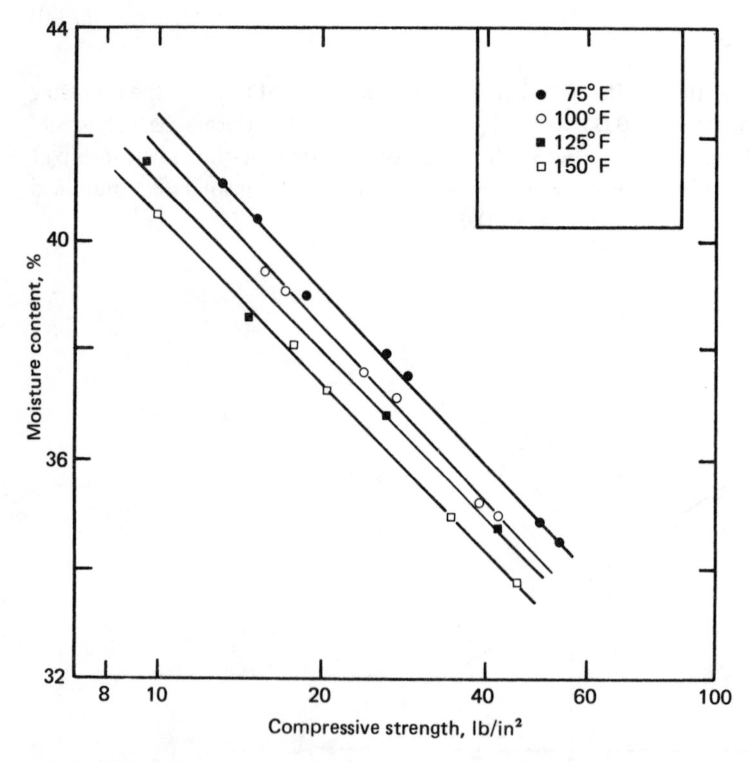

Fig. 7.46 Unconfined compression strength of kaolinite – effect of temperature. (Note: 1 lb/in² = 6.9 kN/m².) *(After M. A. Sherif and C. M. Burrous, Temperature Effect on the Unconfined Shear Strength of Saturated Cohesive Soils,* Highway Research Board, Special Report 103, *1969.)*

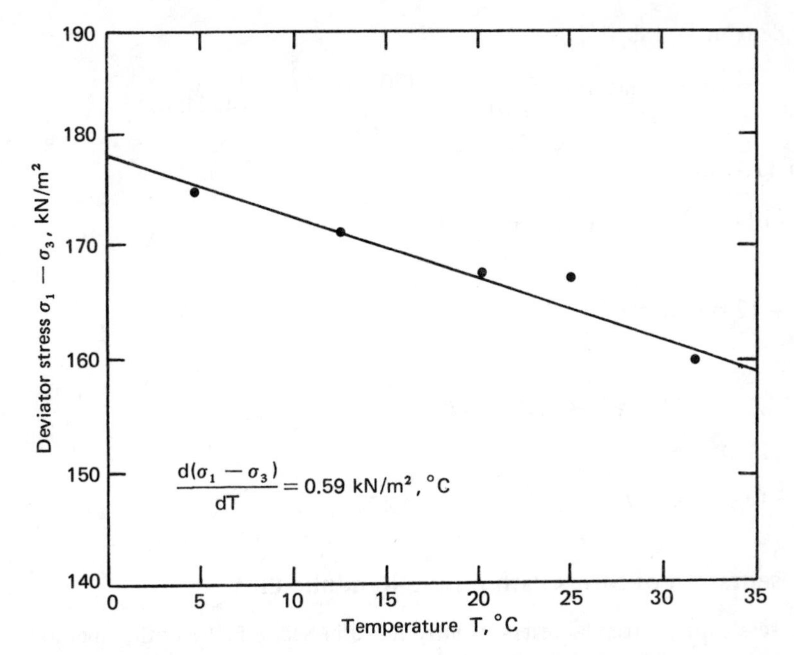

Fig. 7.47 Effect of temperature on shear strength of San Francisco Bay mud. *(Redrawn after J. K. Mitchell, Shearing Resistance of Soils as a Rate Process,* J. Soil Mech. Found. Div., ASCE, *vol. 90, no. SM1, 1964.)*

$$\frac{dS_u}{dT} \approx 0.59 \text{ kN/(m}^2 \cdot {}^\circ\text{C)} \tag{7.41}$$

Kelly (1978) also studied the effect of temperature on the undrained shear strength of some undisturbed marine clay samples and commercial illite and montmorillonite. Undrained shear strengths at 4 and 20°C were determined. Based on the laboratory test results, Kelly proposed the following correlation:

$$\frac{\Delta S_u}{\Delta T} = 0.213 + 0.00747\, S_{u(\text{average})} \tag{7.42}$$

where $S_{u(\text{average})} = (S_{u(4^\circ\text{C})} + S_{u(20^\circ\text{C})})/2$ in lb/ft^2 and T is the temperature in °C.

Example 7.5 The following are the results of an unconsolidated undrained test: $\sigma_3 = 70$ kN/m^2, $\sigma_1 = 210$ kN/m^2. The temperature of the test was 12°C. Estimate the undrained shear strength of the soil at a temperature of 20°C.

Solution

$$S_{u(12^\circ\text{C})} = \frac{\sigma_1 - \sigma_3}{2} = \frac{210 - 70}{2} = 70 \text{ kN/m}^2$$

Since 47.88 $N/m^2 = 1\ lb/ft^2$,

$$S_{u(\text{average})} = \frac{S_{u(4°C)} + S_{u(20°C)}}{2} = S_{u(12°C)} = \frac{(70)\,(1000)}{47.88} = 1462\ lb/ft^2$$

From Eq. (7.42),

$$\Delta S_u = \Delta T[0.213 + 0.00747\ S_{u(\text{average})}].$$

Now,

$$\Delta T = 20 - 12 = 8°C$$

and

$$\Delta S_u = 8\,[0.213 + 0.00747(1462)] = 89.07\ lb/ft^2.$$

Hence,

$$S_{u(20°C)} = 1462 - 89.07 = 1372.93\ lb/ft^2 = 65.74\ kN/m^2$$

7.3.7 Representation of Stress Path on the Rendulic Plot

The use of a stress path for triaxial tests was introduced in Sec. 6.5.1. Another method of representing the stress path for triaxial tests is the plot suggested by Rendulic (1937) and later developed by Henkel (1960). The *Rendulic plot* is the plot of the results of triaxial tests in the stress path of plane *Oabc* as shown in Fig. 7.48. Along *Oa* we plot $\sqrt{2}\,\sigma_r'$, and long *Oc* we plot σ_a' (σ_r' is the effective radial stress and σ_a' is the effective axial stress). Line *Od* in Fig. 7.49 represents the *isotropic stress line.* The direction cosines of this line are $1/\sqrt{3}$, $1/\sqrt{3}$, $1/\sqrt{3}$. Line *Od* in Fig. 7.49 will have a slope of 1 vertical to $\sqrt{2}$ horizontal. Note that the trace of the octahedral plane ($\sigma_1 + \sigma_2 + \sigma_3 =$ constant) will be at right angles to the line *Od*.

In a triaxial equipment, if a soil sample is hydrostatically consolidated (i.e., $\sigma_a' = \sigma_r'$) it may be represented by the point 1 on the line *Od*. If this sample is subjected to a drained axial compression test by increasing σ_a' and keeping σ_r' constant, the stress path can be represented by the line 1-2. Point 2 represents the state of stress at failure. Similarly,

Line 1-3 will represent a drained axial compression test conducted by keeping σ_a' constant and reducing σ_r'.

Line 1-4 will represent a drained axial compression test where the mean principal stress (or $J = \sigma_1' + \sigma_2' + \sigma_3'$) is kept constant.

Line 1-5 will represent a drained axial extension test conducted by keeping σ_r' constant and reducing σ_a'.

Line 1-6 will represent a drained axial extension test conducted by keeping σ_a' constant and increasing σ_r'.

Line 1-7 will represent a drained axial extension test with $J = \sigma_1' + \sigma_2' + \sigma_3'$ constant (i.e. $J = \sigma_a' + 2\sigma_r'$ constant).

Curve 1-8 will represent an undrained compression test.

Curve 1-9 will represent an undrained extension test.

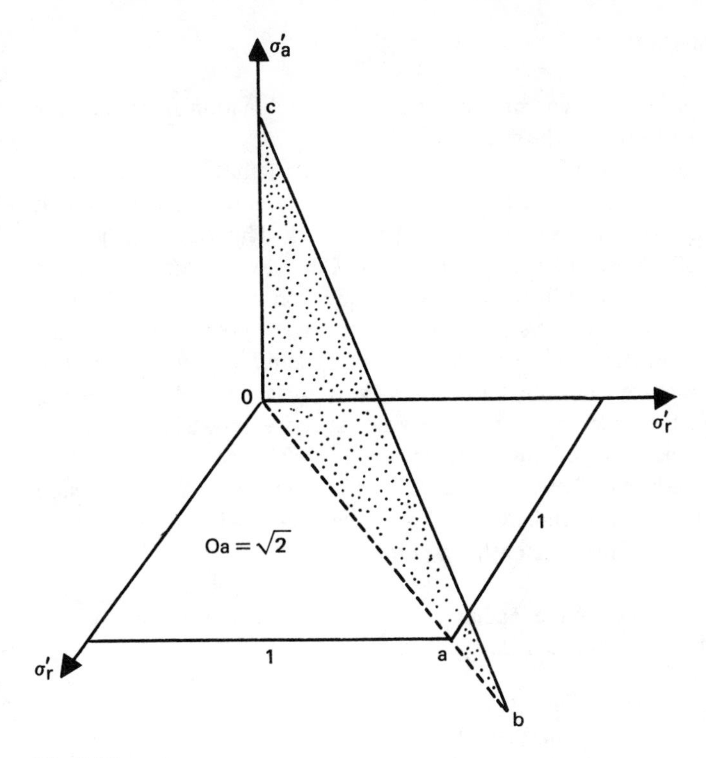

Fig. 7.48

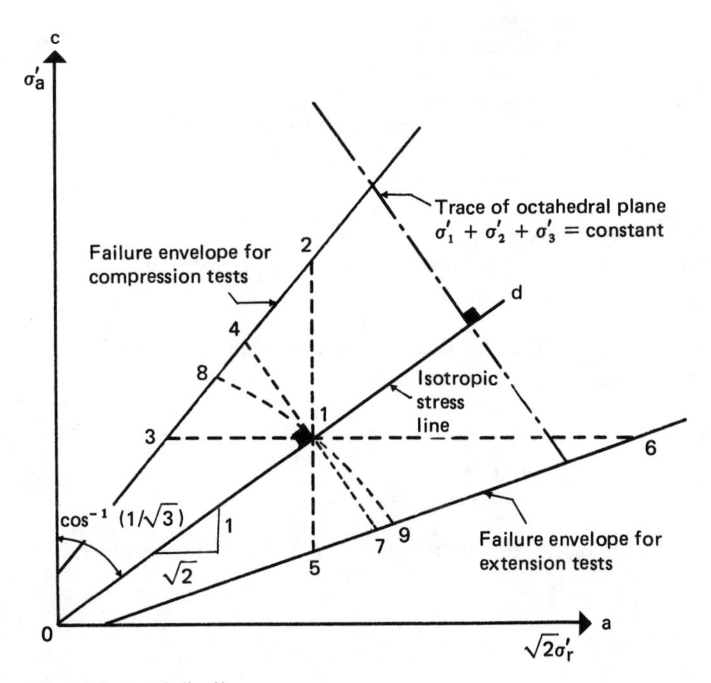

Fig. 7.49 Rendulic diagram.

Curves 1-8 and 1-9 are independent of the total stress combination since the pore water pressure is adjusted to follow the stress path shown.

If the effective stress path from a triaxial test in which failure of the specimen was caused by loading in an undrained condition is given, the pore water pressure at a given state during the loading can be easily determined. This can be explained with the aid of Fig. 7.50. Consider a soil specimen consolidated with an encompassing pressure σ_r and with failure caused in the undrained condition by increasing the axial stress σ_a. Let *acb* be the effective stress path for this test. We are required to find the excess pore water pressures that were generated at points *c* and *b* (i.e., at failure). For this type of triaxial test, we know that the *total stress path* will follow a vertical line such as *ae*. To find the excess pore water pressure at *c*, we draw a line *cf* parallel to the isotropic stress line. Line *cf* intersects line *ae* at *d*. The pore water pressure u_d at *c* is the *vertical distance* between points *c* and *d*. The pore water pressure $u_{d(\text{failure})}$ at *b* can similarly be found by drawing *bg* parallel to the isotropic stress line and measuring the vertical distance between points *b* and *g*.

Example 7.6 Given below are the loading conditions of a number of consolidated *drained* triaxial tests on a remolded clay ($\phi = 25°, c = 0$).

Test no.	Consolidation pressure, kN/m²	Type of loading applied to cause failure
1	400	σ_a increased; σ_r constant
2	400	σ_a constant; σ_r increased
3	400	σ_a decreased; σ_r constant
4	400	σ_a constant; σ_r decreased
5	400	$\sigma_a + 2\sigma_r$ constant; increased σ_a and decreased σ_r
6	400	$\sigma_a + 2\sigma_r$ constant; decreased σ_a and increased σ_r

(*a*) Draw the isotropic stress line.
(*b*) Draw the failure envelopes for compression and extension tests.
(*c*) Draw the stress paths for tests 1 through 6.

SOLUTION *Part (a):* The istropic stress line will make an angle $\theta = \cos^{-1} 1/\sqrt{3}$ with the σ'_a axis; so $\theta = 54.8°$. This is shown in Fig. 7.51 as line *Od*.

Part (b):

$$\sin\phi = \left(\frac{\sigma'_1 - \sigma'_3}{\sigma'_1 + \sigma'_3}\right)_{\text{failure}} \qquad \text{or} \qquad \left(\frac{\sigma'_1}{\sigma'_3}\right)_{\text{failure}} = \frac{1 + \sin\phi}{1 - \sin\phi}$$

where σ'_1 and σ'_3 are the major and minor principal stresses. For *compression tests*, $\sigma'_1 = \sigma'_a$ and $\sigma'_3 = \sigma'_r$. Thus,

$$\left(\frac{\sigma'_a}{\sigma'_r}\right)_{\text{failure}} = \frac{1 + \sin 25°}{1 - \sin 25°} = 2.46 \qquad \text{or} \qquad (\sigma'_a)_{\text{failure}} = 2.46(\sigma'_r)_{\text{failure}}$$

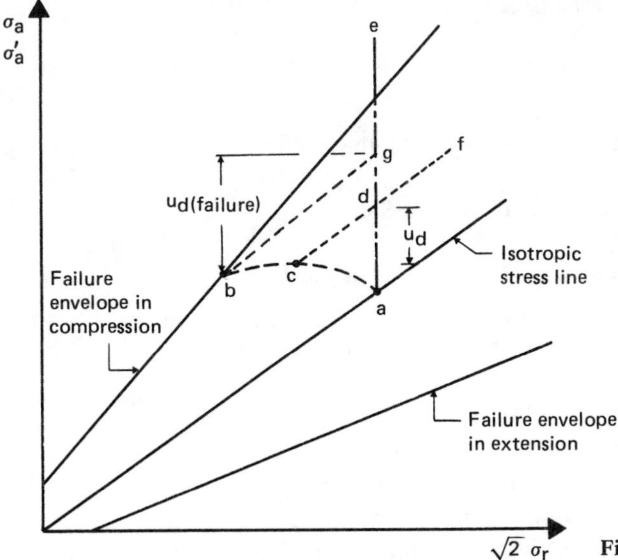

Fig. 7.50 Determination of pore water pressure in a Rendulic plot.

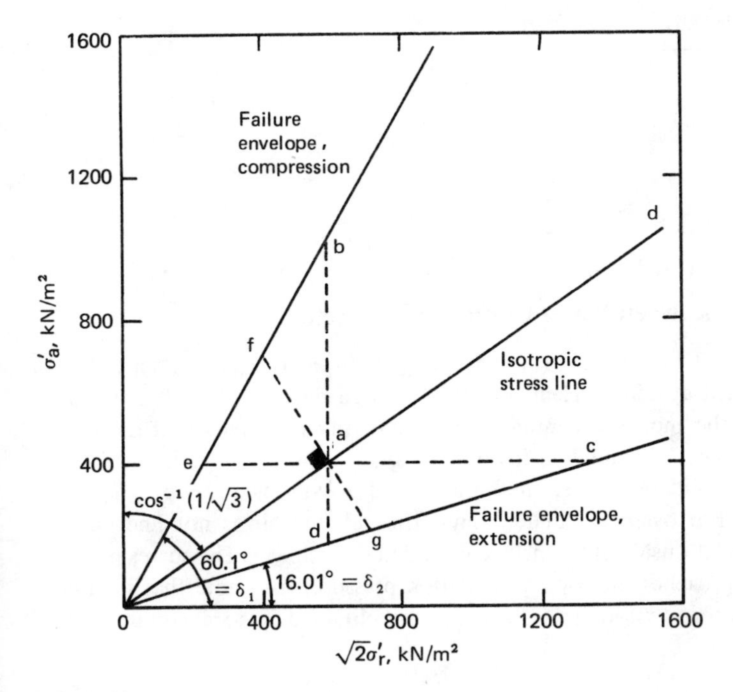

Fig. 7.51

The slope of the failure envelope is

$$\tan \delta_1 = \frac{\sigma'_a}{\sqrt{2}\,\sigma'_r} = \frac{2.46\,\sigma'_r}{\sqrt{2}\,\sigma'_r} = 1.74$$

Hence, $\delta_1 = 60.1°$. The failure envelope for the compression tests is shown in Fig. 7.51.

For *extension tests*, $\sigma'_1 = \sigma'_r$ and $\sigma'_3 = \sigma'_a$. So,

$$\left(\frac{\sigma'_a}{\sigma'_r}\right)_{\text{failure}} = \frac{1 - \sin 25}{1 + \sin 25} = 0.406 \qquad \text{or} \qquad \sigma'_a = 0.406\,\sigma'_r$$

The slope of the failure envelope for extension tests is

$$\tan \delta_2 = \frac{\sigma'_a}{\sqrt{2}\,\sigma'_r} = \frac{0.406\,\sigma'_r}{\sqrt{2}\,\sigma'_r} = 0.287$$

Hence, $\delta_2 = 16.01°$. The failure envelope is shown in Fig. 7.51.

Part (c): Point a on the isotropic stress line represents the point where $\sigma'_a = \sigma'_r$ (or $\sigma'_1 = \sigma'_2 = \sigma'_3$). The stress paths of the test are plotted in Fig. 7.51.

Test no.	Stress path in Fig. 7.51
1	*a–b*
2	*a–c*
3	*a–d*
4	*a–e*
5	*a–f*
6	*a–g*

7.3.8 Relationship between Water Content and Strength

The strength of a soil at failure [i.e., $(\sigma_1 - \sigma_3)_{\text{failure}}$ or $(\sigma'_1 - \sigma'_3)_{\text{failure}}$] is dependent on the moisture content at failure. Henkel (1960) pointed out that there is a unique relationship between the moisture content w at failure and the strength of a clayey soil. This is shown in Figs. 7.52 and 7.53 for Weald clay.

For normally consolidated clays, the variation of w vs. log $(\sigma_1 - \sigma_3)_{\text{failure}}$ is approximately linear. For overconsolidated clays, this relationship is not linear but lies slightly below the relationship of normally consolidated specimens. The curves merge when the strength approaches the overconsolidation pressure. Also note that slightly different relationships for w vs. log $(\sigma_1 - \sigma_3)_{\text{failure}}$ are obtained for axial compression and axial extension tests.

7.3.9 Unique Effective Stress Failure Envelope

When Mohr's envelope is used to obtain the relationship for normal and shear stress at failure (from triaxial test results), separate envelopes need to be drawn for separate

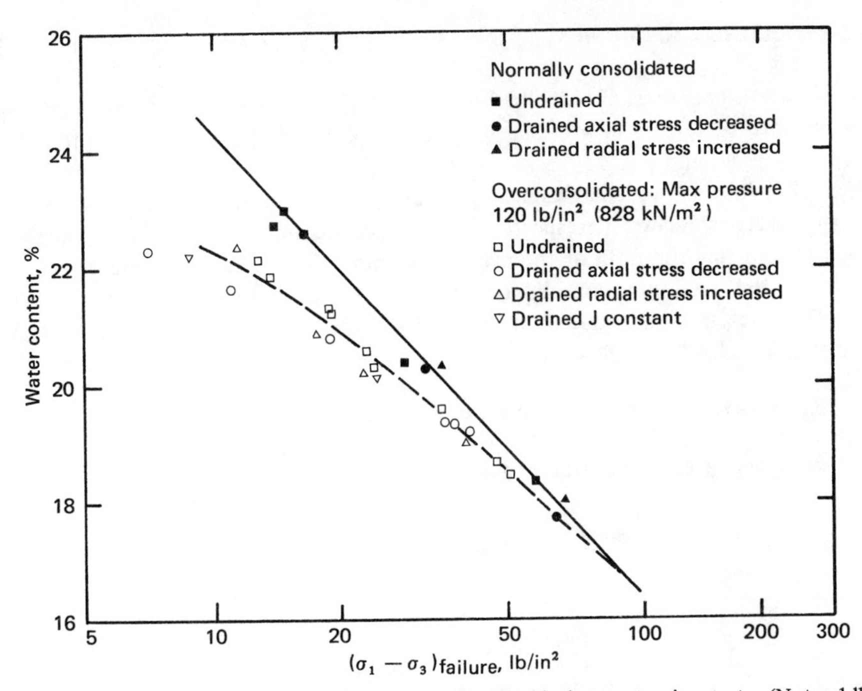

Fig. 7.52 Water content vs. $(\sigma_1 - \sigma_3)_{\text{failure}}$ for Weald clay–extension tests. (Note: 1 lb/in² = 6.9 kN/m².) *(Redrawn after D. J. Henkel, The Shearing Strength of Saturated Remolded Clays,* Proc. Research Conference on Shear Strength of Cohesive Soils, ASCE, *1960.)*

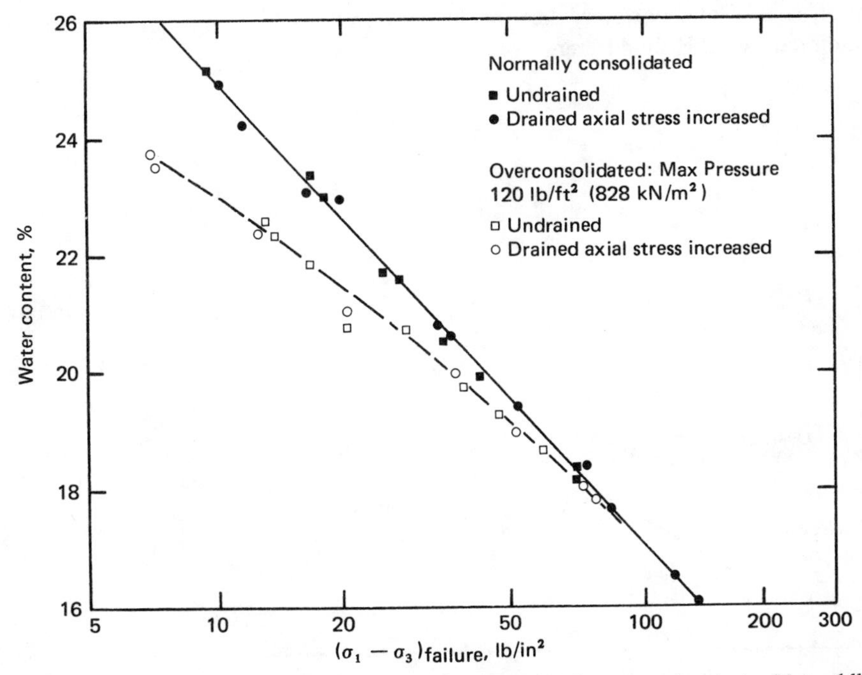

Fig. 7.53 Water content vs. $(\sigma_1 - \sigma_3)_{\text{failure}}$ for Weald clay–compression tests. (Note: 1 lb/in² = 6.9 kN/m².) *(Redrawn after D. J. Henkel, The Shearing Strength of Saturated Remolded Clays,* Proc. Research Conference on Shear Strength of Cohesive Soils, ASCE, *1960.)*

preconsolidation pressures, σ_c'. This is shown in Fig. 7.54. For a soil with a preconsolidation pressure of σ_{c1}', $s = c_1 + \sigma' \tan \phi_{c(1)}$; similarly, for a preconsolidation pressure of σ_{c2}', $s = c_2 + \sigma' \tan \phi_{c(2)}$.

Henkel (1960) showed that a single, general failure envelope for normally consolidated and preconsolidated (irrespective of preconsolidation pressure) soils can be obtained by plotting the ratio of the major to minor effective stress at failure against the ratio of the maximum consolidation pressure to the average effective stress at failure. This fact is demonstrated in Fig. 7.55, which gives the results of triaxial compression tests for Weald clay. In Fig. 7.55,

$$J_m = \text{maximum consolidation pressure} = \sigma_c'$$

$$J_f = \text{average effective stress at failure}$$

$$= \frac{\sigma_{1(\text{failure})}' + \sigma_{2(\text{failure})}' + \sigma_{3(\text{failure})}'}{3}$$

$$= \frac{\sigma_a' + 2\sigma_r'}{3}$$

The results shown in Fig. 7.55 are obtained from normally consolidated specimens and overconsolidated specimens having a maximum preconsolidation pressure of 120 lb/in^2 (828 kN/m^2). Similarly, a unique failure envelope can be obtained from extension tests. Note, however, that the failure envelopes for compression tests and extension tests are slightly different.

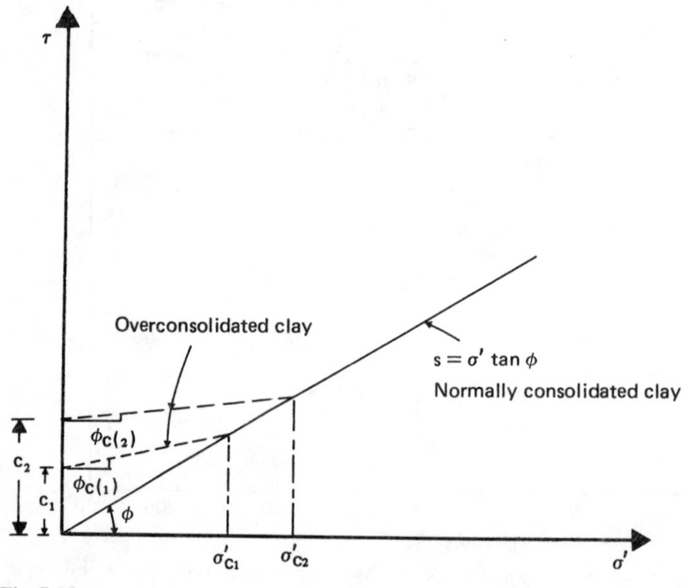

Fig. 7.54

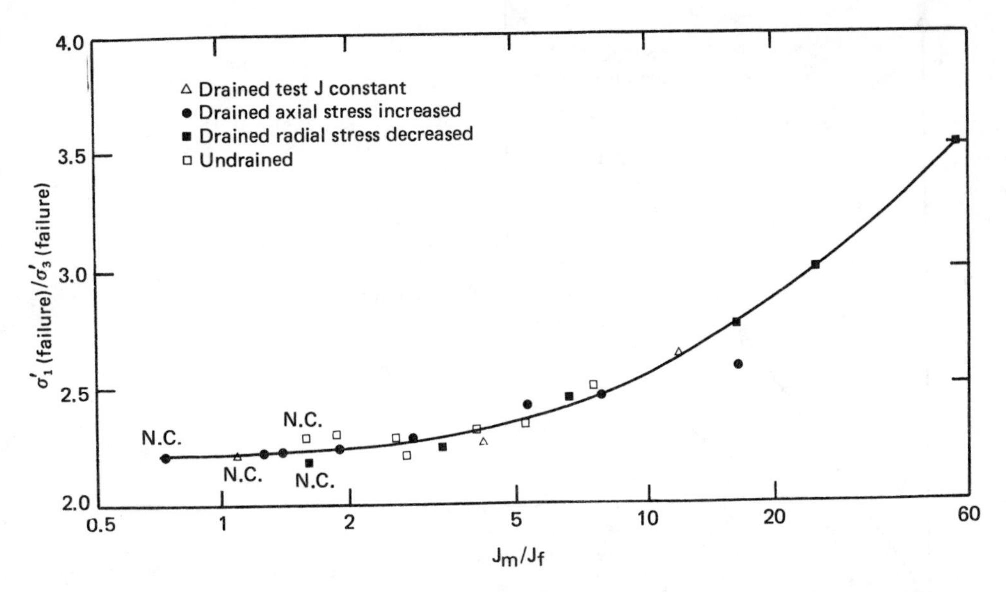

Fig. 7.55 Plot of $\sigma'_{1(\text{failure})}/\sigma'_{3(\text{failure})}$ against J_m/J_f for Weald clay–compression tests. *(After D. J. Henkel, The Shearing Strength of Saturated Remolded Clays,* Proc. Research Conference on Shear Strength of Cohesive Soils, ASCE, *1960.)*

7.3.10 Unique Relationship between Water Content and Effective Stress

There is a unique relationship between the water content of a soil and the effective stresses to which it is being subjected, provided that normally consolidated specimens and specimens with common maximum consolidation pressures are considered separately. This can be explained with the aid of Fig. 7.56, in which a Rendulic plot for a normally consolidated clay is shown. Consider several specimens consolidated at various confining pressures in a triaxial chamber; the states of stress of these specimens are represented by the points *a, c, e, g,* etc. located on the isotropic stress lines. When these specimens are sheared to failure by drained compressions, the corresponding stress paths will be represented by lines such as *ab*, *cd*, *ef*, and *gh*. During drained tests, the moisture contents of the specimens change. We can determine the moisture contents of the specimens during the tests, such as w_1, w_2, . . . , as shown in Fig. 7.56. If these points of equal moisture contents on the drained stress paths are joined, we obtain contours of stress paths of equal moisture contents (for moisture contents w_1, w_2, . . .).

Now, if we take a soil specimen and consolidate it in a triaxial chamber under a state of stress as defined by point *a* and shear it to failure in an undrained condition, it will follow the effective stress path *af* since the moisture content of the specimen during shearing is w_1. Similarly, a specimen consolidated in a triaxial chamber under a state of stress represented by point *c* (moisture content $= w_2$) will follow a stress path *gh* (which is the stress contour of moisture content w_2) when sheared to failure in an undrained state. This means that a unique relationship exists between water content and effective stress.

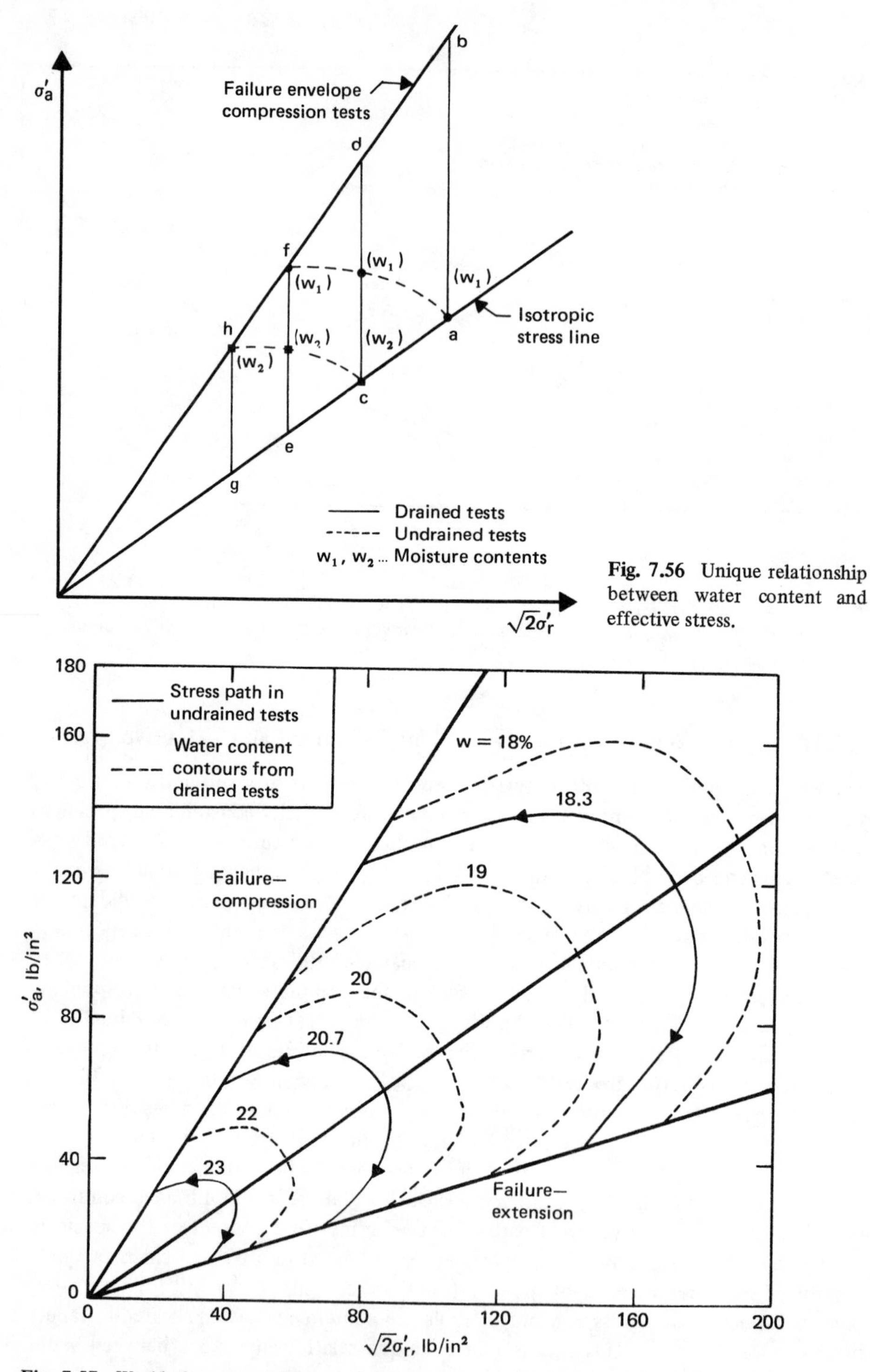

Fig. 7.56 Unique relationship between water content and effective stress.

Fig. 7.57 Weald clay – normally consolidated. (Note: 1 lb/in² = 6.9 kN/m².) *(Redrawn after D. J. Henkel, The Shearing Strength of Saturated Remolded Clays,* Proc. Research Conference on Shear Strength of Cohesive Soils, ASCE, *1960.)*

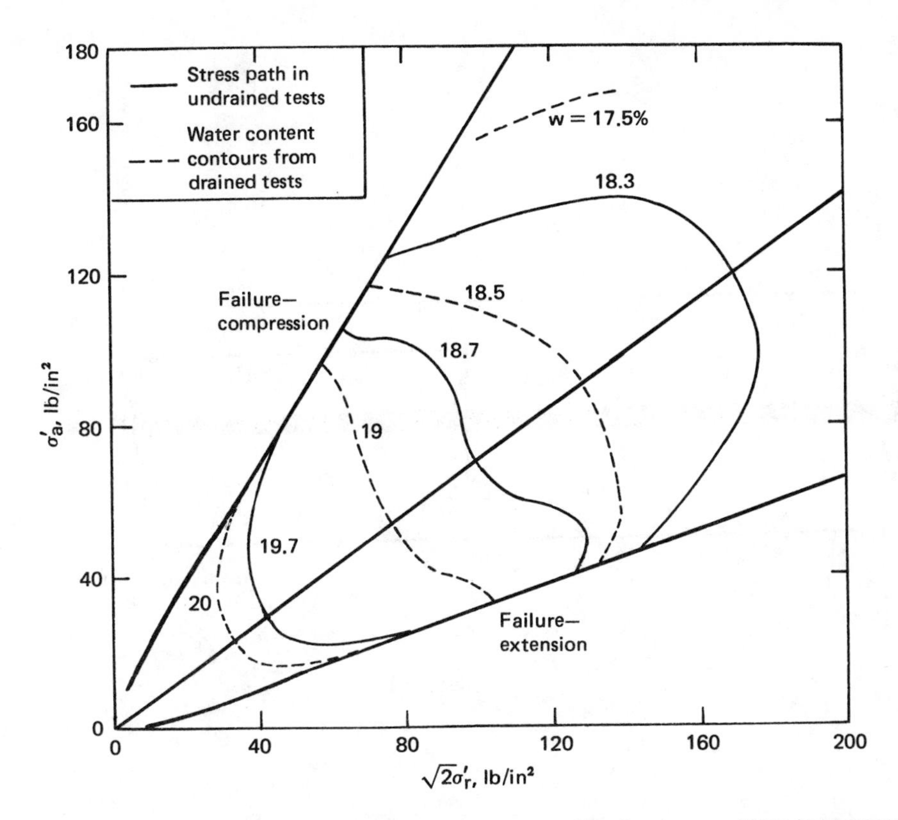

Fig. 7.58 Weald clay—overconsolidated: maximum consolidation pressure 120 lb/in² (828 kN/m²). *(Redrawn after D. J. Henkel, The Shearing Strength of Saturated Remolded Clays,* Proc. Research Conference on Shear Strength of Cohesive Soils, ASCE, *1960.)*

Figures 7.57 and 7.58 show the stress paths for equal water contents for normally consolidated and overconsolidated Weald clay. Note the similarity of shape of the stress paths for normally consolidated clay in Fig. 7.57. For overconsolidated clay, the shape of the stress path gradually changes, depending on the overconsolidation ratio.

7.3.11 Vane Shear Test

The field vane shear test is another method of obtaining the undrained shear strength of cohesive soils. The common shear vane usually consists of four thin steel plates of equal size welded to a steel torque rod (Fig. 7.59*a*). To perform the test, the vane is pushed into the soil and torque is applied at the top of the torque rod. The torque is gradually increased until the cylindrical soil of height H and diameter D fails (Fig. 7.59*b*). The maximum torque T applied to cause failure is the sum of the resisting moment at the top, M_T, and bottom, M_B, of the soil cylinder, plus the resisting moment at the sides of the cylinder, M_S. Thus,

$$T = M_S + M_T + M_B \tag{7.43}$$

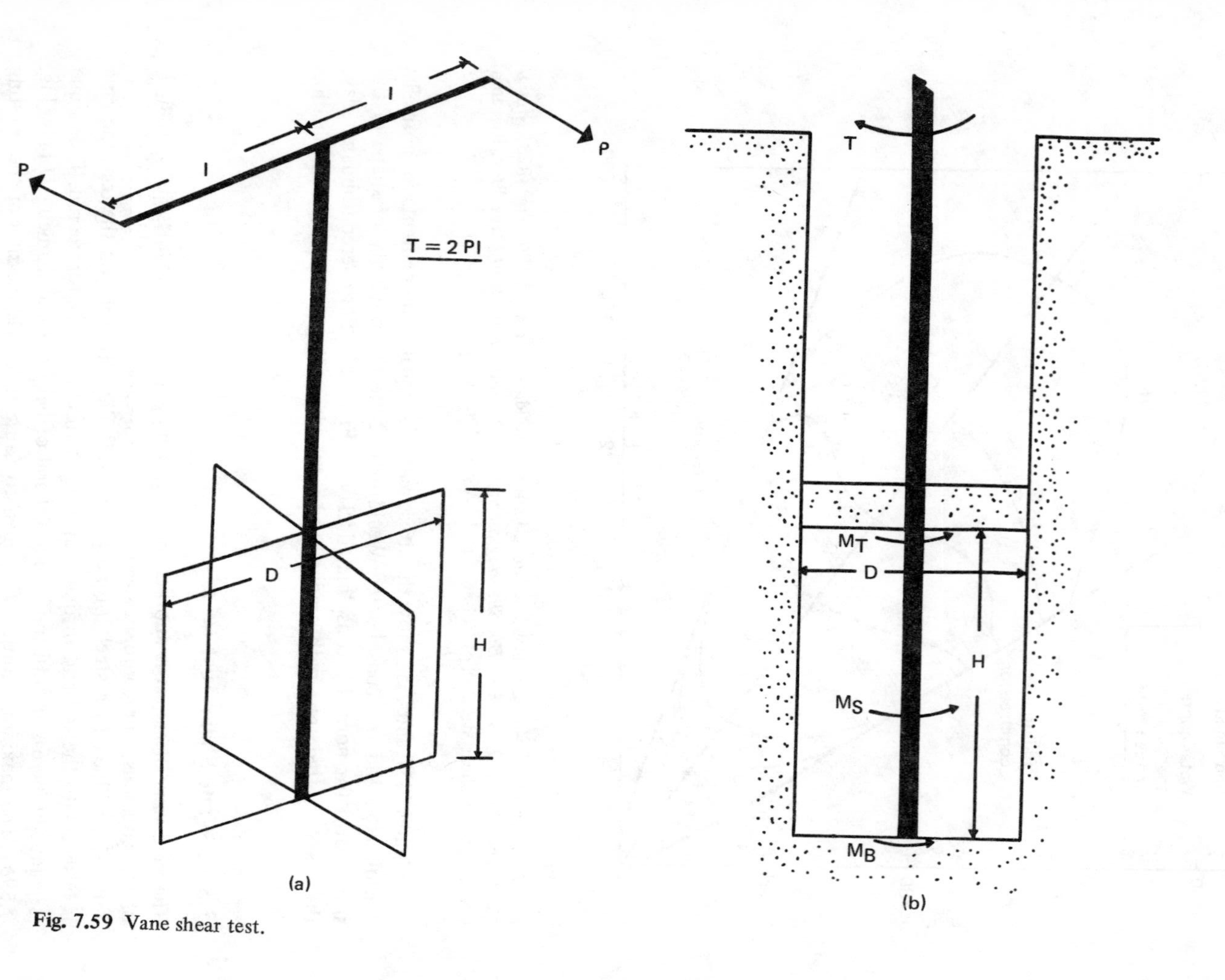

Fig. 7.59 Vane shear test.

But, $M_S = \pi DH \frac{D}{2} S_u$ and $M_T = M_B = \frac{\pi D^2}{4} \frac{2}{3} \frac{D}{2} S_u$

(assuming uniform undrained shear strength distribution at the ends; see Carlson, 1948). So,

$$T = \pi S_u \left[\left(\pi DH \frac{D}{2} \right) + 2 \left(\frac{\pi D^2}{4} \frac{2}{3} \frac{D}{2} \right) \right]$$

or

$$S_u = \frac{T}{\pi(D^2 H/2 + D^3/6)} \tag{7.44}$$

If only one end of the vane (i.e., the bottom) is engaged in shearing the clay, $T = M_S + M_B$. So

$$S_u = \frac{T}{\pi(D^2 H/2 + D^3/12)} \tag{7.45}$$

Laboratory vane shear testing devices are also available for determination of the undrained shear strength of clay soil.

A detailed study comparing the undrained shear strengths of cohesive soils obtained from field vane shear tests, unconfined compression tests, undrained triaxial tests, and laboratory vane shear tests was made by Arman et al. (1975). These tests were carried out at various sites in the Gulf Coast region in Louisiana. The soils varied from stiff pleistocene clays to soft marine deposits, with one site containing highly organic soils. Some of the results of these tests for one site are shown in Fig. 7.60, from which it appears that the field vane shear test consistently gives a higher value of undrained shear strength. The discrepancy can be attributed to factors such as sampling, handling, and testing conditions. A study of the core-slice radiograph by Arman et al. indicated that a thin, partially sheared zone surrounds the failure surface of vane shear tests, as shown in Fig. 7.61. The extent of the partially sheared or reoriented zone depends on the type and cohesiveness of soils. The results obtained by Arman et al. for correlating unconfined compression tests and field vane shear tests in three soils are as follows:

1. Laplace organic clay:

$$S_u = 0.535\, S_{u(\text{field vane})} - 0.007 \tag{7.46}$$

2. Morgan City clay (recent alluvial):

$$S_u = 0.509\, S_{u(\text{field vane})} - 0.066 \tag{7.47}$$

3. Lake Charles clay (pleistocene):

$$S_u = 0.536\, S_{u(\text{field vane})} - 0.0017 \tag{7.48}$$

where $S_u = q_u/2$.

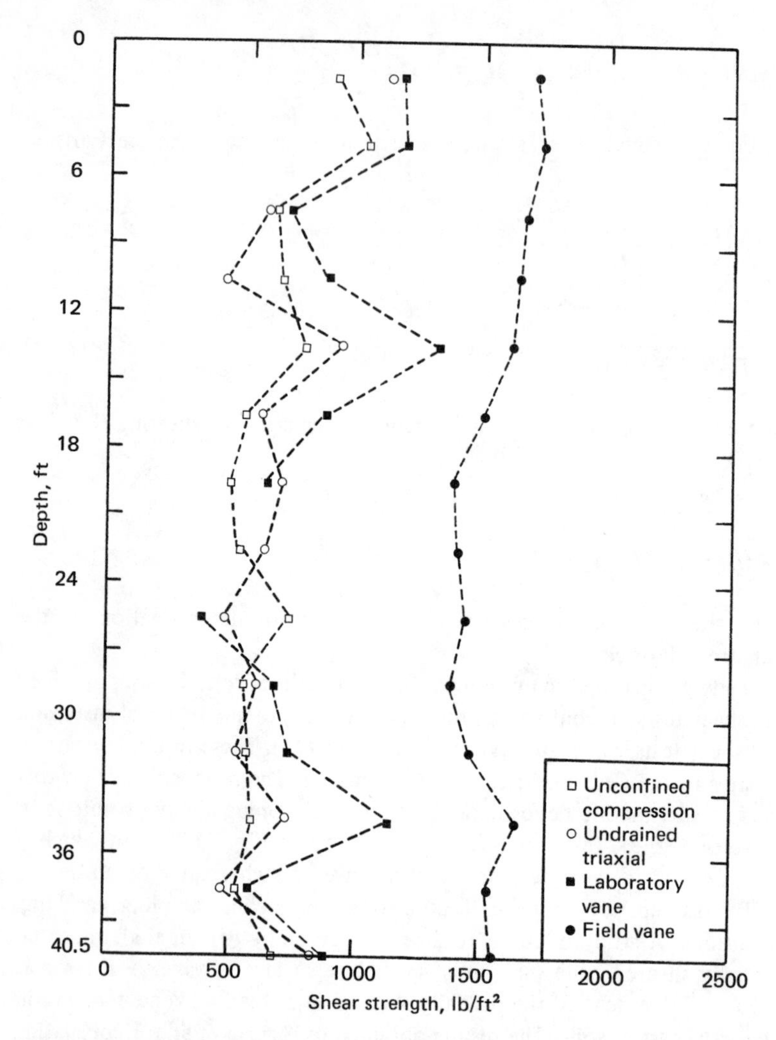

Fig. 7.60 Depth vs. typical undrained shear strength by various test methods for Morgan City recent alluvium. (Note: $1\ \text{ft} = 0.3048\ \text{m}$; $1\ \text{lb/ft}^2 = 47.88\ \text{N/m}^2$.) *(Redrawn after A. Arman, J. K. Poplin, and N. Ahmad, Study of Vane Shear,* Proc. Conference on In Situ Measurement of Soil Properties, ASCE, *vol. 1, 1975.)*

The coefficients of correlation for Eqs. (7.46) to (7.48) were 0.92, 0.81, and 0.87, respectively.

Bjerrum (1972) studied a number of slope failures and concluded that the undrained shear strength obtained by vane shear is too high. He proposed that the vane shear test results obtained from the field should be corrected for the actual design. Thus,

$$S_{u(\text{design})} = \lambda S_{u(\text{field vane})} \tag{7.49}$$

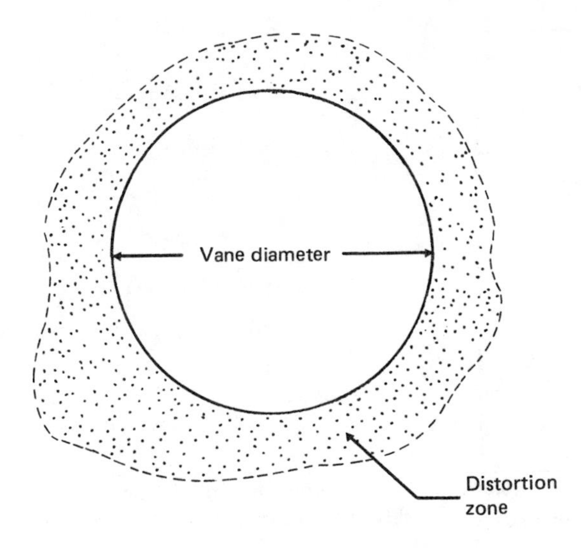

Fig. 7.61 Distorted zone around soil for vane shear test.

where λ is a correction factor. The value of λ decreases with the plasticity index of soil (Fig. 7.62).

As in the case of the unconsolidated undrained test, the vane shear strength is dependent on the rate at which the torque is applied. This is shown in Fig. 7.63 for laboratory vane shear tests for a silty clay. The undrained shear strength, S_u, increases with the increase in the rate of application of the torque.

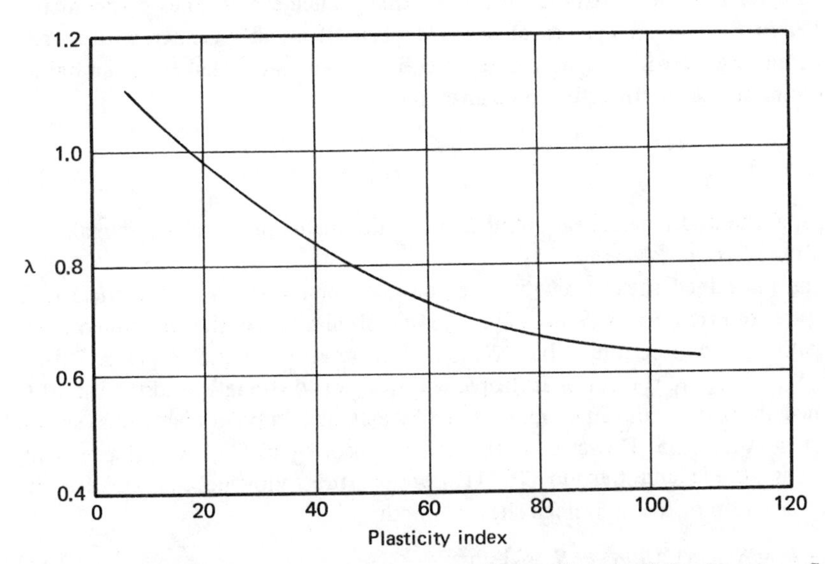

Fig. 7.62 Correction factor for vane shear test. *(After L. Bjerrum, Embankments on Soft Ground,* Proc. Specialty Conference on Performance of Earth and Earth Supported Structures, ASCE, *vol. 2, 1972.)*

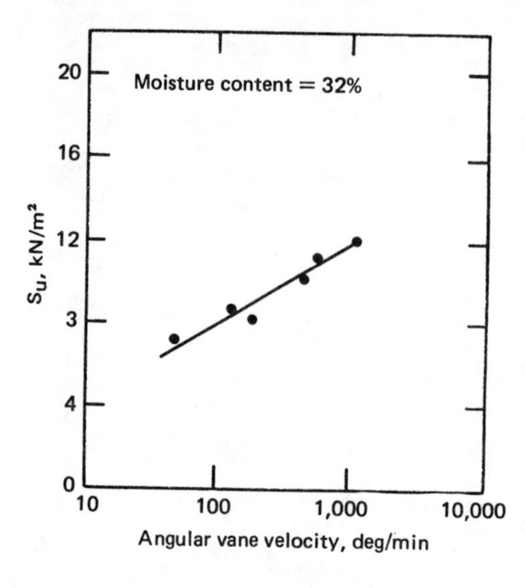

Fig. 7.63 Effect of rate of shear on S_u for a silty clay—laboratory vane shear apparatus.

7.3.12 Undrained Shear Strength of Anisotropic Clay

Due to the nature of the deposition of cohesive soils and subsequent consolidation, clay particles tend to become oriented perpendicular to the direction of the major principal stress. Parallel orientation of clay particles could cause the strength of the clay to vary with direction; or, in other words, the clay could be anisotropic with respect to strength. This fact can be demonstrated with the aid of Fig. 7.64, in which V and H are vertical and horizontal directions that coincide with lines perpendicular and parallel to the bedding planes of a soil deposit. If a soil specimen with its axis inclined at an angle i with the horizontal is collected and subjected to an undrained test, the undrained shear strength can be given by

$$S_{u(i)} = \frac{\sigma_1 - \sigma_3}{2} \tag{7.50}$$

where $S_{u(i)}$ is the undrained shear strength when the major principal stress makes an angle i with the horizontal.

Let the undrained shear strength of a soil specimen with its axis vertical (i.e., $S_{u(i=90°)}$] be referred to as $S_{u(V)}$ (Fig. 7.64*b*); similarly, let the undrained shear strength with its axis horizontal [i.e., $S_{u(i=0°)}$] be referred to as $S_{u(H)}$ (Fig. 7.64*c*). If $S_{u(V)} = S_{u(i)} = S_{u(H)}$, the soil is isotropic with respect to strength, and the variation of undrained shear strength can be represented by a circle in a polar diagram, as shown by curve a in Fig. 7.65. However, if the soil is anisotropic, $S_{u(i)}$ will change with direction. Casagrande and Carrillo (1944) proposed the following equation for the directional variation of the undrained shear strength:

$$S_{u(i)} = S_{u(H)} + [S_{u(V)} - S_{u(H)}] \sin^2 i \tag{7.51}$$

When $S_{u(V)} > S_{u(H)}$, the nature of variation of $S_{u(i)}$ can be represented by curve b in Fig. 7.65. Again, if $S_{u(V)} < S_{u(H)}$, the variation of $S_{u(i)}$ is given by curve c. The coefficient of anisotropy can be defined as

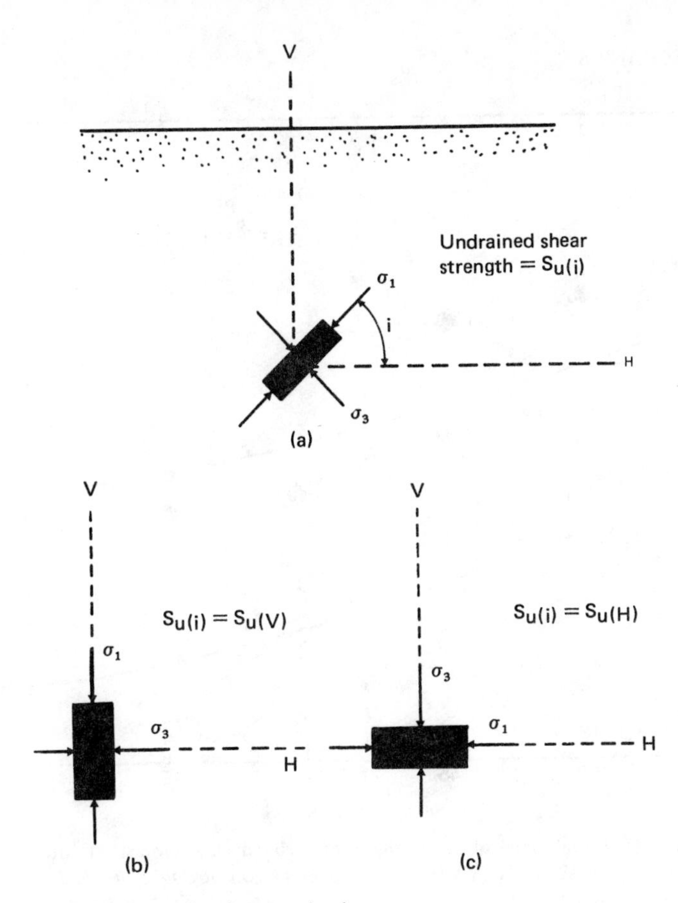

Fig. 7.64 Strength anisotropy in clay.

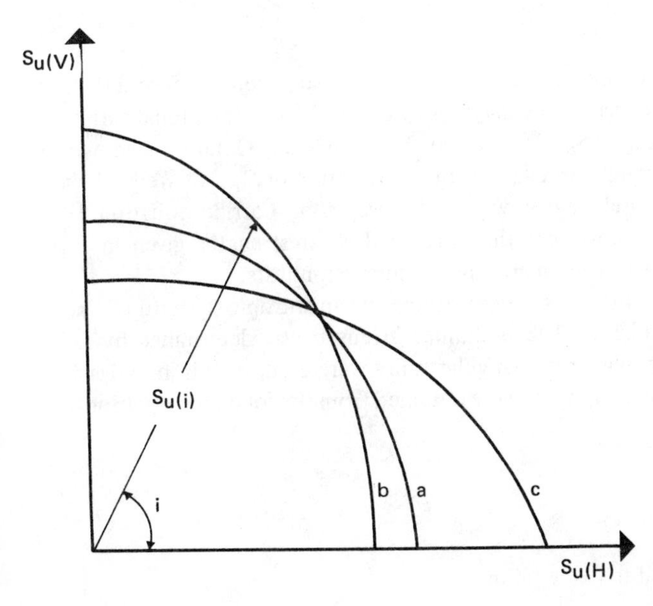

Fig. 7.65 Directional variation of undrained strength of clay.

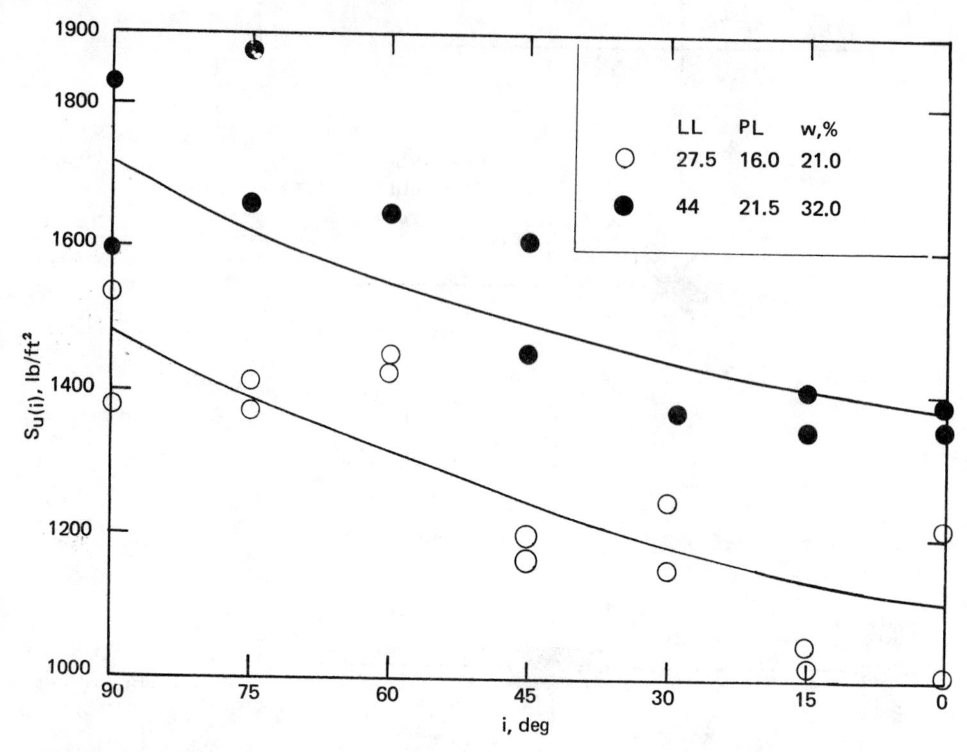

Fig. 7.66 Directional variation of undrained shear strength of Welland clay, Ontario. (Note: 1 lb/ft² = 47.9 N/m².) *(Redrawn after K. Y. Lo, Stability of Slopes in Anisotropic Soils,* J. Soil Mech. Found. Div., ASCE, *vol. 91, no. SM4, 1965.)*

$$K = \frac{S_{u(V)}}{S_{u(H)}} \tag{7.52}$$

In the case of natural soil deposits, the value of K can vary from 0.75 to 2.0. K is generally less than 1 in overconsolidated clays. An example of the directional variation of the undrained shear strength $S_{u(i)}$ for a clay from Welland, Ontario, is shown in Fig. 7.66. Figure 7.67 shows a polar diagram for the variation of $S_{u(i)}$ of Welland clay. The experimental results closely agree with the Casagrande-Carrillo equation [Eq. (7.51)]. It is of interest to know that the undrained shear strengths given in Figs. 7.66 and 7.67 were determined from unconfined compression tests.

Richardson et al. (1975) made a study regarding the anisotropic strength of a soft deposit of marine clay (Thailand). The undrained strength was determined by field vane shear tests. Both rectangular and triangular vanes were used for this investigation (Fig. 7.68). The undrained shear strengths were obtained from the following equations:

1. Triangular vanes:

$$S_{u(i)} = \frac{T}{\frac{4}{3}\pi L^3 \cos^2 i} \tag{7.53}$$

where T = torque applied to cause failure

$$L = \frac{H}{2 \sin i} \text{ (see Fig. 7.68)}$$

H = height of vanes (see Fig. 7.68)

2. Rectangular vanes:

$$T = \pi D^2 \left[\frac{D}{6} S_{u(H)} + \frac{H}{2} S_{u(V)} \right] \tag{7.54}$$

where D = vane diameter

The directional variation of shear strength of this investigation is shown in Fig. 7.69. Based on the experimental results, Richardson et al. concluded that $S_{u(i)}$ can be given by the following relation:

$$S_{u(i)} = \frac{S_{u(H)} S_{u(V)}}{\sqrt{S_H^2 \sin^2 i + S_V^2 \cos^2 i}} \tag{7.55}$$

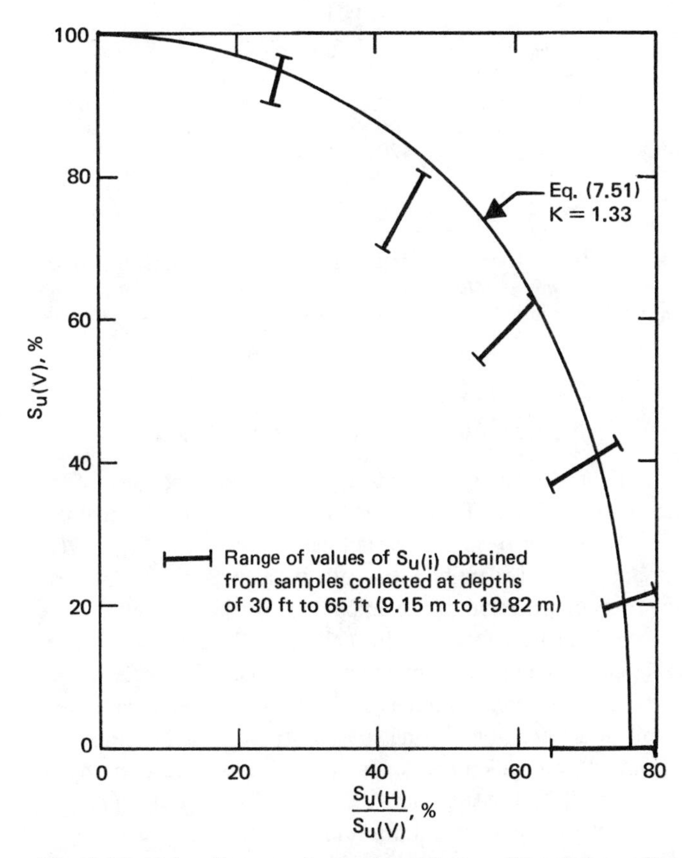

Fig. 7.67 Polar diagram of experimental strength variation with orientation of applied major principal stress of Welland clay, Ontario. *(Redrawn after K. Y. Lo, Stability of Slopes in Anisotropic Soils,* J. Soil Mech. Found. Div., ASCE, *vol. 91, no. SM4, 1965.)*

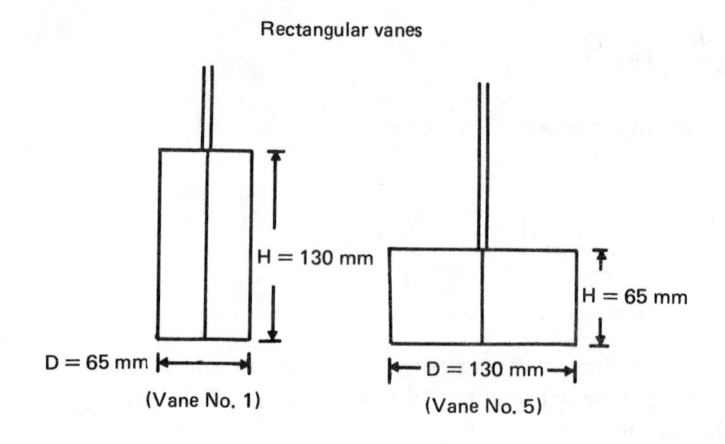

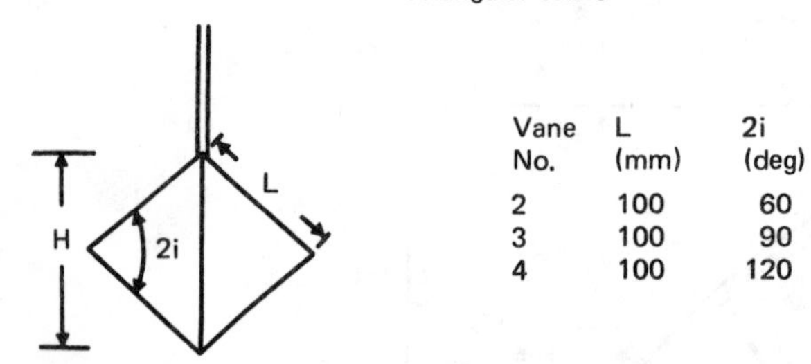

Vane No.	L (mm)	2i (deg)
2	100	60
3	100	90
4	100	120

Fig. 7.68 Configuration of vanes used by Richardson et al. (1975) for use in Eqs. (7.53) and (7.54). *(A. M. Richardson, E. W. Brand, and A. Menon, In Situ Determination of Anisotropy of a Soft Soil,* Proc. Conference on In Situ Measurement of Soil Properties, ASCE, *vol. 1, 1965.)*

7.3.13 Applicability of Drained (c, ϕ) and Undrained (S_u) Shear Strength Parameters for Foundation Design

In conducting the analysis of the stability of various foundations, proper choice of the soil shear strength parameters is necessary. To demonstrate this, consider the problem of the construction of an embankment over a saturated clay as shown in Fig. 7.70*a*. The stability of such foundations is generally analyzed by assuming a cylindrical failure surface. With the gradual construction of the embankment, the induced average shear stress, τ, through a cylindrical surface passing through point P gradually increases with the height of the fill and remains constant from the time of completion of construction (Fig. 7.70*b*). Since the construction is completed in a relatively short time and drainage from the soil is negligible during the construction phase, the pore water pressure in the foundation soil will increase up to the time corresponding to completion of construction (Fig. 7.70*c*). After that, the induced pore water pressure will begin to dissipate and will eventually reach equilibrium.

Up to the time of end of construction, due to the practically undrained condition the shear strength of the clay will remain constant ($\approx S_u$). Hence, the factor of safety of the foundation is lowest at the end of construction (Fig. 7.70*d*). So, during the construction phase, stability analysis using the $\phi = 0$ method (i.e., $s = S_u$) is more applicable. After that, the pore water pressure in the field may be observed and the stability may be reestimated by the drained shear strength parameters (i.e., c and ϕ). Table 7.7 shows several cases of factor of safety from stability calculations for end-of-construction failures of fills on saturated clays. Also shown in Table 7.7 are the factor of safety calculations for stability of footings using the $\phi = 0$ concept. These demonstrate the applicability of $s = S_u$ analysis.

7.3.14 Hvorslev's Parameters

Considering cohesion to be the result of physico-chemical bond forces (thus the interparticle spacing and hence void ratio), Hvorslev (1937) expressed the shear strength of a soil in the form

$$s = c_e + \sigma' \tan \phi_e \tag{7.56}$$

where c_e and ϕ_e are "true cohesion" and "true angle of friction," respectively, which are dependent on void ratio.

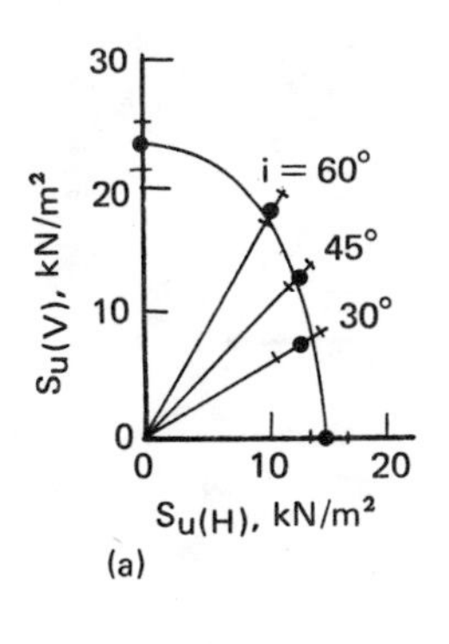

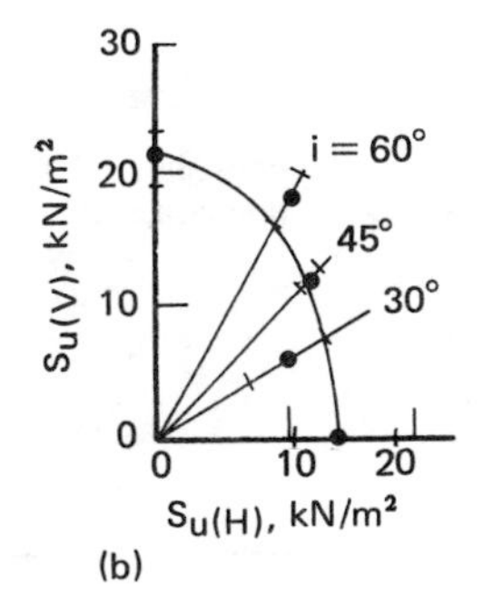

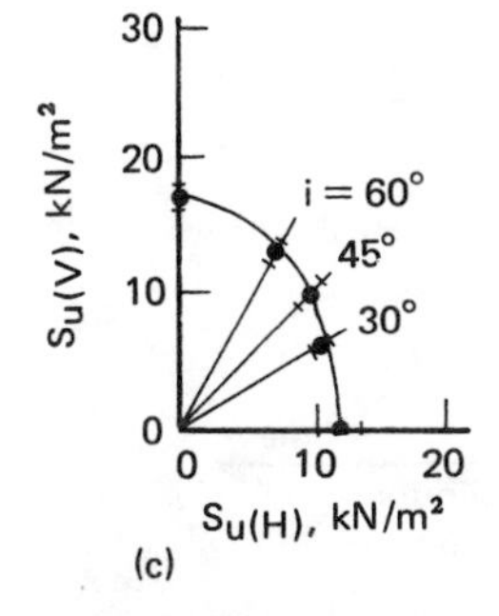

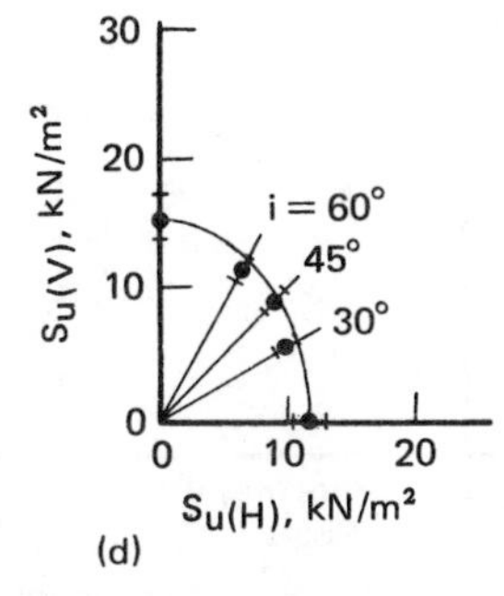

Fig. 7.69 Vane shear-strength polar diagrams for a soft marine clay in Thailand. (*a*) Depth = 1 m; (*b*) depth = 2 m; (*c*) depth = 3 m; (*d*) depth = 4 m. *(After A. M. Richardson, E. W. Brand, and A. Menon, In Situ Determination of Anisotropy of a Soft Clay*, Proc. Conference on In Situ Measurements of Soil Properties, ASCE, *vol. 1, 1975.)*

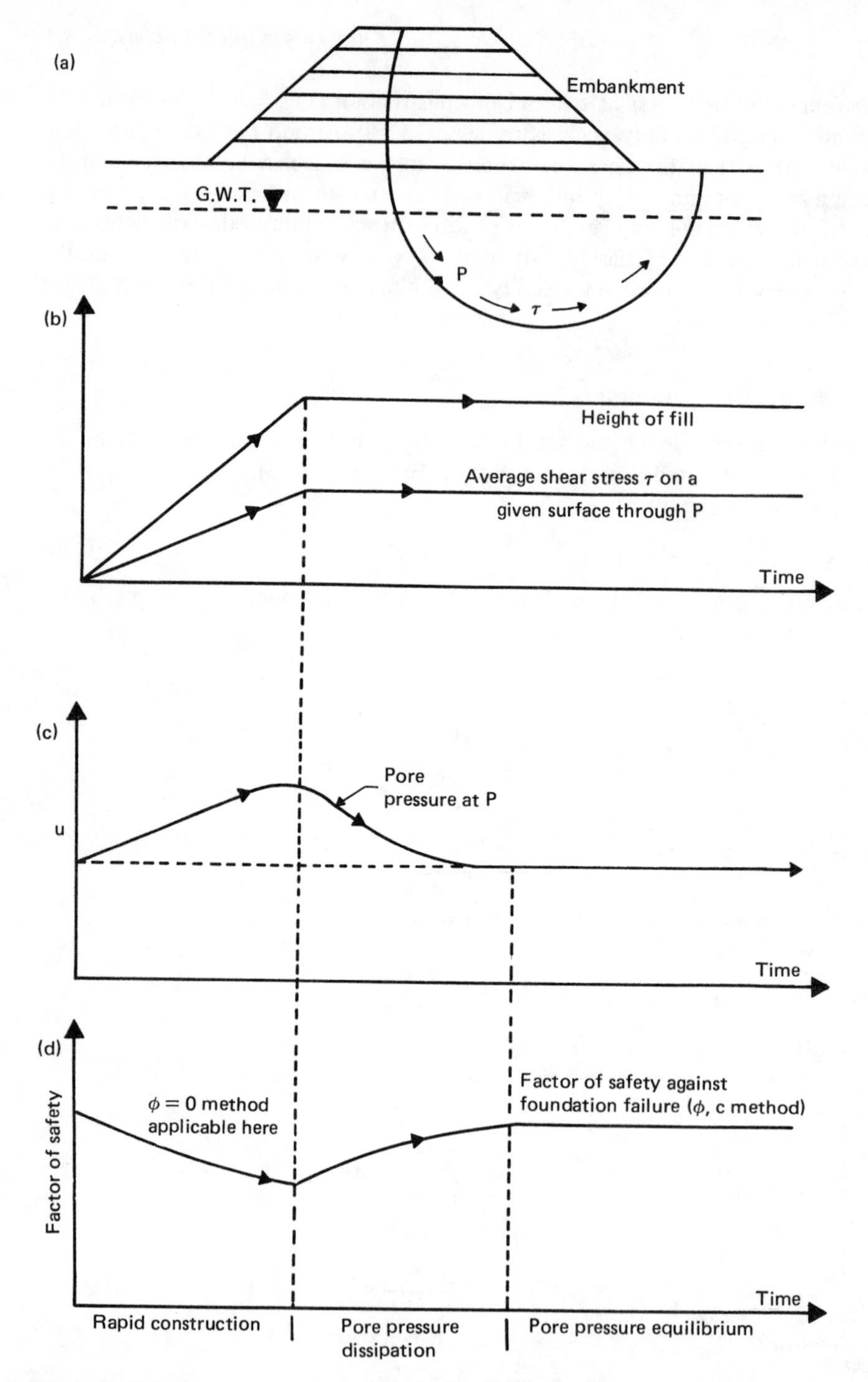

Fig. 7.70 Variation of factor of safety with time for a foundation on a saturated clay beneath a fill. *(Redrawn after A. W. Bishop and L. Bjerrum, The Relevance of the Triaxial Test to the Solution of Stability Problems,* Proc. Research Conference on Shear Strength of Cohesive Soils, ASCE, *1960.)*

The procedure for determination of the above parameters can be explained with the aid of Fig. 7.71, which shows the relation of the moisture content (i.e., void ratio) with effective consolidation pressure. Points 2 and 3 represent normally consolidated stages of a soil and point 1 represents the overconsolidation stage. We now test the soil samples represented by points 1, 2, and 3 in an undrained condition. The effective-stress Mohr's circles at failure are given in Fig. 7.71*b*.

The soil specimens at points 1 and 2 in Fig. 7.71*a* have the same moisture content and, hence, the same void ratio. If we draw a common tangent to the Mohr's circles 1 and 2, the slope of the tangent will give ϕ_e, and the intercept on the shear stress axis will give c_e.

Gibson (1953) found that ϕ_e varies slightly with void ratio. The *true angle of internal friction* decreases with the plasticity index of soil, as shown in Fig. 7.72. The

Table 7.7 End-of-construction failures on fill and footings on saturated clay

Locality	Factor of safety, $\phi = 0$ analysis
(a) Fills	
Chingford	1.05
Gosport	0.93
Panama 2	0.93
Panama 3	0.98
Newport	1.08
Bromma II	1.03
Bocksjon	1.10
Huntington	0.98
(b) Footings: Loading test	
Loading test, Marmorera	0.92
Kensal Green	1.02
Silo, Transcona	1.09
Kippen	0.95
Screw pile, Lock Ryan	1.05
Screw pile, Newport	1.07
Oil tank, Fredrikstad	1.08
Oil tank A, Shellhaven	1.03
Oil tank B, Shellhaven	1.05
Silo, USA	0.98
Loading test, Moss	1.10
Loading test, Hagalund	0.93
Loading test, Torp	0.96
Loading test, Rygge	0.95

After A. W. Bishop and L. Bjerrum, The Relevance of the Triaxial Test to the Solution of Stability Problems, *Proc. Research Conference on Shear Strength of Cohesive Soils, ASCE,* 1960.

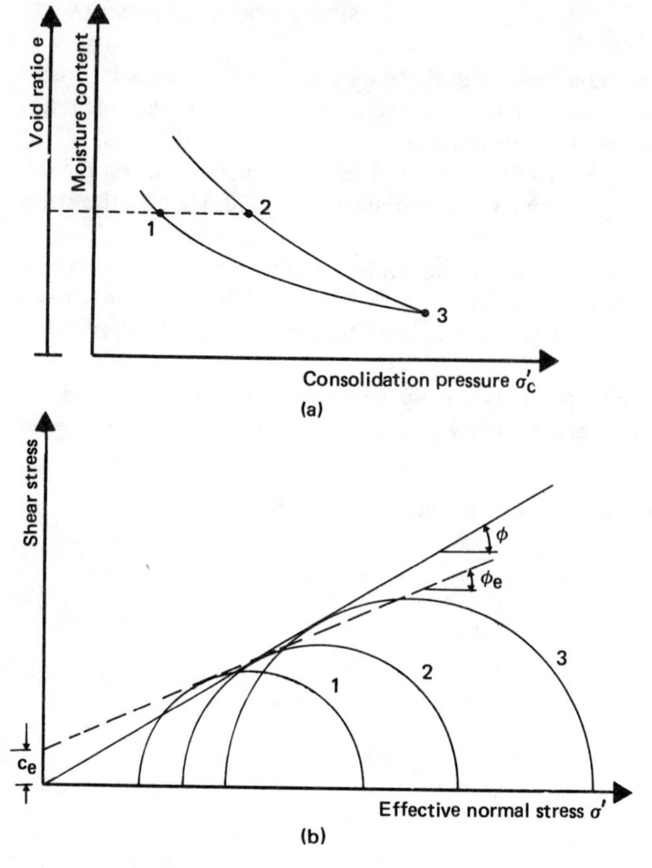

Fig. 7.71 Determination of c_e and ϕ_e.

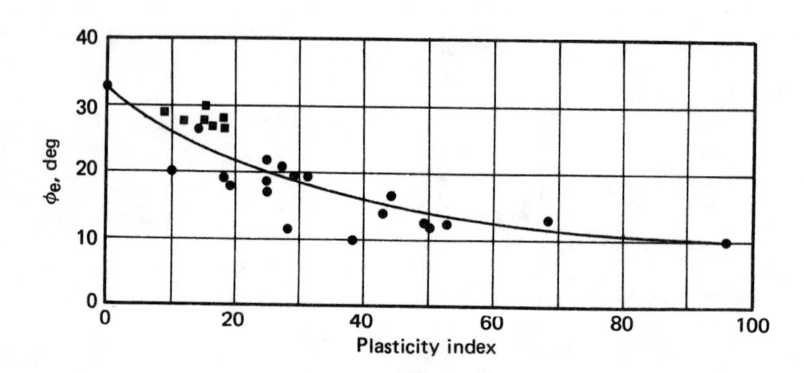

Fig. 7.72 Variation of true angle of friction with plasticity index. *(Redrawn after L. Bjerrum and N. E. Simons, Comparison of Shear Strength Characteristics of Normally Consolidated Clay,* Proc. Research Conference on Shear Strength of Cohesive Soils, ASCE, *1960.)*

variation of the effective cohesion c_e with void ratio may be given by the relation (Hvorslev, 1960)

$$c_e = c_0 \exp(-Be) \tag{7.57}$$

where c_0 = *true cohesion* at zero void ratio
e = void ratio at failure
B = slope of plot of $\ln c_e$ vs. void ratio at failure

Example 7.7 A clay soil specimen was subjected to confining pressures $\sigma_3 = \sigma_3'$ in a triaxial chamber. The moisture content vs. σ_3' relation is shown in Fig. 7.73*a*.

A *normally consolidated* specimen of the same soil was subjected to a consolidated undrained triaxial test. The results are as follows: $\sigma_3 = 440\,\text{kN/m}^2$; $\sigma_1 = 840\,\text{kN/m}^2$; moisture content at failure, 27%; $u_d = 240\,\text{kN/m}^2$.

An overconsolidated specimen of the same soil was subjected to a consolidated undrained test. The results are as follows: overconsolidation pressure, $\sigma_c' = 550\,\text{kN/m}^2$; $\sigma_3 = 100\,\text{kN/m}^2$; $\sigma_1 = 434\,\text{kN/m}^2$; $u_d = -18\,\text{kN/m}^2$; initial and final moisture content, 27%.

Determine ϕ_e, c_e for a moisture content of 27%; also determine ϕ.

SOLUTION For the normally consolidated specimen,

$$\sigma_3' = 440 - 240 = 200\,\text{kN/m}^2$$

$$\sigma_1' = 840 - 240 = 600\,\text{kN/m}^2$$

$$\phi = \sin^{-1}\left(\frac{\sigma_1' - \sigma_3'}{\sigma_1' + \sigma_3'}\right) = \sin^{-1}\left(\frac{600 - 200}{600 + 200}\right) = 30°$$

The failure envelope is shown in Fig. 7.73*b*.

For the overconsolidated specimen,

$$\sigma_3' = 100 - (-18) = 118\,\text{kN/m}^2$$

$$\sigma_1' = 434 - (-18) = 452\,\text{kN/m}^2$$

The Mohr's circle at failure is shown in Fig. 7.73*b*; from this,

$$c_e = 110\,\text{kN/m}^2 \qquad \phi_e = 15°$$

7.3.15 Sensitivity and Thixotropic Characteristics of Clays

Most undisturbed natural clayey soil deposits show a pronounced reduction of strength when they are remolded. This characteristic of saturated cohesive soils is generally expressed quantitatively by a term referred to as *sensitivity*. Thus,

$$\text{Sensitivity} = \frac{S_{u(\text{undisturbed})}}{S_{u(\text{remolded})}} \tag{7.58}$$

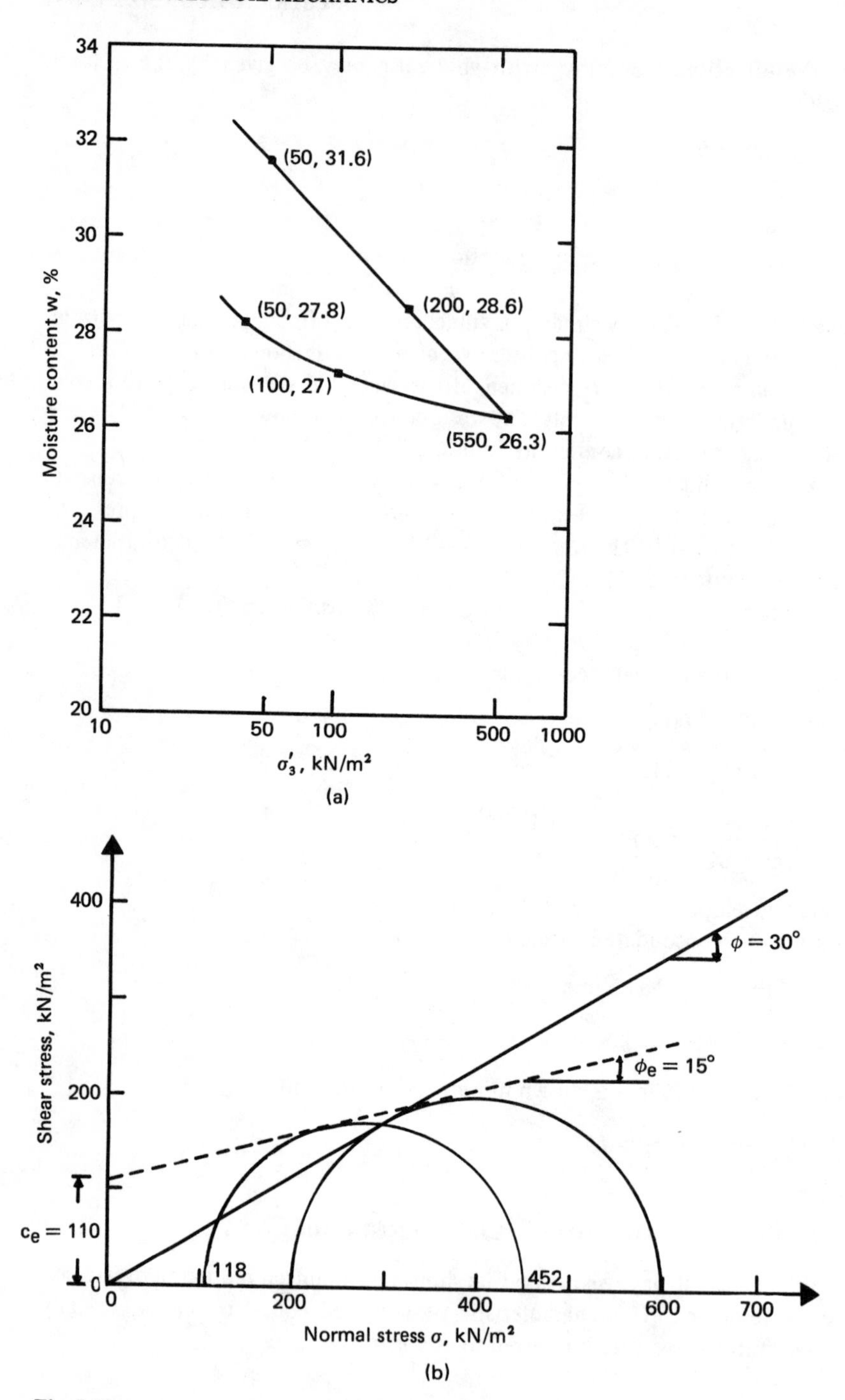
34
32
30
28
26
24
22
20
Moisture content w, %
(50, 31.6)
(200, 28.6)
(50, 27.8)
(100, 27)
(550, 26.3)
10
50
100
500
1000
σ_3', kN/m²
(a)
400
200
Shear stress, kN/m²
$c_e = 110$
$\phi = 30°$
$\phi_e = 15°$
118
452
0
200
400
600
700
Normal stress σ, kN/m²
(b)

Fig. 7.73

The classification of clays based on sensitivity is as follows:

Sensitivity	Clay
≈1	Insensitive
1–2	Low sensitivity
2–4	Medium sensitivity
4–8	Sensitive
8–16	Extra sensitive
>16	Quick

The sensitivity of most clays generally falls in a range of 1 to 8. However, sensitivity as high as 150 for a clay deposit at St. Thurible, Canada, was reported by Peck et al. (1951).

The loss of strength of saturated clays may be due to the breakdown of the original structure of natural deposits and thixotropy. *Thixotropy* is defined as an isothermal, reversible, time-dependent process which occurs under constant composition and volume whereby a material softens, as a result of remolding, and then gradually returns to its original strength when allowed to rest. This is shown in Fig. 7.74. A general review of the thixotropic nature of soils is given by Seed and Chan (1959).

Figure 7.75, which is based on the work of Moretto (1948), shows the thixotropic strength regain of a Laurentian clay with a liquidity index of 0.99 (i.e., the natural water content was approximately equal to the liquid limit). In Fig. 7.76, the acquired sensitivity is defined as

$$\text{Acquired sensitivity} = \frac{S_{u(t)}}{S_{u(\text{remolded})}} \tag{7.59}$$

where $S_{u(t)}$ is the undrained shear strength after a time t from remolding.

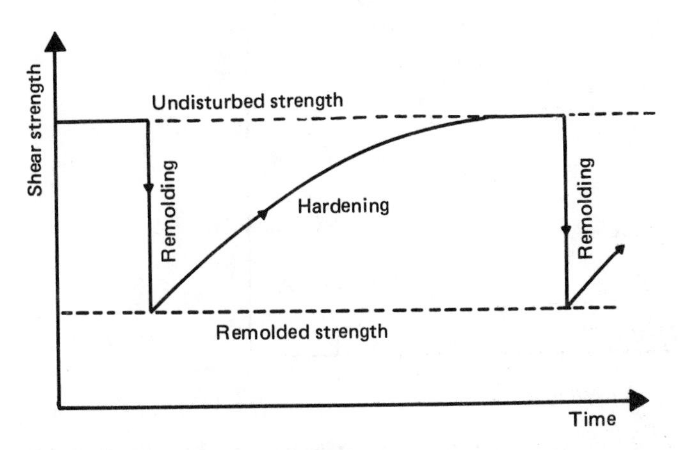

Fig. 7.74 Thixotropy of a material.

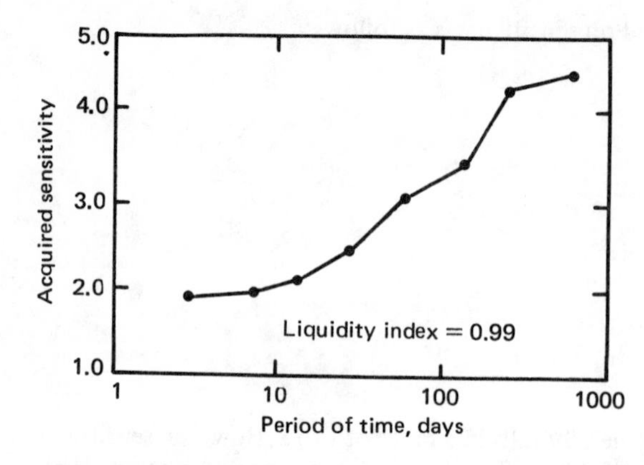

Fig. 7.75 Acquired sensitivity for Laurentian clay. *(Redrawn after H. B. Seed and C. K. Chan, Thixotropic Characteristics of Compacted Clays,* Trans. ASCE, *vol. 124, 1959.)*

Acquired sensitivity generally decreases with the liquidity index (i.e., the natural water content of soil), and this is demonstrated in Fig. 7.76. It can also be seen from this figure that the acquired sensitivity of clays with a liquidity index approaching zero (i.e., natural water content equal to the plastic limit) is approximately one. Thus, thixotropy in the case of overconsolidated clay is very small.

There are some clays that show that sensitivity cannot be entirely accounted for by thixotropy (Berger and Gnaedinger, 1949). This means that only a part of the strength loss due to remolding can be recovered by hardening with time. The other part of the strength loss is due to the breakdown of the original structure of the clay. The general nature of the strength regain of a partially thixotropic material is shown in Fig. 7.77.

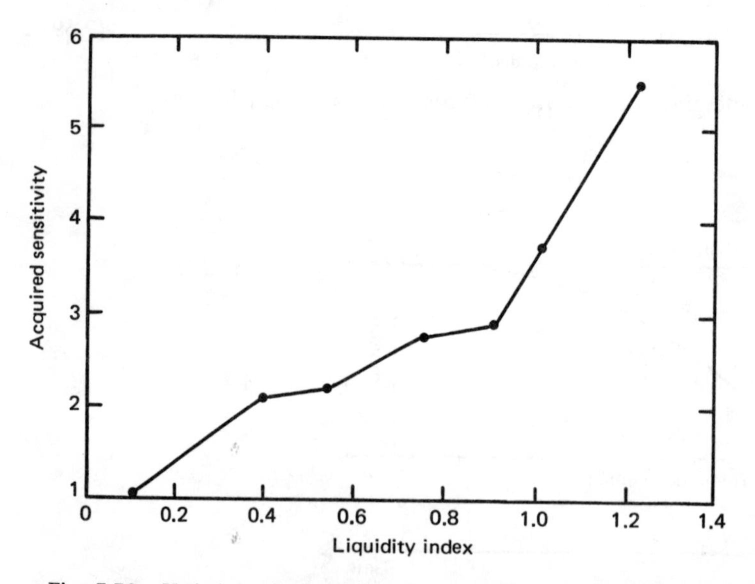

Fig. 7.76 Variation of sensitivity with liquidity index for Laurentian clay. *(Redrawn after H. B. Seed and C. K. Chan, Thixotropic Characteristics of Compacted Clays,* Trans. ASCE, *vol. 124, 1959.)*

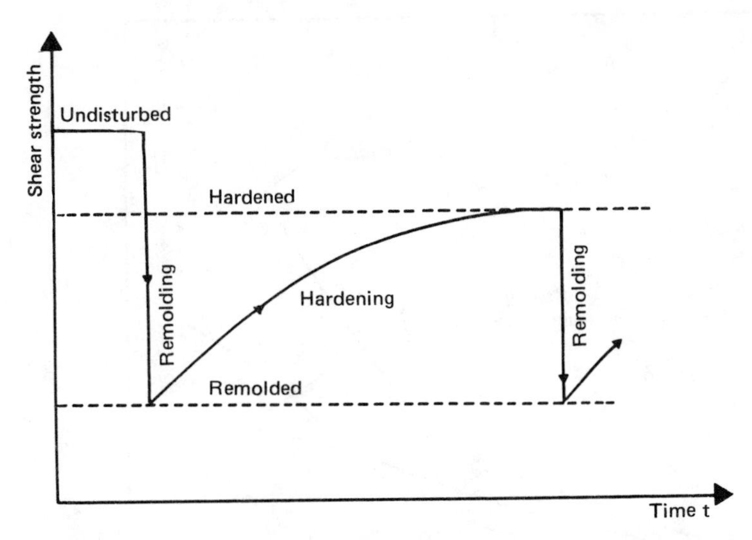

Fig. 7.77 Strength regain of a partially thixotropic material. *(Redrawn after H. B. Seed and C. K. Chan, Thixotropic Characteristics of Compacted Clays,* Trans. ASCE, *vol. 124, 1959.)*

Seed and Chan (1959) conducted several tests on three compacted clays with a water content near or below the plastic limit to study their thixotropic strength-regain characteristics. The properties of these clays are given in Table 7.8, and Fig. 7.78 shows their thixotropic strength ratio with time. The thixotropic strength ratio is defined as follows

$$\text{Thixotropic strength ratio} = \frac{S_{u(t)}}{S_{u(\text{compacted at } t=0)}} \tag{7.60}$$

where $S_{u(t)}$ is the undrained strength at time t after compaction.

These test results demonstrate that thixotropic strength-regain is also possible for soils with a water content at or near the plastic limit.

Table 7.8 Properties of soils shown in Fig. 7.78

Soil	Liquid limit	Plastic limit	Water content, %	Degree of saturation
Vicksburg silty clay	37	23	19.5	95
Pittsburgh sandy clay	35	20	17.4	96
Friant-Kern clay	59	35	22	95

After H. B. Seed and C. K. Chan, Thixotropic Characteristics of Compacted Clays, *Trans. ASCE*, vol. 124, 1959.

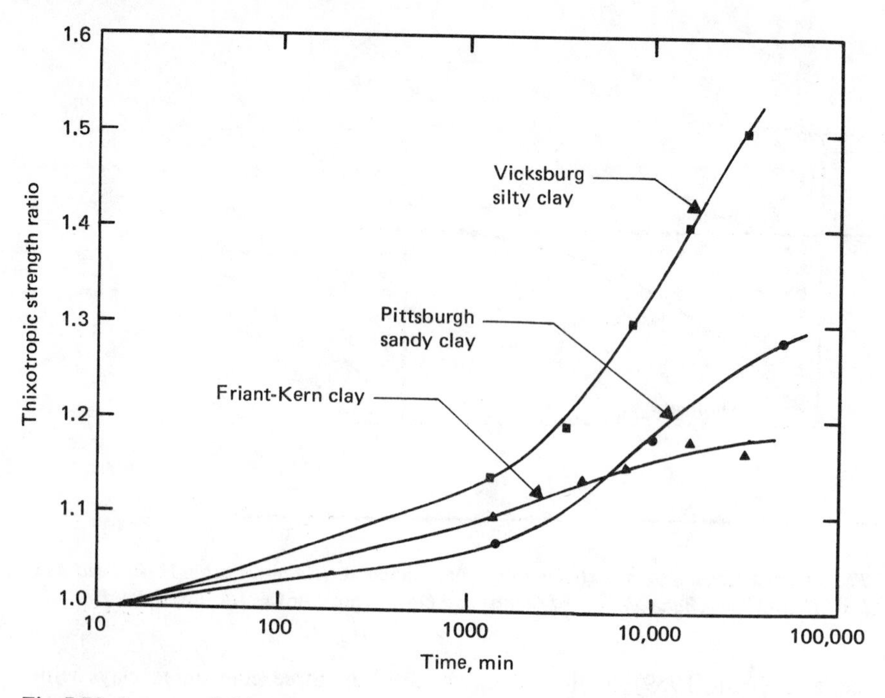

Fig. 7.78 Increase of thixotropic strength with time for three compacted clays. *(After H. B. Seed and C. K. Chan, Thixotropic Characteristics of Compacted Clays,* Trans. ASCE, *vol. 124, 1959.)*

7.3.16 Creep in Soils

Like metals and concrete, most soils exhibit creep, i.e., continued deformation under a sustained loading (Fig. 7.79). In order to understand Fig. 7.79, consider several similar clay specimens subjected to standard undrained loading. For specimen no. 1, if a deviator stress $(\sigma_1 - \sigma_3)_1 < (\sigma_1 - \sigma_3)_{\text{failure}}$ is applied, the strain vs. time (ϵ vs. t) relation will be similar to that shown by curve 1. If specimen no. 2 is subjected to a deviator stress $(\sigma_1 - \sigma_3)_2 < (\sigma_1 - \sigma_3)_1 < (\sigma_1 - \sigma_3)_{\text{failure}}$, the strain vs. time relation may be similar to that shown by curve 2. After the occurrence of a large strain, creep failure will take place in the specimen.

In general, the strain vs. time plot for a given soil can be divided into three parts: primary, secondary, and tertiary. The primary part is the transient stage; this is followed by a steady state, which is secondary creep. The tertiary part is the stage during which there is a rapid strain which results in failure. These three steps are shown in Fig. 7.79. Although the secondary stage is referred to as steady-state creep, in reality a true steady-state creep may not really exist (Singh and Mitchell, 1968). Figure 7.80 shows the plot of creep strain vs. time for a number of soils.

It was observed by Singh and Mitchell (1968) that for most soils (i.e., sand, clay–dry, wet, normally consolidated, and overconsolidated) the logarithm of strain rate has an approximately linear relation with the logarithm of time. This fact is,

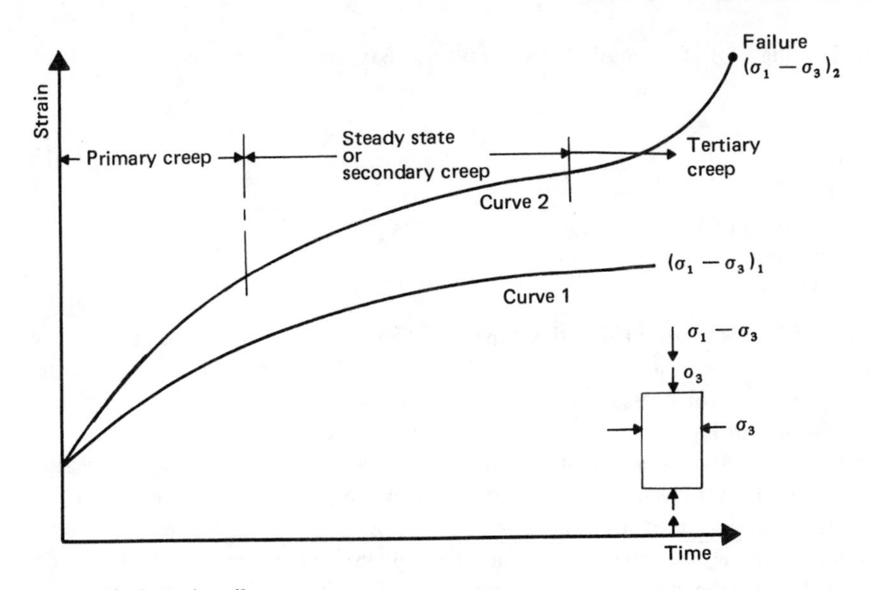

Fig. 7.79 Creep in soils.

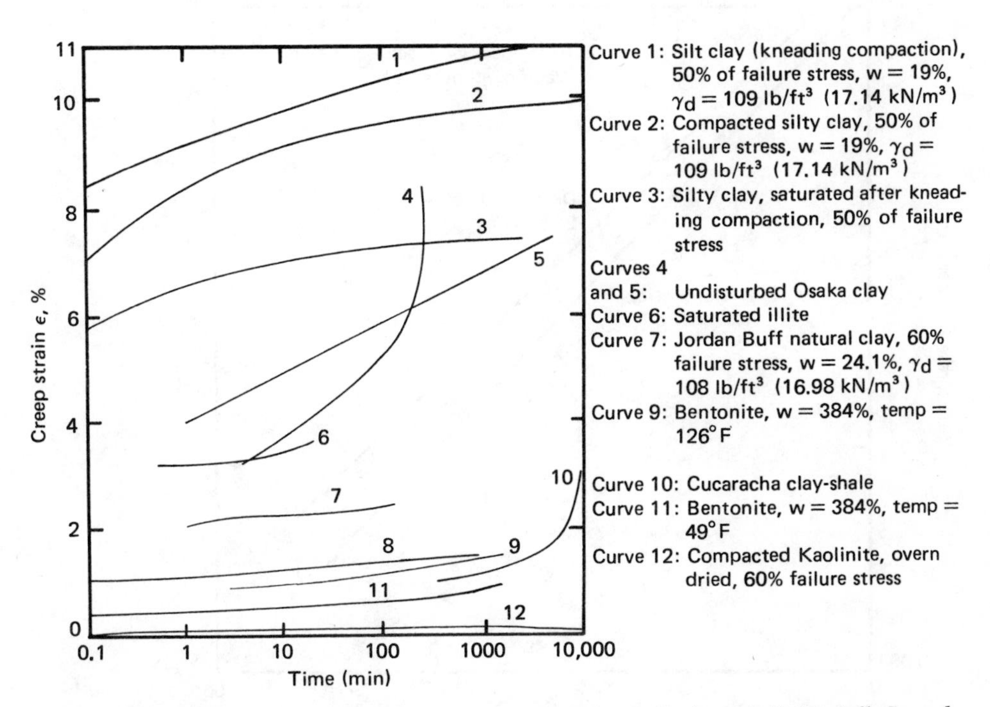

Fig. 7.80 Sustained stress–creep curves for some clays. *(After A. Singh and J. K. Mitchell, General Stress-Strain-Time Functions for Soils,* J. Soil Mech. Found. Div., ASCE, *vol. 94, no. SM1, 1968.)*

illustrated in Fig. 7.81 for remolded San Francisco Bay mud. The strain rate is defined as

$$\dot{\epsilon} = \frac{\Delta \epsilon}{\Delta t} \tag{7.61}$$

where $\dot{\epsilon}$ = strain rate
ϵ = strain
t = time

From Fig. 7.81, it is apparent that the slope of the log $\dot{\epsilon}$ vs. log t plot for a given soil is constant irrespective of the level of the deviator stress. When the failure stage due to creep at a given deviator stress level is reached, the log $\dot{\epsilon}$ vs. log t plot will show a reversal of slope as shown in Fig. 7.82.

Figure 7.83 shows the nature of the variation of the creep strain rate with deviator stress $D = \sigma_1 - \sigma_3$ at a given time t after the start of the creep. For small values of the deviator stress, the curve of log $\dot{\epsilon}$ vs. D is convex upward. Beyond this portion, log $\dot{\epsilon}$ vs. D is approximately a straight line. When the value of D approximately reaches the strength of the soil, the curve takes an upward turn, signalling impending failure.

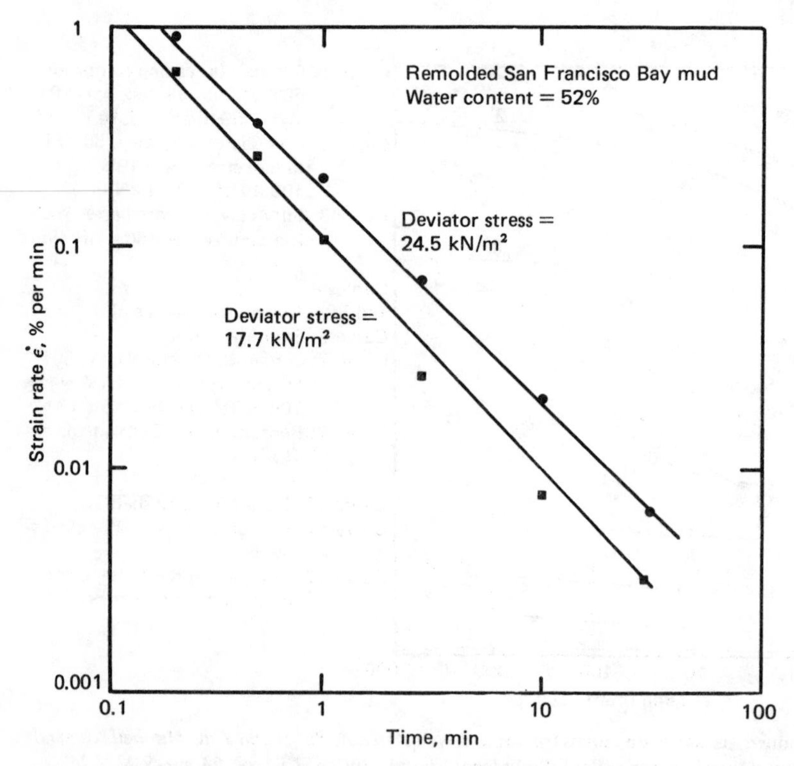

Fig. 7.81 Plot of log $\dot{\epsilon}$ vs. log (t) during undrained creep of remolded San Francisco Bay mud. *(After A. Singh and J. K. Mitchell, General Stress-Strain-Time Functions for Soils,* J. Soil Mech. Found. Div., ASCE, *vol. 94, no. SM1, 1968.)*

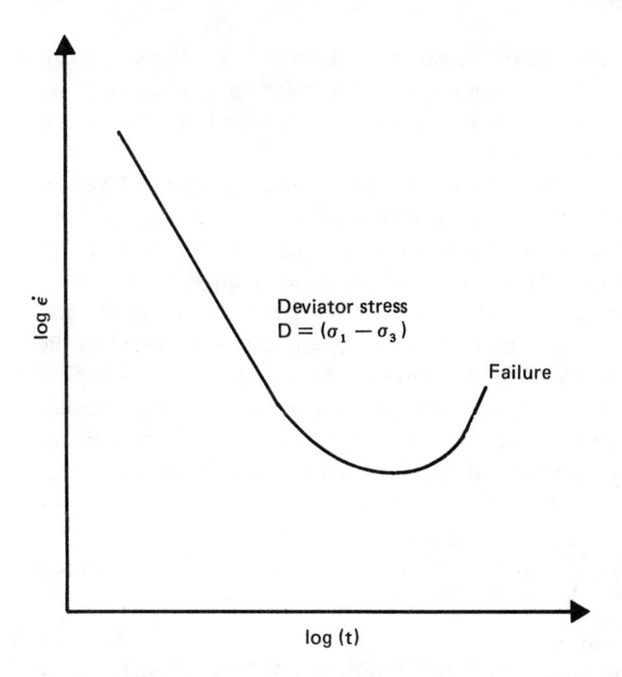

Fig. 7.82 Nature of variation of log $\dot{\epsilon}$ vs. log (t) for a given deviator stress showing the failure stage at large strains.

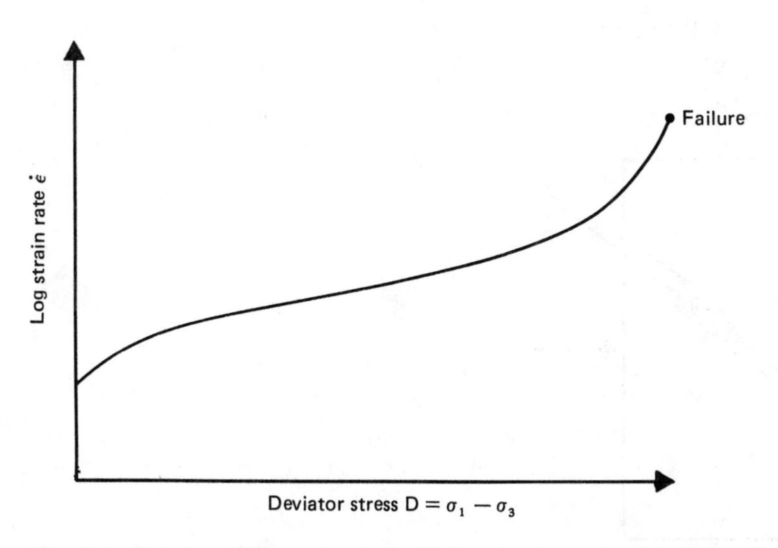

Fig. 7.83 Variation of the strain rate $\dot{\epsilon}$ with deviator stress at a given time t after the start of the test.

For a mathematical interpretation of the variation of strain rate with the deviator stress, several investigators (e.g., Christensen and Wu, 1964; Mitchell et al., 1968) have used the *rate-process theory.* Christensen and Das (1973) also used the rate-process theory to predict the rate of erosion of cohesive soils.

The fundamentals of the rate-process theory can be explained as follows. Consider the soil specimen shown in Fig. 7.84. The deviator stress on the specimen is $D = \sigma_1 - \sigma_3$. Let the shear stress along a plane AA in the specimen be equal to τ. The shear stress is resisted by the bonds at the points of contact of the particles along AA. Due to the shear stress τ the weaker bonds will be overcome, with the result that shear displacement occurs at these localities. As this displacement proceeds, the force carried by the weaker bonds is transmitted partly or fully to stronger bonds. The effect of applied shear stress can thus be considered as making some flow units cross the energy barriers as shown in Fig. 7.85, in which ΔF is equal to the activation energy (in cal/mole of flow unit). The frequency of activation of the flow units to overcome the energy barriers can be given by

$$k' = \frac{kT}{h} \exp\left(-\frac{\Delta F}{RT}\right) = \frac{kT}{h} \exp\left(-\frac{\Delta F}{NkT}\right) \tag{7.62}$$

where k' = frequency of activation
k = Boltzmann's constant = 1.38×10^{-16} erg/K = 3.29×10^{-24} cal/K

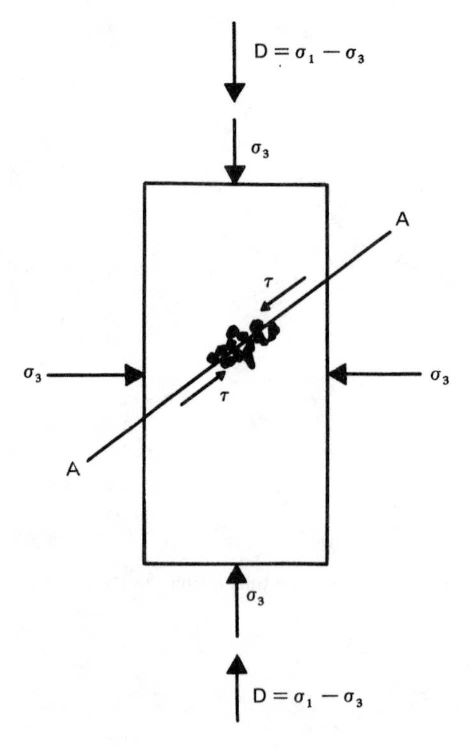

Fig. 7.84

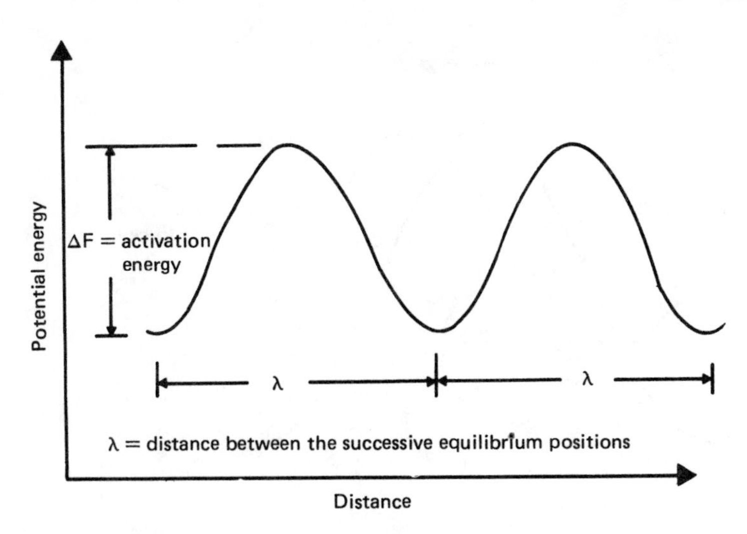

Fig. 7.85 Definition of activation energy.

T = absolute temperature
h = Plank's constant = 6.624×10^{-27} erg/s
ΔF = free energy of activation, cal/mole
R = universal gas constant
N = Avogadro's number = 6.02×10^{23}

Now, referring to Fig. 7.86, when a force f is applied across a flow unit, the energy-barrier height is reduced by $f\lambda/2$ in the direction of the force and increased by $f\lambda/2$ in the opposite direction. By this, the frequency of activation in the direction of the force is

$$\underset{\rightarrow}{k'} = \frac{kT}{h} \exp\left(-\frac{\Delta F/N - f\lambda/2}{kT}\right) \tag{7.63}$$

and, similarly, the frequency of activation in the opposite direction becomes

$$\underset{\leftarrow}{k'} = \frac{kT}{h} \exp\left(-\frac{\Delta F/N + \lambda f/2}{kT}\right) \tag{7.64}$$

where λ is the distance between successive equilibrium positions.

So, the net frequency of activation in the direction of the force is equal to

$$\underset{\rightarrow}{k'} - \underset{\leftarrow}{k'} = \frac{kT}{h}\left[\exp\left(-\frac{\Delta F/N - f\lambda/2}{kT}\right) - \exp\left(-\frac{\Delta F/N + f\lambda/2}{kT}\right)\right]$$

$$= \frac{2kT}{h} \exp\left(-\frac{\Delta F}{RT}\right)\sinh\left(\frac{f\lambda}{2kT}\right) \tag{7.65}$$

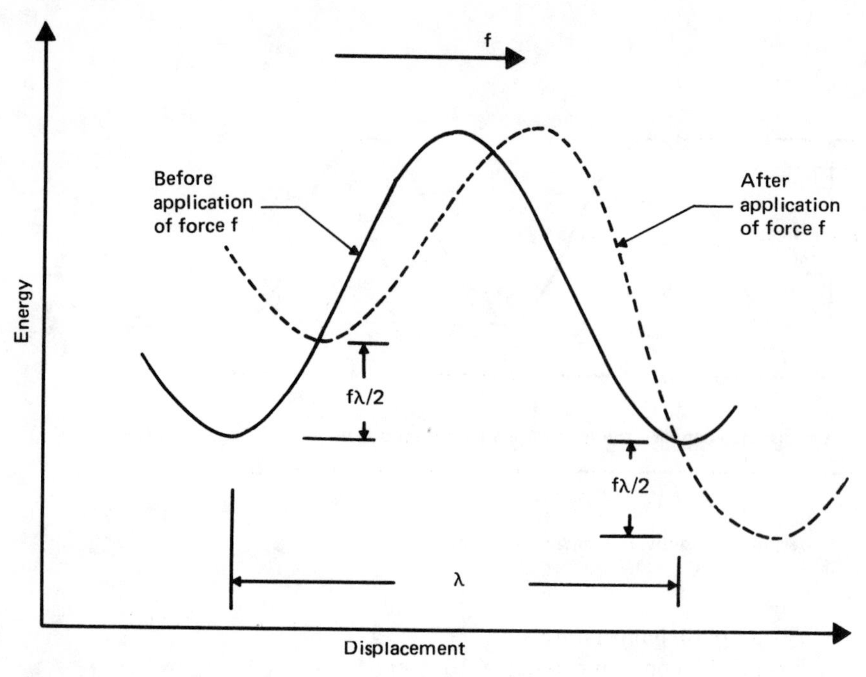

Fig. 7.86 Derivation of Eq. (7.73).

The rate of strain in the direction of the applied force can be given by

$$\dot{\epsilon} = x(\underset{\rightarrow}{k'} - \underset{\leftarrow}{k'}) \tag{7.66}$$

where x is a constant depending on the successful barrier crossings. So,

$$\dot{\epsilon} = 2x \frac{kT}{h} \exp\left(-\frac{\Delta F}{RT}\right) \sinh\left(\frac{f\lambda}{2kT}\right) \tag{7.67}$$

In the above equation,

$$f = \frac{\tau}{S} \tag{7.68}$$

where τ is the shear stress and S is the number of flow units per unit area.

For triaxial shear test conditions as shown in Fig. 7.84,

$$\tau_{\max} = \frac{D}{2} = \frac{\sigma_1 - \sigma_3}{2} \tag{7.69}$$

Combining Eqs. (7.68) and (7,69),

$$f = \frac{D}{2S} \tag{7.70}$$

Substituting Eq. (7.70) into Eq. (7.67), we get

$$\dot{\epsilon} = 2x \frac{kT}{h} \exp\left(-\frac{\Delta F}{RT} \sinh \frac{D\lambda}{4kST}\right) \tag{7.71}$$

For large stresses to cause significant creep—i.e., $D > 0.25 \cdot D_{max} = 0.25(\sigma_1 - \sigma_3)_{max}$ (Mitchell et al., 1968)—$D\lambda/4kST$ is greater than 1. So, in that case,

$$\sinh \frac{D\lambda}{4kST} \approx \frac{1}{2} \exp\left(\frac{D\lambda}{4kST}\right) \tag{7.72}$$

Hence, from Eqs. (7.71) and (7.72),

$$\dot{\epsilon} = x \frac{kT}{h} \exp\left(-\frac{\Delta F}{RT}\right) \exp\left(\frac{D\lambda}{4kST}\right) \tag{7.73}$$

$$= A \exp(BD) \tag{7.74}$$

where $$A = x \frac{kT}{h} \exp\left(-\frac{\Delta F}{RT}\right) \tag{7.75}$$

and $$B = \frac{\lambda}{4kST} \tag{7.76}$$

The quantity A is likely to vary with time because of the variation of x and ΔF with time. B is a constant for a given value of the effective consolidation pressure.

Figure 7.87 shows the variation of the undrained creep rate $\dot{\epsilon}$ with the deviator stress D for remolded illite at elapsed times t equal to 1 min, 10 min, 100 min, and 1000 min. From this, note that at any given time the following apply:

1. For $D < 49$ kN/m^2, the log $\dot{\epsilon}$ vs. D plot is convex upward following the relation given by Eq. (7.71), $\dot{\epsilon} = 2A \sinh(BD)$. For this case, $D\lambda/4SkT < 1$.
2. For 128 kN/m$^2 > D > 49$ kN/m^2, the log $\dot{\epsilon}$ vs. D plot is approximately a straight line following the relation given by Eq. (7.74), $\dot{\epsilon} = Ae^{BD}$. For this case, $D\lambda/4SkT > 1$.
3. For $D > 128$ kN/m^2, the failure stage is reached when the strain rate rapidly increases; this stage cannot be predicted by Eqs. (7.71) and (7.74).

Table 7.9 gives the values of the experimental activation energy ΔF for four different soils.

7.4 OTHER THEORETICAL CONSIDERATIONS

7.4.1 Yield Surfaces in Three Dimension

Comprehensive failure conditions or yield criteria were first developed for metals, rocks, and concrete. In this section, we will examine the application of these theories

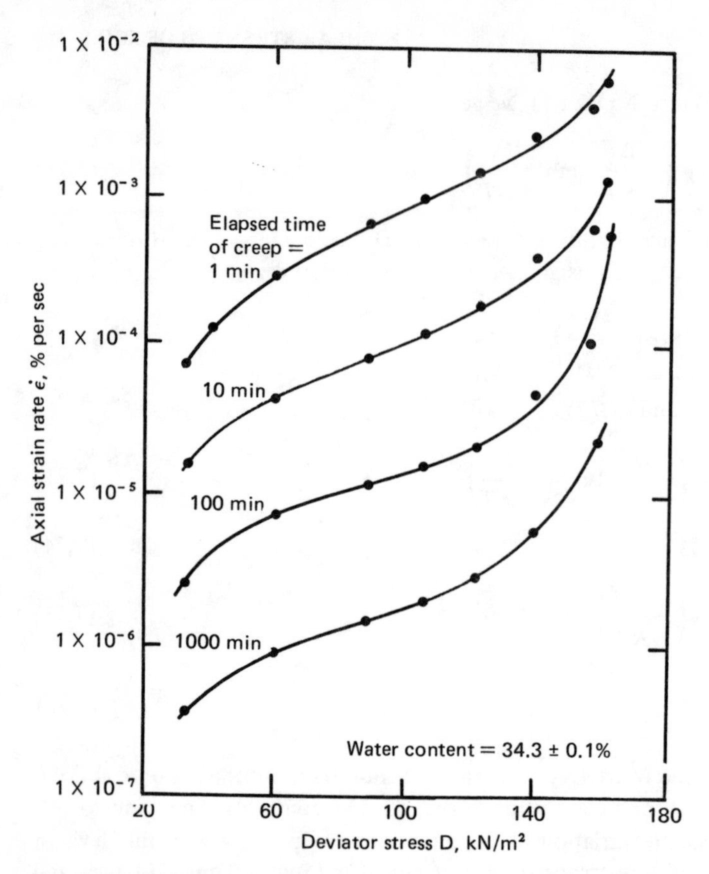

Fig. 7.87 Variation of strain rate with deviator stress for undrained creep of remolded illite. *(Redrawn after J. K. Mitchell, A. Singh, and R. G. Campanella, Bonding, Effective Stresses and Strength of Soils,* J. Soil Mech. Found. Div., ASCE, *vol. 95, no. SM5, 1969.)*

Table 7.9 Values of ΔF for some soils

Soil	ΔF, kcal/mole
Saturated, remolded illite; water content 30 to 43%	25 to 40
Dried illite, samples air-dried from saturation then evacuated over dissicant	37
Undisturbed San Francisco Bay mud	25 to 32
Dry Sacramento River sand	~ 25

After J. K. Mitchell, A. Singh, and R. G. Campanella, Bonding, Effective Stresses, and Strength of Soils, *J. Soil Mech. Found. Div., ASCE,* vol. 95, no. SM5, 1969.)

to soil and determine the yield surfaces in the principal stress space. The notations σ_1', σ_2', and σ_3' will be used for effective principal stresses without attaching an order of magnitude—i.e., σ_1', σ_2', and σ_3' are not necessarily, major, intermediate, and minor principal stresses, respectively.

Von Mises (1913) proposed a simple yield function which may be stated as

$$F = (\sigma_1' - \sigma_2')^2 + (\sigma_2' - \sigma_3')^2 + (\sigma_3' - \sigma_1')^2 - 2Y^2 = 0 \tag{7.77}$$

where Y is the yield stress obtained in axial tension. However, the octahedral shear stress can be given by the relation

$$\tau_{\text{oct}} = \tfrac{1}{3}\sqrt{(\sigma_1' - \sigma_2')^2 (\sigma_2' - \sigma_3')^2 (\sigma_3' - \sigma_1')^2}$$

Thus, Eq. (7.77) may be written as

$$3\tau_{\text{oct}}^2 = 2Y^2$$

or

$$\tau_{\text{oct}} = \sqrt{\tfrac{2}{3}}\, Y \tag{7.78}$$

Equation (7.78) means that failure will take place when the octahedral shear stress reaches a constant value equal to $\sqrt{2/3}\ Y$. Let us plot this on the octahedral plane ($\sigma_1' + \sigma_2' + \sigma_3' = \text{constant}$), as shown in Fig. 7.88. The locus will be a circle with a radius equal to $\tau_{\text{oct}} = \sqrt{2/3}\ Y$ and with its center at point a. In Fig. 7.88a, Oa is the octahedral normal stress $(\sigma_1' + \sigma_2' + \sigma_3')/3 = \sigma_{\text{oct}}'$; also, $ab = \tau_{\text{oct}}$, and $Ob = \sqrt{\sigma_{\text{oct}}'^2 + \tau_{\text{oct}}^2}$. Note that the locus is unaffected by the value of σ_{oct}'. Thus, various values of σ_{oct}' will generate a circular cylinder coaxial with the hydrostatic axis, which is a yield surface (Fig. 7.88b).

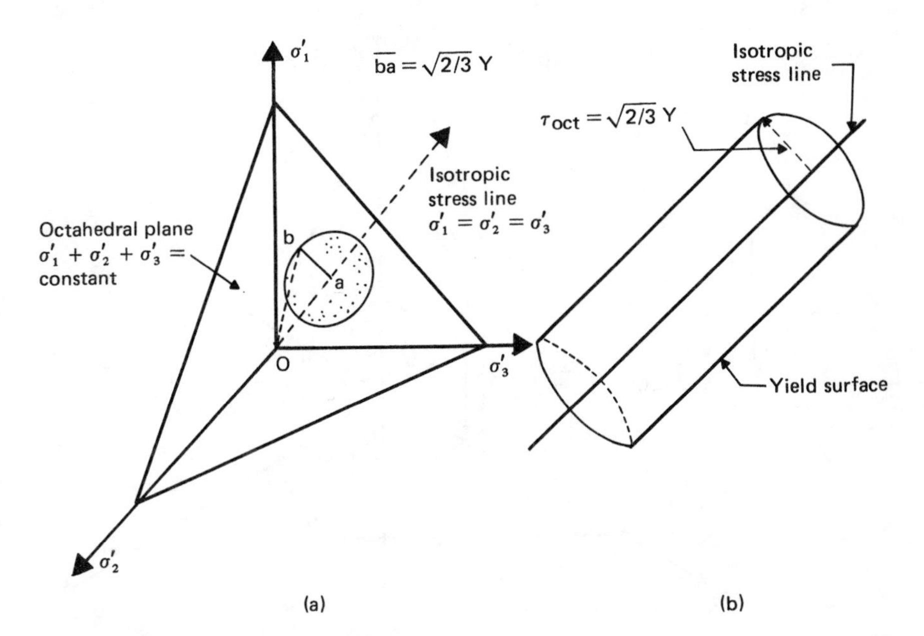

Fig. 7.88 Yield surface—Von Mises criteria.

Another yield function suggested by Tresca (1868) can be expressed in the form

$$\sigma_{max} - \sigma_{min} = 2k \tag{7.79}$$

Equation (7.79) assumes that failure takes place when the maximum shear stress reaches a constant critical value. The factor k of Eq. (7.79) is defined for the case of simple tension by the Mohr's circle shown in Fig. 7.89. Note that for soils this is actually the $\phi = 0$ condition. In Fig. 7.89, the yield function is plotted on the octahedral plane ($\sigma_1' + \sigma_2' + \sigma_3' =$ constant). The locus is a regular hexagon. Point a is the point of intersection of the *hydrostatic axis* or *isotropic stress line* with octahedral plane, and so it represents the octahedral normal stress. The point b represents the failure condition in compression for $\sigma_1' > \sigma_2' = \sigma_3'$, and point e represents the failure condition in extension with $\sigma_2' = \sigma_3' > \sigma_1'$. Similarly, the point d represents the failure condition for $\sigma_3' > \sigma_1' = \sigma_2'$, point g for $\sigma_1' = \sigma_2' > \sigma_3'$, point f for $\sigma_2' > \sigma_3' = \sigma_1'$, and point c for $\sigma_3' = \sigma_1' > \sigma_2'$. Since the locus is unaffected by the value of σ_{oct}', the yield surface will be a hexagonal cylinder.

We have seen from Eq. (7.20) that, for the Mohr–Coulomb condition of failure, $(\sigma_1' - \sigma_3') = 2c \cos\phi + (\sigma_1' + \sigma_3') \sin\phi$, or $(\sigma_1' - \sigma_3')^2 = [2c \cos\phi + (\sigma_1' + \sigma_3')^2 \sin\phi]^2$. In its most general form, this can be expressed as

$$\begin{aligned} &\{(\sigma_1' - \sigma_2')^2 - [2c \cos\phi + (\sigma_1' + \sigma_2') \sin\phi]\}^2 \\ &\quad \times \{(\sigma_2' - \sigma_3')^2 - [2c \cos\phi + (\sigma_2' + \sigma_3') \sin\phi]\}^2 \\ &\quad \times \{(\sigma_3' - \sigma_1') - [2c \cos\phi + (\sigma_3' + \sigma_1') \sin\phi]\}^2 = 0 \end{aligned} \tag{7.80}$$

When the yield surface defined by Eq. (7.80) is plotted on the octahedral plane,

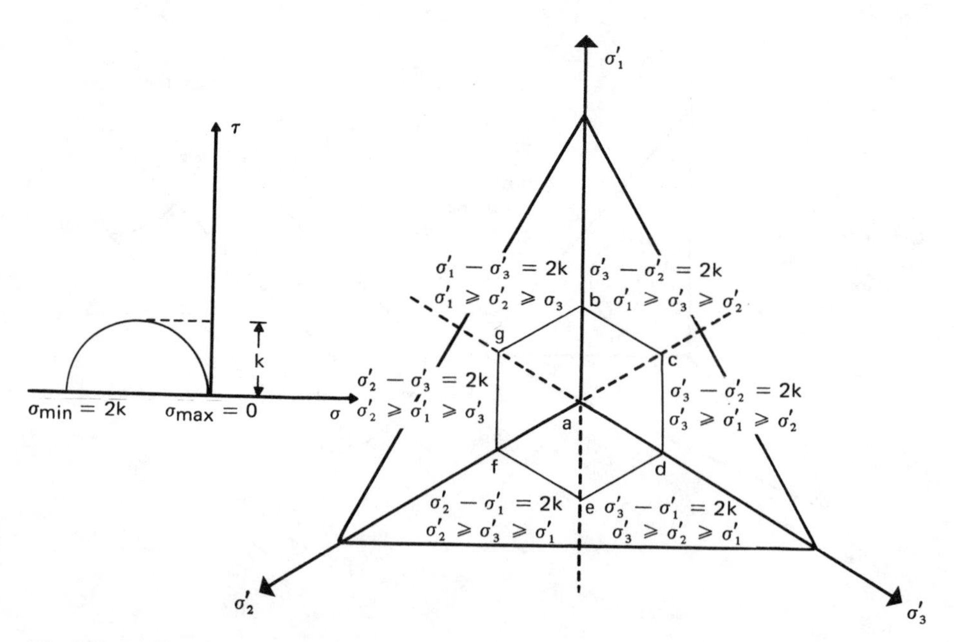

Fig. 7.89 Yield surface – Tresca criteria.

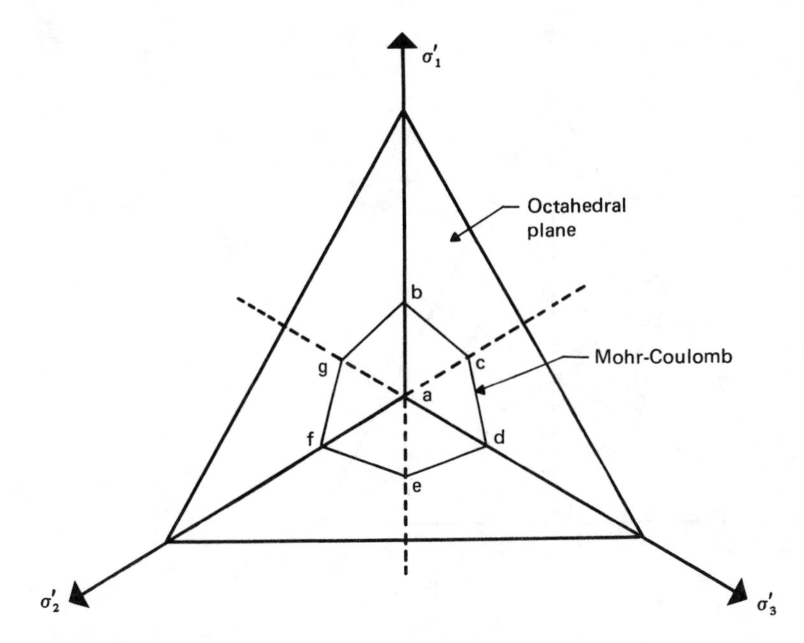

Fig. 7.90 Mohr–Coulomb failure criteria.

it will appear as shown in Fig. 7.90. This is an irregular hexagon in section with non-parallel sides of equal length. Point a in Fig. 7.90 is the point of intersection of the hydrostatic axis with the octahedral plane. Thus, the yield surface will be a hexagonal cylinder coaxial with the isotropic stress line.

Figure 7.91a shows a comparison of the three yield functions described above. In a Rendulic-type plot, the failure envelopes will appear in a manner as shown in Fig. 7.91b. At point a, $\sigma'_1 = \sigma'_2 = \sigma'_3 = \sigma'$ (say). At point b, $\sigma'_1 = \sigma' + \overline{ba'} = \sigma' + ab \sin\theta$, where $\theta = \cos^{-1}(1/\sqrt{3})$. Thus,

$$\sigma'_1 = \sigma' + \sqrt{\tfrac{2}{3}}\,\overline{ab} \tag{7.81}$$

$$\sigma'_2 = \sigma'_3 = \sigma' - \frac{aa'}{\sqrt{2}} = \sigma' - \frac{\overline{ab}\cos\theta}{\sqrt{2}} = \sigma' - \frac{1}{\sqrt{6}}\overline{ab} \tag{7.82}$$

For the Mohr–Coulomb failure criterion, $\sigma'_1 - \sigma'_3 = 2c\cos\phi + (\sigma'_1 + \sigma'_3)\sin\phi$. Substituting Eqs. (7.81) and (7.82) in the preceding equation, we obtain

$$\left(\sigma' + \sqrt{\frac{2}{3}}\,\overline{ab} - \sigma' + \frac{1}{\sqrt{6}}\overline{ab}\right) = 2c\cos\phi$$

$$+ \left(\sigma' + \sqrt{\frac{2}{3}}\,\overline{ab} + \sigma' - \frac{1}{\sqrt{6}}\overline{ab}\right)\sin\phi$$

$$\overline{ab}\left[\left(\sqrt{\frac{2}{3}} + \frac{1}{\sqrt{6}}\right) - \left(\sqrt{\frac{2}{3}} - \frac{1}{\sqrt{6}}\right)\sin\phi\right] = 2(c\cos\phi + \sigma'\sin\phi)$$

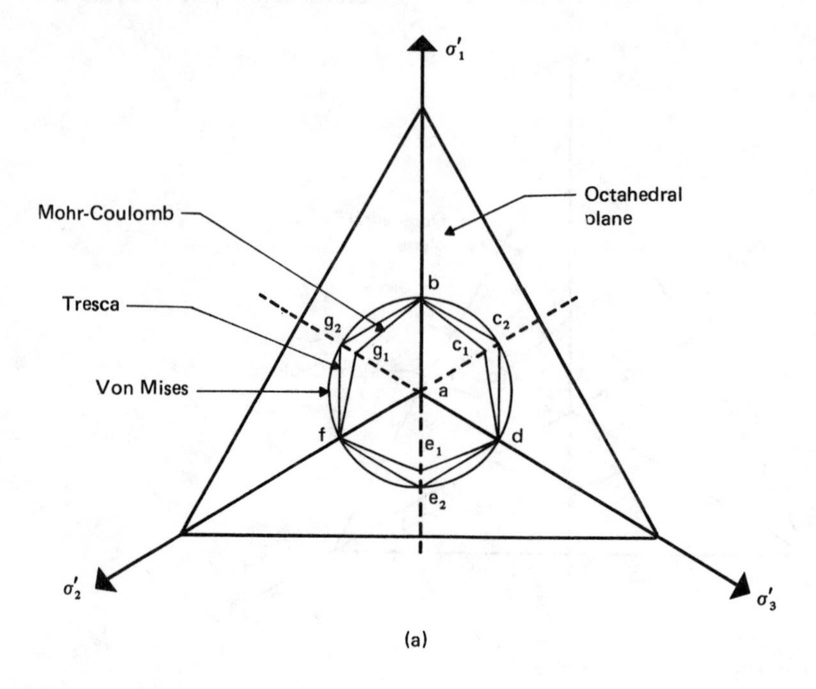

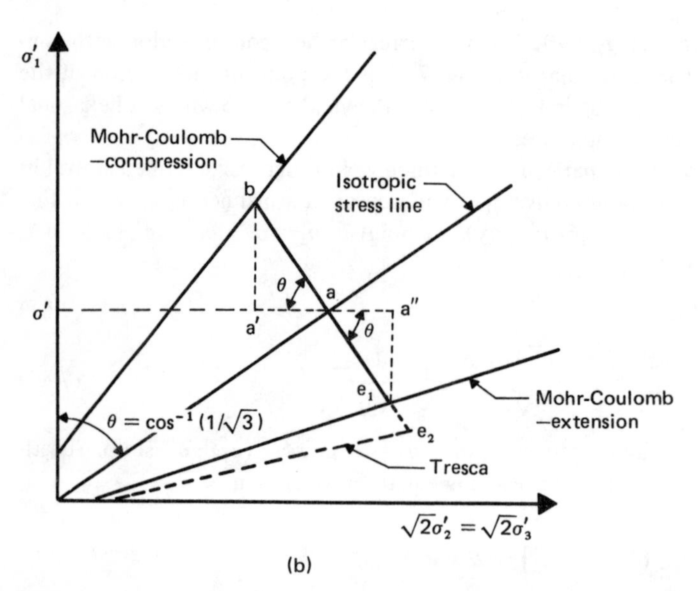

Fig. 7.91 Comparison of Von Mises, Tresca, and Mohr–Coulomb yield functions.

or $$\overline{ab}\,\frac{3}{\sqrt{6}}\left(1-\frac{1}{3}\sin\phi\right)=2(c\cos\phi+\sigma'\sin\phi) \tag{7.83}$$

Similarly, for *extension* (i.e., at point e_1),

$$\sigma_1'=\sigma'-\overline{e_1a''}=\sigma'-\overline{ae_1}\sin\theta=\sigma'-\sqrt{\frac{2}{3}}\,\overline{ae_1} \tag{7.84}$$

$$\sigma_2'=\sigma_3'=\sigma'+\frac{\overline{aa''}}{\sqrt{2}}=\sigma'+\frac{\overline{ae_1}\cos\theta}{\sqrt{2}}=\sigma'+\frac{1}{\sqrt{6}}ae_1 \tag{7.85}$$

Now, $\sigma_3'-\sigma_1'=2c\cos\phi+(\sigma_3'+\sigma_1')\sin\phi$. Substituting Eqs. (7.84) and (7.85) into the preceding equation, we get

$$ae_1\left[\left(\sqrt{\frac{2}{3}}+\frac{1}{\sqrt{6}}\right)+\left(\sqrt{\frac{2}{3}}-\frac{1}{\sqrt{6}}\right)\sin\phi\right]=2(c\cos\phi+\sigma'\sin\phi) \tag{7.86}$$

or $$\overline{ae_1}\,\frac{3}{\sqrt{6}}\left(1+\frac{1}{3}\sin\phi\right)=2(c\cos\phi+\sigma'\sin\phi) \tag{7.87}$$

Equating Eqs. (7.83) and (7.87),

$$\frac{\overline{ab}}{\overline{ae_1}}=\frac{1+\frac{1}{3}\sin\phi}{1-\frac{1}{3}\sin\phi} \tag{7.88}$$

Table 7.10 gives the ratios of $\overline{ab}$ to $\overline{ae_1}$ for various values of ϕ. Note that this ratio is not dependent on the value of cohesion, c.

It can be seen from Fig. 7.91*a* that the Mohr-Coulomb and the Tresca yield functions coincide for the case $\phi=0$.

Von Mises' yield function [Eq. (7.77)] can be modified to the form

$$(\sigma_1'-\sigma_2')^2+(\sigma_2'-\sigma_3')^2+(\sigma_3'-\sigma_1')^2=\left[c+\frac{k_2}{3}(\sigma_1'+\sigma_2'+\sigma_3')\right]^2$$

or $$(\sigma_1'-\sigma_2')^2+(\sigma_2'-\sigma_3')^2+(\sigma_3'-\sigma_1')^2=(c+k_2\sigma_{\text{oct}}')^2 \tag{7.89}$$

where k_2 is a function of $\sin\phi$, and c = cohesion. Eq. (7.89) is called the extended Von Mises' yield criterion.

Table 7.10 Ratio of $\overline{ab}$ to $\overline{ae_1}$ [Eq. (7.88)]

ϕ	ab/ae_1
40	0.647
30	0.715
20	0.796
10	0.889
0	1.0

Similarly, Tresca's yield function [Eq. (7.79)] can be modified to the form

$$[(\sigma_1' - \sigma_2')^2 - (c + k_3\sigma_{oct}')^2] \times [(\sigma_2' - \sigma_3') - (c + k_3\sigma_{oct}')^2] \times [(\sigma_3' - \sigma_1')^2 - (c + k_3\sigma_{oct}')^2] = 0 \tag{7.90}$$

where k_3 is a function of $\sin\phi$ and c = cohesion. Equation (7.90) is generally referred to as the extended Tresca criterion.

7.4.2 Experimental Results to Compare the Yield Functions

Kirkpatrick (1957) devised a special shear test procedure for soils called the *hollow cylinder test,* which provides the means for obtaining the variation in the three principal stresses. The results from this test can be used to compare the validity of the various yield criteria suggested in the preceding section.

A schematic diagram of the laboratory arrangement for the hollow cylinder test is shown in Fig. 7.92*a*. A soil specimen in the shape of a hollow cylinder is placed inside a test chamber. The specimen is encased by both an inside and an outside membrane. As in the case of a triaxial test, radial pressure on the soil specimen can be applied through water. However, in this type of test, the pressures applied to the inside and outside of the specimen can be controlled separately. Axial pressure on the specimen is applied by a piston. In the original work of Kirkpatrick, the axial pressure was obtained from load differences applied to the cap by the fluid on top of the specimen [i.e., piston pressure was not used; see Eq. (7.97)].

The relations for the principal stresses in the soil specimen can be obtained as follows (see Fig. 7.92*b*). Let σ_o and σ_i be the outside and inside fluid pressures, respectively. For *drained tests,* the total stresses σ_o and σ_i are equal to the effective stresses, σ_o' and σ_i'. For an axially symmetrical case, the equation of continuity for a given point in the soil specimen can be given by (see any text on theory of elasticity)

$$\frac{d\sigma_r'}{dr} + \frac{\sigma_r' - \sigma_\theta'}{r} = 0 \tag{7.91}$$

where σ_r' and σ_θ' are the radial and tangential stresses, respectively, and r is the radial distance from the center of the specimen to the point.

We will consider a case where the failure in the specimen is caused by *increasing σ_i', keeping σ_o' constant.* Let

$$\sigma_\theta' = \lambda\sigma_r' \tag{7.92}$$

Substituting Eq. (7.92) in Eq. (7.91), we get

$$\frac{d\sigma_r'}{dr} + \frac{\sigma_r'(1-\lambda)}{r} = 0$$

or

$$\frac{1}{\lambda - 1}\int \frac{d\sigma_r'}{\sigma_r'} = \int \frac{dr}{r}$$

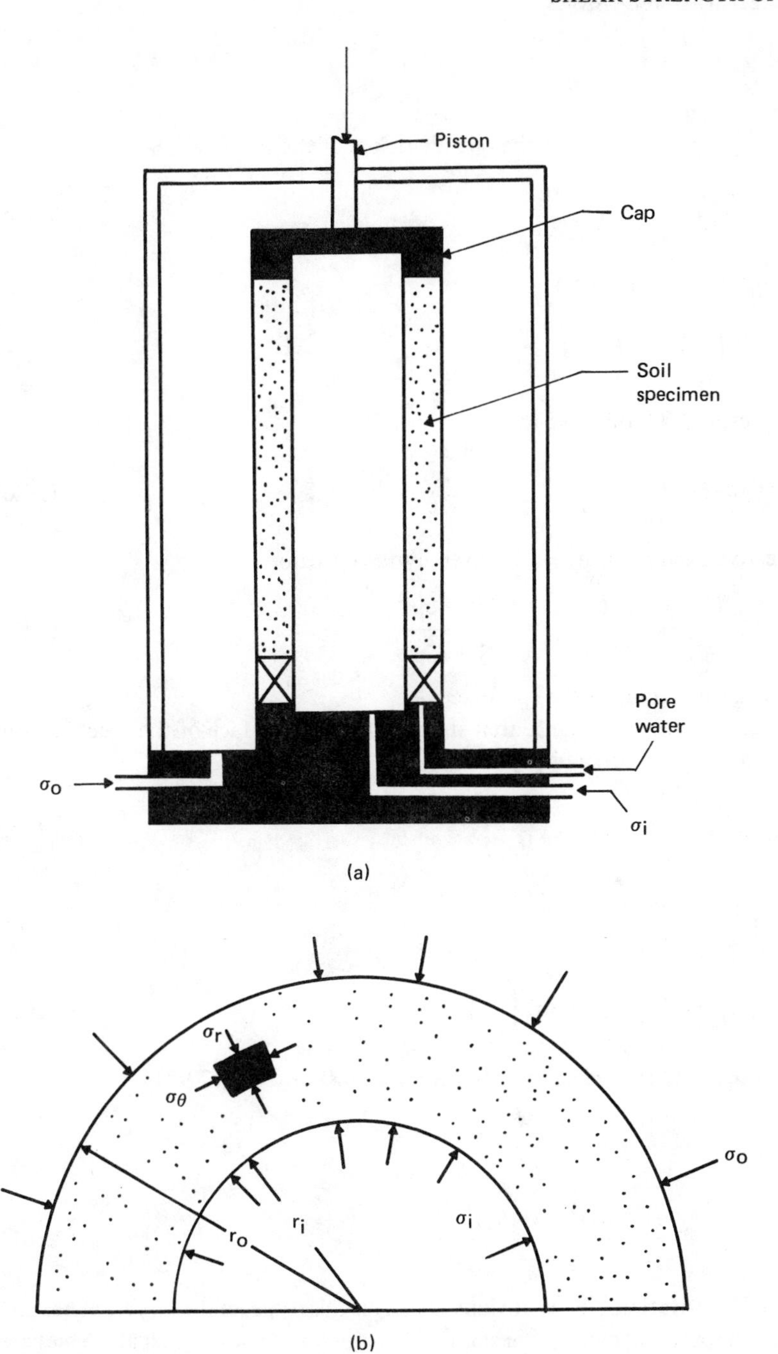

Fig. 7.92 Hollow cylinder test.

$$\sigma_r' = Ar^{\lambda-1} \tag{7.93}$$

where A is a constant.

But $\sigma_r' = \sigma_o'$ at $r = r_o$, which is the outside radius of the specimen. So

$$A = \frac{\sigma_o'}{r_o^{\lambda-1}} \tag{7.94}$$

Combining Eqs. (7.93) and (7.94)

$$\sigma_r' = \sigma_o' \left(\frac{r}{r_o}\right)^{\lambda-1} \tag{7.95}$$

Again, from Eqs. (7.92) and (7.95),

$$\sigma_\theta' = \lambda\sigma_o' \left(\frac{r}{r_o}\right)^{\lambda-1} \tag{7.96}$$

The effective axial stress σ_a' can be given by the equation

$$\sigma_a' = \frac{\sigma_o'(\pi r_o^2) - \sigma_i'(\pi r_i^2)}{\pi r_o^2 - \pi r_i^2} = \frac{\sigma_o' r_o^2 - \sigma_i' r_i^2}{r_o^2 - r_i^2} \tag{7.97}$$

where r_i is the inside radius of the specimen.

At failure, the radial and tangential stresses at the inside face of the specimen can be obtained from Eqs. (7.95) and (7.96):

$$\sigma_{r(\text{inside})}' = (\sigma_i')_{\text{failure}} = \sigma_o' \left(\frac{r_i}{r_o}\right)^{\lambda-1} \tag{7.98}$$

or

$$\left(\frac{\sigma_i'}{\sigma_o'}\right)_{\text{failure}} = \left(\frac{r_i}{r_o}\right)^{\lambda-1} \tag{7.99}$$

$$\sigma_{\theta(\text{inside})}' = (\sigma_\theta')_{\text{failure}} = \lambda\sigma_o' \left(\frac{r_i}{r_o}\right)^{\lambda-1} \tag{7.100}$$

To obtain σ_a' at failure, we can substitute Eq. (7.98) into Eq. (7.97):

$$(\sigma_a')_{\text{failure}} = \frac{\sigma_o'[(r_o/r_i)^2 - (\sigma_i'/\sigma_o')]}{(r_o/r_i)^2 - 1}$$

$$= \frac{\sigma_o'[(r_o/r_i)^2 - (r_o/r_i)^{1-\lambda}]}{(r_o/r_i)^2 - 1} \tag{7.101}$$

From the above derivations, it is obvious that for this type of test (i.e., increasing σ_i' to cause failure and keeping σ_o' constant) the major and minor principal stresses are σ_r' and σ_θ'. The intermediate principal stress is σ_a'. For granular soils, the value of the cohesion c is 0; and, from the Mohr–Coulomb failure criterion,

$$\left(\frac{\text{Minor principal stress}}{\text{Major principal stress}}\right)_{\text{failure}} = \frac{1 - \sin\phi}{1 + \sin\phi}$$

Table 7.11 Results of Kirkpatrick's hollow cylinder test on a sand

Test no.	$(\sigma_i')_{failure}$,* lb/in²	σ_o',† lb/in²	λ [from Eq. (7.99)] ‡	σ_θ' (inside) at failure, § lb/in²	σ_θ' (outside) at failure, ¶ lb/in²	σ_a' [from Eq. 7.97), lb/in²
1	21.21	14.40	0.196	4.16	2.82	10.50
2	27.18	18.70	0.208	5.65	3.89	13.30
3	44.08	30.60	0.216	9.52	6.61	22.30
4	55.68	38.50	0.215	11.95	8.28	27.95
5	65.75	45.80	0.192	12.61	8.80	32.30
6	68.63	47.92	0.198	13.60	9.48	34.05
7	72.88	50.30	0.215	15.63	10.81	35.90
8	77.16	54.02	0.219	16.90	11.83	38.90
9	78.43	54.80	0.197	15.4	10.80	38.20

*$(\sigma_i')_{failure} = \sigma_{r(inside)}'$ at failure.
†$(\sigma_o') = \sigma_{r(outside)}'$ at failure.
‡For these tests, $r_o = 2$ in (50.8 mm) and $r_i = 1.25$ in (31.75 mm).
§$\sigma_\theta'(\text{inside}) = \lambda(\sigma_i')_{failure}$.
¶$\sigma_\theta'(\text{outside}) = \lambda(\sigma_o')_{failure}$.
Note: 1 lb/in² = 6.9 kN/m²

or

$$\left(\frac{\sigma_\theta'}{\sigma_r'}\right)_{\text{failure}} = \frac{1-\sin\phi}{1+\sin\phi} \tag{7.102}$$

Comparing Eqs. (7.92) and (7.102),

$$\frac{1-\sin\phi}{1+\sin\phi} = \tan^2\left(45° - \frac{\phi}{2}\right) = \lambda \tag{7.103}$$

The results of some hollow cylinder tests conducted by Kirkpatrick (1957) on a sand are given in Table 7.11, together with the calculated values of λ, $(\sigma_a')_{failure}$, $(\sigma_r')_{failure}$, and $(\sigma_\theta')_{failure}$.

A comparison of the yield functions on the octahedral plane and the results of Kirkpatrick is given in Fig. 7.93. The results of triaxial compression and extension tests conducted on the same sand by Kirkpatrick are also shown in Fig. 7.93. The experimental results indicate that the Mohr–Coulomb criterion gives a better representation for soils than the extended Tresca and Von Mises criteria. However, the hollow cylinder tests produced slightly higher values of ϕ than those from the triaxial tests.

Hollow cylinder test results of Wu et al. Wu et al. (1963) also conducted a type of hollow cylinder shear test with sand and clay specimens. In these tests, failure was produced by increasing the inside, outside, and axial stresses on the specimens in various combinations. The axial stress increase was accomplished by the application of a force P on the cap through the piston as shown in Fig. 7.92. Triaxial compression and extension tests were also conducted. Out of a total of six series of tests, there were two in which failure was caused by increasing the outside pressure. For those two series of tests, $\sigma_\theta' > \sigma_a' > \sigma_r'$. Note that this is opposite to Kirkpatrick's tests, in which $\sigma_r' > \sigma_a' > \sigma_\theta'$. Based on the Mohr–Coulomb criterion, we can write [see Eq.

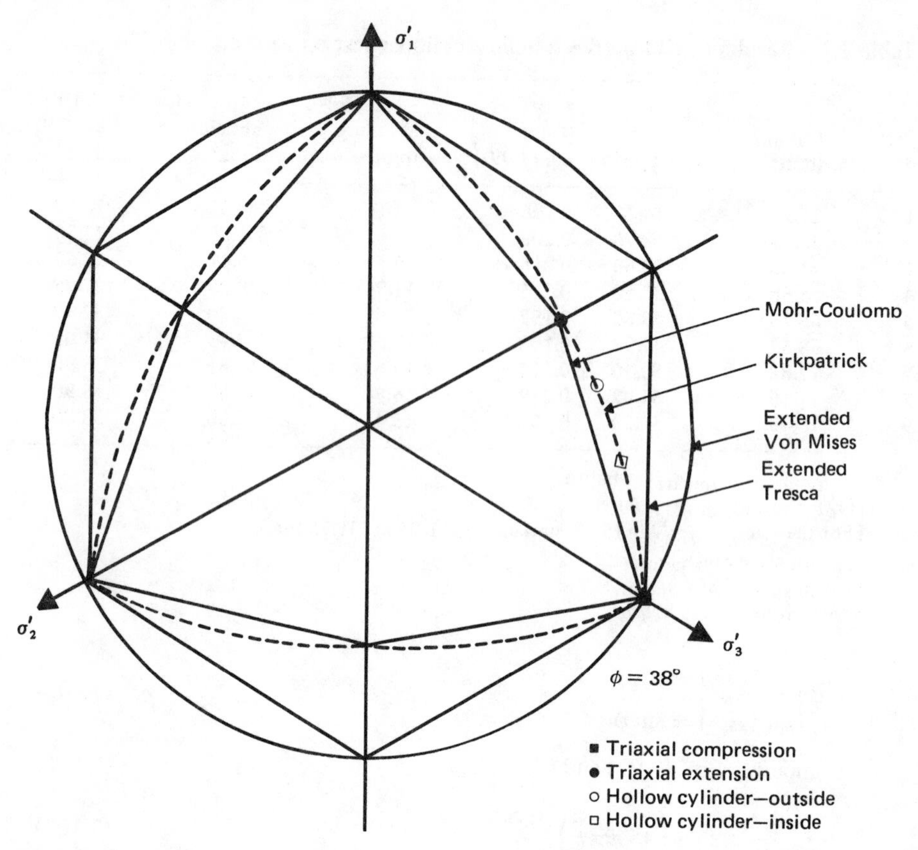

Fig. 7.93 Comparison of the yield functions on the octahedral plane along with the results of Kirkpatrick.

(7.21)], $\sigma'_{\max} = \sigma'_{\min} N + 2cN^{1/2}$. So, for the case where $\sigma'_\theta > \sigma'_a > \sigma'_r$,

$$\sigma'_\theta = \sigma'_r N + 2cN^{1/2} \tag{7.104}$$

The value of N in the above equation is $\tan^2(45° + \phi/2)$, and so the λ in Eq. (7.92) is equal to $1/N$. From Eq. (7.91),

$$\frac{d\sigma'_r}{dr} = \frac{\sigma'_\theta - \sigma'_r}{r}$$

Combining the preceding equation and Eq. (7.104), we get

$$\frac{d\sigma'_r}{dr} = \frac{1}{r}\left[\sigma'_r(N-1) + 2cN^{1/2}\right] \tag{7.105}$$

Using the boundary condition that, at $r = r_i$, $\sigma'_r = \sigma'_i$, Eq. (7.105) gives the following relation

$$\sigma'_r = \left(\sigma'_i + \frac{2cN^{1/2}}{N-1}\right)\left(\frac{r}{r_i}\right)^{N-1} - \frac{2cN^{1/2}}{N-1} \tag{7.106}$$

Also, combining Eqs. (7.104) and (7.106),

$$\sigma'_\theta = \left(\sigma'_i N + \frac{2cN^{3/2}}{N-1}\right)\left(\frac{r}{r_i}\right)^{N-1} - \frac{2cN^{1/2}}{N-1} \tag{7.107}$$

At failure, $\sigma'_{r(\text{outside})} = (\sigma'_o)_{\text{failure}}$. So,

$$(\sigma'_o)_{\text{failure}} = \left(\sigma'_i + \frac{2cN^{1/2}}{N-1}\right)\left(\frac{r_o}{r_i}\right)^{N-1} - \frac{2cN^{1/2}}{N-1} \tag{7.108}$$

For granular soils and normally consolidated clays, $c = 0$. So, at failure, Eqs. (7.106) and (7.107) simplify to the form

$$(\sigma'_r)_{\text{outside, at failure}} = (\sigma'_o)_{\text{failure}} = \sigma'_i\left(\frac{r_o}{r_i}\right)^{N-1} \tag{7.109}$$

and $$(\sigma'_\theta)_{\text{outside, at failure}} = \sigma'_i N\left(\frac{r_o}{r_i}\right)^{N-1} \tag{7.110}$$

Hence, $$\left(\frac{\sigma'_r}{\sigma'_\theta}\right)_{\text{failure}} = \frac{\text{minor principal effective stress}}{\text{major principal effective stress}} = \frac{1}{N} = \lambda \tag{7.111}$$

Compare Eqs. (7.92) and (7.111).

Wu et al. also derived equations for σ'_r and σ'_θ for the case $\sigma'_a > \sigma'_\theta > \sigma'_r$.

Figure 7.94 shows the results of Wu et al. plotted on the octahedral plane

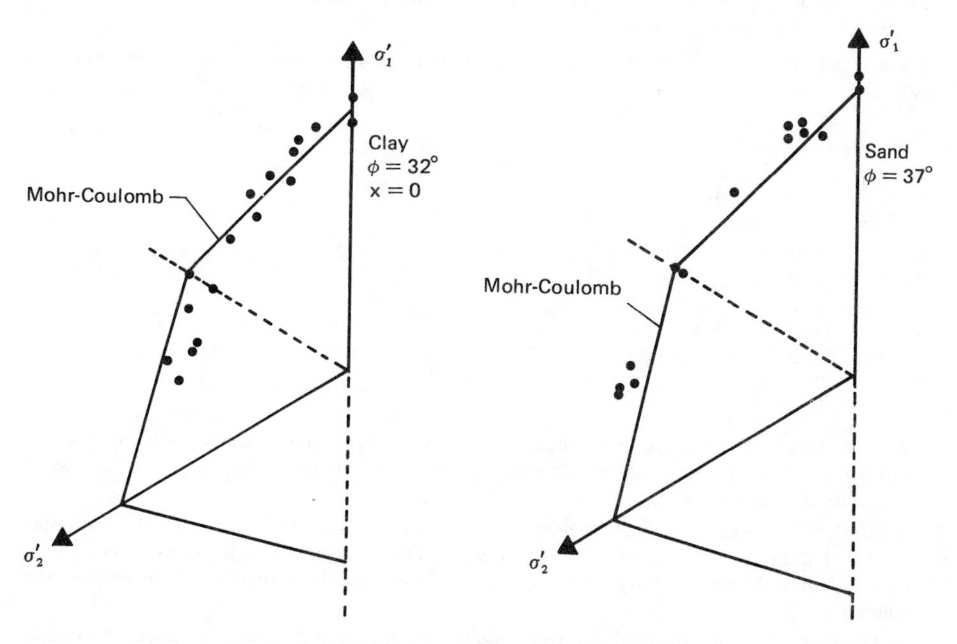

Fig. 7.94 Results of hollow cylinder tests plotted on octahedral plane $\sigma'_1 + \sigma'_2 + \sigma'_3 = 1$. *(After T. W. Wu, A. K. Lok, and L. E. Malvern, Study of Failure Envelope of Soils,* J. Soil Mech. Found. Div., ASCE, *vol. 89, no. SM1, 1963.)*

$\sigma_1' + \sigma_2' + \sigma_3' = 1$. The Mohr–Coulomb yield criterion has been plotted by using the triaxial compression and extension test results. The results of other hollow cylinder tests are plotted as points. In general, there is good agreement between the experimental results and the yield surface predicted by the Mohr–Coulomb theory. However, as in Kirkpatrick's test, hollow cylinder tests indicated somewhat higher values of ϕ than triaxial tests in the case of sand. In the case of clay, the opposite trend is generally observed.

PROBLEMS

7.1 The results of a direct shear test on a dry sand are as follows: normal stress, 96.6 kN/m^2; shear stress at failure, 67.7 kN/m^2. By means of a Mohr's diagram, find the direction and magnitude of the principal stresses acting on a soil element in the zone of failure.

7.2 The results of two consolidated drained triaxial tests are as follows:

Test no.	σ_3, kN/m^2	$\Delta\sigma_f$, kN/m^2
1	66	134.77
2	91	169.1

Determine c and ϕ. Also determine the magnitudes of the normal and shear stress on the planes of failure for the two specimens used in the tests.

7.3 For the following consolidated drained triaxial tests on a clay, draw a p' vs. q' diagram and determine c and ϕ.

Test no.	p', lb/ft^2	q', lb/ft^2
1	600	740
2	800	780
3	1540	1040
4	2120	1340
5	2800	1600

7.4 A specimen of normally consolidated clay ($\phi = 28°$) was consolidated under a chamber confining pressure of 40 lb/in^2. For a drained test, by how much does the axial stress have to be reduced to cause failure by axial extension?

7.5 A normally consolidated clay specimen ($\phi = 31°$) was consolidated under a chamber confining pressure of 132 kN/m^2. Failure of the specimen was caused by an added axial stress of 158.1 kN/m^2 in an undrained condition. Determine ϕ_{cu}, A_f, and the pore water pressure in the specimen at failure.

7.6 The results of a consolidated undrained test, in which $\sigma_3 = 392$ kN/m^2, on a normally consolidated clay are given below:

Axial strain, %	$\Delta\sigma$, kN/m^2	u_d, kN/m^2
0	0	0
0.5	156	99
0.75	196	120
1	226	132
1.3	235	147
2	250	161
3	245	170
4	240	173
4.5	235	175

Draw the K_f line in a p' vs. q' diagram. Also draw the stress path for this test in that diagram.

7.7 For the test results given in Prob. 7.6, draw a stress path in a Rendulic-type diagram.

7.8 A normally consolidated clay is consolidated under a triaxial chamber confining pressure of 495 kN/m^2, and $\phi = 29°$. In a Rendulic-type diagram, draw the stress path the specimen would follow if sheared to failure in a *drained condition* in the following ways:

(*a*) By increasing the axial stress and keeping the radial stress constant.
(*b*) By reducing the radial stress and keeping the axial stress constant.
(*c*) By increasing the axial stress and reducing the radial stress such that $\sigma'_a + 2\sigma'_r =$ constant.
(*d*) By reducing the axial stress and keeping the radial stress constant.
(*e*) By increasing the radial stress and keeping the axial stress constant.
(*f*) By reducing the axial stress and increasing the radial stress such that $\sigma'_a + 2\sigma'_r =$ constant.

7.9 Repeat Prob. 7.8 for a triaxial chamber confining pressure of 120 lb/in^2.

7.10 A specimen of soil was collected from a depth of 12 m in a deposit of clay. The groundwater table coincides with the ground surface. For the soil, $LL = 68$, $PL = 29$, and $\gamma_{sat} = 17.8\ kN/m^3$. Estimate the undrained shear strength, S_u, of this clay for the following cases:

(*a*) If the clay is normally consolidated.
(*b*) If the preconsolidation pressure is 191 kN/m^2.

7.11 The undrained shear strength, S_u, of a saturated clay specimen was determined in the laboratory (UU test). The results of the test were $\sigma_3 = 1500\ lb/ft^2$ and $\Delta\sigma_f = 1521\ lb/ft^2$ and the temperature of the test was 12°C. Estimate the undrained shear strength of this soil at a temperature of 4°C.

7.12 A specimen of clay was collected from the field from a depth of 16 m (Fig. P7.1). A consolidated undrained triaxial test yielded the following results: $\phi = 32°$, $A_f = 0.8$. Estimate the undrained shear strength S_u of the clay.

7.13 Refer to Fig. 7.65. For an anisotropic clay deposit, the results from unconfined compression tests were $S_{u(i=30°)} = 2125\ lb/ft^2$ and $S_{u(i=60°)} = 2575\ lb/ft^2$. Find the anisotropy coefficient, K, of the soil based on the Casagrande–Carrillo equation. Plot a polar diagram showing the variation of S_u.

7.14 The stress path for a normally consolidated clay is shown in Fig. P7.2 (Rendulic plot). The stress path is for a consolidated undrained triaxial test where failure was caused by increasing the axial stress while keeping the radial stress constant. Determine:

(*a*) ϕ for the soil.
(*b*) The pore water pressure induced at A.
(*c*) The pore water pressure at failure.
(*d*) The value of A_f.

7.15 A sand specimen was subjected to a drained shear test using hollow cylinder test equipment. Failure was caused by increasing the inside pressure while keeping the outside pressure constant.

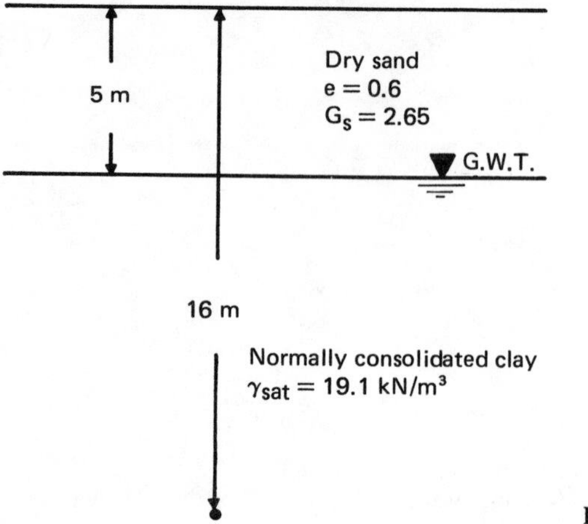

Fig. P7.1

At failure, $\sigma_o = 28\ \text{lb/in}^2$ and $\sigma_i = 38.3\ \text{lb/in}^2$. The inside and outside radii of the specimen were 40 and 60 mm, respectively.

(*a*) Calculate the soil friction angle.

(*b*) Calculate the axial stress on the specimen at failure (Eq. 97).

7.16 For the soil specimen described in Prob. 7.15, if failure had been induced by keeping $\sigma_i = 195\ \text{kN/m}^2$ constant and increasing the outside pressure, what would have been the value of σ_o at failure?

7.17 The results of some drained triaxial tests on a clay soil are given below. Failure of each specimen was caused by increasing the axial stress while the radial stress was kept constant.

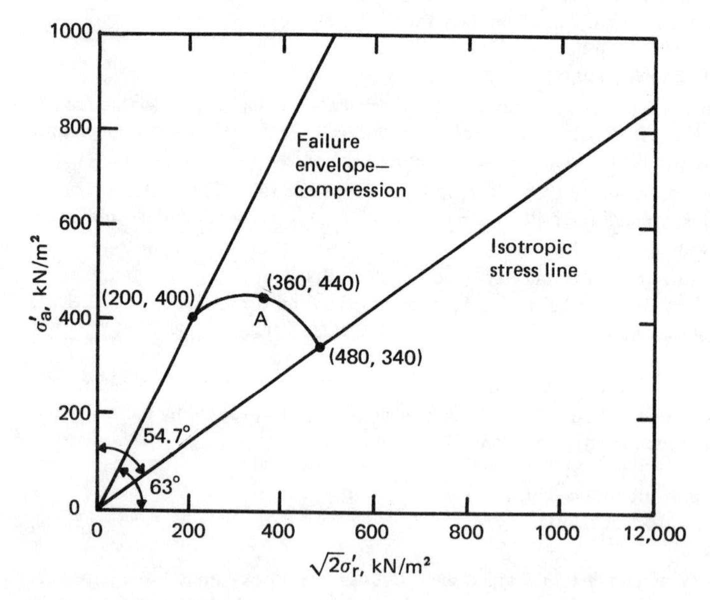

Fig. P7.2

Test no.	Chamber consolidation pressure σ_3', kN/m²	σ_3' at failure, kN/m²	$\Delta\sigma_f$, kN/m²	Moisture content of specimen at failure, %
1	105	105	154	24.2
2	120	120	176	22.1
3	162	162	237	18.1
4	250	35	109	24.2
5	250	61	137	22.1
6	250	140	229	18.1

(*a*) Determine ϕ for the soil.
(*b*) Determining Hvorslev's parameters ϕ_e and c_e at moisture contents of 24.2, 22.1, and 18.1%.

REFERENCES

Arman, A., J. K. Poplin, and N. Ahmad, Study of Vane Shear, *Proc. Conf. In-Situ Measure. Soil Prop., ASCE,* vol. 1, pp. 93–120, 1975.

Berger, L., and J. Gnaedinger, Strength Regain of Clays, *Bull. ASTM,* Sept. 1949.

Bishop, A. W., Correspondence on Shear Characteristics of Saturated Soil, Measured in Triaxial Compression, *Geotechnique,* vol. 4, pp. 43–55, 1954.

Bishop, A. W., Discussion on Soil Properties and Their Measurement, *Proc. 4th Int. Conf. Soil Mech. Found. Eng.,* vol. 3, pp. 103–104, 1957.

Bishop, A. W., Discussion on Soil Properties and Their Measurement, *Proc. 5th Int. Conf. Soil Mech. Found. Eng.,* vol. 3, pp. 91–100, 1961.

Bishop, A. W., and L. Bjerrum, The Relevance of the Triaxial Test to the Solution of Stability Problems, *Proc. Res. Conf. Shear Strength Cohesive Soils, ASCE,* pp. 437–501, 1960.

Bishop, A. W., and D. J. Henkel, "The Measurement of Soil Properties in the Triaxial Test," 2d ed., Edward Arnold, London, 1969.

Bjerrum, L., Embankments on Soft Ground, *Proc. Specialty Conf. Perform. Earth and Earth-Supported Struct., ASCE,* vol. 2, pp. 1–54, 1972.

Bjerrum, L., and O. Kummeneje, "Shearing Resistance of Sand Samples with Circular and Rectangular Cross Sections," Norwegian Geotechnical Institute, Publication No. 44, 1961.

Bjerrum, L., and N. E. Simons, Comparison of Shear Strength Characteristics of Normally Consolidated Clay, *Proc. Res. Conf. Shear Strength Cohesive Soils, ASCE,* pp. 711–726, 1960.

Bjerrum, L., N. Simons, and I. Torblaa, The Effect of Time on Shear Strength of a Soft Marine Clay, *Proc. Brussels Conf. Earth Press. Prob.,* vol. 1, pp. 148–158, 1958.

Bromwell, L. G., "The Friction of Quartz in High Vacuum," Sc.D. thesis, Massachusetts Institute of Technology, 1966.

Carlson, L., Determination In Situ of the Shear Strength of Undisturbed Clay by Means of a Rotating Auger, *Proc. 2d. Int. Conf. Soil Mech. Found. Eng., Rotterdam,* vol. 1, pp. 265–270, 1948.

Casagrande, A., and N. Carrillo, Shear Failure of Anisotropic Materials, *in* "Contribution to Soil Mechanics 1941–1953," Boston Society of Civil Engineers, Boston, 1944.

Casagrande, A., and S. D. Wilson, "Investigation of the Effects of the Long-time Loading on the Strength of Clays and Shales at Constant Water Content," Report to the U.S. Waterways Experiment Station, Harvard University, 1949.

Casagrande, A., and S. D. Wilson, Effect of the Rate of Loading on the Strength of Clays and Shales at Constant Water Content, *Geotechnique,* vol. 1, pp. 251–263, 1951.

Christensen, N. H., "Model Tests on Plane Active Earth Pressure in Sand," Danish Geotechnical Institute, Copenhagen, Bulletin No. 10., 1961.

Christensen, R. W., and B. M. Das, "Hydraulic Erosion of Remolded Cohesive Soils." Highway Research Board, Special Report 135, pp. 9–19, 1973.

Christensen, R. W., and T. H. Wu, Analysis of Clay Deformation as a Rate Process, *J. Soil Mech. Found. Eng. Div., ASCE,* vol. 90, no. SM6, pp. 125–157, 1964.

Cornforth, D. H., Some Experiments on the Influence of Strain Conditions on the Strength of Sand, *Geotechnique,* vol. 14, no. 2, pp. 143–167, 1964.

Coulomb, C. A., Essai Sur une Application des regles des Maximis et Minimis a Quelques Problemes des Statique Relatifs a L'Architecture, *Mem. Acad. Roy. Pres. Divers Savants, Paris,* vol. 7, 1776.

D'Appolonia, D. J., E. D'Appolonia, and R. F. Brissette, Settlement of Spread Footings on Sand, *J. Soil Mech. Found. Eng. Div., ASCE,* vol. 94, no. SM3, pp. 735–760, 1968.

Dickey, J. W., "Frictional Characteristics of Quartz," S. B. Thesis, Massachusetts Institute of Technology, 1966.

Duncan, J. M., and H. B. Seed, Anisotropy and Stress Reorientation in Clay, *J. Soil Mech. Found. Eng. Div., ASCE,* vol. 92, no. SM5, pp. 21–50, 1966*a*.

Duncan, J. M., and H. B. Seed, Strength Variation Along Failure Surfaces in Clay, *J. Soil Mech. Found. Eng. Div., ASCE,* vol. 92, no. SM6, pp. 81–104, 1966*b*.

Finn, W. D., and H. K. Mittal, "Shear Strength in a General Stress Field," ASTM STP 361, pp. 42–51, 1964.

Gibson, R. E., Experimental Determination of True Cohesion and True Angle of Internal Friction in Clay, *Proc. 3d Int. Conf. Soil Mech. Found. Eng., Zurich,* vol. 1, p. 126, 1953.

Healey, K. A., "The Dependence of Dilatation in Sand on Rate of Shear Strain," Report No. 13, U.S. Army Engineers, Massachusetts Institute of Technology, 1963.

Henkel, D. J., The Shearing Strength of Saturated Remolded Clays, *Proc. Res. Conf. Shear Strength Cohesive Soils, ASCE,* pp. 533–554, 1960.

Henkel, D. J., and N. H. Wade, Plane Strain Tests on a Saturated Remolded Clay, *J. Soil Mech. Found. Eng. Div., ASCE,* vol. 92, no. SM6, pp. 67–80, 1966.

Hennes, R. G., "The Strength of Gravel in Direct Shear," Symposium on Direct Shear Testing of Soils, ASTM, STP 131, pp. 52–62, 1952.

Horne, H. M., and D. U. Deere, Frictional Characteristics of Minerals, *Geotechnique,* vol. 12, pp. 319–335, 1962.

Hvorslev, J., Physical Component of the Shear Strength of Saturated Clays, *Proc. Res. Conf. Shear Strength Cohesive Soils, ASCE,* pp. 169–173, 1960.

Hvorslev, M. J., Uber Die Festigheitseigen-schaften Gestorter Bindinger Boden, *in* "Ingeniorvidenskabelige Skrifter," no. 45, Danmarks Naturvidenskabelige Samfund, Kovenhavn, 1937.

Karlsson, R., and L. Viberg, Ratio *c/p'* in Relation to Liquid Limit and Plasticity Index with Special Reference to Swedish Clays, *Proc. Geotech. Conf., Oslo, Norway,* vol. 1, pp. 43–47, 1967.

Kelly, W. E., Correcting Shear Strength for Temperature, *J. Geotech. Eng. Div., ASCE,* vol. 104, no. GT5, pp. 664–667, 1978.

Kenney, T. C., Discussion, *Proc. ASCE,* vol. 85, no. SM3, pp. 67–79, 1959.

Kirkpatrick, W. M., The Condition of Failure of Sands, *Proc. 4th Int. Conf. Soil Mech. Found. Eng.,* vol. 1, pp. 172–178, 1957.

Kjellman, W., Report on an Apparatus for Consummate Investigation of Mechanical Properties of Soils, *Proc. 1st Int. Conf. Soil Mech. Found. Eng., Cambridge, Mass.,* vol. 1, pp. 16–20, 1936.

Ladd, C. C., and R. Foot, New Design Procedure for Stability of Soft Clays, *J. Geotech. Eng. Div., ASCE,* vol. 100, no. GT7, pp. 763–786, 1974.

Lee, I. K., Stress-Dilatancy Performance of Feldspar, *J. Soil Mech. Found. Eng. Div., ASCE,* vol. 92, no. SM2, pp. 79–103, 1966.

Lee, K. L., Comparison of Plane Strain and Triaxial Tests on Sand, *J. Soil Mech. Found. Eng. Div., ASCE,* vol. 96, no. SM3, pp. 901–923, 1970.

Leonards, G. A., "Foundation Engineering," McGraw-Hill, New York, 1962.

Leussink, H., and W. Wittke, "Difference in Triaxial and Plane Strain Shear Strength," ASTM, STP 361, pp. 77–89, 1964.

Lo, K. Y., Stability of Slopes in Anisotropic Soils, *J. Soil Mech. Found. Eng. Div., ASCE,* vol. 91, no. SM4, pp. 85–106, no. 1, Proc. paper 866, pp. 1–19, 1965.

Meyerhof, G. G., Penetration Test and Bearing Capacity of Cohesionless Soils, *J. Soil Mech. Found. Eng. Div., ASCE,* vol. 82, no. 1, Proc. paper 866, pp. 1–19, 1956.

Meyerhof, G. G., Some Recent Research on the Bearing Capacity of Foundations, *Can. Geotech. J.,* vol. 1, no. 1, pp. 16–26, 1963.

Mitchell, J. K., Shearing Resistance of Soils as a Rate Process, *J. Soil Mech. Found. Eng. Div., ASCE,* vol. 90, no. SM1, pp. 29–61, 1964.

Mitchell, J. K., "Fundamentals of Soil Behavior," Wiley, New York, 1976.

Mitchell, J. K., R. G. Campanella, and A. Singh, Soil Creep as a Rate Process, *J. Soil Mech. Found. Eng. Div., ASCE,* vol. 94, no. SM1, pp. 231–253, 1968.

Mitchell, J. K., A. Singh, and R. G. Campanella, Bonding, Effective Stresses, and Strength of Soils, *J. Soil Mech. Found. Eng. Div., ASCE,* vol. 95, no. SM5, pp. 1219–1246, 1969.

Moretto, O., Effect of Natural Hardening on the Unconfined Strength of Remolded Clays, *Proc. 2d Int. Conf. Soil Mech. Found. Eng.,* vol. 1, pp. 218–222, 1948.

Nash, K. L., The Shearing Resistance of a Fine Closely Graded Sand, *Proc. 3d Int. Conf. Soil Mech. Found. Eng.,* vol. 1, pp. 161–164, 1953.

Olson, R. E., Shearing Strength of Kaolinite, Illite, and Montmorillonite, *J. Geotech. Eng. Div., ASCE,* vol. 100, no. GT11, pp. 1215–1230, 1974.

Peck, R. B., W. E. Hanson, and T. H. Thornburn, "Foundation Engineering" 2d ed., Wiley, New York, 1974.

Peck, R. B., H. O. Ireland, and T. S. Fry, Studies of Soil Characteristics: The Earth Flows of St. Thuribe, Quebec, *Soil Mechanics Series No. 1, Univ. Illinois, Urbana,* 1951.

Rendulic, L., Ein Grundgesetzder tonmechanik und sein experimentaller beweis, *Bauingenieur,* vol. 18, pp. 459–467, 1937.

Richardson, A. M., Jr., and R. V. Whitman, Effect of Strain Rate upon Undrained Shear Resistance of Saturated Remolded Fat Clay, *Geotechnique,* vol. 13, no. 4, pp. 310–346, 1964.

Richardson, A. M., E. W. Brand, and A. Menon, "In-Situ Determination of Anisotropy of a Soft Clay, *Proc. Conf. In-Situ Measure. Soil Proper., ASCE,* vol. 1, pp. 336–349, 1975.

Roscoe, K. H., A. N. Schofield, and A. Thurairajah, An Evaluation of Test Data for Selecting a Yield Criterion for Soils, *Proc. Symp. Lab. Shear Test. Soils, ASTM Special Tech. Pub. No. 361,* pp. 111–133, 1963.

Rowe, P. W., The Stress-Dilatency Relation for Static Equilibrium of an Assembly of Particles in Contact, *Proc. Roy. Soc.,* A269, pp. 500–527, 1962.

Seed, H. B., and C. K. Chan, Thixotropic Characteristics of Compacted Clays, *Trans. ASCE,* vol. 124, pp. 894–916, 1959.

Seed, H. B., and K. L. Lee, Liquefaction of Saturated Sands During Cyclic Loading, *J. Soil Mech. Found. Eng. Div., ASCE,* vol. 92, no. SM6, pp. 105–134, 1966.

Selig, E. T., and K. E. McKee, Static and Dynamic Behavior of Small Footings, *J. Soil Mech. Found. Eng. Div., ASCE,* vol. 87, no. SM6, pp. 29–47, 1961.

Sherif, M. A., and C. M. Burrous, "Temperature Effect on the Unconfined Shear Strength of Saturated Cohesive Soils," Highway Research Board, Special Report 103, pp. 267–272, 1969.

Simons, N. E., The Effect of Overconsolidation on the Shear Strength Characteristics of an Undisturbed Oslo Clay, *Proc. Res. Conf. Shear Strength Cohesive Soils, ASCE,* pp. 747–763, 1960.

Singh, A., and J. K. Mitchell, General Stress-Strain-Time Functions for Soils, *J. Soil Mech. Found. Eng. Div., ASCE,* vol. 94, no. SM1, pp. 21–46, 1968.

Skempton, A. W., Discussion: The Planning and Design of New Hong Kong Airport, *Proc. Inst. Civil Eng.,* vol. 7, pp. 305–307, 1957.

Skempton, A. W., Correspondence, *Geotechnique,* vol. 10, no. 4, p. 186, 1960.

Skempton, A. W., Effective Stress in Soils, Concrete and Rock, *in* "Pore Pressure and Suction in Soils," Butterworths, London, pp. 4-16, 1961.

Skempton, A. W., Long-Term Stability of Clay Slopes, *Geotechnique,* vol. 14, p. 77, 1964.

Sultan, H. A., and H. B. Seed, Stability of Sloping Core Earth Dam, *J. Soil Mech. Found. Eng. Div., ASCE,* vol. 93, no. SM4, pp. 45-67, 1967.

Taylor, D. W., A Comparison of Results of Direct Shear and Cylindrical Compression Test, *Proc. Symp. Shear Test. Soils, ASTM,* vol. 39, pp. 1059-1070, 1939.

Taylor, D. W., "Fundamentals of Soil Mechanics," Wiley, New York, 1948.

Tresca, H., Memoire sur L'Ecoulement des Corps Solids, *Men. Pres. Par Div. Sav,* vol. 18, 1868.

Von Mises, R., Mechanik der festen Korper in Plastichdeformablen Zustand, *Goettinger-Nachr. Math-Phys. K1,* 1913.

Wade, N. H., "Plane Strain Characteristics of a Saturated Clay," Ph.D. thesis, University of London, 1963.

Whitman, R. V., The Behavior of Soils under Transient Loading, *Proc. 4th Int. Conf. Soil Mech. Found. Eng., London,* vol. 1, p. 207, 1957.

Whitman, R. V., and K. A. Healey, Shear Strength of Sand During Rapid Loadings, *Trans. ASCE,* vol. 128, pp. 1553-1594, 1963.

Whitman, R. V., and W. Luscher, Basic Experiments into Soil-Structure Interaction, *J. Soil Mech. Found. Eng. Div., ASCE,* vol. 88, no. SM6, pp. 135-167, 1962.

Wu, T. H., A. K. Loh, and L. E. Malvern, Study of Failure Envelope of Soils, *J. Soil Mech. Found. Eng. Div., ASCE,* vol. 89, no. SM1, pp. 145-181, 1963.

INDEX